T0253034

Combinatorics of
Permutations

Discrete Mathematics and Its Applications
Series editors:
Miklós Bóna, Donald L. Kreher, Douglas B. West

Algorithmics of Nonuniformity
Tools and Paradigms
Micha Hofri, Hosam Mahmoud

Handbook of Geometric Constraint Systems Principles
Edited by Meera Sitharam, Audrey St. John, Jessica Sidman

Introduction to Chemical Graph Theory
Stephan Wagner, Hua Wang

Extremal Finite Set Theory
Daniel Gerbner, Balazs Patkos

The Mathematics of Chip-Firing
Caroline J. Klivans

Computational Complexity of Counting and Sampling
Istvan Miklos

Volumetric Discrete Geometry
Karoly Bezdek, Zsolt Langi

The Art of Proving Binomial Identities
Michael Z. Spivey

Combinatorics and Number Theory of Counting Sequences
Istvan Mezo

Applied Mathematical Modeling
A Multidisciplinary Approach
Douglas R. Shier, K.T. Wallenius

Analytic Combinatorics
A Multidimensional Approach
Marni Mishna

50 years of Combinatorics, Graph Theory, and Computing
Edited By Fan Chung, Ron Graham, Frederick Hoffman, Ronald C. Mullin, Leslie Hogben, Douglas B. West

Fundamentals of Ramsey Theory
Aaron Robertson

Methods for the Summation of Series
Tian-Xiao He

The Lambert W Function
and its Generalizations and Applications
István Mező

Combinatorics of Permutations, Third Edition
Miklos Bona

https://www.routledge.com/Discrete-Mathematics-and-Its-Applications/book-series/CHDISMTHAPP

Combinatorics of Permutations

Third Edition

Miklós Bóna

CRC Press
Taylor & Francis Group
Boca Raton London New York

CRC Press is an imprint of the
Taylor & Francis Group, an **informa** business
A CHAPMAN & HALL BOOK

Third edition published 2022
by CRC Press
6000 Broken Sound Parkway NW, Suite 300, Boca Raton, FL 33487-2742

and by CRC Press
2 Park Square, Milton Park, Abingdon, Oxon, OX14 4RN

First edition published by CRC Press 2004
Second edition published by CRC Press 2012

CRC Press is an imprint of Taylor & Francis Group, LLC

ISBN: 9780367222581 (hbk)
ISBN: 9781032223506 (pbk)
ISBN: 9780429274107 (ebk)

DOI: 10.1201/9780429274107

Typeset in NimbusSanL-Regu font
by KnowledgeWorks Global Ltd.

Publisher's note: This book has been prepared from camera-ready copy provided by the authors

Dedication

To Linda
To Mikike, Benny, and Vinnie

To the Mathematicians whose relentless and brilliant efforts throughout the centuries unearthed the gems that we call Combinatorics of Permutations.

The Tribute of the Current to the Source.

Robert Frost, *West Running Brook*

Contents

Foreword

Permutations have a remarkably rich combinatorial structure. Part of the reason for this is that a permutation of a finite set can be represented in many equivalent ways, including as a word (sequence), a function, a collection of disjoint cycles, a matrix, etc. Each of these representations suggests a host of natural invariants (or "statistics"), operations, transformations, structures, etc., that can be applied to or placed on permutations. The fundamental statistics, operations, and structures on permutations include descent set (with numerous specializations), excedance set, cycle type, records, subsequences, composition (product), partial orders, simplicial complexes, probability distributions, etc. How is the newcomer to this subject able to make sense of and sort out these bewildering possibilities? Until now it was necessary to consult a myriad of sources, from textbooks to journal articles, in order to grasp the whole picture. Now, however, Miklós Bóna has provided us with a comprehensive, engaging, and eminently readable introduction to all aspects of the combinatorics of permutations. The chapter on pattern avoidance is especially timely and gives the first systematic treatment of this fascinating and active area of research.

This book can be utilized at a variety of levels, from random samplings of the treasures therein to a comprehensive attempt to master all the material and solve all the exercises. In whatever direction the reader's tastes lead, a thorough enjoyment and appreciation of a beautiful area of combinatorics is certain to ensue.

Richard Stanley

Cambridge, Massachusetts

Preface to the First Edition

A few years ago, I was given the opportunity to teach a graduate combinatorics class on a special topic of my choice. I wanted the class to focus on the combinatorics of permutations. However, I instantly realized that while there were several excellent books that discussed some aspects of the subject, there was no single book that would have contained all, or even most, areas that I wanted to cover. Many areas were not covered in any book, which was easy to understand as the subject is developing at a breathtaking pace, producing new results faster than textbooks are published. Classic results, while certainly explained in various textbooks of very high quality, seemed to be scattered in numerous sources. This was again no surprise; indeed, permutations are omnipresent in modern combinatorics, and there are quite a few ways to look at them. We can consider permutations as linear orders; we can consider them as elements of the symmetric group; we can model them by matrices; or by graphs. We can enumerate them according to countless interesting statistics; we can decompose them in many ways, and we can bijectively associate them to other structures. One common feature of these activities is that they all involve factual knowledge, new ideas, and serious fun. Another common feature is that they all evolve around permutations, and quite often, the remote-looking areas are connected by surprising results. Briefly, they do belong to one book, and I am very glad that now you are reading such a book.

***

As I have mentioned, there are several excellent books that discuss various aspects of permutations. Therefore, in this book, I cover these aspects less deeply than the areas that have previously not been contained in any book. Chapter 1 is about descents and runs of permutations. While Eulerian numbers have been given plenty of attention during the last 200 years, most of the research was devoted to analytic concepts. Nothing shows this better than the fact that I was unable to find published proofs of two fundamental results of the area using purely combinatorial methods. Therefore, in this chapter, I concentrated on purely combinatorial tools dealing with these issues. By and large, the same is true for Chapter 2, whose subject is inversions in permutations, and in permutations of multisets. Chapter 3 is devoted to permutations as products of cycles, which is probably the most-studied of all areas covered in this book. Therefore, while there were many classic results we had to include there for the sake of completeness, nevertheless we still managed to squeeze in less well-known topics, such as applications of Darroch's theorem, or transpositions and trees.

The area of pattern avoidance is a young one, and has not been given significant space in textbooks before. Therefore, we devoted two full chapters to it. Chapter 4 walks the reader through the quest for the solution of the Stanley-Wilf conjecture, ending with the recent spectacular proof of Marcus and Tardos for this 23-year-old problem. Chapter 5 discusses aspects of pattern avoidance other than upper bounds or exact formulae. Chapter 6 looks at random permutations and Standard Young Tableaux, starting with two classic and difficult proofs of Greene, Nijenhaus, and Wilf. Standard techniques for handling permutation statistics are presented. A relatively new concept, that of min-wise independent families of permutations, is discussed in the Exercises. Chapter 7, Algebraic Combinatorics of Permutations, is the one in which we had to be very selective. Each of the three sections of that chapter covers an area that is sufficiently rich to be the subject of an entire book. Our goal with that chapter is simply to raise interest in these topics and prepare the reader for the more detailed literature that is available in those areas. Chapter 8 is about combinatorial sorting algorithms, many of which are quite recent. This is the first time many of these algorithms (or at least, most aspects of them) are discussed in a textbook, so we treated them in depth.

Besides the Exercises, each chapter ends with a selection of Problems Plus. These are typically more difficult than the exercises, and they are meant to raise interest in some questions for further research, and to serve as reference material of what is known. Some of the Problems Plus are not classified as such because of their level of difficulty, but because they are less tightly connected to the topic at hand. A solution manual for the even-numbered Exercises is available for instructors teaching a class using this book, and it can be obtained from the publisher.

Preface to the Third Edition

It has been nine years since the second edition of *Combinatorics of Permutations* was published. Many areas of the subject went through significant progress during those years.

The youngest area, *permutation patterns*, that is the content of our Chapters 4 and 5, was a major contributor to that progress. Several new methods were discovered to prove upper bounds on the size of some permutation classes, most of all the extremely challenging class of 1324-avoiding permutations. We know now that most principal permutation classes have a nonrational generating function, but a conjecture on which patterns of a given size are the easiest to avoid proved to be false. Records have fallen in just about every version of the superpattern problem.

The related area of *stack sorting* also has much stronger, and more numerous, results than it had nine years ago. This also means that some of the long-standing conjectures of the field are very likely to be false, though we do not always know which ones. Chapter 8 describes some of this progress, and the new open problems that it brought.

The *enumeration of vertices of a given rank* in many varieties of rooted trees has been the subject of vigorous research in the last decade. As many of these tree varieties correspond to some kind of permutations, these results will appear throughout the book. Section 6.3 is entirely devoted to this topic.

As the Combinatorics of Permutations keeps expanding, books about the subject must become more and more selective. While it is impossible to cover all, or even, most, of the important results in the field, we tried to discuss at least those that we find particularly rewarding, and that we want the readers to enjoy as well.

Gainesville, FL
February 2022

Acknowledgments

This book grew out of various graduate combinatorics courses that I taught at the University of Florida. I am indebted to the authors of the books I used in those courses, for shaping my vision, and for teaching me facts and techniques. These books are *The Art of Computer Programming* by D. E. Knuth, *Enumerative Combinatorics* by Richard Stanley, *The Probabilistic Method* by Noga Alon and Joel Spencer, *The Symmetric Group* by Bruce Sagan, and *Enumerative Combinatorics* by Charalambos Charalambides. For my knowledge of biologically motivated sorting algorithms, which is the topic of a new chapter in the second edition, I am indebted to the book *Combinatorics of Genome Rearrangements* by Guillaume Fertin, Anthony Labarre, Irena Rusu, Éric Tannier, and Stéphane Vialette.

Needless to say, I am grateful to all the researchers whose results made possible a textbook devoted exclusively to the combinatorics of permutations. I am sure that new discoveries will follow.

I am thankful to my former research advisor, Richard Stanley, for having introduced me to this fascinating field, and to Doron Zeilberger, and the late Herb Wilf, who kept asking intriguing questions attracting scores of young mathematicians like myself to the subject.

Some of the presented material was part of my own research, sometimes in collaboration. I would like to say thanks to my co-authors, Richard Ehrenborg, Andrew MacLennan, Bruce Sagan, Rodica Simion, Ryan Flynn, Daniel Spielman, Vincent Vatter, and Dennis White. I also owe thanks to Michael Atkinson, who introduced me to the history of stack sorting algorithms.

I am deeply indebted to Aaron Robertson for an exceptionally thorough and knowledgeable reading of the first edition, and to the anonymous referees of the second edition. I am also deeply appreciative for manuscript reading by my colleague Andrew Vince, and by Rebecca Smith. I feel grateful to the many mathematicians who pointed out various typos in the first edition.

A significant part of the book was written during the summer of 2003. In the first half of that summer, I enjoyed the stimulating professional environment at LABRI, at the University of Bordeaux I, in Bordeaux, France. The hospitality of colleagues Olivier Guibert and Sylvain Pelat-Alloin made it easy for me to keep writing during my one-month visit. In the second half of the summer, I enjoyed the hospitality of my parents, Miklós and Katalin Bóna, at Lake Balaton in Hungary. In 2005, I spent a sabbatical semester at the University of Pennsylvania, where I learned from Herb Wilf and Robin Pemantle.

My gratitude is extended to Tina Freebody for preparing the cover page.

Last, but not least, I must be thankful to my wife, Linda, my first reader and critic, who keeps tolerating my book-writing endeavors. I will not forget how much she helped me, and neither will she.

No Way around It. Introduction.

This book is devoted to the study of permutations. While the overwhelming majority of readers already know what they are, we are going to define them for the sake of completeness. Note that this is by no means the only definition possible.

DEFINITION 0.1   *A linear ordering of the elements of the set $[n] = \{1, 2, 3, \cdots, n\}$ is called a* permutation, *or, if we want to stress the fact that it consists of n entries, an n-permutation.*

In other words, a permutation lists all elements of $[n]$ so that each element is listed exactly once.

Example 0.2
If $n = 3$, then the n-permutations are 123, 132, 213, 231, 312, 321.   ◻

There is nothing magic about the set $[n]$; other sets having n elements would be just as good for our purposes, but working with $[n]$ will simplify our discussion. In Chapter 2, we will extend the definition of permutations to multisets, and in Chapter 3, we will consider permutations from a different perspective. The set of all n-permutations will be denoted by S_n, and the reason for that will become clear in Chapter 3.

For now, we will denote an n-permutation by $p = p_1 p_2 \cdots p_n$, with p_i being the ith entry in the linear order given by p.

The following simple statement is probably the best-known fact about permutations.

PROPOSITION 0.3
The number of n-permutations is $n!$.

PROOF   When building an n-permutation $p = p_1 p_2 \cdots p_n$, we can choose n entries to play the role of p_1, then $n - 1$ entries for the role of p_2, and so on.   ∎

We promise the rest of the book will be less straightforward.

1

In One Line and Close. Permutations as Linear Orders.

1.1 Descents

The "most orderly" of all n-permutations is obviously the increasing permutation $123\cdots n$. All other permutations have at least some "disorder" in them; for instance, it happens that an entry is immediately followed by a *smaller* entry in them. This simple phenomenon is at the center of our attention in this Section.

1.1.1 Definition of Descents

DEFINITION 1.1 *Let $p = p_1 p_2 \cdots p_n$ be a permutation, and let $i < n$ be a positive integer. We say that i is a* descent *of p if $p_i > p_{i+1}$. Similarly, we say that i is an* ascent *of p if $p_i < p_{i+1}$.*

Example 1.2
Let $p = 3412576$. Then 2 and 6 are descents of p, while 1, 3, 4, and 5 are ascents of p. ▯

Note that the descents denote the *positions* within p, and not the entries of p. The set of all descents of p is called the *descent set of p* and is denoted by $D(p)$. The cardinality of $D(p)$, that is, the number of descents of p, is denoted by $d(p)$, though certain authors prefer des(p).

This very natural notion of descents raises some obvious questions for the enumerative combinatorialist. How many n-permutations are there with a given number of descents? How many n-permutations are there with a given descent set? If two n-permutations have the same descent set, or the same number of descents, what other properties do they share?

The answers to these questions are not always easy, but are always interesting. We would like to start with the problem of finding the number of permutations with a given descent set S. However, it turns out that it is even easier to find the number of permutations whose descent set is *contained* in S, so we start with that result.

DOI: 10.1201/9780429274107-1

LEMMA 1.3

Let $S = \{s_1, s_2, \cdots, s_k\} \subseteq [n-1]$, where the $s_1 < s_2 < \cdots < s_k$, and let $\alpha(S)$ be the number of n-permutations whose descent set is contained in S. Then the identity

$$\alpha(S) = \binom{n}{s_1}\binom{n-s_1}{s_2-s_1}\binom{n-s_2}{s_3-s_2}\cdots\binom{n-s_k}{n-s_k}$$

holds.

PROOF   The crucial idea of the proof is the following. We arrange our n entries into $k+1$ segments so that the first i segments together have s_i entries for each i. Then, within each segment, we put our entries in increasing order. Then the only places where the resulting permutation has a chance to have a descent is where two segments meet, that is, at $s_1, s_2, \cdots, s_k$. Therefore, the descent set of the resulting permutation is contained in S.

How many ways are there to arrange our entries in these segments? The first segment has to have length s_1, and therefore the entries that go there can be chosen in $\binom{n}{s_1}$ ways. The second segment has to be of length $s_2 - s_1$, and has to be disjoint from the first one. Therefore, its entries can be chosen in $\binom{n-s_1}{s_2-s_1}$ ways. In general, the ith segment must have length $s_i - s_{i-1}$ if $i < k + 1$, and its entries have to be chosen from the remaining $n - s_{i-1}$ entries, in $\binom{n-s_{i-1}}{s_i-s_{i-1}}$ ways. There is only one choice for the last segment as all remaining $n - s_k$ entries have to go there. This completes the proof. ∎

Now we are in a position to state and prove the formula for the number of n-permutations with a given descent set.

THEOREM 1.4

Let $S \subseteq [n-1]$. Then the number of n-permutations with descent set S is

$$\beta(S) = \sum_{T \subseteq S} (-1)^{|S-T|}\alpha(T). \tag{1.1}$$

PROOF   This is a direct conclusion of the Principle of Inclusion and Exclusion. (See any textbook on introductory combinatorics, such as [74], for this principle.) Note that permutations with a given h-element descent set $H \subseteq S$ are counted $a_h = \sum_{i=0}^{|S-H|}(-1)^i\binom{|S-H|}{i} = (1+(-1))^{|S-H|}$ times on the right–hand side of (1.1). The value of a_h is 0 except when $|S - H| = 0$, that is, when $S = H$. So the right-hand side counts precisely the permutations with descent set S. ∎

1.1.2   Eulerian Numbers

Let $A(n, k)$ be the number of n-permutations with $k - 1$ descents. You may be wondering what the reason for this shift in the parameter k is. If p has $k - 1$ descents, then p is the union of k increasing subsequences of consecutive entries. These are called the *ascending runs* of p. (Some authors call them just "runs," others call something else "runs." This is why we add the adjective "ascending" to avoid confusion.) Also note that in some papers, $A(n, k)$ is used to denote the number of permutations with k descents.

Example 1.5
The three ascending runs of $p = 2415367$ are 24, 15, and 367.   ▯

Example 1.6
There are four permutations of length three with one descent, namely 132, 213, 231, and 312. Therefore, $A(3, 2) = 4$. Similarly, $A(3, 3) = 1$ corresponding to the permutation 321, and $A(3, 1) = 1$, corresponding to the permutation 123.
▯

Thus the permutations with k ascending runs are the same as permutations with $k - 1$ descents, providing one answer for the notation $A(n, k)$. We note that some authors use the notation $\left\langle {n \atop k} \right\rangle$ for $A(n, k)$.

The numbers $A(n, k)$ are called the *Eulerian numbers*, and have several beautiful properties. Several authors provided extensive reviews of this field, including Leonard Carlitz [110], Dominique Foata and Marcel-Paul Schützenberger [172], Donald E. Knuth [230], and Charalambos Charalambides [109]. The most recent, and most comprehensive, coverage of the subject is the book [261] by T. Kyle Petersen. In our treatment of the Eulerian numbers, we will make an effort to be as combinatorial as possible, and avoid the analytic methods that probably represent a majority of the available literature. We start by proving a simple recurrence relation.

THEOREM 1.7
For all positive integers k and n satisfying $k \leq n$, the identity

$$A(n, k + 1) = (k + 1)A(n - 1, k + 1) + (n - k)A(n - 1, k)$$

holds.

PROOF   There are two ways we can get an n-permutation p with k descents from an $(n - 1)$-permutation p' by inserting the entry n into p'. Either p' has k descents, and the insertion of n does not form a new descent, or p' has $k - 1$ descents, and the insertion of n does form a new descent.

In the first case, we have to put the entry n at the end of p', or we have to insert n between two entries that form one of the k descents of p'. This means we have $k + 1$ choices for the position of n. As we have $A(n - 1, k + 1)$ choices for p', the first term of the -hand side is explained.

In the second case, we have to put the entry n at the front of p', or we have to insert n between two entries that form one of the $(n-2)-(k-1) = n-k-1$ ascents of p'. This means that we have $n - k$ choices for the position of n. As we have $A(n - 1, k)$ choices for p', the second part of the right-hand side is explained, and the theorem is proved. ∎

We note that $A(n, k + 1) = A(n, n - k)$; in other words, the Eulerian numbers are symmetric. Indeed, if $p = p_1 p_2 \cdots p_n$ has k descents, then its reverse $p^r = p_n p_{n-1} \cdots p_1$ has $n - k - 1$ descents.

The following theorem shows some additional significance of the Eulerian numbers. In fact, the Eulerian numbers are sometimes *defined* using this relation.

THEOREM 1.8

Set $A(0,0) = 1$, and $A(n,0) = 0$ for $n > 0$. Then for all nonnegative integers n, and for all complex numbers z, the equality

$$z^n = \sum_{k=1}^{n} A(n, k) \binom{z + n - k}{n} \tag{1.2}$$

holds.

Example 1.9

Let $n = 3$. Then we have $A(3, 1) = 1$, $A(3, 2) = 4$, and $A(3, 3) = 1$, enumerating the sets of permutations $\{123\}$, $\{132, 213, 231, 312\}$, and $\{321\}$. And indeed,

$$z^3 = \binom{z + 2}{3} + 4 \binom{z + 1}{3} + \binom{z}{3}.$$

⬚

PROOF   (of Theorem 1.8) Let us assume first that z is a positive integer. Then the left-hand side counts the n-element sequences in which each digit comes from the set $[z]$. We will show that the right-hand side counts these same sequences. Let $a = a_1 a_2 \cdots a_n$ be such a sequence. Rearrange entries of a into a nondecreasing order $a' = a_{i_1} \leq a_{i_2} \leq \cdots \leq a_{i_n}$, with the extra condition that identical digits appear in a' in the increasing order of their indices. Then $i = i_1 i_2 \cdots i_n = p(a)$ is an n-permutation that is uniquely determined by a. Note that i_1 tells from which position of a the first entry of

a' comes, i_2 tells from which position of a the second entry of a' comes, and so on.

For instance, if $a = 311243$, then the rearranged sequence is $a' = 112334$, leading to the permutation $i = 234165$.

If we can show that each permutation i having $k-1$ descents is obtained from exactly $\binom{z+n-k}{n}$ sequences a in this way, then we will have proved the theorem.

The crucial observation is that if $a_{i_j} = a_{i_{j+1}}$ in a', then $i_j < i_{j+1}$ in i. Taking contrapositives, if j is a descent in $i = i_1 i_2 \cdots i_n$, then $a_{i_j} < a_{i_{j+1}}$. This means that the sequence a' has to be *strictly increasing* whenever j is a descent of $i_1 i_2 \cdots i_n$. The reader should verify that in our running example, i has descents at 3 and 5, and indeed, a' is strictly increasing in those positions (but not only there).

How many sequences a can lead to the permutation $i = 234165$? It follows from the above argument that in sequences with that property, we must have

$$1 \leq a_2 \leq a_3 \leq a_4 < a_1 \leq a_6 < a_5 \leq x,$$

as strict inequality is required in the third and fifth positions. The above chain of inequalities is obviously equivalent to

$$1 \leq a_2 < a_3 + 1 < a_4 + 2 < a_1 + 2 < a_6 + 3 < a_5 + 3 \leq z + 3,$$

and, therefore, the number of such sequences is clearly

$$\binom{z+3}{6}.$$

So this is the number of sequences a for which $a' = 234165$. Generalizing this argument for any n and for permutations i with $k-1$ descents, we get that each n-permutation with $k-1$ descents will be obtained from $\binom{z+(n-1)-(k-1)}{n} = \binom{z+n-k}{n}$ sequences.

If z is not a positive integer, note that the two sides of the equation to be proved can both be viewed as polynomials in the variable z. As they agree for infinitely many values (the positive integers), they must be identical. ∎

Exercise 7 gives a more mechanical proof that simply uses Theorem 1.7.

COROLLARY 1.10

For all positive integers n, the identity

$$z^n = \sum_{k=1}^{n} A(n,k) \binom{z+k-1}{n}$$

holds.

PROOF   Replace z by $-z$ in the result of Theorem 1.8, to get

$$z^n(-1)^n = \sum_{k=1}^{n} A(n,k)\binom{-z+n-k}{n}.$$

Now note that

$$\binom{-z+n-k}{n} = \frac{(-z+n-k)(-z+n-k-1)\cdots(-z+1-k)}{n!}$$

$$= (-1)^n \binom{z+k-1}{n}.$$

Comparing these two identities yields the desired result.   ∎

The obvious question that probably crossed the mind of the reader by now is whether there exists an *explicit formula* for the numbers $A(n,k)$. The answer to that question is in the affirmative, though the formula contains a summation sign. This formula is more difficult to prove than the previous formulae in this section.

THEOREM 1.11
For all nonnegative integers n and k satisfying $k \leq n$, the identity

$$A(n,k) = \sum_{i=0}^{k}(-1)^i \binom{n+1}{i}(k-i)^n \tag{1.3}$$

holds.

While this theorem is a classic (it is more than 100 years old), in 2003 we could not find an immaculately direct proof for it in the literature. Proofs we did find used generating functions, or manipulations of double sums of binomial coefficients, or inversion formulae to obtain (1.3). Therefore, we solicited simple, direct proofs at the problem session of the 15th Formal Power Series and Algebraic Combinatorics conference, which took place in Vadstena, Sweden. The proof we present here was contributed by Richard Stanley. A similar proof was proposed by Hugh Thomas.

PROOF   (of Theorem 1.11) Let $k > 0$, and let us write down $k-1$ bars with k compartments in between. (For $k = 0$, the statement of the theorem is obvious.) Place each element of $[n]$ in a compartment. There are k^n ways to do this, the term in the above sum indexed by $i = 0$. Arrange the numbers in each compartment in increasing order. For example, if $k = 4$ and $n = 9$, then one arrangement is

$$237||19|4568. \tag{1.4}$$

Ignoring the bars, we get a permutation (in the above example, it is 237194568) with *at most* $k-1$ descents.

There are several issues to take care of. There could be empty compartments, or there could be neighboring compartments with no descents in between. We will show how to sieve out permutations having either of these problems, and therefore, less than $k-1$ descents, at the same time.

Let us say that a bar is a *wall* if it is not immediately followed by another bar. Let us say that a wall is *extraneous* if by removing it we still get a legal arrangement, that is, an arrangement in which each compartment consist of integers in *increasing* order.

For instance, in (1.4), the second bar is an extraneous wall. Our goal is to enumerate the arrangements with *no extraneous walls*, as these are clearly in bijection with permutations with $k-1$ descents.

In order to do this, we will apply the Principle of Inclusion and Exclusion. Let us call the spaces between consecutive entries of a permutation, as well as the space preceding the first entry and the space following the last entry a *position*. So we associate $n+1$ positions to an n-permutation. Let $S \subseteq [n+1]$, and let A_S be the set of arrangements in which there is an extraneous wall in each position belonging to S.

Let $i \leq k-1$ be the size of S. Then we claim that

$$|A_S| = (k-i)^n.$$

In order to see this, first take any legal arrangement that contains $k-i-1$ bars. There are $(k-i)^n$ such arrangements as we can proceed from the entry 1 to the entry n in an increasing order, and choose a compartment for each entry. Now insert i extra bars by inserting one to each position that belongs to S. (If there is already a bar in such a position, then put the new bar immediately on the right of that bar.) This results in an arrangement that belongs to A_S. Conversely, each arrangement belonging to A_S will be obtained exactly once in this way. Indeed, if $a \in A_S$, then removing one bar from each of the i positions that belong to S, we get the unique original arrangement with $k-i-1$ bars that leads to a.

As there are $\binom{n+1}{i}$ choices for the set S, the proof of our theorem is now immediate by the Principle of Inclusion and Exclusion. ∎

For the sake of completeness, we include a more computational proof that does not need a clever idea as the previous one did.

First, we recall a lemma from the theory of binomial coefficients.

LEMMA 1.12

[Chu - Vandermonde Convolution Formula] Let x and y be real numbers, and

let z be a positive integer. Then the identity

$$\binom{x+y}{z} = \sum_{d=0}^{z} \binom{x}{d}\binom{y}{z-d}$$

holds.

PROOF   Let us assume first that x and y are positive integers. Then the left-hand side enumerates the z-element subsets of the set $[x + y]$, while the right-hand side enumerates these same objects, according to the size of their intersection with the set $[x]$.

For general x and y, note that both sides can be viewed as polynomials in x and y, and they agree for infinitely many values (the positive integers). Therefore, they have to be identical. ∎

PROOF   (of Theorem 1.11) As a first step, consider formula (1.2) with $x = 1$, then with $x = 2$, and then for $x = i$ for $i \leq k$. We get

$$1 = A(n, 1) \cdot \binom{n}{n},$$

$$2^n = A(n, 2) \cdot \binom{n}{n} + A(n, 1) \cdot \binom{n+1}{n},$$

and so on, the hth equation being

$$h^n = \sum_{j=0}^{h-1} A(n, k - j)\binom{n+j-1}{n}, \tag{1.5}$$

and the last equation being

$$k^n = \sum_{j=0}^{k-1} A(n, k - j)\binom{n+j-1}{n} \tag{1.6}$$

We will now add certain multiples of our equations to the last one, so that the left-hand side becomes the right-hand side of formula (1.3) that we are trying to prove.

To start, let us add $(-1)\binom{n+1}{1}$ times the $(k-1)$st equation to the last one. Then add $\binom{n+1}{2}$ times the $(k-2)$nd equation to the last one. Continue this way, that is, in step i, add $(-1)^i\binom{n+1}{i}$ times the $(k-i)$th equation to the last one. This gives us

$$\sum_{i=0}^{k}(-1)^i\binom{n+1}{i}(k-i)^n = \sum_{j=1}^{k} A(n, j)\sum_{i=0}^{k-j}\binom{n+k-i-j}{n}\binom{n+1}{i}(-1)^i. \tag{1.7}$$

The left-hand side of (1.7) agrees with the right-hand side of (1.3). There-fore, (1.3) will be proved if we can show that the coefficient $a(n,j)$ of $A(n,j)$ on the right-hand side above is 0 for $j < k$. It is obvious that $a(n,k) = 1$ as $A(n,k)$ occurs in the last equation only.

Set $b = k - j$. Then $a(n,k)$ can be transformed as follows.

$$a(n,k) = \sum_{i=0}^{b}(-1)^i \binom{n+1}{i}\binom{n-i+b}{n}.$$

Recalling that for a positive real number x, we have $\binom{-x}{a} = \binom{x+a-1}{a}(-1)^a$, and noting that $(-1)^b = (-1)^{b-2i}$, this yields

$$(-1)^b a(n,k) = \sum_{i=0}^{b}(-1)^{b-i}\binom{n+1}{i}\binom{n-i+b}{n}$$

$$= \sum_{i=0}^{b}(-1)^{b-i}\binom{n+1}{i}\binom{n-i+b}{b-i} = \sum_{i=0}^{b}\binom{n+1}{i}\binom{-1-n}{b-i} = \binom{0}{b} = 0,$$

where the last step holds as $b = k - j > 0$, and the next-to-last step is a direct application of Lemma 1.12.

This shows that the right-hand side of (1.7) simplifies to $A(n,k)$, and proves our theorem. ∎

1.1.3 Stirling Numbers and Eulerian Numbers

DEFINITION 1.13 *A partition of the set $[n]$ into r blocks is a distri-bution of the elements of $[n]$ into r disjoint nonempty sets, called blocks, so that each element is placed into exactly one block.*

In Section 2.1, we will define the different concept of partitions of an *integer*. If there is a danger of confusion, then partitions of the set $[n]$ will be called *set partitions*, to distinguish them from partitions of the integer n.

Example 1.14
Let $n = 7$ and $r = 4$. Then $\{1,2,4\},\{3,6\},\{5\},\{7\}$ is a partition of $[7]$ into four blocks. ∎

Note that neither the order of blocks nor the order of elements within each block matters. That is, $\{4,1,2\},\{6,3\},\{5\},\{7\}$ and $\{4,1,2\},\{6,3\},\{7\},\{5\}$ are considered the same partition as the one in Example 1.14.

DEFINITION 1.15 *The number of partitions of $[n]$ into k blocks is denoted by $S(n,k)$ and is called a Stirling number of the second kind.*

n=0				1			
n=1				0	1		
n=2			0	1	1		
n=3		0	1	3	1		
n=4	0	1	7	6	1		
n=5	0	1	15	25	10	1	

FIGURE 1.1
The values of $S(n, k)$ for $n \leq 5$. Note that the Northeast–Southwest diagonals contain values of $S(n, k)$ for fixed k. Row n starts with $S(n, 0)$.

By convention, we set $S(n, 0) = 0$ if $n > 0$, and $S(0, 0) = 1$. The next chapter will explain what the Stirling numbers of the first kind are.

Example 1.16
The set $[4]$ has six partitions into three blocks, each consisting of one doubleton and two singletons. Therefore, $S(4, 3) = 6$.   □

Whereas Stirling numbers of the second kind do not directly count permutations, they are inherently related to two different sets of numbers that do. One of them is the set of Eulerian numbers, and the other one is the aforementioned set of Stirling numbers of the first kind. Therefore, exploring some properties of the numbers $S(n, k)$ in this book is well-motivated. See Figure 1.1 for the values of $S(n, k)$ for $n \leq 5$.

See Exercises 8 and 14 for two simple recurrence relations satisfied by the numbers $S(n, k)$. It turns out that an explicit formula for these numbers can be proved without using any recurrence relations.

LEMMA 1.17
For all positive integers n and r, the identity

$$S(n, r) = \frac{1}{r!} \sum_{i=0}^{r} (-1)^i \binom{r}{i} (r - i)^n$$

holds.

PROOF   An *ordered partition* of $[n]$ into r blocks is a partition of $[n]$ into r blocks in which the set of blocks is totally ordered. So $\{1, 3\}, \{2, 4\}$ and $\{2, 4\}, \{1, 3\}$ are different ordered partitions of $[4]$ into two blocks. Note that an ordered partition of $[n]$ into r blocks is just the same as a surjection from $[n]$ to $[r]$. In order to enumerate all such surjections, let A_i be the set of functions from $[n]$ into $[r]$ whose image *does not* contain i. The function $f : [n] \to [r]$ is a surjection if and only if it is not contained in $A_1 \cup A_2 \cup \cdots \cup A_r$, and our claim can be proved by a standard application of the Principle of Inclusion and Exclusion. ∎

Stirling numbers of the second kind and Eulerian numbers are closely related, as shown by the following theorem.

THEOREM 1.18
For all positive integers n and r, the identity

$$S(n,r) = \frac{1}{r!} \sum_{k=0}^{r} A(n,k) \binom{n-k}{r-k} \qquad (1.8)$$

holds.

PROOF   Multiplying both sides by $r!$ we get

$$r!S(n,r) = \sum_{k=0}^{r} A(n,k) \binom{n-k}{r-k}.$$

Here the left-hand side is obviously the number of ordered partitions of $[n]$ into r blocks. We will now show that the right-hand side counts the same objects. Take a permutation p counted by $A(n,k)$. The k ascending runs of p then naturally define an ordered partition of $[n]$ into k parts. If $k = r$, then there is nothing left to do. If $k < r$, then we will split up some of the ascending runs into several blocks of consecutive elements, in order to get an ordered partition of r blocks. As we currently have k blocks, we have to increase the number of blocks by $r - k$. This can be achieved by choosing $r - k$ of the $n - k$ "gap positions" (gaps between two consecutive entries within the same block).
    This shows that we can generate $\sum_{k=0}^{r} A(n,k) \binom{n-k}{r-k}$ ordered partitions of $[n]$ that consist of r blocks each by the above procedure. It is straightforward to show that each such partition will be obtained exactly once. Indeed, if we write the elements within each block of the partition in increasing order, we can just read the entries of the ordered partition left to right and get the unique permutation having at most r ascending runs that led to it. We can then recover the gap positions used. This completes the proof. ∎

Inverting this result leads to a formula expressing the Eulerian numbers by the Stirling numbers of the second kind.

COROLLARY 1.19
For all positive integers n and k, the identity

$$A(n,k) = \sum_{r=1}^{k} S(n,r)r! \binom{n-r}{k-r}(-1)^{k-r} \qquad (1.9)$$

holds.

PROOF   Let us consider formula (1.8) for each $r \leq k$, and multiply each equality by $r!$. We get the equations

$$1! \cdot S(n,1) = A(n,1)\binom{n-1}{0},$$

$$2! \cdot S(n,2) = A(n,1)\binom{n-1}{1} + A(n,2)\binom{n-2}{0},$$

the equation for general r being

$$r! \cdot S(n,r) = \sum_{i=1}^{r} A(n,i)\binom{n-i}{r-i}, \tag{1.10}$$

and the last equation being

$$k! \cdot S(n,k) = \sum_{i=1}^{k} A(n,i)\binom{n-i}{r-i}. \tag{1.11}$$

Our goal is to eliminate each term from the right-hand side of (1.11), except for the term $A(n,k)\binom{n-k}{k-k} = A(n,k)$. We claim that this can be achieved by multiplying (1.10) by $(-1)^{k-r}\binom{n-r}{k-r}$, doing this for all $r \in [k-1]$, then adding these equations to (1.11).

To verify our claim, look at the obtained equation

$$\sum_{r=1}^{k} S(n,r)r!(-1)^{k-r}\binom{n-r}{k-r} = \sum_{r=1}^{k}(-1)^{k-r}\binom{n-r}{k-r}\sum_{i=1}^{r} A(n,i)\binom{n-i}{r-i}, \tag{1.12}$$

or, after changing the order of summation,

$$\sum_{r=1}^{k} S(n,r)r!(-1)^{k-r}\binom{n-r}{k-r} = \sum_{i=1}^{r} A(n,i)\binom{n-i}{r-i}\sum_{r=1}^{k}(-1)^{k-r}\binom{n-r}{k-r} \tag{1.13}$$

whose left-hand side is identical to the right-hand side of (1.9).

It is obvious that the coefficient of $A(n,k)$ on the right-hand side is $\binom{n-k}{k-k} = 1$. Therefore, our statement will be proved if we can show that the coefficient $t(n,i)$ of $A(n,i)$ in the last expression is equal to zero if $i < k$.

Note that $\binom{n-i}{r-i} = 0$ if $r < i$. Therefore, for any fixed $i < k$, we have

$$t(n,i) = \sum_{r=i}^{k} \binom{n-i}{r-i}\binom{n-r}{k-r}(-1)^{k-r} = \sum_{r=i}^{k}\binom{n-i}{r-i}\binom{k-n-1}{k-r}$$

$$= \binom{k-i-1}{k-i} = 0.$$

We used Cauchy's convolution formula (Lemma 1.12) in the last step. This proves that if $i < k$, then $A(n, i)$ vanishes on the right-hand side of (1.13). We have discussed that $A(n, k)$ will have coefficient 1 there. (This can be seen again by setting $k = i$ in the last expression, leading to $t(n, i) = \binom{-1}{0} = 1$.) So (1.13) implies the claim of this corollary. ∎

1.1.4 Generating Functions and Eulerian Numbers

There are several ways one can define a generating function whose coefficients are certain Eulerian numbers. Let us start with a "horizontal" version.

DEFINITION 1.20 *For all nonnegative integers n, the polynomial*

$$A_n(z) = \sum_{k=1}^{n} A(n, k) z^k$$

is called the nth Eulerian polynomial.

The Eulerian polynomials have several interesting properties that can be proved by purely combinatorial means. We postpone the study of those properties until the next subsection. For now, we will explore the connection between these polynomials and some infinite generating functions.

THEOREM 1.21
For all positive integers n, the nth Eulerian polynomial has the alternative description

$$A_n(z) = (1 - z)^{n+1} \sum_{i \geq 0} i^n z^i.$$

Note that Euler first defined the polynomials $A_n(z)$ in the above form.

Example 1.22
For $n = 1$, we have

$$A_1(z) = (1 - z)^2 \sum_{i \geq 0} i z^i = (1 - z)^2 \cdot \frac{z}{(1 - z)^2} = z,$$

and for $n = 2$, we have

$$A_2(z) = (1 - z)^3 \sum_{i \geq 0} i^2 z^i = (1 - z)^3 \cdot \left(\frac{2z^2}{(1 - z)^3} + \frac{z}{(1 - z)^2} \right) = z + z^2.$$

◻

PROOF   (of Theorem 1.21) Let us use (1.3) to write the Eulerian polynomials as

$$\sum_{k=1}^{n} A(n,k)z^k = \sum_{k=1}^{n} \sum_{0 \le i \le k} (-1)^i \binom{n+1}{i}(k-i)^n z^k$$

$$= \sum_{k=1}^{n} \left(\sum_{0 \le i \le k} (-1)^{k-i} \binom{n+1}{k-i} i^n z^k \right).$$

Changing the order of summation, and noting that the sum in parentheses, being equal to $A(n,k)$, vanishes for $k > n$, we get

$$\sum_{i \ge 0} i^n z^i \cdot \sum_{k \ge i} \binom{n+1}{k-i}(-z)^{k-i} = (1-z)^{n+1} \sum_{i \ge 0} i^n z^i.$$

∎

It is often useful to collect all Eulerian numbers $A(n,k)$ for all n and all k in a master generating function. This function turns out to have the following simple form.

THEOREM 1.23

Let

$$r(t,z) = \sum_{n \ge 0} \sum_{k \ge 0} A(n,k) t^k \frac{z^n}{n!}.$$

Then the equality

$$r(t,z) = \frac{1-t}{1 - te^{z(1-t)}}$$

holds.

PROOF   Using the result of Theorem 1.21, we see that

$$r(t,z) = \sum_{n \ge 0} \left((1-t)^{n+1} \sum_{i \ge 0} i^n t^i \right) \frac{z^n}{n!} = (1-t) \sum_{i \ge 0} t^i \sum_{n \ge 0} \frac{(iz(1-t))^n}{n!} =$$

$$(1-t) \sum_{i \ge 0} t^i e^{iz(1-t)} = \frac{1-t}{1 - te^{z(1-t)}}.$$

∎

n=1				1			
n=2			1		1		
n=3			1	4	1		
n=4		1	11		11	1	
n=5	1		26	66	26		1
n=6	1	57	302	302	57	1	

FIGURE 1.2

Eulerian numbers for $n \leq 6$. Again, the NE–SW diagonals contain the values of $A(n, k)$ for fixed k. Row n starts with $A(n, 1)$.

1.1.5 Sequences of Eulerian Numbers

Let us take a look at the numerical values of the Eulerian numbers for small n, and $k = 0, 1, \cdots, n - 1$. The nth row of Figure 1.2 contains the values of $A(n, k)$, for $1 \leq k \leq n$, up to $n = 6$.

We notice several interesting properties. As we pointed out before, the sequence $A(n, k)$ is symmetric for any fixed n. Moreover, it seems that these sequences first increase steadily, then decrease steadily. This property is so important in combinatorics that it has its own name.

DEFINITION 1.24 *We say that the sequence of positive real numbers $a_1, a_2, \cdots, a_n$ is* unimodal *if there exists an index k such that $1 \leq k \leq n$, and $a_1 \leq a_2 \cdots \leq a_k \geq a_{k+1} \geq \cdots \geq a_n$.*

The sequences $A(n, k)_{\{1 \leq k \leq n\}}$ seem to be unimodal for any fixed n. In fact, they seem to have a stronger property.

DEFINITION 1.25 *We say that the sequence of positive real numbers $a_1, a_2, \cdots, a_n$ is* log-concave *if $a_{k-1} a_{k+1} \leq a_k^2$ holds for all indices k.*

PROPOSITION 1.26

If the sequence $a_1, a_2, \cdots, a_n$ of positive real numbers is log-concave, then it is also unimodal.

PROOF The reader should find the proof first, then check the proof that we provide as a solution for Exercise 5. ∎

The conjecture suggested by our observations is in fact correct. This is the content of the following theorem.

THEOREM 1.27

For any positive integer n, the sequence $A(n, k)_{\{1 \leq k \leq n\}}$ of Eulerian numbers is log-concave.

While this result has been known for a long time, it was usually shown as a corollary to a stronger, analytical result that we will discuss shortly, in Theorem 1.34. Direct combinatorial proofs of this fact are more recent. The proof we present here was given by Bóna and Ehrenborg [51] who built on an idea of Vesselin Gasharov [189].

If a path on a square grid uses steps $(1, 0)$ and $(0, 1)$ only, we will call it a *northeastern lattice path.*

Before proving the theorem, we need to set up some tools, which will be useful in the next section as well. We will construct a bijection from the set $\mathcal{A}(n, k)$ of n-permutations with k descents onto that of labeled northeastern lattice paths with n edges, exactly k of which are vertical. (Note the shift in parameters: $|\mathcal{A}(n, k)| = A(n, k + 1)$, but this will not cause any confusion.)

If a path on a square grid uses steps $(1, 0)$ and $(0, 1)$ only, we will call it a *northeastern lattice path.*

Let $\mathcal{P}(n)$ be the set of labeled northeastern lattice paths that have edges $a_1, a_2, \ldots, a_n$ and that corresponding positive integers $e_1, e_2, \ldots, e_n$ as labels, so that the following hold:

(i) the edge a_1 is horizontal and $e_1 = 1$,

(ii) if the edges a_i and a_{i+1} are both vertical, or both horizontal, then $e_i \geq e_{i+1}$,

(iii) if a_i and a_{i+1} are perpendicular to each other, then $e_i + e_{i+1} \leq i + 1$.

The starting point of a path in $\mathcal{P}(n)$ has no additional significance. Let $\mathcal{P}(n, k)$ be the set of all lattice paths in $\mathcal{P}(n)$ which have k vertical edges, and let $P(n, k) = |\mathcal{P}(n, k)|$.

PROPOSITION 1.28

The following two properties of paths in $\mathcal{P}(n)$ are immediate from the definitions.

- *For all $i \geq 2$, the inequality $e_i \leq i - 1$ holds.*

- *Fix the label e_i. If e_{i+1} can take value v, then it can take all positive integer values $w \leq v$.*

Also note that all restrictions on e_{i+1} are given by e_i, independently of preceding e_j, $j < i$. Now we are going to explain how we will encode our permutations by these labeled lattice paths.

LEMMA 1.29

The following description defines a bijection from $\mathcal{S}(n)$ onto $\mathcal{P}(n)$, where $\mathcal{S}(n)$ is the set of all n-permutations. Let $p \in \mathcal{S}(n)$. To obtain the edge a_i and the label e_i for $2 \leq i \leq n$, restrict the permutation p to the i first entries and relabel the entries to obtain a permutation $q = q_1 \cdots q_i$ of $[i]$. Then proceed as follows.

1. *If the position $i - 1$ is a descent of the permutation p (equivalently, of the permutation q), let the edge a_i be vertical and the label e_i be equal to q_i.*

2. *If the position $i - 1$ is an ascent of the permutation p, let the edge a_i be horizontal and the label e_i be $i + 1 - q_i$.*

Moreover, this bijection restricts naturally to a bijection between $\mathcal{A}(n, k)$ and $\mathcal{P}(n, k)$ for $0 \leq k \leq n - 1$.

PROOF

The described map is clearly injective. Let us assume that $i - 1$ and i are both descents of the permutation p. Let q, respectively r, be the permutation when restricted to the i, respectively $i + 1$, first elements. Observe that q_i is either r_i or $r_i - 1$. Since $r_i > r_{i+1}$ we have $q_i \geq r_{i+1}$ and condition (ii) is satisfied in this case. By similar reasoning, the three remaining cases (based on $i - 1$ and i being ascents or descents) are shown, hence the map is into the set $\mathcal{P}(n)$.

To see that this is a bijection, we show that we can recover the permutation p from its image. To that end, it is sufficient to show that we can recover p_n, and then use induction on n for the rest of p. To recover p_n from its image, simply recall that p_n is equal to the label ℓ of the last edge if that edge is vertical, and to $n + 1 - \ell$ if that edge is horizontal. Conditions (ii) and (iii) assure that this way we always get a number between 1 and n for p_n. $\blacksquare$

See Figure 1.3 for an example of this bijection.

Now we are in position to prove that the Eulerian numbers are log-concave.

PROOF (of Theorem 1.27). We construct an *injection*

$$\Phi : \mathcal{P}(n, k - 1) \times \mathcal{P}(n, k + 1) \longrightarrow \mathcal{P}(n, k) \times \mathcal{P}(n, k).$$

This injection Φ will be defined differently on different parts of the domain.

Let $(P, Q) \in \mathcal{P}(n, k - 1) \times \mathcal{P}(n, k + 1)$. Place the initial points of P and Q at $(0, 0)$ and $(1, -1)$, respectively. Then the endpoints of P and Q are $(n - k + 1, k - 1)$ and $(n - k, k)$, respectively, so while Q starts "below" P, it ends "above" P.

Let X be the *first* (most southwestern) common point of P and Q. It then follows that P arrives at X by an east step, and Q arrives at X by a north

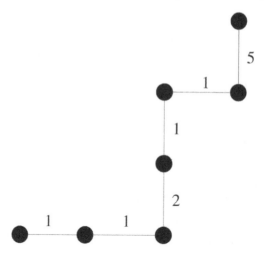

FIGURE 1.3
The image of the permutation 243165.

step. We will now show how to proceed if neither P nor Q changes directions at X, that is, P leaves X by an east step, and Q leaves X by a north step. The other cases are very similar and are left to the reader. Essentially, in all of the other cases, one of the transformations discussed below will have the desired effect when applied appropriately. See Problem Plus 17 for details.

Decompose $P = P_1 \cup P_2$ and $Q = Q_1 \cup Q_2$, where P_1 is a path from $(0,0)$ to X, P_2 is a path from X to $(n - k, k)$, Q_1 is a path from $(1, -1)$ to X, and Q_2 is a path from X to $(n - k + 1, k - 1)$. Let a, b, c, d be the labels of the four edges adjacent to X as shown in Figure 1.5, the edges AX and XB originally belonging to P and the edges CX and XD originally belonging to Q. Then by condition (ii) we have $a \geq b$ and $c \geq d$. Let $P' = P_1 \cup Q_2$ and let $Q' = Q_1 \cup P_2$.

1. If P' and Q' are valid paths, that is, if their labeling fulfills conditions (i)–(iii), then we set $\Phi(P, Q) = (P', Q')$. See Figure 1.4 for this construction. This way we have defined Φ for pairs $(P, Q) \in \mathcal{P}(n, k) \times \mathcal{P}(n, k)$ in which $a + d \leq i$ and $b + c \leq i$, where $i - 1$ is the sum of the two coordinates of X. We also point out that we have not changed any labels, therefore, in (P', Q') we still have $a \geq b$ and $c \geq d$, though that is no longer required, as the edges in question are no longer parts of the same path.

   It is clear that $\Phi(P, Q) = (P', Q') \in \mathcal{P}(n, k) \times \mathcal{P}(n, k)$, (in particular, (P', Q') belongs to the subset of $\mathcal{P}(n, k) \times \mathcal{P}(n, k)$ consisting of *intersecting* pairs of paths), and that Φ is one-to-one.

2. There are pairs $(P, Q) \in \mathcal{P}(n, k - 1) \times \mathcal{P}(n, k + 1)$ for which $\Phi(P, Q)$

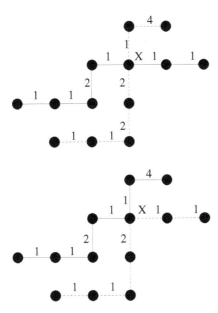

FIGURE 1.4
The new pair of paths.

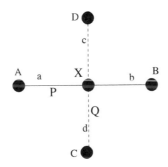

FIGURE 1.5
Labels around the point X.

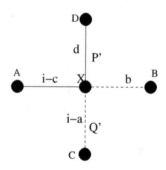

FIGURE 1.6
New labels around the point X.

cannot be defined in the way it was defined in the previous case, that is, when either $a + d > i$ or $b + c > i$ holds. The reader is invited to verify that such pairs (P, Q) actually exist. One example is when P is the path belonging to the permutation 1237654, and Q is the path belonging to the permutation 4567123.

Change the label of the edge AX to $i - c$ and change the label of the edge CX to $i - a$ as seen in Figure 1.6, then proceed as in the previous case to get $\Phi(P, Q) = (P', Q')$, where $P' = P_1 \cup Q_2$ and $Q' = Q_1 \cup P_2$.

We claim that P' and Q' are valid paths. Indeed we had at least one of $a + d > i$ and $b + c > i$, so we must have $a + c > i$ as $a \geq b$ and $c \geq d$. Therefore, $i - a < c$ and $i - c < a$, so we have decreased the values of the labels of edges AX and CX, and that is always possible as shown in Proposition 1.28. Moreover, no constraints are violated in P' and Q' by the edges adjacent to X as $i - c + d \leq i$ and $i - a + b \leq i$. It is also clear that Φ is one-to-one on this part of the domain, too. Finally, we have to show that the image of this part of the domain is disjoint from that of the previous part. This is true because in this part of the domain we have at least one of $a + d > i$ and $b + c > i$, that is, at least one of $i - c < b$ and $i - a < d$, so in the image, at least one of the pairs of edges AX, XB and CX, XD does not have the property that the label of the first edge is at least as large as that of the second one. And, as pointed out in the previous case, all elements of the image of the previous part of the domain do have that property.

Given $\Phi(P, Q) = (P', Q')$, the vertex X can be uniquely determined as the most southwestern point of the intersection of P' and Q'.

It then follows that the map Φ we created is an injection. This proves the inequality

$$A(n, k - 1)A(n, k + 1) \leq A(n, k)^2,$$

so our theorem is proved. ∎

There is a property of sequences of positive real numbers that is even stronger than log-concavity.

DEFINITION 1.30 *Let $a_1, a_2, \cdots, a_n$ be a sequence of positive real numbers. We say that this sequence has real roots only or real zeros only if the polynomial $\sum_{i=1}^{n} a_i z^i$ has real roots only.*

We note that sometimes the sequence can be denoted $a_0, a_1, \cdots, a_n$, and sometimes it is better to look at the polynomial $\sum_{i=0}^{n} a_i z^i$ (which, of course, has real roots if and only if $\sum_{i=0}^{n} a_i z^{i+1}$ does).

Example 1.31
For all positive integers n, the sequence $a_0, a_1, \cdots, a_n$ defined by $a_i = \binom{n}{i}$ has real zeros only. □

SOLUTION We have $\sum_{i=0}^{n} a_i z^i = \sum_{i=0}^{n} \binom{n}{i} z^i = (1+z)^n$, so all roots of our polynomial are equal to -1. ∎

Having real zeros is a stronger property than being log-concave, as is shown by the following theorem of Newton.

THEOREM 1.32
If a sequence of positive real numbers has real roots only, then it is log-concave.

PROOF Let $a_0, a_1, \cdots, a_n$ be our sequence, and let $P(z) = \sum_{k=0}^{n} a_k z^k$. Then for all roots (z, y) of the polynomial $Q(x, y) = \sum_{k=0}^{n} a_k z^k y^{n-k}$, the ratio (z/y) must be real. (Otherwise z/y would be a non-real root of $P(z)$). Therefore, by Rolle's Theorem, this also holds for the partial derivatives $\partial Q/\partial z$ and $\partial Q/\partial y$. Iterating this argument, we see that the polynomial $\partial^{a+b} Q/\partial z^a \partial y^b$ also has real zeros, if $a + b \leq n - 1$. In particular, this is true in the special case when $a = j - 1$, and $b = n - j - 1$, for some fixed j. This implies that the quadratic polynomial $R(z, y) = \partial^{n-2} Q/\partial z^{j-1} \partial y^{n-j-1}$ has real roots only, and therefore the discriminant of $R(z, y)$ is non-negative. On the other hand, we can compute $R(z, y)$ by computing the relevant partial derivatives. Note that we only have to look at the values of k ranging from $j - 1$ to $j + 1$ as all other summands of $Q(z, y)$ vanish after derivation. We get

$$R(z, y) = a_{j-1}(j-1)!\frac{1}{2}(n-j+1)!y^2 + a_j j!(n-j)!zy + a_{j+1}(n-j-1)!\frac{1}{2}(j+1)!z^2$$

As we said, this polynomial has to have a non-negative discriminant, meaning that

$$a_j^2 \geq \frac{j+1}{j} \cdot \frac{n-j+1}{n-j} \cdot a_{j-1}a_{j+1}, \tag{1.14}$$

which is stronger than our original claim, $a_j^2 \geq a_{j-1}a_{j+1}$. ∎

The alert reader has probably noticed that by (1.14), a log-concave sequence does not necessarily have real zeros only. For instance, the sequence 1, 1, 1 is certainly log-concave, but the polynomial $P(z) = 1 + z + z^2$ has two complex roots.

One might ask why we would want to know whether a combinatorially defined sequence has real zeros or not. In certain cases, proving the real zeros property is the only, or the easiest, way to prove log-concavity and unimodality. In some cases, unimodality and log-concavity can be proved by other means, but that does not always tell us where the maximum or maxima of a given sequence is, or just how many maxima the sequence has. Note that a constant sequence is always log-concave, so a log-concave sequence could possibly have any number of maxima. The following Proposition shows that in a sequence with real zeros only, the situation is much simpler.

PROPOSITION 1.33
If the sequence $\{a_k\}_{0 \leq k \leq n}$ has real zeros only, then it has either one or two maximal elements.

PROOF Formula (1.14) shows that in such a sequence, the ratio a_{j+1}/a_j strictly decreases, so it can be equal to 1 for at most one index j. ∎

Theorem 3.25 will show how to find the maximum (or maxima) of a sequence with real zeros.

The following theorem shows that Eulerian numbers have this last, stronger property as well.

THEOREM 1.34
For any fixed n, the sequence $\{A(n,k)\}_k$ of Eulerian numbers has real roots only. In other words, all roots of the polynomial

$$A_n(z) = \sum_{k=1}^{n} A(n,k)z^k$$

are real.

Recall that the polynomials $A_n(z)$ of Theorem 1.34 are called the *Eulerian polynomials*. This theorem is a classic result, but surprisingly, it is not easy to find a full, self-contained proof for it in the literature. The ideas of the proof we present here are due to Herb Wilf and Aaron Robertson.

PROOF (of Theorem 1.34) Theorem 1.7 implies

$$A_n(z) = (z - z^2)A'_{n-1}(z) + nzA_{n-1}(z) \qquad (n \geq 1; A_0(z) = z).$$

Indeed, the coefficient of z^k on the left-hand side is $A(n, k)$, while the coefficient of z^k on the right-hand side is

$$kA(n - 1, k) - (k - 1)A(n - 1, k - 1) + nA(n - 1, k - 1) =$$

$$kA(n - 1, k) + (n - k + 1)A(n - 1, k - 1) = A(n, k).$$

Now note that the right-hand side closely resembles the derivative of a product. This suggests the following rearrangement:

$$A_n(z) = z(1 - z)^{n+1}\frac{d}{dz}\left\{(1 - z)^{-n}A_{n-1}(z)\right\} \qquad (1.15)$$

with $n \geq 1$ and $A_0(z) = z$.

The Eulerian polynomial $A_0(z) = z$ vanishes only at $z = 0$. Suppose, inductively, that $A_{n-1}(z)$ has $n - 1$ distinct real zeros, one at $z = 0$, and the others negative. From (1.15), or otherwise, $A_n(z)$ vanishes at the origin. Further, by Rolle's Theorem, (1.15) shows that $A_n(z)$ has a root between each pair of consecutive roots of $A_{n-1}(z)$. This accounts for $n - 1$ of the roots of $g_n(z)$. Since we have accounted for all but one root, the remaining last root must be real since complex roots of polynomials with real coefficients come in conjugate pairs. ∎

We mention that an elementary survey of unimodal, log-concave, and real-roots-only sequences can be found in [75]. The articles [98] and [295] are earlier, high-level survey papers. A recent, high-level summary of the subject is Chapter 7 of [73], by Petter Brändén.

Eulerian numbers can count permutations according to properties other than descents. Let $p = p_1p_2 \cdots p_n$ be a permutation. We say that i is an *excedance* of p if $p_i > i$. (Note that for this definition, it is important to require that the entries of p are the elements of $[n]$ and not some other n-element set.)

Example 1.35
The permutation 24351 has three excedances, 1, 2, and 4. Indeed, $p_1 = 2 > 1$, $p_2 = 4 > 2$, and $p_4 = 5 > 4$. ▯

THEOREM 1.36
The number of n-permutations with $k - 1$ excedances is $A(n, k)$.

We postpone the proof of this theorem until Section 3.3.2, where it will become surprisingly easy, due to a different way of looking at permutations. However, we mention that if $f : S_n \to \mathbf{N}$ is a function associating natural numbers to permutations, then it is often called a *permutation statistic*. (Recall from the Introduction that S_n denotes the set of all n-permutations.) If a permutation statistic f has the same distribution as the statistic "number of descents", that is, if for all $k \in [n]$, we have

$$| \{p \in S_n : f(p) = k\} | = | \{p \in S_n : d(p) = k\} |, \qquad (1.16)$$

then we say that f is an *Eulerian* statistic. So Theorem 1.36 says that "number of excedances," sometimes denoted by *exc*, is an Eulerian statistic. We will see further Eulerian statistics in the Exercises section.

1.2 Alternating Runs

Let us modify the notion of ascending runs that we discussed in the last section. Let $p = p_1 p_2 \cdots p_n$ be a permutation. We say that p changes direction at position i if either $p_{i-1} < p_i > p_{i+1}$, or $p_{i-1} > p_i < p_{i+1}$. In other words, p changes directions when p_i is either a *peak* or a *valley*.

DEFINITION 1.37 *We say that p has k alternating runs if there are $k - 1$ indices i so that p changes direction at these positions.*

For example, $p = 3561247$ has 3 alternating runs as p changes direction when $i = 3$ and when $i = 4$. A geometric way to represent a permutation and its alternating runs by a diagram is shown in Figure 1.7. The alternating runs are the line segments (or edges) between two consecutive entries where p changes direction. So a permutation has k alternating runs if it can be represented by k line segments so that the segments go "up" and "down" exactly when the entries of the permutation do.

The origins of this line of work go back to the nineteenth century. More recently, D. E. Knuth [230] has discussed the topic in connection to sorting and searching.

Let $G(n, k)$ denote the number of n-permutations having k alternating runs. There are significant similarities between these numbers and the Eulerian numbers. For instance, for fixed n, both sequences have real zeros only, and both satisfy similar recurrence relations. However, the sequence of the $G(n, k)$ is not symmetric. On the other hand, almost half of all roots of the generating function $G_n(z) = \sum_{p \in S_n} z^{r(p)} = \sum_{k \geq 1} G(n, k) z^k$ are equal to -1. Here $r(p)$ denotes the number of alternating runs of p.

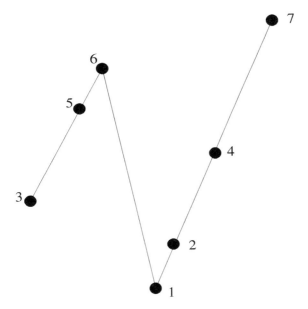

FIGURE 1.7
Permutation 3561247 has three alternating runs.

First we prove a simple recurrence relation on the numbers $G(n,k)$, which was first proved by André in 1883.

LEMMA 1.38
For positive integers n and k we have

$$G(n,k) = kG(n-1,k) + 2G(n-1,k-1) + (n-k)G(n-1,k-2), \quad (1.17)$$

where we set $G(1,0) = 1$, and $G(1,k) = 0$ for $k > 0$.

PROOF Let p be an $(n-1)$-permutation having k alternating runs, and let us try to insert n into p without increasing the number of alternating runs. We can achieve that by inserting n at one of k positions. These positions are right before the beginning of each descending run, and right after the end of each ascending run. This gives us $kG(n-1,k)$ possibilities.

Now let q be an $(n-1)$-permutation having $k-1$ alternating runs. We want to insert n into q so that it increases the number of alternating runs by 1. We can achieve this by inserting n into one of two positions. These two positions are very close to the beginning and the end of q. Namely, if q starts in an ascending run, then insert n to the front of q, and if q starts in a descending run, then insert n right after the first entry of q. Proceed dually at the end of the permutation.

$$2$$
$$2 \quad 4$$
$$2 \quad 12 \quad 10$$
$$2 \quad 28 \quad 58 \quad 32$$
$$2 \quad 60 \quad 236 \quad 300 \quad 122$$

FIGURE 1.8
The values of $G(n,k)$ for $n \le 6$. The first value of row n is $G(n,1)$. The NE–SW diagonals contain the values of $G(n,k)$ for fixed k.

Finally, let r be an $(n-1)$-permutation having $k-2$ alternating runs, and observe that by inserting n into any of the remaining $n - (k-2) - 2 = n - k$ positions, we increase the number of alternating runs by two. This completes the proof. ∎

The first values of $G(n,k)$ are shown in Figure 1.8 for $n \le 6$.

Looking at these values of $G(n,k)$, we note they are all even. This is easy to explain as p and its reverse always have the same number of alternating runs.

Taking a second look at the polynomials $G_n(z)$, we note that $G_4(z) = (z+1)(10z^2 + 2z)$, and that

$$G_5(z) = 32z^4 + 58z^3 + 28z^2 + 2z = (z+1)(32z^3 + 26z^2 + 2z).$$

Further analysis shows that $G_6(z)$ and $G_7(z)$ are divisible by $(z+1)^2$, and that $G_8(z)$ and $G_9(z)$ are divisible by $(z+1)^3$, and so on. In general, it seems that for any positive integer $n \ge 4$, the polynomial $G_n(z)$ is divisible by $(z+1)^{\lfloor (n-2)/2 \rfloor}$. We will show a recent proof of this fact that comes from [82].

Let $p = p_1 p_2 \cdots p_n$ be a permutation. The *complement* of p is the permutation $p^c = n+1-p_1 \; n+1-p_2 \; \cdots n+1-p_n$. For instance, the complement of 425613 is 352164. It is clear that p and $\bar{p}$ have the same number of alternating runs, since the diagram of p^c is just the diagram of p reflected through a horizontal line. In what follows, we will say *flipped* instead of *reflected through a horizontal line*. Note that this symmetry implies that all coefficients of $G_n(z)$ are even for $n \ge 2$.

Let s be a subsequence of entries in p that are in consecutive positions. Let S be set of entries that occur in s, in other words, the *underlying set* of s. Then the *complement of s relative to S* is the subsequence obtained from s so that for each j, the jth smallest entry of S is replaced by the jth largest entry of S. For example, if $s=24783$, then the complement of s relative to its underlying set S is 84327.

We will use a similar notion for sets. Let $T \subseteq U$. be finite sets. Let $T = \{t_1, t_2, \cdots, t_j\}$, where the t_i are listed in increasing order. Let us say that t_i is the a_ith smallest element of U. Then the *vertical complement* of

T with respect to U is the set consisting of the a_ith largest elements of U, for all i. For instance, if $T = \{1, 4, 6\}$, and $U = \{1, 2, 3, 4, 6, 8, 9\}$, then the vertical complement of T with respect to U is $\{3, 4, 9\}$. Indeed, T consists of the first, fourth, and fifth smallest elements of U, so its vertical complement with respect to U consists of the first, fourth, and fifth largest elements of U.

DEFINITION 1.39 *For $1 \leq i \leq n$, let c_i be the transformation on the set of all permutations $p = p_1 p_2 \cdots p_n$ that leaves the string $p_1 p_2 \cdots p_{i-1}$ unchanged, and replaces the string $p_i p_{i+1} \cdots p_n$ by its complement relative to its underlying set.*

Note that $c_1(p) = p^c$ and $c_n(p) = p$ for all p.

Example 1.40
Let $p = 315462$. Then $c_3(p) = 314526$, while $c_5(p) = 315426$. ⬜

PROPOSITION 1.41
Let $n \geq 4$, and let $3 \leq i \leq n-1$. Let p be any permutation of length n. Then one of p and $c_i(p)$ has exactly one more alternating run than the other.

PROOF As c_i does not change the number of runs of the string $p_1 p_2 \cdots p_{i-1}$ or the number of runs of the string $p_i p_{i+1} \cdots p_n$ and its image, all changes occur within the four-element string $p_{i-2} p_{i-1} p_i p_{i+1}$. There are only 24 possibilities for the pattern of these four entries, and it is routine to verify the statement for each of the possible 12 pairs. In fact, checking the six pairs in which $p_{i-2} < p_{i-1}$ is sufficient, for symmetry reasons.

As it is only the relations among the entries that matter, not the actual values of the entries, we can assume that those four entries are 1, 2, 3, and 4. Then the six pairs to check are

- 1234 and 1243,

- 1324 and 1342,

- 1423 and 1432,

- 2134 and 2143,

- 2314 and 2341, and

- 3412 and 3421.

Indeed, each of these pairs consists of two permutations so that one of them has one alternating run more than the other. ∎

In other words, $z^{\mathrm{run}(p)} + z^{\mathrm{run}(c_i(p))}$ is divisible by $1 + z$.

The following lemma is crucial for our purposes.

LEMMA 1.42

Let $1 \leq i \leq j - 2 \leq n - 2$. Then for all permutations p of length n, the identity $c_i(c_j(p)) = c_j(c_i(p))$ holds.

PROOF   Neither c_i nor c_j acts on any part of the initial segment $p_1 p_2 \cdots p_{i-1}$, so that segment, unchanged, will start both $c_i(c_j(p))$ and $c_j(c_i(p))$. The ending segment $p_j p_{j+1} \cdots p_n$ gets flipped twice by both $c_i c_j$ and $c_j c_i$, so in the end, the pattern of the last $n - j + 1$ entries will be the same in $c_i(c_j(p))$ and $c_j(c_i(p))$, because in both permutations, it will be the same pattern as it was in p. The middle segment $p_i p_{i+1} \cdots p_j$ will get flipped once by both $c_i c_j$ and $c_j c_i$, so on both sides, the pattern of entries in positions $i, i+1, \cdots, j$ will be the complement of the pattern of $p_i p_{i+1} \cdots p_j$.

Finally, the *set* of entries in the last $n - j + 1$ positions is the same in both $c_i(c_j(p))$ and $c_j(c_i(p))$, since both sets are equal to the vertical complement of the set $\{p_j, p_{j+1}, \cdots, p_n\}$ with respect to the set $\{p_i, p_{j+1}, \cdots, p_n\}$. ∎

Note that $c_i(c_j(p)) \neq p$, since the segment $p_i \cdots p_{j-1}$ is of length at least two and gets flipped exactly once. Note that this is where we use the fact that $i < j - 1$. Indeed, if $i = j - 1$, then it could happen that $c_i c_j(p) = p$. An example for this is when $p = 34251$, $i = 3$, and $j = 4$.

Example 1.43

Continuing Example 1.40, we get that $c_3(c_5(315462)) = c_5(c_3(315462)) = 314562.$ ▯

Now let $n \geq 4$ be any integer. If n is even, let

$$\mathcal{C}_n = \{c_3, c_5, c_7, \cdots, c_{n-1}\}.$$

If n is odd, let

$$\mathcal{C}_n = \{c_3, c_5, c_7, \cdots, c_{n-2}\}.$$

In both cases, $\mathcal{C}_n$ consists of $m = \lfloor (n - 2)/2 \rfloor$ operations. Each of these operations is an involution, and by Lemma 1.42, these involutions pairwise commute. None of these operations can be obtained as a product of the others, since only c_i changes turns the pattern $p_{i-1} p_i p_{i+1}$ into something other than itself or its complement. Therefore, the operations in $\mathcal{C}_n$ define a group $H_m \cong \mathbf{Z}_2^m$ that acts on the set of all permutations of length n.

Note that with this set of generators, $c_i(c_j(p)) \neq p$, since the segment $p_i \cdots p_{j-1}$ is of length at least two and gets flipped exactly once. Also note that the action of the individual c_i on the set of all permutations of length

n is independent in the following sense. If p is a permutation, and $c_i(p)$ has one more alternating run than p, then $c_j(c_i(p))$ also has one more alternating run than $c_j(p)$. Loosely speaking, the increase or decrease caused by c_j depends only on the permutation p that c_j is applied to, not on whether other transformations c_i are applied to p or not.

The action of H_m on the set of all permutations of length n creates orbits of size 2^m.

LEMMA 1.44
Let A be any orbit of H_m on the set of all permutations of length n. Then the equality

$$\sum_{p \in A} z^{\text{run(p)}} = z^a(1 + z)^m$$

holds, where a is a nonnegative integer.

PROOF Let $p \in A$, and going through p left to right, let us apply or not apply each element of C_n so as to minimize the number of alternating runs of the obtained permutation. That is, if applying c_i increases the number of alternating run, then do not apply it, if it decreases the number of alternating runs, then apply it. Let q be the obtained permutation. Then we call q the *minimal* permutation in A, since among all permutations in A, it is q that has the smallest number of alternating runs. Now elements of A with i more alternating runs than q can be obtained from q by applying exactly i elements of C_n to q. As there are $\binom{m}{i}$ ways to choose i such elements, our statement is proved by summing over i. ∎

Now we are ready to prove our conjecture about the high multiplicity of -1 as a root of $G_n(z)$.

THEOREM 1.45
For $n \geq 4$, the equality

$$G_n(z) = (1 + z)^m \sum_q z^{\text{run(q)}}$$

holds, where the summation is over permutations q that are minimal in their orbit under the action of H_m. Here $m = \lfloor (n - 2)/2 \rfloor$.

PROOF This follows from Lemma 1.44 by summing over all orbits. ∎

So almost half of the roots of $G_n(z)$ are equal to -1; in particular, they are real numbers. This raises the question whether the other half are real numbers as well. That question has recently been answered in the affirmative by Herb Wilf [330]. In his proof, he used the rather close connections between Eulerian polynomials, and the generating functions $G_n(z) = \sum_{k \geq 1} G(n, k) z^k$. This connection, established in [131], and given in a more concise form in [230], can be described by

$$G_n(z) = \left(\frac{1+z}{2}\right)^{n-1} (1+w)^{n+1} A_n \left(\frac{1-w}{1+w}\right), \tag{1.18}$$

where $w = \sqrt{\frac{1-z}{1+z}}$. The proof of (1.18) uses the similarities between the recurrence relations for $A_n(z)$ and $G_n(z)$ to get a differential equation satisfied by certain generating functions in two variables. The details can be found in [131], pages 157–162.

THEOREM 1.46

(H. Wilf [330].) For any fixed n, the polynomial $G_n(z)$ has real roots only.

PROOF From (1.18) it follows that $G_n(z)$ can vanish only if either $z = -1$ or $z = 2y/(1+y^2)$, where y is a zero of A_n. Indeed, if y is a root of $A_n(z)$ and $y = \frac{1-w}{1+w}$, then $w = \frac{1-y}{1+y}$. Therefore,

$$\sqrt{\frac{1-z}{1+z}} = \frac{1-y}{1+y}.$$

Squaring both sides and solving for z, we get our claim. As we know that the roots of A_n are real, our statement is proved. ∎

1.3 Alternating Subsequences

1.3.1 Definitions and a Recurrence Relation

The concept of alternating subsequences in permutations was introduced by Richard Stanley in [299]. In this section, we describe some of the major results of this recently developed subject, and we explain the very close connection between alternating subsequences and alternating runs.

DEFINITION 1.47 *An* alternating subsequence *in a permutation $p = p_1 p_2 \cdots p_n$ is a subsequence $p_{i_1} p_{i_2} \cdots p_{i_k}$ so that*

$$p_{i_1} > p_{i_2} < p_{i_3} > p_{i_4} < \cdots.$$

Similarly, *a* reverse alternating subsequence *in p is a subsequence* $p_{j_1} p_{j_2} \cdots p_{j_k}$ so that

$$p_{j_1} < p_{i_2} > p_{i_3} < p_{i_4} > \cdots .$$

The length of the longest alternating subsequence of p is denoted by $as(p)$.
For instance, if $p = 3416527$, then 3165, 31657, and 427 are examples of alternating subsequences of p.

Example 1.48

If $p = 35714268$, then $as(p) = 5$. Indeed, 31426 is an alternating subsequence of p of length 5. On the other hand, no alternating subsequence of p can contain more than one of the first three entries of p, and no alternating subsequence can contain more than two of the last three entries of p. Therefore, no alternating subsequence of p can be longer than five. ☐

Interestingly, the parameter $as(p)$ is much easier to handle in many aspects than its older and more studied brother, the length of the longest *increasing* subsequence of p. (We will study the latter in Chapter 4 and various later chapters.) The reason for this is that $as(p)$ is more conducive to arguments using recurrences, due to the following fact.

PROPOSITION 1.49

Let p be an n-permutation. Then p contains an alternating subsequence of length $as(p)$ that contains the entry n.

PROOF   Let us assume the contrary, that is, that there exists an n-permutation p so that each maximum-length alternating subsequence of p avoids the entry n. This means that all maximum-length alternating subsequences of p start on the right of n, or end on the left of n, or "skip" n; that is, contain an entry on the left of n that is followed by an entry on the right of n. It is easy to see that each of these three kinds of subsequences can be transformed into an alternating subsequence of the same length that contains n. For instance, if s is an alternating subsequence of p so that $a < b$ are two consecutive entries in s, and a is on the left of n and b is on the right of n, then we can replace b by n in s, and obtain the alternating subsequence s' that is still of maximum length. If $b < a$, then we can replace a by n to get the maximum length alternating subsequence s'' that contains n. Similarly, if t is a maximum length alternating subsequence that starts on the right of n, then the first entry of t can simply be replaced by n to create a maximum length alternating subsequence containing n. Finally, if u is a maximum length alternating subsequence ending on the left of n, then u has to end in an ascent; otherwise, n could be appended to the end of u to get a longer alternating subsequence, which would be a contradiction. So u ends in an ascent, but then the last entry of u can be replaced by n. ∎

The following definitions set the framework for our efforts to count permutations according to their alternating subsequences.

DEFINITION 1.50 *Let $a_k(n)$ be the number of permutations p of length n whose longest alternating subsequence is of length k. Set $a_0(0) = 1$. Furthemore, let $b_k(n) = \sum_{i=1}^{k} a_k(n)$ be the number of permutations of length n with no alternating subsequences longer than k.*

Note that $a_k(n) = b_k(n) - b_{k-1}(n)$.

Our first theorem on the subject of alternating subsequences turns the observation of Proposition 1.49 into an enumeration formula.

THEOREM 1.51

[299] Let $1 \leq k \leq n+1$. Then

$$a_k(n+1) = \sum_{j=0}^{n} \binom{n}{j} \sum_{2r+s=k-1} (a_{2r}(j) + a_{2r+1}(j))a_s(n-j),$$

where r and s range non-negative *integers satisfying $2r + s = k - 1$.*

PROOF We are going to build an $(n+1)$-permutation p that has a longest alternating subsequence of length k. Let us place the entry $n+1$ into the $(j+1)$st position of p; let L be the subsequence of the first j entries of p (since these are the entries on the left of p), and let R be the subsequence of the last $n - j$ entries of p.

It goes without saying that there are $\binom{n}{j}$ ways to choose the set of entries in L.

Let us now focus on a longest alternating subsequence *long* of $p = L(n+1)R$ containing $n+1$. Then *long* intersects L in an alternating subsequence *alts* of length $2r$ and R in a reverse alternating subsequence *rev* of length $s = k - 1 - 2r$. Clearly, no reverse alternating subsequence of R can be longer than *rev*. Therefore, as(rev) = s, so there are $a_s(n-j)$ choices for R. The situation for L is a little bit more complicated. Indeed, *alts* may or may not be the longest alternating subsequence of L. If *alts* is of even length, then it is possible that *alts* can be extended by another entry within L, but then that entry cannot be followed by an ascent (and so by the entry $n+1$) in any alternating subsequences. Therefore, as(alts) = 2r or as(alts) = 2r+1, and so the number of choices for L is $a_{2r}(j) + a_{2r+1}(j)$, completing the proof. ∎

Theorem 1.51 can be turned into generating function identities, which in turn can be turned into explicit formulae. The interested reader can find the details in [299].

THEOREM 1.52
For all positive integers $k \leq n$ we have

$$b_k(n) = \frac{1}{2^{k-1}} \sum_{\substack{r+2s \leq k \\ r \equiv k \ (mod 2)}} (-2)^s \binom{k-s}{(k+r)/2} \binom{n}{s} r^n.$$

As $a_k(n) = b_k(n) - b_{k-1}(n)$, formulae for $a_k(n)$ can be obtained from the preceding theorem.

For small values of k, Theorem 1.52 yields simple formulae such as

1. $b_1(n) = 1$,

2. $b_2(n) = 2^{n-1}$,

3. $b_3(n) = \frac{3^n - (2n-3)}{4}$,

4. $b_4(n) = \frac{4^n - 2(n-2)2^n}{8}$,

5. $b_5(n) = \frac{5^n - (2n-5)3^n + 2(n^2 - 5n + 5)}{16}$,

6. $b_6(n) = \frac{6^n - 2(n-3)4^n + (2n^2 - 12n + 15)2^n}{32}$.

1.3.2 Alternating Runs and Alternating Subsequences

The length of the longest alternating subsequence of a permutation is closely connected to the number of alternating runs as shown by the following proposition.

PROPOSITION 1.53
Let $n \geq 2$. Then $a_k(n) = \frac{1}{2}(G(n, k-1) + G(n, k))$.

PROOF If an n-permutation p has i alternating runs and starts in a descent, then as$(p) = i+1$ as can be seen by considering the entries of p that are peaks or valleys, as well as the first and last entry of p. It follows from the pigeon-hole principle that p cannot contain a longer alternating subsequence. If p starts in an ascent, then as$(p) = i$ by similar considerations.

Therefore, the n-permutations p satisfying as$(p) = k$ are precisely the n-permutations with $k - 1$ alternating runs starting in a descent and the n-permutations with k alternating runs starting in an ascent. ∎

Proposition 1.53 implies that if $T_n(z) = \sum_{p \in S_n} z^{\text{as}(p)}$, then

$$T_n(z) = \frac{1}{2}(1+z)G_n(z).$$

So the polynomials $T_n(z)$ have real roots only, and $\lfloor n/2 \rfloor$ of their roots are equal to -1.

The first few polynomials $T_n(z)$ are shown below.

1. $T_1(z) = z$,

2. $T_2(z) = z + z^2$,

3. $T_3(z) = z + 3z^2 + 2z^3$,

4. $T_4(z) = z + 7z^2 + 11z^3 + 5z^4$,

5. $T_5(z) = z + 15z^2 + 43z^3 + 45z^4 + 16z^5$,

6. $T_6(z) = z + 31z^2 + 148z^3 + 268z^4 + 211z^5 + 61z^6$,

7. $T_7(z) = z + 63z^2 + 480z^3 + 1344z^4 + 1767z^5 + 1113z^6 + 272z^7$.

1.3.3   Alternating Permutations

Sometimes an entire permutation is an alternating sequence, leading to the following definition.

DEFINITION 1.54   *We say that the n-permutation p is* alternating *if the longest alternating subsequence of p is of length n. Similarly, we say that p is* reverse alternating *if the longest reverse alternating subsequence of p is of length n.*

For instance, 312 and 5241736 are alternating permutations. Clearly, p is alternating if and only if its *complement*, that is, the n-permutation whose ith entry is $n + 1 - p_i$, is reverse alternating.

The number of alternating n-permutations is called an *Euler* number (not to be confused with the Eulerian numbers $A(n, k)$) and is denoted by E_n. The reader is invited to verify that $E_2 = 1$, $E_3 = 2$, $E_4 = 5$, and $E_5 = 16$. The Euler numbers have a very interesting exponential generating function. This is the content of the next theorem.

THEOREM 1.55
Set $E_0 = 1 = E_1$. *Then the equality*

$$E(z) = \sum_{n \geq 0} E_n \frac{z^n}{n!} = \sec z + \tan z$$

holds.

PROOF Let $L(n+1)R$ be an alternating or reverse alternating permutation of length $n+1$. So L is the string on the left of the maximal entry, and R is the string on the right of the maximal entry. Then R is reverse alternating, and so is L^r, that is, the reverse of L.

This observation leads to the recurrence relation

$$2E_{n+1} = \sum_{k=0}^{n} \binom{n}{k} E_k E_{n-k},$$

for $n \geq 1$. In terms of generating functions, this is equivalent to

$$2E'(z) = E^2(z) + 1,$$

with $E(0) = 1$.

We can easily solve the last displayed differential equation, and obtain the solution

$$
\begin{aligned}
y(z) &= \tan\left(\frac{z}{2} + \frac{\pi}{4}\right) \\
&= \frac{1 + \tan(z/2)}{1 - \tan(z/2)} = \frac{1 + \tan^2(z/2)}{1 - \tan^2(z/2)} + \frac{2\tan(z/2)}{1 - \tan^2(z/2)} \\
&= \sec z + \tan(z).
\end{aligned}
$$

∎

We point out that $\sec z$ is an even function and $\tan z$ is an odd function, so $\tan z$ is the exponential generating function for the Euler numbers with odd indices, and $\sec z$ is the exponential generating function for the Euler numbers with even indices. The numbers E_{2n} are often called the *secant numbers* and the numbers E_{2n+1} are often called the *tangent numbers*. There are various other combinatorial objects that are counted by the Euler numbers. Exercises 15 and 46 show two of them.

Alternating permutations have many fascinating properties, and we will return to those in upcoming chapters.

Exercises

1. (−) Simplify the formula obtained for $\alpha(S)$ in Lemma 1.3.

2. Let $p+1$ be a prime. What can be said about $A(p, k)$ modulo $p+1$?

3. (–) Find an alternative proof for the fact that $A(n, k+1) = A(n, n-k)$.

4. (–) What is the value of $A'_n(1)$? Here $A_n(z)$ denotes the nth Eulerian polynomial.

5. Prove Proposition 1.26.

6. We have n boxes numbered from 1 to n. We run an n-step experiment as follows. In step i, we drop one ball into a box, chosen randomly from boxes labeled 1 through i. So during the entire experiment, n balls will be dropped. Let $B(n, k)$ be the number of experiments in which at the end, $k - 1$ boxes are left empty. Prove that $B(n, k) = A(n, k)$.

7. Deduce Theorem 1.8 from Theorem 1.7.

8. (–) Prove that for all positive integers $k \leq n$, the equality

$$S(n, k) = S(n - 1, k - 1) + kS(n - 1, k)$$

holds.

9. Prove that

$$A(n, k) = \sum_{h=k-1}^{n} (-1)^{h-k+1} \binom{h}{k-1} S(n, n-h) \cdot (n - h)!.$$

10. Let $p = p_1 p_2 \cdots p_n$ be a permutation, and let b_i be the number of indices $j < i$ so that $p_j > p_i$. Find a formula for the number $C(n, k)$ of arrays $(b_1, b_2, \cdots, b_n)$ obtained this way in which exactly k different integers occur. Note that the permutation statistic defined as above is often called the *Dumont statistic*, and its value on the permutation p is denoted by $dmc(p)$.

11. Prove that

$$A_n(z) = z \sum_{k=1}^{n} k! S(n, k)(z - 1)^{n-k}.$$

12. (–) Prove that

$$\sum_{i=1}^{k} A(n, k) \leq k^n.$$

13. We say that i is a *weak excedance* of $p = p_1 p_2 \cdots p_n$ if $p_i \geq i$. Assuming Theorem 1.36, prove that the number of n-permutations with k weak excedances is $A(n, k)$.

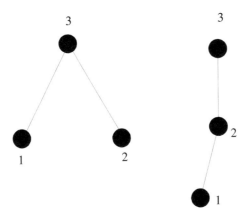

FIGURE 1.9
The two decreasing non-plane 1-2 trees on vertex set $[3]$.

14. Prove that for all positive integers n, the equality

$$S(n+1, k+1) = \sum_{m=k}^{n} \binom{n}{m} S(m, k)$$

holds.

15. A *decreasing non-plane 1-2 tree* is a rooted nonplane tree on vertex set $[n]$ in which each non-leaf vertex has at most two children, and the label of each vertex is smaller than that of its parent. See Figure 1.9 for the two decreasing non-plane trees on vertex set $[3]$. Let T_n denote the number of decreasing non-plane trees on vertex set $[n]$. Prove that $T_n = E_n$.

16. Find a combinatorial proof for Corollary 1.19.

17. Let r be a positive integer, and let us say that $i \in [n-1]$ is an *r-fall* of the permutation $p = p_1 p_2 \cdots p_n$ if $p(i) \geq p(i+1) + r$. Let $A(n, k, r)$ denote the number of all n-permutations with $k-1$ such r-falls. The numbers $A(n, k, r)$ are called the *r-Eulerian numbers*. Prove that

$$A(n, k, r) = (k + r - 1)A(n - 1, k, r) + (n + 2 - k - r)A(n - 1, k - 1, r).$$

18. Let $A(n, k, r)$ be defined as in the previous exercise. Prove that

$$A(n+r-1, k, r) = (k-1)! \sum_{i=0}^{n-k} (-1)^i \binom{n+r}{i} \binom{n+r-k-i}{r-1} (n-k-i+1)^n.$$

19. Are the r-Eulerian numbers defined in Exercise 17 and the ℓ-Takács-Eulerian numbers defined in Problem Plus 3 identical? (Try to give a very short solution.)

20. Let k be a fixed positive integer. Find the explicit form of the ordinary generating function $F_k(z) = \sum_{n \geq k} S(n, k) z^n$.

21. Prove that for all positive integers n, the equality

$$z^n = \sum_{m=0}^{n} S(n, m)(z)_m \tag{1.19}$$

holds. Recall that $(z)_m = z(z - 1) \cdots (z - m + 1)$.

22. Let $P(z)$ and $Q(z)$ be two polynomials with log-concave and positive coefficients. Prove that the polynomial $P(z)Q(z)$ also has log-concave coefficients.

23. Is it true that if $P(z)$ and $Q(z)$ are two polynomials with unimodal coefficients, then $P(z)Q(z)$ also has unimodal coefficients?

24. Is it true that if $P(z)$ and $Q(z)$ are two polynomials with *symmetric and unimodal* coefficients, then $P(z)Q(z)$ also has *symmetric and unimodal* coefficients?

25. (–)

    (a) Find an explicit formula for $G(n, 2)$.
    (b) Find an explicit formula for $G(n, 3)$.

    Here $G(n, k)$ is the number of n-permutations with k alternating runs.

26. (–) Let n be a fixed positive integer. For what pairs (k, m) does there exist an n-permutation with k descents and m alternating runs?

27. It follows from Exercise 21 that if $n \geq 1$, then

$$\sum_{m=1}^{n} (-1)^m m! S(n, m) = (-1)^n.$$

Find a direct combinatorial proof for this identity.

28. Prove that for n sufficiently large, the chain of inequalities $(k - 1)^n < G(n, k) < k^n$ holds, for all $k \geq 2$.

29. A sequence $f : \mathbf{N} \to \mathbf{C}$ is called P-recursive if there exist polynomials $P_0, P_1, \cdots, P_k \in \mathbf{Q}[n]$, with $P_k \neq 0$ so that

$$P_k(n + k)f(n + k) + P_{k-1}(n + k - 1)f(n + k - 1) + \cdots + P_0(n)f(n) = 0 \tag{1.20}$$

for all natural numbers n. Here P-recursive stands for "polynomially recursive". For instance, the function f defined by $f(n) = n!$ is P-recursive as $f(n+1) - (n+1)f(n) = 0$. Prove that for any fixed k, $A(n,k)$ is a P-recursive function of n.

30. Prove that for any fixed k, the function $S(n,k)$ is a polynomially recursive function of n.

31. A *decreasing binary tree* is a rooted binary plane tree that has vertex set $[n]$ and root n, and in which each vertex has 0, 1, or 2 children, and each child is smaller than its parent. Prove that the number of decreasing binary trees is $n!$.

32. Prove that the number of decreasing binary trees on $[n]$ in which $k-1$ vertices have a left child is $A(n,k)$.

33. Let $2 \le i \le n-1$. Recall that we say that p_i is a *peak* of the permutation $p = p_1 p_2 \cdots p_n$ if p_i is larger than both of its neighbors, that is $p_{i-1} < p_i$ and $p_i > p_{i+1}$. Let $n \ge 4$, and let $k \ge 0$. Find a formula for the number $Peak(n,k)$ of n-permutations having exactly k peaks.

34. How many decreasing binary trees are there on n vertices in which exactly one vertex has two children?

35. Use decreasing binary trees to prove that for any fixed n, the sequence of Eulerian numbers $\{A(n,k)\}_{1 \le k \le n}$ is symmetric and unimodal.

36. Which is the stronger requirement for two permutations, to have the same set of descents, or to have decreasing binary trees that are identical as unlabeled trees?

37. The *minmax tree* of a permutation $p_1 p_2 \cdots p_n$ is defined as follows. Let $p = umv$ where m is the *leftmost* of the minimum and maximum letters of p, u is the subword preceding m, and v is the subword following m. The *minmax tree* T_p^m has m as its root. The right subtree of T_p^m is obtained by applying the definition recursively to v. Similarly, the left subtree of T_p^m is obtained by applying the definition recursively to u. See Figure 1.10 for an example.

   (a) Prove that if $1 \le i \le n-2$, then there are $n!/3$ permutations p so that p_i is a leaf in T_p^m.

   (b) How many n-permutations p are there so that p_{n-1} (resp. p_n) is a leaf in T_p^m?

38. (+) Let $p = p_1 p_2 \cdots p_n$ be a permutation, and let $x \in [n]$. Define the x-factorization of p into the set of strings $u\lambda(x)x\gamma(x)v$ as follows. The string $\lambda(x)$ is the longest string of consecutive entries that are larger than x and are immediately on the left of x, and the string $\gamma(x)$ is the

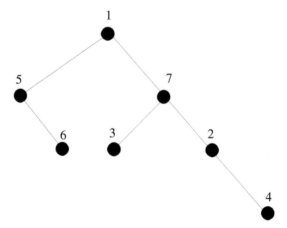

FIGURE 1.10
The minmax tree of $p = 5613724$.

longest string of of consecutive entries that are larger than x and are immediately on the right of x. Finally, $u(x)$ and $v(x)$ are the leftover strings at the beginning and end of p. Note that each of $\lambda(x)$, $\gamma(x)$, u, and v can be empty.

For instance, if $p = 31478526$ and $x = 4$, then $u = 31$, $\lambda(x) = \emptyset$, $\gamma(x) = 785$, and $v = 26$.

The notion of *André permutations* proved to be useful in various areas of algebraic combinatorics. We say that p is an *André permutation of the first kind* if

(a) there is no i so that $p_i > p_{i+1} > p_{i+2}$, and

(b) $\gamma(x) = \emptyset \Longrightarrow \lambda(x) = \emptyset$, and

(c) if $\gamma(x)$ and $\lambda(x)$ are both nonempty, then $\max \lambda(x) < \max \gamma(x)$.

Prove that p is an André permutation of the first kind if and only if all non-leaf nodes of the minmax tree T_p^m are chosen because they are minimum (and not maximum) nodes.

39. Attach labels to the edges of a $k \times (n - k + 1)$ square grid of points as shown in Figure 1.11. That is, both the edges of column i and row i get label i.

Take a northeastern lattice path s from the southwest corner to the northeast corner of the grid. This path s will consist of $n - 1$ steps. Define the weight P_s of s as the product of the labels of all edges of s. Prove that

$$A(n, k) = \sum_s P_s,$$

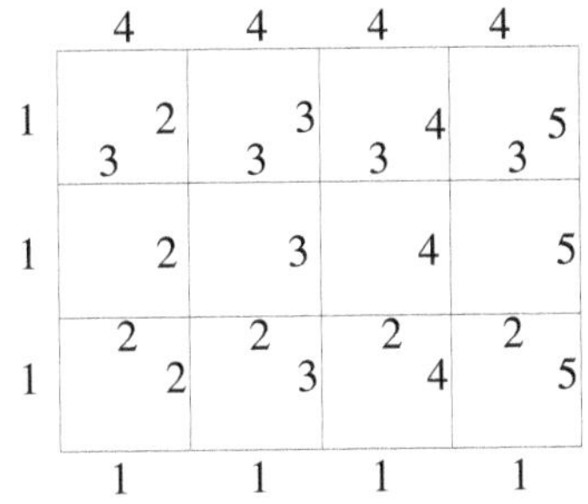

FIGURE 1.11
The labeled grid for $k = 4$ and $n = 8$.

where the sum is taken over all $\binom{n-1}{k-1}$ northeastern lattice paths s from the southwest corner to the northeast corner.

40. Prove, preferably by a combinatorial argument, that if $k < (n-1)/2$, then the inequality $G(n, k) \leq G(n, k+1)$ holds.

41. Let r be a positive integer, and modify the labeling of the vertical edges in the previous exercise so that the label of the edges in column i is $i + r - 1$ instead of i. Prove that

$$A(n, k, r) = r! \sum_{s,r} P_s,$$

where $A(n, k, r)$ is an r-Eulerian number as defined in Exercise 17, the weight P_s of a path s is still the product of the labels of its edges, and the sum is taken on all $\binom{n-r}{k-1}$ northeastern lattice paths from $(0, 0)$ to $(k - 1, n - r - k)$.

42. (−) Let Q_n be the set of permutations of the multiset $\{1, 1, 2, 2, \cdots, n, n\}$ in which for all i, all entries between the two occurrences of i are larger than i. For instance, Q_2 has three elements, namely 1122, 1221, and 2211. Note that the elements of Q_n are called *Stirling permutations* of length $2n$. Prove that for $n \geq 2$, the set Q_n has $(2n - 1)!!$ elements.

43. Let $p = p_1 p_2 \cdots p_{2n} \in Q_n$ be a Stirling permutation as defined in the previous exercise. Let us say that $i \in \{1, \cdots, 2n\}$ is a descent of p if $a_i > a_{i+1}$ or $i = 2n$. Let us say that $i \in \{1, 2, \cdots, 2n - 1\}$ is a *plateau* of p if $a_i = a_{i+1}$.

   (a) Let $C_{n,i}$ be the number of elements of Q_n with i descents. Prove that then for all positive integers $n, i \geq 2$, the recurrence relation $C_{n,i} = iC_{n-1,i} + (2n - i)C_{n-1,i-1}$ holds.

   (b) Let $c_{n,i}$ be the number of elements of Q_n with i plateaux. Prove that then for all positive integers $n, i \geq 2$, the recurrence relation $c_{n,i} = ic_{n-1,i} + (2n - i)c_{n-1,i-1}$ holds.

(c) Conclude that $C_{n,i} = c_{n,i}$ for all positive integers i and n, with $i \leq n$.

44. (+) Keep the notation of the previous exercise, and let

$$C_n(z) = \sum_{i=1}^{n} C_{n,i} z^i.$$

Prove that for all positive integers n, the roots of the polynomial $C_n(z)$ are all real, distinct, and non-positive.

45. (−) Find a direct combinatorial proof (no generating functions, no alternating runs) for the fact that for $n \geq 2$, the number as(p) is even for exactly half of all n-permutations p.

46. A permutation $p = p_1 p_2 \cdots p_n$ is called a *simsun* permutation if there exists no $k \leq n$ so that removing the entries larger than k from p, the remaining permutation has two descents in consecutive positions. For instance, $p = 35241$ is not simsun, since selecting $k = 3$ leads to the substring 321. Prove that the number of simsun permutations of length n is E_{n+1}.

Problems Plus

1. A *simplicial complex* is a collection Δ of subsets of a given set with the property that if $E \in \Delta$, and $F \subseteq E$, then $F \in \Delta$. The sets that belong to the collection Δ are called the *faces* of Δ. If $S \in \Delta$ has i elements, then we call S an $(i - 1)$-dimensional face. The dimension of Δ is, by definition, the dimension of its maximal faces.

   Prove that there exists a simplicial complex Δ whose set of $(i - 1)$-dimensional faces is in natural bijection with the set of n-permutations having exactly $i - 1$ descents.

2. (a) Let T be a rooted tree with root 0 and non-root vertex set $[n]$. Define a vertex of T to be a descent if it is greater than at least one of its children. Prove that the number of forests of rooted trees on a given vertex set with $i + 1$ leaves and j descents is the same as the number of forests of rooted trees with $j + 1$ leaves and i descents.

   (b) Why is the above notion of descents a generalization of the notion of descents in permutations?

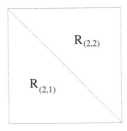

FIGURE 1.12
The regions $R_{n,k}$ for $n = 2$.

3. Define the ℓ-Stirling numbers of the second kind by the recurrence

$$S(n+1, k, \ell) = S(n, k-1, \ell) + k^\ell \ell S(n, k, \ell),$$

and the initial conditions $S(0,0,\ell) = 1$, $S(n,0,\ell) = 0$ for $n \geq 1$, and $S(0,k,\ell) = 0$ for $k \geq 1$. Note that for $\ell = 1$, these are just the Stirling numbers of the second kind as shown in Exercise 8. Define the ℓ-Takács-Eulerian numbers by

$$A_t(n, k, \ell) = \sum_{r=k-1}^{n} (-1)^{r-k+1} \binom{r}{k-1} S(n, n-r, \ell)[(n-r)!]^\ell.$$

Note that in the special case of $\ell = 1$, we get the Eulerian numbers, as shown in Exercise 9. Prove that these numbers generalize Eulerian numbers in the following sense. Modify the experiment of Exercise 6 so that in each step, ℓ balls are distributed, independently from each other. Prove that $\frac{A_t(n,k,\ell)}{n!^\ell}$ is the probability that after n steps, exactly k boxes remain empty.

4. Let $k \leq n$ be fixed positive integers. Compute the volume of the region $R_{n,k}$ of the hypercube $[0,1]^n$ contained between the two hyperplanes $\sum_{i=1}^{n} x_i = k - 1$ and $\sum_{i=1}^{n} x_i = k$. See Figure 1.12 for an illustration.

5. (a) Let $p = p_1 p_2 \cdots p_n$ be a permutation, and define

$$\delta_p = \sum_{1 \leq i < j \leq n} ||i - j| - |p_i - p_j||.$$

Prove that the smallest possible *positive* value of δ_p is $2n - 4$.

(b) Which graph theoretical problem contains part (a) as a special case?

6. Let G be a graph. A k-*coloring* of G is the number of ways to color the vertices of G using only the colors $1, 2, \cdots, k$ so that adjacent vertices have different colors. Let $P(n)$ be the number of n-colorings of G. It

is then well-known that $P(n)$ is a polynomial function of n, called the *chromatic polynomial* of G.

Now let

$$F_G(z) = \sum_{n \geq 0} P(n)z^n.$$

It is proved in [244] that $F_G(x) = \frac{Q(z)}{(1-z)^{m+1}}$, where $Q(z)$ is a polynomial of degree m, and with nonnegative integer coefficients.

So we can set $Q(z) = \sum_{i=k}^{m} w_i z^i$, where k is the smallest number for which G has a k-coloring, called the *chromatic number* of G.

(a) Find a combinatorial interpretation for the numbers w_p in terms of permutations.

(b) Explain why the polynomial $Q(z)$ is a generalization of the Eulerian polynomials.

7. Let n, i, and j be fixed positive integers, and set

$$S(i, j, n) = \sum_{0 \leq k \leq n} k^i (n-k)^j.$$

Prove that

$$S(i, 0, n) = \sum_{r=0}^{i} \binom{n+1}{r+1} r! S(i, r).$$

8. Let us say that a permutation p contains a *very tight ascending run* of length k if it has k consecutive entries $p_i p_{i+1} \cdots p_{i+k-1}$ so that $p_{i+j} = p_i + j - 1$ for $0 \leq j \leq k-1$. In other words, the sequence $p_i p_{i+1} \cdots p_{i+k-1}$ is a sequence of *consecutive integers*.

Find a formula for the number of permutations of length $r+k$ containing a tight ascending run of length at least k, if $k > r$.

9. The *Bessel number* $B(n, k)$ is defined as the number of partitions of $[n]$ into k nonempty blocks of size at most two. Prove that for any fixed n, the sequence $B(n, 1), B(n, 2), \cdots, B(n, n)$ is unimodal.

10. (a) Prove that if $n = 2^m - 1$ for some positive integer m, then all Eulerian numbers $A(n, k)$ with $1 \leq k \leq n$ are odd.

(b) Generalize the statement of part (a).

11. Let $A(n, k)_i$ denote the number of n-permutations with $k - 1$ descents that begin with i.

Prove that

$$A(n, k)_i = \sum_{j \geq 0} (-1)^{k-1-j} \binom{n}{k-1-j} j^{i-1} (j+1)^{n-i}.$$

12. John selects an n-permutation p at random. Jane must predict the descent set of p. What is Jane's best bet?

13. Prove that the Euler numbers E_n satisfy the formula

$$\frac{E_n}{n!} = 2 \left(\frac{2}{\pi}\right)^{n+1} \sum_{k \geq 0} (-1)^{k(n+1)} \frac{1}{(2k+1)^{n+1}}.$$

14. Let P be a finite partially ordered set whose vertices are bijectively labeled with the elements of $[n]$. The *Jordan–Hölder set* of P, denoted by $L(P)$, is the set of linear extensions of P. That is, $L(P)$ is the set of permutations of $p = p_1 p_2 \cdots p_n$ so that if $p_i <_P p_j$, then $i < j$ as integers.

Let $W(P, z) = \sum_{p \in L(P)} z^{d(p)}$. The *Stanley–Neggers conjecture* claimed that for all finite posets P, the polynomial $W(P, z)$ has real roots only.

(a) Let us call a finite poset P a *forest* if each element of P is covered by at most one other element. That is, P is a forest if for each $x \in P$, there is at most one $y \in P$ so that $x < y$ but there is no u so that $x < u < y$. Prove that the Stanley–Neggers conjecture is true for forests.

(b) Deduce that the Eulerian polynomials $A_n(z)$ have real roots only.

15. (a) Determine the polynomial $W(P, z)$ if P is the disjoint union of an m-element chain and an n-element chain.

(b) Let $P_{m,n}$ be the poset that consists of the disjoint union of an m-element chain and an n-element chain so that the elements of the first chain are labeled $1, 2, \cdots, m$ from the bottom up, and the elements of the second chain are labeled $m + 1, m + 2, \cdots, m + n$ from the bottom up, with the extra relation $m + 1 < m$ added. See Figure 1.13 for an illustration.

Compute $W(P_{m,n}, z)$.

(c) Prove that the Stanley–Neggers conjeture is false.

16. Let a_n be the number of $(n + 1)$-permutations that do not contain a very tight ascending run of length two. (See Problem Plus 8 for the definition of a very tight ascending run.) Find a closed formula for the exponential generating function $A(z) = \sum_{n \geq 0} a_n \frac{z^n}{n!}$.

17. In the proof of Theorem 1.27, we discussed the case when neither the path P nor the path Q changes directions at their first common point X. Complete the proof for the other cases.

18. *Stirling permutations* are defined in Exercise 42 and are the subject of several exercises after that. Let $p = p_1 p_2 \cdots p_{2n}$ be a permutation

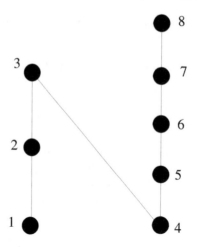

FIGURE 1.13
The poset $P_{3,5}$.

of the multiset $\{1, 1, 2, 2, \cdots, n, n\}$. We say that p is a *quasi-Stirling* permutation if there do not exist indices $i < j < k < \ell$ so that $p_i = p_k$ and $p_j = p_\ell$.

(a) Find a formula for the number of quasi-Stirling permutations of length $2n$.

(b) Find a formula for the number of quasi-Stirling permutations of length $2n$ that have a maximum number of descents.

19. Prove that there is a unique way to write the Eulerian polynomials in the form
$$A_n(z) = \sum_{k=1}^{\lfloor (n+1)/2 \rfloor} \gamma_{n,k} z^k (1+z)^{n-2k+1}.$$

20. Find a combinatorial interpretation for the numbers $\gamma_{n,k}$ in the previous Problem Plus.

Solutions to Problems Plus

1. This result is due to Vesselin Gasharov [189], who used the same lattice path model in his solution as he used to injectively prove that the Eulerian polynomials have log-concave coefficients.

2. (a) This result is due to Ira Gessel [195]. Let $d(F)$ be the number of descents of a forest F and let $\ell(F)$ be the number of leaves of F. Then let $u_n(\alpha, \beta)$ be the bivariate generating function

$$u_n(\alpha, \beta) = \sum_F \alpha^{d(F)} \beta^{\ell(F)-1},$$

where the sum is over all rooted forests on $[n]$. Then Gessel shows that the trivariate generating function

$$U(z, \alpha, \beta) = \sum_{n \geq 1} u_n(\alpha, \beta) \frac{z^n}{n!}$$

is symmetric in α and β by proving that it satisfies the functional equation

$$1 + U = (1 + \alpha U)(1 + \beta U)e^{z(1-\alpha-\beta-\alpha\beta U)}.$$

(b) If the number of leaves is one, then the tree consists of one line, and the sequence of the vertices corresponds to an n-permutation. The notion of descents of the tree then simplifies to that of descent in this permutation.

3. This result is due to Lajos Takács [308], though note that his paper denoted the Eulerian number $A(n, k)$ by $A(n, k - 1)$. The main idea of the proof is the following. Let

$$B_r(n) = \sum_{k=r}^{n} \binom{k}{r} P(n, k),$$

where $P(n, k)$ is the probability that at the end of the trials there are k empty boxes. Then it can be proved that

$$B_r(n) = S(n, n - r, \ell) \left(\frac{(n-r)!}{n!} \right)^{\ell}$$

by showing that both sides satisfy the same recurrence relations. Then, by the formula $P(k, n) = \sum_{r=k}^{n} (-1)^{r-k} \binom{r}{k} B_r(n)$, our claim follows.

4. The volume of $R_{n,k}$ is equal to $A(n, k)/k!$. A nice combinatorial proof was given by Richard Stanley [291]; though the result was probably known by Laplace. The main element of Stanley's proof is the following measure-preserving map. It is straightforward that $A(n, k)/k!$ is the volume of the set $S_{n,k}$ of all points $(x_1, x_2, \cdots, x_n) \in [0, 1]^n$ for which $x_{i-1} < x_i$ for exactly k values of i. (This includes $i = 0$, where we set $x_0 = 0$.) Let $f(x_1, x_2, \cdots, x_n) = (y_1, y_2, \cdots, y_n)$ where

$$y_i = \begin{cases} x_{i-1} - x_i \text{ if } x_{i-1} > x_i, \\ \\ 1 + x_{i-1} - x_i \text{ if } x_{i-1} < x_i \end{cases}$$

Note that f is not defined on the set of points where $x_{i-1} = x_i$ for some i, but that is not a problem as the set of those points has volume zero. Apart from that, however, f maps the rest of $S_{n,k}$ into $R(n,k)$ as $k - 1 \le \sum_{i=1}^{n} = k - x_n \le k$. Stanley then shows that apart from a subset of volume zero of $R(n,k)$, the map f has an inverse, and that f is an affine transformation of determinant $(-1)^n$, implying that f is order-preserving.

5. (a) This result is due to W. Aitken [1]. He called δ_p the *total relative displacement* of p.

   (b) Let G be a graph with n vertices, and let $d(x,y)$ be the graph-theoretical distance (number of edges in the shortest path) between x and y. Then, for a permutation p of the vertices of G, one can define
   $$\delta_{G,p} = \sum_{1 \le i < j \le n} |d(p_x, p_y) - d(x,y)|.$$

   Then part (a) corresponds to the special case when G is the path $12 \cdots n$. Also note that δ_p of part (a) is equal to 0 if and only if $p = 123 \cdots n$ or $p = n \cdots 321$, which is also a special case of the general fact that $\delta_{G,p} = 0$ if and only if p is an automorphism of G.

6. This result is due to I. Tomescu [311]. In that paper, various formulae are proved for the numbers w_k.

   (a) Let I be an acyclic orientation of G, and let G have m vertices. The transitive closure I' is then a partial ordering of $[m]$. Let f be a *bijective* coloring of the vertices of I that is compatible with I'. In other words, if $x < y$, then $f(x) <_{I'} f(y)$. Finally, let $T(I)$ be the set of all total orders that extend I'. In other words, the $T(I)$ are all the possible choices for the bijective coloring f.

   Now for any $f \in T(I)$, note that f in fact defines a permutation of $[m]$. Let $U(I)$ be the set of all these permutations. Finally, let $M(G)$ be the *multiset* obtained by taking the union of all $T(I)$, for all acyclic orientations I of G, preserving the multiplicities.

   It is then proved in [311] that for any graph G, the coefficient w_k is the number of permutations in $M(G)$ that have k ascents.

   (b) If G is the empty graph on m vertices, then $M(G)$ contains all $m!$ permutations of length m, and then $Q(z) = A_m(z)/z$.

7. There are several papers that are devoted to exploring connections between powers of integers and Stirling numbers of the second kind, or Eulerian numbers. See, for instance, [203] for this result.

8. It is proved in [219] that this number is $r!(r^2 + r + 1)$.

9. It is easy to prove that

$$B(n,k) = \frac{n!}{2^{n-k}(n-k)!(2k-n)!},$$

and then the result follows by checking that the sequence $B(n+1,k)/B(n,k)$ is decreasing, therefore if it dips below 1, it has to stay below 1. This result was published in [115].

10. (a) It is well known (see, for example, [74], Exercise 14 of Chapter 4) that if t is a power of 2, then $\binom{t}{k}$ is even, except when $k = t$ or $k = 1$. In our case, this means that $\binom{n+1}{k}$ is always even, except in those special cases, and the claim follows from Theorem 1.11.

    (b) Similarly, if $n = p^m - 1$ where p is a prime, then all Eulerian numbers $A(n,k)$ are congruent to 1 modulo p.

    A combinatorial proof of these facts not using Theorem 1.11 can be found in [309].

11. This result is due to Mark Conger, and was published in [124].

12. Jane's best bet is that p will be alternating or reverse alternating, that is, that $D(p) = \{1, 3, \cdots\} \cap [n-1]$ or $D(p) = \{2, 4, \cdots\} \cap [n-1]$. This is a classic result that has several proofs. Richard Stanley's survey paper [300] gives a modern proof using non-commuting variables, and mentions earlier proofs.

    Note that the answer is what one would intuitively expect. If $i \in D(p)$, then one would expect that $i+1$ is more likely to be an ascent than a descent, and if it is an ascent, then $i+2$ is more likely to be a descent, and so on.

13. A comprehensive treatment of the analytical tools needed to prove this can be found in *Analytic Combinatorics* by Philippe Flajolet and Robert Sedgewick [169]. In particular, a special case of Theorem IV.10 of that book states that if $f(z) = \sum_{n \geq 0} f_n z^n$ is a function all of whose singularities in the closed disc $|z| \leq R$ are simple poles, these poles are at points $\alpha_1, \alpha_2, \cdots$ in increasing order of their distance from 0, and p_i is the residue of f at α_i, then

$$f_n = \sum_{j \geq 1} p_j \alpha_j^{-n} + O(R^{-n}).$$

All singularities of $\sum_{n \geq 0} E_n \frac{z^n}{n!} = \tan z + \sec z$ are simple poles. They are at $\pi/2$, at $-3\pi/2$, at $5\pi/2$, and so on. The residue at each pole can be computed, and each turns out to be -2. As R goes to infinity, the sum on the right-hand side converges. This proves our claim.

14. (a) This result is due to David Wagner and can be found in [322].

    (b) Let P be an antichain of n elements. Then P is a forest, so the result of part (a) applies. On the other hand, $L(P) = S_n$, since the restriction that if $p_i <_P p_j$, then $i < j$ as integers is vacuous. So $W(P, z) = A_n(z)/z$, and the claim is proved.

15. Results in this exercise are due to Petter Brändén [92].

    (a) We claim that in this case, $W(P, z) = \sum_{k=0}^{\min(m,n)} \binom{m}{k}\binom{n}{k} z^k$. Indeed, let $p = p_1 p_2 \cdots p_n \in L(P)$. It follows from the structure of P that all descents of p must be formed by one element from each chain. That is, if $p_j > p_{j+1}$, then p_j ("the descent top") is one of the n elements of $[m+n]$ that are larger than n, and p_{j+1} ("the descent bottom") is one of the m elements of $[m]$. If p is to have i descents, then any i-element subset of $[m+n] \setminus [m]$ and any i-element subset of $[m]$ can play the role of descent tops and descent bottoms in exactly one (increasing) order, proving the claim.

    (b) We claim that $W_{P_{m,n}, z} = \sum_{k=1}^{\min(m,n)} \binom{m}{k}\binom{n}{k} z^k$. Indeed, the only permutation that is not in $L(P_{m,n})$ but is in $L(P)$ where P is as in part (a) is the increasing permutation.

    (c) See [92] for a proof of the result that if M is a positive integer, then for m and n such that $\min(m, n)$ is sufficiently large, the polynomial $W(P_{m,n}, z)$ has more than M non-real zeros.

Note that the Stanley–Neggers conjecture is still open for *naturally labeled posets*, that is, for posets in which if $x > y$, then the label of x is larger than the label of y.

16. First, we claim that $a_n = na_{n-1} + (n-1)a_{n-2}$, with $a_0 = a_1 = 1$. Indeed, we can insert $n+1$ into any position of any permutation counted by a_{n-1} except immediately on the right of n. Furthermore, in any permutation counted by a_{n-2}, we can replace any entry i by the string $i\,(n+1)\,(i+1)$. This recurrence relation leads to the generating function $A(z) = e^{-z}/(1-z)^2$.

17. We will describe all cases in a fairly uniform manner, in that there is always an earliest intersection of P and Q with the property that before that intersection, P was above Q, and immediately after the intersection, Q was above P. In the most generic case, discussed in the text, that intersection is just one point X, but in general, it can be a sequence of edges.

So, let us assume that the two edges of P that are attached to X are EE, with labels a and b, while the two edges of Q that are attached to X are NE, with labels c and d. This implies that

$$a \geq b, \qquad (1.21)$$

and

$$c + d \leq i. \tag{1.22}$$

Now proceed as follows.

(a) If we can simply swap the parts of P and Q that start at X, do so. That happens if and only if $a \geq d$ and $b + c \leq i$.

Note that the image will be a pair of paths so that we can swap their parts starting at X and get a pair of valid paths.

(b) If we cannot, proceed as above because $b + c > i$, then $a + c > i$ by (1.21). Replace the label a by $i - c$, and replace the label c by $i - a$. Both of these labels decrease, so they will not violate any constraints with the edges preceding them. Now swap the parts of paths that start at X. We get an EE path with labels $i - c$ and d, which is fine, by (1.22), and a NE path with labels $i - a$ and b, which is fine, by (1.21).

Note that the image of (P, Q) will be a pair of paths so that we *cannot* just swap the parts starting at X and get a pair of valid paths, because we would get an EE path with labels $i - c$ and b, and that is not allowed, since $i - c < b$ in this case.

So in these two cases, our map Φ is injective. Now we discuss the remaining case, that is, when $b + c \leq i$, but $a < d$.

In this case, we take a look at how P and Q continue *after* their coinciding edges adjacent to X. Recall that before and after X, the edges of P are EE, with labels a and b, while the edges of Q are NE, with labels c and d. The paths P and Q may coincide for a while after their edges labeled b and d, but when they first separate, Q will have an N step, and P will have an E step. (This follows from how we defined X in this solution.)

Let $b, b_1, \cdots, b_k$ be the labels of edges of P on that coinciding segment, and let $d, d_1, \cdots, d_k$ be the labels of edges of Q on that coinciding segment.

Furthermore, let us say that b_k is followed by the E edge labeled b', and d_k is followed by the N edge labeled d'.

Now we are in a position to define our map Φ in the current case.

(a) If there is an index j so that $b_j \geq d_{j+1}$ and $d_j \geq b_{j+1}$, then find the smallest such j, and swap the parts of P and Q that start with b_{j+1} and d_{j+1}. Importantly, this is reversible, so we can recover the original pair of paths from their image.

(b) If there is no such j, that means that for all j, we have $d_j > b_j$. This is easy to prove from the fact that $d_0 = d > a \geq b_0 = b$. So $d > b_1$, and therefore, the fact that $j = 1$ does not work implies that $d_1 > b \geq b_1$. Then we can similarly prove that $d_2 > b_2$, and so on. In this case, we swap the parts of P and Q that start with b' and d'.

In this way, we get valid paths, because

- the inequality $b_k + d' < d_k + d'$ holds, and even the right-hand side was small enough to not violate the constraint in Q, and
- the inequality $d_k \geq b_k \geq b'$ also holds.

18. (a) This number is $(2n)!/(n+1)!$, as is proved in [16].

   (b) The number of such quasi-Sterling permutations is $(n+1)^{n-1}$. See [161] for a proof.

19. This is a classic result of Dominique Foata and Marcel-Paul Schützenberger [172].

20. The number $\gamma_{n,k}$ is the number of all permutations of length n that have k descents and no *double descents* [172]. A double descent is a descent that is either followed by another descent, or by the last entry of the permutation.

2

In One Line and Anywhere. Permutations as Linear Orders. Inversions.

2.1 Inversions

2.1.1 Generating Function of Permutations by Inversions

In Section 1.1, we looked at descents of permutations. That is, we studied instances in which an entry in a permutation was larger than *the entry directly following it*. A more "global" permutation statistic is that of *inversions*. This statistic will look for instances in which an entry of a permutation is smaller than *some entry following it* (not necessarily directly).

DEFINITION 2.1 *Let $p = p_1 p_2 \cdots p_n$ be a permutation. We say that (p_i, p_j) is an inversion of p if $i < j$ but $p_i > p_j$.*

Example 2.2
Permutation 31524 has four inversions, namely $(3, 1)$, $(3, 2)$, $(5, 2)$, and $(5, 4)$.
□

This line of research started as early as 1901 [254]. In this section, we survey some of the most interesting results in this area. The number of inversions of p will be denoted by $i(p)$, though some authors prefer $inv(p)$. It is clear that $0 \leq i(p) \leq \binom{n}{2}$ for all n-permutations, and that the two extreme values are attained by permutations $12 \cdots n$ and $n(n-1) \cdots 1$, respectively. It is relatively easy to find the generating function enumerating all permutations of length n with respect to their number of inversions.

THEOREM 2.3
For all positive integers $n \geq 2$,

$$\sum_{p \in S_n} z^{i(p)} = I_n(z) = (1 + z)(1 + z + z^2) \cdots (1 + z + z^2 + \cdots + z^{n-1}).$$

DOI: 10.1201/9780429274107-2

PROOF   We prove the statement by induction on n. In fact, we prove that each of the $n!$ expansion terms of the product $I_n(z)$ corresponds to exactly one permutation in S_n. Moreover, the expansion term $z^{a_1} z^{a_2} \cdots z^{a_{n-1}}$ will correspond to the unique permutation in which, for each $i \in [n]$, the entry $i + 1$ precedes exactly a_i entries that are smaller than itself.

If $n = 2$, then there are two permutations to count, $p = 12$ has no inversions, and $p' = 21$ has one inversion. So $\sum_{p \in S_2} z^{i(p)} = 1 + z$ as claimed. Furthermore, $p = 12$ is represented by the expansion term 1, and $p' = 21$ is represented by the expansion term z.

Now let us assume that we know that the statement is true for $n - 1$, and prove it for n. Let p be a permutation of length $n - 1$. Insert the entry n into p to get the new permutation q. If we insert n into the last position, we create no new inversions. If we insert n into the next-to-last position of p, we create one new inversion as n will be larger than the last element of q. In general, if we insert n into p so that it precedes exactly i entries of p, we create i new inversions as n will form an inversion with each entry on its right, and with no entry on its left. Therefore, depending on where we inserted n, the new permutation q has 0 or 1 or 2, etc., or $n - 1$ more inversions than p did. If p was represented by the expansion term $z^{a_1} z^{a_2} \cdots z^{a_{n-2}}$, and n is inserted so that it precedes i entries, then q is represented by the new expansion term $z^{a_1} z^{a_2} \cdots z^{a_{n-2}} z^i$. This argument works for all p, proving that

$$I_n(z) = (1 + z + \cdots + z^{n-1}) I_{n-1}(z) = (1 + z)(1 + z + z^2) \cdots (1 + z + \cdots + z^{n-1}).$$

∎

Later in this chapter we will have the techniques to write Theorem 2.3 in a much more compact form.

Therefore, the number $b(n, k)$ of n-permutations with k inversions is the coefficient of z^k in $I_n(z)$. The fact that the polynomial $I_n(z)$ can be decomposed into a product of factors enables us to prove the following result on these numbers.

COROLLARY 2.4
For any fixed n, the sequence $b(n, 0), b(n, 1), \cdots, b(n, \binom{n}{2})$ is log-concave.

PROOF   Let us call a polynomial log-concave if its coefficients form a log-concave sequence. It is then not hard to prove (see Exercise 22 of Chapter 1) that the product of log-concave polynomials is log-concave. The previous theorem shows that the generating function of our sequence is the product of several log-concave polynomials (of the form $1 + z + z^2 + \cdots + z^t$). Therefore, our sequence itself is log-concave.   ∎

The first few values of the numbers $b(n, k)$ are shown in Figure 2.1.

n=1						1						

```
n=1                            1
n=2                       1        1
n=3                   1    2    2    1
n=4                1    3    5    6    5    3    1
n=5           1   4  9  15  20  22  20  15  9  4  1
```

FIGURE 2.1

The values of $b(n, k)$ for $n \leq 5$. Row n starts with $b(n, 0)$.

We would like to point out that it is not true that the polynomial $I_n(z)$ has real roots only. Indeed, if $n \geq 3$, then $I_n(z)$ is divisible by $1 + z + z^2$, and therefore has some complex roots.

We have found, with not much effort, the generating function of the numbers $b(n, k)$ for fixed n. An enumerative combinatorialist will certainly ask next whether it is possible to find a recursive, or even better, an explicit formula for these numbers, just as we did for the numbers $A(n, k)$ in Section 1.1. As we will see, the latter is a somewhat more difficult task.

To start, we prove a recurrence relation.

LEMMA 2.5

Let $n \geq k$. Then we have

$$b(n + 1, k) = b(n + 1, k - 1) + b(n, k). \tag{2.1}$$

PROOF   Let $p = p_1 p_2 \cdots p_{n+1}$ be an $(n + 1)$-permutation with k inversions, where $k \leq n$. If $p_{n+1} = n + 1$, then we can remove the entry $n + 1$ from the end of p and get an n-permutation with k inversions. If $p_i = n + 1$ for $i \leq n$, then let us interchange $n + 1$ and the entry immediately following it. This results in an $(n + 1)$-permutation with $k - 1$ inversions in which the entry $n+1$ is not in the first position. However, all $(n+1)$-permutations with $k - 1$ inversions have that property (that $n + 1$ is not in the first position) as putting $n + 1$ to the first position would result in at least $n \geq k > k - 1$ inversions. This completes the proof, and also shows why the condition $n \geq k$ is needed.   ∎

Even without going into details, it is obvious that (2.1) does not hold in general, that is, when $k > n$. For instance, if $k > \binom{n}{2}/2$, then $b(n + 1, k) < b(n+1, k-1)$, which makes it impossible for (2.1) to hold as $b(n, k) \geq 0$. See Exercise 28 for a recurrence relation for the case when $k > n$.

Finding an explicit formula for the numbers $b(n, k)$ is significantly more difficult, even if we assume $n \geq k$. A little examination of the polynomial

$I_n(z) = \sum_{k=0}^{\binom{n}{2}} b(n,k) z^k$ shows that

$$b(n,0) = 1 = \binom{n}{0},$$

$$b(n,1) = n - 1 = \binom{n}{1} - \binom{n}{0} \qquad n \geq 1,$$

$$b(n,2) = \binom{n}{2} - \binom{n}{0}, \qquad n \geq 2,$$

$$b(n,3) = \binom{n+1}{3} - \binom{n}{1} \qquad n \geq 3,$$

$$b(n,4) = \binom{n+2}{4} - \binom{n+1}{2} \qquad n \geq 4.$$

In order to see how these results are obtained, and to obtain a general formula, we need some notions that are probably well-known to most readers.

DEFINITION 2.6   *Let n be a positive integer. If $a_1 + a_2 + \cdots + a_k = n$, and the a_i are all positive integers, then we say that the k-tuple $(a_1, a_2, \cdots, a_k)$ is a composition of n into k parts. If the a_i are all nonnegative integers, then we say that the k-tuple $(a_1, a_2, \cdots, a_k)$ is a weak composition of n into k parts.*

In the unlikely event that the reader has not met compositions before, the reader should take a moment to prove that the number of compositions of n into k parts is $\binom{n-1}{k-1}$, whereas the number of weak compositions of n into k parts is $\binom{n+k-1}{k-1}$.

For instance, to get $b(n,2)$ as the coefficient of z^2 in $I_n(z)$, one has to count the weak compositions of 2 into $n-1$ nonnegative parts, the first of which is at most 1. Indeed, one needs to find the coefficient of z^2 in the generating function $I_n(z) = (1+z)(1+z+z^2)\cdots(1+z+\cdots+z^{n-1})$. The number of all weak compositions of 2 into $n-1$ parts is $\binom{2+n-1-1}{n-1} = \binom{n}{n-1} = n$, one of which consists of a first part equal to 2. This proves that $b(n,2) = n-1$. See Exercises 3 and 4 for proofs in the cases of $k=3$ and $k=4$.

This line of formulae suggests that maybe the formula for $b(n,k)$ will be obtained by taking the difference of two suitably chosen binomial coefficients. However, this conjecture is false as we have

$$b(n,5) = \binom{n+3}{5} - \binom{n+2}{3} + 1.$$

Further conjectures claiming $b(n,k)$ to satisfy a formula of comparable simplicity also turn out to be false. The truth is a bit more complicated than that.

FIGURE 2.2
The Ferrers shape of $p = (5, 2, 1)$.

Our main tool in finding the correct formula comes, remarkably, from the theory of *integer partitions*. Many readers are probably familiar with the following definition.

DEFINITION 2.7 *Let $a_1 \geq a_2 \geq \cdots \geq a_m \geq 1$ be integers so that $a_1 + a_2 + \cdots + a_m = n$. Then the array $a = (a_1, a_2, \cdots, a_m)$ is called a* partition *of the* integer n, *and the numbers a_i are called the* parts *of the partition a. The number of all partitions of n is denoted by $p(n)$.*

Partitions of the integer n are not to be confused with partitions of the set $[n]$. If there is a danger of confusion, we may refer to the objects we have just defined as *integer partitions* and to the objects we defined in Section 1.1 as *set partitions*.

Example 2.8
The integer 5 has seven partitions, namely (5), $(4, 1)$, $(3, 2)$, $(3, 1, 1)$, $(2, 2, 1)$, $(2, 1, 1, 1)$, and $(1, 1, 1, 1, 1)$. Therefore, $p(5) = 7$. ☐

The topic of integer partitions has been extensively researched for several centuries, from combinatorial, number theoretical, and analytic aspects. See [13] for a survey.

We will use the following simple, but extremely useful, representation of partitions by diagrams. A *Ferrers shape* of a partition $p = (a_1, a_2, \cdots, a_k)$ is a set of n square boxes with sides parallel to the coordinate axes so that in the ith row we have a_i boxes and all rows start at the same vertical line. The Ferrers shape of the partition $p = (5, 2, 1)$ is shown in Figure 2.2. Clearly, there is an obvious bijection between partitions of n and Ferrers shapes of size n.

We will need some basic facts about the generating functions of various partitions.

PROPOSITION 2.9
The ordinary generating function of the sequence of the numbers $p(n)$, where

$n = 0, 1, \cdots,$ *is*

$$\sum_{n \geq 0} p(n) z^n = \prod_{i=1}^{\infty} \frac{1}{1 - z^i}. \tag{2.2}$$

PROOF   We can decompose the right-hand side as

$$(1 + z + z^2 + \cdots)(1 + z^2 + z^4 + \cdots) \cdots (1 + z^i + z^{2i} + \cdots).$$

It is now clear that the coefficient of x^n in this product is equal to the number of vectors $(c_1, c_2, \cdots)$ with nonnegative integer coefficients for which $\sum_{i=1}^{\infty} i c_i = n$. Note that such a vector can have only a finite number of nonzero coordinates. Finally, there is a natural bijection between these vectors and the partitions of n. This bijection maps $(c_1, c_2, \cdots)$ into the partition that has c_i parts equal to i. So the coefficient of z^n on the right-hand side is $p(n)$. ∎

COROLLARY 2.10
Let $p_d(n)$ be the number of partitions of n into distinct parts. Then the equality

$$\sum_{n \geq 0} p_d(n) z^n = \prod_{i=1}^{\infty} (1 + z^i) \tag{2.3}$$

holds.

COROLLARY 2.11
Let $p(n, m)$ be the number of partitions of n into at most m parts. Then the equality

$$\sum_{n \geq 0} p(n, m) z^n = \prod_{i=1}^{m} \frac{1}{1 - z^i} \tag{2.4}$$

holds.

A *pentagonal number* is a non-negative integer n satisfying $n = \frac{1}{2}(3j^2 \pm j)$ for some non-negative integer j. So the first few pentagonal numbers are $0, 1, 2, 5, 7, \cdots$.

The following partition identity is the most interesting one in our quest for the formula for the numbers $b(n, k)$. It is not nearly as easy as the preceding two corollaries.

LEMMA 2.12
Let $p_{o,d}(n)$ (resp. $p_{e,d}(n)$) be the number of partitions of n into an odd number

p

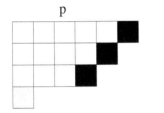

q

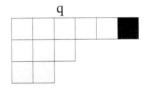

FIGURE 2.3

We have $b(p) = 3$, $g(p) = 1$, and $b(q) = 1$, $g(q) = 2$.

of distinct parts (resp. an even number of distinct parts). Then

$$p_{e,d}(n) - p_{o,d}(n) = \begin{cases} 0 \text{ if } n \text{ is not pentagonal,} \\ (-1)^j \text{ if } n = \frac{1}{2}(3j^2 \pm j). \end{cases}$$

Example 2.13

Let $n = 6$, which is not a pentagonal number. Then $p_{e,d}(6) = 2$, since the suitable partitions are $(5, 1)$ and $(4, 2)$. On the other hand, $p_{o,d}(6) = 2$, since the suitable partitions are (6) and $(3, 2, 1)$.

Now let $n = 7$, which is a pentagonal number. Then $p_{e,d}(7) = 3$, since the suitable partitions are $(6, 1)$, $(5, 2)$, and $(4, 3)$. On the other hand, $p_{o,d}(7) = 2$, since the suitable partitions are (7) and $(4, 2, 1)$. ◻

PROOF We will define a map ϕ from the set of partitions of n into an even number of distinct parts to the set of partitions of n into an odd number of distinct parts. We will then show that if n is not a pentagonal number, then our map is a bijection, proving the first part of our theorem. The second part of the theorem will be proved by analyzing why ϕ is not a bijection if n is pentagonal.

Let $q = (q_1, q_2, \cdots, q_j)$ be any partition of n. We define two parameters for q. The simpler one, $g(q)$ (for gray) is just the size of the smallest part of q. In other words, $g(q) = q_j$. The other one, $b(q)$ (for black) is the length of the longest strictly decreasing subsequence $(q_1, q_2, \cdots)$ of parts of q in which each part is exactly one less than the part immediately preceding it. Note that by definition, this strictly decreasing subsequence *must* start with q_1. So in particular, if $q_2 \neq q_1 - 1$, then $b(q) = 1$.

See Figure 2.3 for an illustration.

The set of black boxes will be called the *outer rim* of a Ferrers shape.

Now let p be a partition of n into an even number of distinct parts. We distinguish two cases.

1. If $b(q) \geq g(q)$, then we remove the last part of q, and add one to each of the first $g(q)$ parts of q, to get the partition $\phi(q)$. This operation

decreases the number of parts by one, and always keeps all parts distinct. The procedure can always be carried out *except* in the case when each part of q, including the last one, is one less than the preceding one, and $b(q) = g(q)$, such as in the case of $q = (3, 2)$ or $q = (7, 6, 5, 4)$. Indeed, it is precisely in that case that we have no row left into which we could place the last gray box. We do not define ϕ in that exceptional case.

Note that at the end of this procedure, $b(\phi(q)) < g(\phi(q))$ holds. This is because $b(\phi(q)) = g(q)$, and $g(\phi(q))$ is at least as large as the *next-to-last* row of q.

2. If $b(q) < g(q)$, then we decrease each of the first $b(q)$ parts of q by one, then affix a last part of size $b(q)$ to the end of q to get the partition $\phi(q)$. This operation increases the number of parts by one, and keeps the parts all distinct, *except* in the case when each part of q, including the last one, is one less than the preceding one, and $b(q) = g(q) - 1$, such as in the case of $q = (4, 3)$. Indeed, it is precisely in that case that the new last part that we create is just as long as the part immediately preceding it, that is, of length $g(q) - 1$. We do not define ϕ in that exceptional case.

Note that at the end of this procedure, $b(\phi(q)) \geq g(\phi(q))$ holds. Indeed, the procedure does not decrease b and does not increase g.

So the only times when $\phi(q)$ is not defined are as follows.

(A) When $b(q) = g(q)$, and the last part of q and the outer rim of the Ferrers shape of q both consist of j boxes. Therefore, the partition q is of the form $q = (2j - 1, 2j - 2, \cdots, j)$, so $n = \frac{(3j-1)j}{2}$.

(B) When $b(q) = g(q) - 1$, then the outer rim of q consists of j boxes, and the last part of q is $j + 1$. Therefore, in this case, $q = (2j, 2j - 1, \cdots, j + 1)$, so $n = \frac{(3j+1)j}{2}$.

Note that in both of these exceptional cases, the integer j has to be even, to assure that q has an even number of parts.

So in all cases but the exceptional cases (A) and (B), our function ϕ maps into the set of partitions of n into an odd number of distinct parts. It is clear that ϕ is one-to-one since it is one-to-one on both parts of its domain, and the images of the two parts are disjoint, as explained in the last sentence of the definition of $\phi(q)$ in each case. Let us examine whether ϕ is surjective.

Let r be a partition of n into an odd number of distinct parts, and let us try to find the preimage of r under ϕ. If $g(r) \leq b(r)$, then this preimage can be found by removing the last part of r and adding one to each of the first $g(r)$ parts of r. This can always be done, unless $g(r) = b(r)$, and each part of r, including the last one, is one less than the preceding one, as in $q = (5, 4, 3)$. If $g(r) > b(r)$, then this preimage can be found by decreasing each of the first $b(r)$ parts of r by one, and creating a new, last part of r that is of size $b(r)$. This procedure does not yield a partition into distinct parts only if $b(r) = g(r) - 1$, and each part of r is one less than the preceding one, as in $q = (6, 5, 4)$. So we can find the unique preimage $\phi^{-1}(r)$ of r unless

(A') $b(r) = g(r)$, so $r = (2j - 1, 2j - 2, \cdots, j)$, and therefore, $n = \frac{(3j-1)j}{2}$, or

(B') $b(r) = g(r) - 1$, so $r = (2j, 2j - 1, \cdots, j + 1)$, and therefore, $n = \frac{(3j+1)j}{2}$.

Also note that in cases (A') and (B'), the number j has to be odd to ensure that r has an odd number of parts.

In other words, if n is not a pentagonal number, then ϕ is a bijection from the set of partitions enumerated by $p_{e,d}(n)$ onto the set of partitions enumerated by $p_{o,d}(n)$.

In exceptional cases (A) and (B), (which occur when $n = \frac{(3j+1)j}{2}$ for some even positive integer j), there is one partition in the domain of ϕ that does not get mapped into a partition consisting of an odd number of parts, showing that $p_{e,d}(n) - p_{o,d}(n) = 1 = (-1)^j$ as j is even.

In exceptional cases (A') and (B'), (which occur when $n = \frac{(3j-1)j}{2}$ for some odd positive integer j), there is one partition of n into an odd number of distinct parts that does not have a preimage under ϕ, proving that $p_{e,d}(n) - p_{o,d}(n) = -1 = (-1)^j$ as j is odd. ∎

Now let $p_d(n, m)$ be the number of partitions of n into m distinct parts. As a consequence of the previous lemma, note that if n is not of the form $(3j^2 + j)/2$ or $(3j^2 - j)/2$ for some nonnegative integer j, then we have

$$\sum_{m=1}^{n} (-1)^m p_d(n, m) = 0.$$

Otherwise, we have

$$\sum_{m=1}^{n} (-1)^m p_d(n, m) = (-1)^j.$$

The following remarkable Corollary links the pentagonal numbers to the enumeration of permutations according to their number of inversions.

COROLLARY 2.14
[Euler's formula.] The identity

$$f(z) = (1 - z)(1 - z^2)(1 - z^3) \cdots = 1 - z - z^2 + z^5 + z^7 - z^{12} - \cdots$$
$$= \sum_{j \in \mathbf{Z}} (-1)^j z^{(3j^2 + j)/2}$$

holds.

The reader should consider for a minute how fundamentally interesting this result is. Even if someone has no interest in, or knowledge of, integer partitions and inversions in permutations, he or she could still very well ask what the power series form of the product $\prod_{i \geq 1}(1 - z^i)$ is. And the interesting answer is that most coefficients in that power series are 0, and the nonzero coefficients all have absolute value 1, and they all occur at exponents corresponding to pentagonal numbers.

PROOF   The left-hand side is similar to the generating function of the numbers $p_d(n)$ as given in (2.3), except for the negative sign within each term. This implies that the coefficient of z^n on the left-hand side is not simply the sum of all the numbers $p_d(n, m)$, but their *signed sum* $\sum_m (-1)^m p_d(n, m)$. We know from Lemma 2.12 that this sum is 0, except when n is of the form $(3j^2 + j)/2$ or $(3j^2 - j)/2$, in which case this sum is equal to $(-1)^j$. This completes the proof.   ■

We mention that the rather unusual summation $\sum_{j \in \mathbf{Z}}$ in Euler's formula is used to include pentagonal numbers of the form $(3j^2 + j)/2$ and $(3j^2 - j)/2$ in the same sum. One can think of the sum $\sum_{j \in \mathbf{Z}}(-1)^j z^{(3j^2+j)/2}$ as the sum in which j ranges through all integers in order $0, -1, 1, -2, 2, -3, 3, \cdots$. For $j \in \mathbf{Z}$, let us set $d_j = (3j^2 + j)/2$.

Recall that by Theorem 2.3, the polynomial $I_n(z)$ can be rearranged as

$$I_n(z) = \prod_{i=1}^{n}(1 + z + \cdots + z^{i-1}) = \prod_{i=1}^{n} \frac{1 - z^i}{1 - z}.$$

Let $k \leq n$. While $I_n(z)$ is a polynomial and $\frac{f(z)}{(1-z)^n}$ is an infinite product, their factors of degree at most k agree; therefore, their coefficients for terms of degree at most k also agree. So our task is reduced to finding the coefficient of z^k in

$$f(z) \cdot (1 - z)^{-n} = f(z) \cdot \sum_{h \geq 0} \binom{n + h - 1}{h} z^h,$$

where we set $\binom{-1}{0} = 1$. In order to get a term with coefficient k in the product $f(z) \cdot (1 - z)^{-n}$, we have to multiply the term $(-1)^j z^{(3j^2+j)/2} = (-1)^j z^{d_j}$ of $f(z)$ by the term of $(1 - z)^{-n}$ that has exponent $k - d_j$, that is, in which $h = k - d_j$. Therefore, we have proved the following theorem.

THEOREM 2.15
Let $n \geq k$. Then the coefficient of z^k in $I_n(z)$, or, equivalently, the number of n-permutations with k inversions, is

$$b(n, k) = \sum_{j}(-1)^j \binom{n + k - d_j - 1}{k - d_j} \tag{2.5}$$

j	0	1		2	
	-	+	-	+	
d_j	0	1	2	5	7

FIGURE 2.4
The first five pentagonal numbers.

where j is such that the pentagonal number d_j is at most as large as k.

The first few pentagonal numbers are shown in Figure 2.4.
Expanding (2.5), we see that if $n \geq k$, then the formula for $b(n, k)$ starts as follows.

$$b(n, k) = \binom{n+k-1}{k} - \binom{n+k-2}{k-1} - \binom{n+k-3}{k-2}$$
$$+ \binom{n+k-6}{k-5} + \binom{n+k-8}{k-7} - \cdots .$$

2.1.2　Major Index

There are other permutation statistics that are *equidistributed* with the number of inversions. That is, there exist other permutation statistics *stat* so that for all non-negative integers n and k, the number of n-permutations p satisfying $stat(p) = k$ is equal to $b(n, k)$. The most famous of these statistics is the major index, which was named after the rank of its inventor, Percy MacMahon, in the British Army.

DEFINITION　2.16　*Let $p = p_1 p_2 \cdots p_n$ be a permutation, and define the* major index *or* greater index *$maj(p)$ of p to be the sum of the descents of p. That is, $maj(p) = \sum_{i \in D(p)} i$.*

Example 2.17
If $p = 352461$, then $D(p) = \{2, 5\}$; therefore, $maj(p) = 7$.　☐

In 1916, MacMahon showed [248] the following surprising theorem by proving that the two relevant generating functions were identical. It was not until 1968 that a bijective proof was found by Dominique Foata [170], who worked in a more general setup. Another proof that can be turned into a bijective proof is given in Exercises 31 and 32. We present Foata's proof in the simplified language of permutations.

THEOREM 2.18
For all positive integers n and all nonnegative integers k, there are as many n-permutations with k inversions as there are n-permutations with major index k.

In other words, the permutation statistics "number of inversions," which we denoted by i, and "major index," which we denoted by maj, are *equidistributed* on S_n. If a permutation statistic s has the same distribution on S_n as i, then s is called *Mahonian*.

PROOF   (of Theorem 2.18) For any permutation $p = p_1 p_2 \cdots p_n$, we call the entry p_i large if $p_i > p_n$, and we call p_i small if $p_i < p_n$.

We are going to prove our statement by recursively defining a bijection $\phi : S_n \to S_n$ so that for all $p \in S_n$, the equality $maj(p) = i(\phi(p))$ holds. Our map ϕ will have the additional feature of keeping the last element of p fixed.

It will not surprise the reader that we define $\phi(1) = 1$ for the initial case of $n = 1$, and $\phi(12) = 12$ and $\phi(21) = 21$ for the case of $n = 2$.

Now let us assume that we have defined ϕ for all $(n-1)$-permutations, and the ϕ fixes the last entry of each such permutation. In order to define ϕ for all n-permutations, we distinguish two cases. Let $p = p_1 p_2 \cdots p_n$ be any n-permutation.

1. First we consider the case when p_{n-1} is a small entry. In this case, take $w_p = \phi(p_1 p_2 \cdots p_{n-1}) = q_1 q_2 \cdots q_{n-1}$. Let $q_{i_1}, q_{i_2}, \cdots, q_{i_j}$ be the small entries of p in w_p, that is, those that are less than p_n. Note that $p_{n-1} = q_{n-1}$. Set $i_0 = 0$. Let $Q_j = q_{i_{j-1}+1} \cdots q_{i_j}$. In other words, the Q_j provide the unique decomposition of w into subwords that contain exactly one small entry, and contain that small entry in the last position. For instance, if $q_1 \cdots q_6 = 425613$, then there are two small entries, 1 and 2, and therefore, $Q_1 = 42$, and $Q_2 = 561$. Now define

$$f(Q_j) = \begin{cases} Q_j \text{ if } Q_j \text{ is of length at most } 1, \\ \\ x_m x_1 x_2 \cdots x_{m-1} \text{ if } Q_j = x_1 x_2 \cdots x_m, \text{ with } m \geq 2. \end{cases}$$

Finally, define

$$f(w_p) = f(Q_1) f(Q_2) \cdots f(Q_k),$$

and

$$\phi(p) = f(w_p) p_n.$$

Example 2.19

Let $n = 5$, and $p = 54213$. Then we have $w_p = \phi(5421) = 5421$, and $Q_1 = 542$, $Q_2 = 1$. Therefore, $f(w_p) = 2541$, and so $\phi(p) = 25413$. ☐

2. When p_{n-1} is a large entry, the procedure is very similar. The only difference is in the definition of the strings Q_j. In this case, the Q_j provide the unique decomposition of w into subwords that contain exactly one *large* entry, and contain that *large* entry in the last position.

Example 2.20

Let $n = 5$, and let $p = 13452$. Then we have $w_p = \phi(1345) = 1345$, and $Q_1 = 13$, $Q_2 = 4$, and $Q_3 = 5$. Therefore, $f(w_p) = 3145$, and so $\phi(p) = 31452$. ☐

It is easy to see that $\phi : S_n \to S_n$ is a bijection. Indeed, verifying both cases, one sees that the first rule was used to create $\phi(p)$ if and only if the last element of $\phi(p)$ is larger than the first element of $\phi(p)$. Once we know which rule was used to create $\phi(p)$, we can recover w_p from $f(w_p)$. Indeed, if the first (resp. second) rule was used, then the $f(Q_i)$ are the subwords that contain only one small (resp. large) entry, and contain that small (resp. large) entry in the *first* position. As f is a bijection, recovering the $f(Q_i)$ this way allows us to recover the Q_i, and therefore, w_p itself. Finally, $\phi : S_{n-1} \to S_{n-1}$ is a bijection by induction, so we recover $p_1 p_2 \cdots p_{n-1}$ from $f(p_1 p_2 \cdots p_{n-1}) = w_p$.

We still need to prove that $\phi : S_n \to S_n$ has the desired property, that is, it maps a permutation with major index k into a permutation with k inversions. We accomplish this by considering the two above cases separately.

1. When p_{n-1} is a small entry, then

$$maj(p) = maj(p_1 p_2 \cdots p_{n-1}) = i(\phi(p_1 p_2 \cdots p_{n-1})) = i(w_p). \quad (2.6)$$

How does the map f change the number of inversions of $w(p)$? It does not change the order among the small entries, or among the large entries. If a small entry belongs to the subword Q_j of length $t > 1$, then it jumps forward and passes all $t - 1$ large entries of Q_j, decreasing the number of inversions by $t - 1$.

As each large entry will be passed by one small entry, the total decrease in inversions is equal to the number of large entries, that is, to $n - p_n$. However, affixing p_n to the end of $f(w_p)$ will create precisely $n - p_n$ new inversions. Therefore,

$$i(\phi(p)) = i(f(w_p)p_n) = i(w_p),$$

which, compared to (2.6), shows that $maj(p) = i(\phi(p))$ as claimed.

2. When p_{n-1} is a large entry, then

$$maj(p) = maj(p_1 p_2 \cdots p_{n-1}) + (n-1) \tag{2.7}$$
$$= i(\phi(p_1 p_2 \cdots p_{n-1})) + n - 1 = i(w_p) + n - 1. \tag{2.8}$$

When f is applied to w_p, each large entry belonging to a subword of length $t > 1$ jumps forward, passes all $t - 1$ small entries of its subword, and increases the number of inversions by $t - 1$. Each small entry is passed by one large entry, so the total increase in the number of inversions is equal to the number of small entries, that is, $p_n - 1$. On the other hand, affixing p_n to the end of $f(w_p)$ will create precisely $n - p_n$ new inversions. Therefore,

$$i(\phi(p)) = i(f(w_p)p_n) = i(w_p) + (p_n - 1) + (n - p_n) = i(w_p) + n - 1, \tag{2.9}$$

which, compared to (2.7), shows that again, $maj(p) = i(\phi(p))$ as claimed.

∎

Other examples of Mahonian statistics can be found among the exercises.

2.1.3 Application: Determinants and Graphs

2.1.3.1 Explicit Definition of Determinants

There are several undergraduate mathematics courses and textbooks that only give a recursive definition of the *determinant* of a square matrix. That is, $\det \begin{pmatrix} a & b \\ c & d \end{pmatrix}$ is defined to be equal to $ad - bc$, and then the determinant of the $n \times n$ matrix $A = (a_{ij})$ is defined to be

$$\det A = \sum_{j=1}^{n} (-1)^{j-1} a_{1j} A_{1j} \tag{2.10}$$

where A_{1j} is the $(n-1) \times (n-1)$ matrix obtained from A by removing the first row and the jth column.

If that is the only definition of determinants the reader has seen, he may find the following result interesting.

THEOREM 2.21
Let $A = (a_{ij})$ be an $n \times n$ matrix. Then we have

$$\det A = \sum_{p \in S_n} (-1)^{i(p)} a_{1p_1} a_{2p_2} \cdots a_{np_n}. \tag{2.11}$$

That is, $\det A$ is obtained by taking all $n!$ possible n-tuples of entries so that there is exactly one of the n entries in each row and each column, multiplying the elements of each such n-tuple together, finally taking a signed sum of these $n!$ products, where the sign is determined by the parity of $i(p)$, and p is the permutation determined by each chosen n-tuple.

In other words, the n-tuples correspond to all possible placements of n rooks on an $n \times n$ chessboard so that no two of them hit each other.

Example 2.22

Let $n = 3$. Then there are three 3-permutations with an even number of inversions, namely 123, 312, and 231, and there are three 3-permutations with an odd number of inversions, namely 132, 213, and 321. Therefore, we have

$$\det A = a_{11}a_{22}a_{33} + a_{13}a_{21}a_{32} + a_{12}a_{23}a_{31} - a_{11}a_{23}a_{32} - a_{12}a_{21}a_{33} - a_{13}a_{22}a_{31}.$$

⬜

PROOF (of Theorem 2.21). We prove the statement by induction on n, the initial cases of $n = 1$ and $n = 2$ being obvious. Let us assume that the statement is true for $(n-1) \times (n-1)$ matrices. That means that

$$\det A_{1j} = \sum_{q} (-1)^{i(q)} a_{2q_2} a_{3q_3} \cdots a_{nq_n}, \tag{2.12}$$

where $q = q_2 q_3 \cdots q_n$ is a *partial permutation*, that is, a list of the integers $1, 2, \cdots, j-1, j+1, \cdots, n$ in some order.

Therefore, $a_{1j} \det A_{1j}$ will contribute the products of all n-tuples starting with a_{1j} to $\det A$. In other words, $a_{1j} \det A_{1j}$ will correspond to all non-hitting rook placements in which there is a rook in position j of the first row. This argument can be applied for each j. So our theorem will be proved if we can show that the signs of these products are what they should be.

Substituting the expression provided for A_{1j} by formula (2.12) into formula (2.10), we see that the sign of the n-tuple that belongs to q becomes $(-1)^{j-1+i(q)}$. And indeed, the permutation $p = jq_2q_3 \cdots q_n$ has precisely $j-1$ more inversions than the partial permutation $q = q_2q_3 \cdots q_n$ as j is larger than $j-1$ other elements. This shows that the contribution of this n-tuple is indeed counted with sign $(-1)^{j-1}$. ∎

2.1.3.2 Perfect Matchings in Bipartite Graphs

The explicit definition of the determinant has some surprising applications in graph theory. A simple graph is a graph with no loops or multiple edges. A *perfect matching* M in the simple graph G is a set of pairwise vertex-disjoint edges covering all vertices of G. In other words, each vertex of G belongs to exactly one edge in M. The graph G is called bipartite if the vertex set of G can be cut into two parts X and Y so that all edges of G have one vertex in X and one vertex in Y. The *truncated adjacency matrix* of a simple bipartite graph G is the matrix $B(G) = (b_{ij})$ in which $b_{ij} = 1$ if there is an edge between $i \in X$ and $j \in Y$, and $b_{ij} = 0$ otherwise. Note that the rows of B represent the vertices of X, and the columns of B represent the vertices of Y.

Whether a bipartite graph has a perfect matching is an interesting and well-studied question. A sufficient and necessary condition for this existence problem is the well-known Marriage Theorem, which is included in most elementary graph theory books, such as [74].

The concept of truncated adjacency matrices provides us with a sufficient condition for the existence of a perfect matching that is very easy to verify.

THEOREM 2.23

Let G be a simple bipartite graph with $|X| = |Y| = n$ that does not have a perfect matching. Then $\det B(G) = 0$.

In other words, if $\det B(G) \neq 0$, then G has a perfect matching.

PROOF We prove that $\det B(G) = 0$ by showing that all $n!$ summands in the explicit definition (2.11) of $B(G)$ are equal to 0. This is because the existence of a nonzero term $b_{1p_1} b_{2p_2} \cdots b_{np_n}$ would be equivalent to the existence of a perfect matching, namely the perfect matching in which $i \in X$ is matched to $p_i \in Y$. ∎

We also note that the *number* of all perfect matchings of G can be obtained by computing the *permanent* of $B(G)$ that is defined by

$$\operatorname{per} B(G) = \sum_{p \in S_n} b_{1p_1} b_{2p_2} \cdots b_{np_n}.$$

That is, $\operatorname{per} B(G)$ is defined just like $\det B(G)$, except that each term is added with a positive sign.

2.2 Inversions in Permutations of Multisets

Instead of permuting the elements of our favorite set, $[n]$, in this section we are going to permute elements of *multisets*. We will use the notation $\{1^{a_1}, 2^{a_2}, \cdots, k^{a_k}\}$ for the multiset consisting of a_i copies of i, for all $i \in [k]$.

For our purposes, a *permutation* of a multiset is just a way of listing all its elements. It is straightforward to see, and is proved in most undergraduate textbooks on enumerative combinatorics, that the number of all permutations of the multiset $K = \{1^{a_1}, 2^{a_2}, \cdots, k^{a_k}\}$ is

$$\frac{n!}{a_1! a_2! \cdots a_k!},$$

where $n = a_1 + a_2 + \cdots + a_k$.

An inversion of a permutation $p = p_1 p_2 \cdots p_n$ of a multiset is defined similarly to the way in which it was for permutations of sets, that is, (i, j) is an inversion if $i < j$, but $p_i > p_j$.

Example 2.24
The multiset-permutation 1322 has two inversions: $(2, 3)$ and $(2, 4)$.   ⬚

If we want to generalize Theorem 2.3 for permutations of multisets, that is, if we want to count permutations of multisets according to their inversions, we encounter exciting and surprising connections between the objects at hand, and a plethora of remote-looking areas of combinatorics.

Our goal is to find a closed expression for the sum

$$\sum_{p \in S_K} q^{i(p)}, \tag{2.13}$$

where S_K denotes the set of all permutations of the multiset K. We cannot reasonably expect something quite as simple as the result of Theorem 2.3 as the formula to be found will certainly depend on each of the a_i, and not just their sum n. Therefore, the reader will hopefully understand that we need some new notions before we can find the desired closed formula for (2.13).

Let $[\mathbf{n}] = 1 + q + q^2 + \cdots + q^{n-1}$, the polynomial whose importance we know from Theorem 2.3, and let $[\mathbf{n}]! = [\mathbf{1}] \cdot [\mathbf{2}] \cdots [\mathbf{n}]$. Do not confuse $[n] = \{1, 2, \cdots, n\}$, which is a set, and $[\mathbf{n}] = 1 + q + q^2 + \cdots + q^{n-1}$, which is a polynomial. One way to avoid the danger of confusion is to use the notation $[n]_q$ instead of $[\mathbf{n}]$, but that notation can result in crowded formulas. Note that if we substitute $q = 1$, then $[\mathbf{i}] = i$, and therefore $[\mathbf{n}]! = n!$, so this concept generalizes the concept of factorials. The crucial definition of this section is the following.

DEFINITION 2.25   *Let k and n be positive integers so that $k \leq n$. Then the (n, k)-Gaussian coefficient or q-binomial coefficient is denoted by $\begin{bmatrix} n \\ k \end{bmatrix}$, and is given by*

$$\begin{bmatrix} \mathbf{n} \\ \mathbf{k} \end{bmatrix} = \frac{[\mathbf{n}]!}{[\mathbf{k}]![(\mathbf{n} - \mathbf{k})]!}.$$

Note that $\begin{bmatrix} n \\ k \end{bmatrix} = \begin{bmatrix} n \\ n-k \end{bmatrix}$ clearly holds. Also note that substituting $q = 1$ reduces this definition to that of the usual binomial coefficients. This, and other connections between binomial and q-binomial coefficients, will be further explored shortly. Finally, we can define q-multinomial coefficients accordingly.

DEFINITION   2.26   *Let $a_1, a_2, \cdots, a_k$ be positive integers satisfying $\sum_{i=1}^{k} a_i = n$. Then the $(a_1, a_2, \cdots, a_k)$-Gaussian coefficient, or q-multinomial coefficient is denoted by $\begin{bmatrix} n \\ a_1, a_2, \cdots, a_k \end{bmatrix}$, and is given by*

$$\begin{bmatrix} n \\ a_1, a_2, \cdots, a_k \end{bmatrix} = \frac{[n]!}{[a_1]! [a_2]! \cdots [a_k]!}.$$

We point out that similarly to multinomial coefficients, the q-multinomial coefficients satisfy the identity

$$\begin{bmatrix} n \\ a_1, a_2, \cdots, a_k \end{bmatrix} = \begin{bmatrix} n \\ a_1 \end{bmatrix} \begin{bmatrix} n - a_1 \\ a_2 \end{bmatrix} \begin{bmatrix} n - a_1 - a_2 \\ a_3 \end{bmatrix} \cdots \begin{bmatrix} a_k \\ a_k \end{bmatrix}. \qquad (2.14)$$

Note that using this terminology, Theorem 2.3 can be written as

$$\sum_{p \in S_n} q^{i(p)} = [n]!.$$

The Gaussian coefficients look like rational functions of q, but, as we will soon see, it is not difficult to prove that they are in fact *polynomials* in q. Even more strongly, they are polynomials with *positive integer* coefficients. This is why sometimes they are called *Gaussian polynomials*.

2.2.1   Application: Gaussian Polynomials and Subset Sums

Before we start applying Gaussian polynomials to obtain generating functions of multiset permutations, it seems beneficial to take a look at one of their several natural occurrences. The advantage of this will be that the reader will see in what sense the Gaussian coefficients $\begin{bmatrix} n \\ k \end{bmatrix}$ are generalizations of the binomial coefficients $\binom{n}{k}$. That, in turn, will be helpful in putting into context the recurrence relations of Gaussian coefficients that we are going to use.

THEOREM 2.27
Let n and k be fixed non-negative integers so that $k \leq n$. Let a_i denote the number of k-element subsets of $[n]$ whose elements have sum $i + \binom{k+1}{2}$, that is, i larger than the minimum. Then

$$\begin{bmatrix} n \\ k \end{bmatrix} = \sum_{i=0}^{k(n-k)} a_i q^i. \qquad (2.15)$$

In other words, $\begin{bmatrix} n \\ k \end{bmatrix}$ is the ordinary generating function of the k-element subsets of $[n]$ according to the sum of their elements.

Example 2.28

Let $n = 4$ and $k = 2$. Then, among the six 2-element subsets of $[4]$, two, namely $\{1, 4\}$ and $\{2, 3\}$, have sum 5, and all other sums from 3 to 7 are attained by exactly one subset. Therefore, the right-hand side of (2.15) becomes $1 + q + 2q^2 + q^3 + q^4$, which is indeed equal to

$$\begin{bmatrix} 4 \\ 2 \end{bmatrix} = \frac{(q^4 - 1)(q^4 - q)}{(q^2 - 1)(q^2 - q)} = (q^2 + 1)(q^2 + q + 1).$$

⬚

PROOF (of Theorem 2.27) We prove the statement by induction on n, the initial case of $n = 1$ being obvious. Let us assume that the statement is true for $n - 1$ and prove it for n. Exercise 21 shows that

$$\begin{bmatrix} n \\ k \end{bmatrix} = q^{n-k} \cdot \begin{bmatrix} n-1 \\ k-1 \end{bmatrix} + \begin{bmatrix} n-1 \\ k \end{bmatrix}. \tag{2.16}$$

Therefore, our induction step will be complete if we can show that the polynomials $\sum_{i=0}^{k(n-k)} a_i q^i$ satisfy the same recurrence relation. That is, let b_i be the number of k-element subsets of $[n-1]$ whose sum of elements is $i + \binom{k+1}{2}$ and let c_i be the number of $(k-1)$-element subsets of $[n-1]$ whose sum of elements is $i + \binom{k}{2}$; we then need to show that

$$\sum_{i=0}^{k(n-k)} a_i q^i = \left(\sum_{i=0}^{k(n-k-1)} b_i q^i \right) + \left(q^{n-k} \cdot \sum_{i=0}^{(k-1)(n-k)} c_i q^i \right).$$

This is the same as showing that $a_i = b_i + c_i q^{n-k}$ for all i, where undefined coefficients are to be treated as zero. However, the last equation is clearly true as a k-subset of $[n]$ either does not contain n, and then it is accounted for by b_i, or it does, and then it is accounted for by $c_{i-(n-k)}$, because of the shift in the definition of c_i. ∎

2.2.2 Inversions and Gaussian Coefficients

Now we are ready to announce and prove the result describing the generating function of multiset-permutations according to the number of their inversions.

THEOREM 2.29

Let $K = \{1^{a_1}, 2^{a_2}, \cdots, k^{a_k}\}$ be a multiset so that $\sum_{i=1}^{k} a_i = n$, and let S_K denote the set of all permutations of K. Then the equality

$$\sum_{p \in S_K} q^{i(p)} = \begin{bmatrix} n \\ a_1, a_2, \cdots, a_k \end{bmatrix} \tag{2.17}$$

holds.

PROOF   First we prove the statement in the special case of $k = 2$. In this case, we will write K' instead of K, to remind the reader that K' is a multiset consisting of a_1 copies of 1 and a_2 copies of 2, where that $a_1 + a_2 = n$, and an inversion is an occurrence of a 2 on the left of a 1. We need to prove that in this special case,

$$\sum_{p \in S_{K'}} q^{i(p)} = \begin{bmatrix} n \\ a_1 \end{bmatrix}. \tag{2.18}$$

We prove this statement by induction on n. For $n = 1$, the statement is trivially true as $\begin{bmatrix} 1 \\ 0 \end{bmatrix} = \begin{bmatrix} 1 \\ 1 \end{bmatrix} = 1$. Now let us assume the statement is true for $n - 1$, and prove it for n. A multiset permutation of K' either ends in a 2, and then its last entry is not involved in any inversion, or it ends in a 1, and then its last entry is involved in exactly $a_2 = n - a_1$ inversions. By the induction hypothesis, this means that

$$\sum_{p \in S_{K'}} q^{i(p)} = \begin{bmatrix} n-1 \\ a_1 \end{bmatrix} + q^{n-a_1} \cdot \begin{bmatrix} n-1 \\ a_1 - 1 \end{bmatrix}.$$

By (2.16), it is now easy to see that the right-hand side is in fact equivalent to $\begin{bmatrix} n \\ a_1 \end{bmatrix}$, completing the induction proof of (2.18).

We are now in a position to prove our theorem in its general form. We will do this by induction on k, the case of $k = 1$ being trivial, and the case of $k = 2$ being solved above. Let us assume that the statement of the theorem is true for $K = \{1^{a_1}, 2^{a_2}, \cdots, k^{a_k}\}$, and prove that then it is also true for $K^+ = \{1^{a_1}, 2^{a_2}, \cdots, k^{a_k}, (k+1)^{a_{k+1}}\}$.

Note that any permutation of K^+ is completely determined by the pair (p', p''), where p' is the multiset-permutation obtained from p by replacing all entries less than $k + 1$ by 1, and p'' is the permutation obtained from p by removing all copies of $k + 1$. It is then clear that

$$i(p) = i(p') + i(p''),$$

and that p' and p'' are independent of each other.

Then the problem of finding $\sum_{p'} q^{i(p')}$ is clearly equivalent to the previous special case, and therefore we get that $\sum_{p'} q^{i(p')} = \begin{bmatrix} n \\ a_{k+1} \end{bmatrix}$.

Now let us find $\sum_{p''} q^{i(p'')}$. If we remove all copies of $k + 1$, we can apply the induction hypothesis, and see that

$$\sum_{p''} q^{i(p'')} = \begin{bmatrix} n - a_{k+1} \\ a_1 \end{bmatrix} \cdot \begin{bmatrix} n - a_{k+1} - a_1 \\ a_2 \end{bmatrix} \cdots \begin{bmatrix} a_k \\ a_k \end{bmatrix}.$$

Finally, as any p' (consisting of $a_1 + a_2 + \cdots + a_k$ copies of 1, and a_{k+1} copies of $k + 1$) can be paired with any p'' (consisting of a_i copies of i for

$i \in [k]$), it follows that

$$\sum_{p \in S_{K'}} q^{i(p)} = \sum_{p'} q^{i(p')} \cdot \sum_{p''} q^{i(p'')}$$

$$= \begin{bmatrix} n \\ a_{k+1} \end{bmatrix} \cdot \begin{bmatrix} n - a_{k+1} \\ a_1 \end{bmatrix} \cdot \begin{bmatrix} n - a_{k+1} - a_1 \\ a_2 \end{bmatrix} \cdots \begin{bmatrix} a_k \\ a_k \end{bmatrix}$$

$$= \begin{bmatrix} n \\ a_1, a_2, \cdots, a_{k+1} \end{bmatrix}.$$

Here the last equation is immediate from (2.14). This completes our induction proof. ∎

2.2.3   Major Index and Permutations of Multisets

Recall that for permutations of the set $[n]$, we found in Theorem 2.18 that the statistics i and maj were equidistributed. We would like to see whether something similar is true for permutations of multisets. In order to be able to do that, we need to define the major index of multiset permutations. As a first step to that end, we need to define descents of multiset permutations.

Fortunately, both of these definitions are what one expects them to be. If $p = p_1 p_2 \cdots p_n$ is a permutation of a multiset, then we say that i is a descent of p if $p_i > p_{i+1}$. Similarly, the major index of the multiset permutation p is defined by $maj(p) = \sum_{i \in D(p)} i$.

Now we are ready to state the q-generalization of Theorem 2.18.

THEOREM 2.30
Let $K = \{1^{a_1}, 2^{a_2}, \cdots, k^{a_k}\}$ be a multiset so that $a_1 + a_2 + \cdots + a_k = n$. Then the statistics i and maj are equidistributed on the set S_K of all permutations of K. In other words,

$$\sum_{p \in S_K} q^{i(p)} = \sum_{p \in S_K} q^{maj(p)} = \begin{bmatrix} n \\ a_1, a_2, \cdots, a_k \end{bmatrix}.$$

PROOF   Theorem 2.29 shows that the far-left side and the far-right side are equal. Therefore, it suffices to show that the sum in the middle is also equal to the far-left side. That, in turn, can be proved injectively, in a very similar manner Theorem 2.18 that covers the case in which $a_i = 1$ for all i.

The maps f and ϕ, and the permutation w_p can be defined just as in the proof of Theorem 2.18. The only differences are the following.

1. An entry p_i is *large* if it $p_i > p_n$, and an entry p_i is *small* if $i \neq n$, but $p_i \leq p_n$.

2. When p_{n-1} is a small entry, then the application of ϕ causes each large entry of p to be passed by small entry. So, the decrease of the number

of inversions caused by these movements is equal to the number of large entries, that is, $\sum_{j=i+1}^{k} a_j = n - \sum_{i=1}^{j} a_i$. On the other hand, affixing p_n to the end of $f(w_p)$ creates the same number of non-inversions, proving that

$$i(\phi(p)) = i(f(w_p)p_n) = i(w_p)$$

holds again.

3. When p_{n-1} is a large entry, then the application of ϕ causes each small entry of p to be passed by one large entry. So, the increase of the number of inversions caused by these movements is equal to the number of small entries, that is, $\left(\sum_{i=1}^{j} a_i\right) - 1$, where $p_n = j$. On the other hand, affixing p_n to the end of $f(w_p)$ creates $\sum_{j=i+1}^{k} a_j = n - \sum_{i=1}^{j} a_i$ additional inversions. Therefore,

$$i(\phi(p)) = i(f(w_p)p_n) = i(w_p) + \left(\sum_{i=1}^{j} a_i\right) - 1 + n - \sum_{i=1}^{j} a_i = i(w_p) + n - 1,$$

just as in (2.9).

∎

As we have mentioned, there are many interesting occurrences of Gaussian coefficients in combinatorics. Perhaps the most direct one is the following.

THEOREM 2.31

Let q be a power of a prime number, and let V be an n-dimensional vector space over the q-element field. Then the number of k-dimensional subspaces of V is $\begin{bmatrix} n \\ k \end{bmatrix}$.

PROOF   First, let us choose a k-tuple of vectors in V that form an (ordered) basis for a k-dimensional subspace. For this, we have to choose k linearly independent vectors from our vector space V. For the first basis vector v_1, we can choose any vector in V except the zero vector, so we have $q^n - 1$ choices. For the second basis vector, we cannot choose any multiples of v_1, therefore we have only $q^n - q$ choices. For the third vector, we cannot choose any of the q^2 possible linear combinations of v_1 and v_2, yielding $q^n - q^2$ choices, and so on. Iterating this argument, we see that we have

$$(q^n - 1)(q^n - q) \cdots (q^n - q^{k-1}) \tag{2.19}$$

choices for an ordered basis of a k-dimensional subspace of V. It goes without saying that any such subspace has many ordered bases. In fact, repeating the above argument with k playing the role of n shows that the number of ordered bases of a k-dimensional subspace is

$$(q^k - 1)(q^k - q)\cdots(q^k - q^{k-1}).$$

So this is how many times each k-dimensional subset of V is counted by (2.19). Therefore, the number of such subspaces is

$$\frac{(q^n - 1)(q^n - q)\cdots(q^n - q^{k-1})}{(q^k - 1)(q^k - q)\cdots(q^k - q^{k-1})} = \frac{(q^n - 1)(q^{n-1} - 1)\cdots(q^{n-k+1} - 1)}{(q^k - 1)(q^{k-1} - 1)\cdots(q - 1)}$$

$$= \frac{[n][n-1]\cdots[n-k+1]}{[k]!} = \begin{bmatrix} n \\ k \end{bmatrix}.$$

∎

We will see some alternative interpretations of the Gaussian coefficients in the exercises.

Exercises

1. (+) Let us generalize the notion of Eulerian polynomials as follows. Let

$$B_n(z) = \sum_{p \in S_n} (-1)^{i(p)} z^{1+d(p)}.$$

That is, the only difference between this definition and that of $A_n(z)$ is that here the parity of the number of inversions is taken into account. Prove that

$$B_{2n}(z) = (1-z)^n A_n(z),$$

and

$$B_{2n+1}(z) = (1-z)^n A_{n+1}(z).$$

2. Following the line of thinking found in Exercise 17 of Chapter 1, define the r-*major index* of p, denoted by $rmaj(p)$ as

$$rmaj(p) = \left(\sum_{i \in RD(p)} i\right) + |\{(i,j): 1 \le i < j \le n,\ p_i > p_j > p_i - r\}|,$$

where $RD(p)$ denotes the set of all r-falls of p, as defined in the mentioned exercise.

(a) Explain why the r-major index is a generalization of both the number of inversions and the major index.

(b) Prove that for any positive integers r, the r-major index is a mahonian statistic.

3. Prove (without using the general formula for $b(n, k)$) that $b(n, 3) = \binom{n+1}{3} - \binom{n}{1}$ if $n \geq 3$.

4. Prove (without using the general formula for $b(n, k)$) that $b(n, 4) = \binom{n+2}{4} - \binom{n+1}{2}$ if $n \geq 4$.

5. (-) Let $p = p_1 p_2 \cdots p_n$, and let us call (p_i, p_j) a *2-inversion* if $i < j$, and $p_i > p_j + 1$. What is the total number of all 2-inversions in all permutations of length n?

6. Let $m > 1$ be a positive integer, and let j be a nonnegative integer, with $j < m$. Prove that if n is large enough, then the number of n-permutations p for which $i(p) \equiv j \bmod(m)$ is independent of j.

7. Let T be a rooted tree with root 0 and non-root vertex set $[n]$. Define an *inversion* of T to be a pair (i, j) of vertices so that $i > j$, and the unique path from 0 to j goes through i. How many such trees have zero inversions?

8. Let $p \in S_n$ have $n - 2$ descents. What is the minimal and maximal possible value of $i(p)$?

9. It follows from Lemma 2.12 that

$$\sum_{j \text{ even}} p(n - a(j)) = \sum_{j \text{ odd}} p(n - a(j)),$$

where $a(j) = (3j^2 + j)/2$. Find a direct bijective proof of this identity.

10. Let $p = p_1 p_2 \cdots p_n$ be a permutation, and let our goal be to eliminate all four-tuples of entries (p_a, p_b, p_c, p_d) in which $a < b < c < d$ and $p_a < p_c < p_b < p_d$. In order to achieve that goal, we use the following algorithm. We choose a four-tuple F with the above property at random, and interchange its two middle entries. By doing that, we took away the undesirable property of F, but we may have created new four-tuples with that property. Then pick another four-tuple with that property, and repeat the procedure.

Prove that no matter what p is, and how we choose our four-tuples, this algorithm will always stop, that is, it will eliminate all four-tuples with the undesirable property.

11. Is it true that $b(n, k)$ is a polynomially recursive function of n for any fixed k? (Polynomially recursive functions are defined in Exercise 29 of Chapter 1.)

12. Let $p(n, k)$ be the number of partitions of n into k parts. Let $P(z) = \sum_{i=1}^{n} p(n, k)z^k$. Does there exist an integer $n > 2$ so that $P(z)$ has real zeros only?

13. Let B_n be the set of all n-tuples $(b_1, b_2, \cdots, b_n)$ of non-negative integers that satisfy $b_i \leq i - 1$ for all i. How many elements of B_n satisfy $\sum_{i=1}^{n} b_i = k$?

14. Let B_n be defined as in Exercise 13, and let $B(n, k)$ be the number of n-tuples in B_n that have exactly k different entries. Find a formula for $B(n, k)$.

15. Express $b(n, k)$ using summands of the type $b(n - 1, i)$.

16. Compute the value of $\sum_{k=0}^{\binom{n}{2}} (-1)^k b(n, k)$.

17. Let $p \in S_n$ have $n-1$ alternating runs, and let us assume that $n = 2k+1$. What is the minimal and maximal possible value of $i(p)$?

18. (a) (+) Let $A = \{1^{a_1}, 2^{a_2}\}$, and let us assume that a_1 and a_2 are relative primes to each other, with $a_1 + a_2 = n$. Let $I(A, k)$ be the number of permutations p of A so that

$$i(p) \equiv k \mod n.$$

Prove that $I(A, k) = \frac{1}{n}\binom{n}{a_1}$ for all k.

(b) (+) What can we say about $I(A, k)$ if a_1 and a_2 have largest common divisor $d > 1$?

19. (+) The *Denert* statistic, denoted by den, is defined on S_n as follows. Let $p \in S_n$, then $den(p)$ is the number of pairs (ℓ, k) of integers satisfying $1 \leq \ell < k \leq n$, and one of the conditions listed below

$$p_k < p_\ell \leq k,$$

$$p_\ell \leq k < p_k.$$

$$k < p_k < p_\ell.$$

So for instance, $den(132) = 2$ as the pair $(2, 3)$ satisfies the first condition, and the pair $(1, 2)$ satisfies the second condition. Prove that the Denert statistic is Mahonian.

20. We know that $\binom{n}{k}$ is the number of northeastern lattice paths from $(0, 0)$ to $(k, n - k)$. Extend this correspondence to one that provides an interpretation for $\begin{bmatrix} n \\ k \end{bmatrix}$.

21. Prove by way of computation that

$$\begin{bmatrix} n \\ k \end{bmatrix} = \begin{bmatrix} n - 1 \\ k \end{bmatrix} + q^{n-k} \cdot \begin{bmatrix} n - 1 \\ k - 1 \end{bmatrix}.$$

22. Prove the identity of the previous exercise by a combinatorial argument.

23. Prove that

$$\begin{bmatrix} m \\ k \end{bmatrix} = q^{m-k} \cdot \begin{bmatrix} m-1 \\ k-1 \end{bmatrix} + q^{m-k-1} \cdot \begin{bmatrix} m-2 \\ k-1 \end{bmatrix} + \cdots + q \cdot \begin{bmatrix} k \\ k-1 \end{bmatrix} + \begin{bmatrix} k-1 \\ k-1 \end{bmatrix}.$$

24. Prove by way of computation that $\begin{bmatrix} n \\ k \end{bmatrix}$ is always a polynomial with non-negative integers as coefficients.

25. Prove that

$$\begin{bmatrix} i+k \\ k \end{bmatrix} = \sum_{n \geq 0} q^n p(i, k, n),$$

where $p(i, k, n)$ is the number of partitions of the integer n into at most i parts of size at most k each.

26. For what values of n and k will $\begin{bmatrix} n \\ k \end{bmatrix}$ have log-concave coefficients?

27. Let $m \leq n$, and let

$$A_m(n) = \{1, 1, 2, 2, \cdots, m, m, m+1, m+2, \cdots, n\}.$$

Let $a_m(n)$ be the number of all permutations of the multiset $A_m(n)$ in which $12 \cdots n$ occurs as a subword. (The letters of this subword do not have to be consecutive entries of the permutation.) Prove that

$$a_{m+1}(n) = (n + 2m)a_m(n) - m(n + m)a_{m-1}(n).$$

28. Let $n < k \leq \binom{n}{2}$. Prove that

$$b(n+1, k) = b(n+1, k-1) + b(n, k) - b(n, k-n-1).$$

29. (+) Find a formula for

$$\sum_{k=0}^{n} (-1)^k \begin{bmatrix} n \\ k \end{bmatrix}.$$

30. (+) Consider the following refinement of the Eulerian polynomials. Let

$$A_{n,k}(q) = \sum_{p} q^{maj(p)},$$

where the sum is taken over all n-permutations having $k - 1$ descents. These polynomials are often called the q-Eulerian polynomials. Prove that

$$[z]^n = \sum_{k=1}^{n} A_{n,k}(q) \begin{bmatrix} z+n-k \\ n \end{bmatrix}.$$

31. Let p be a permutation of length $n - 1$. Insert the entry n into all p in all possible ways. This yields n distinct permutations of length n. Compute the major index of each of these permutations. Prove that all these n major indices will be distinct, and that their set will be the set of integers in the interval $[maj(p), maj(p) + n - 1]$.

32. Use the result of the previous exercise to give an induction proof of Theorem 2.18, that is, of the fact that i and maj are equidistributed.

33. A simple graph G on vertex set $[n]$ is called the *inversion* graph of the n-permutation p if ij is an edge of G if and only if (i, j) is an inversion of G. Find an example for an unlabeled graph U that is not an inversion graph of any permutation. Try to find an example with as few vertices as possible.

34. Let $p = p_1 p_2 \cdots p_n$ be an n-permutation, and let $G(p)$ be the inversion graph of p as defined in the previous exercise. Let $i < j < k$ be three elements of $[n]$, and interchange the strings $p_i \cdots p_{j-1}$ and $p_j \cdots p_{k-1}$ of p, to get the permutation

$$p' = p_1 \cdots p_{i-1} p_j \cdots p_{k-1} p_i \cdots p_{j-1} p_k \cdots p_n.$$

Describe $G(p')$ in terms of $G(p)$.

35. A graph G on vertex set $\{a_1, a_2, \cdots, a_n\}$ is called a *comparability graph* if there exists a poset P on vertex set $\{a_1, a_2, \cdots, a_n\}$ so that (a_i, a_j) is an edge in G if and only if a_i and a_j are comparable elements in P. Prove that all inversion graphs are comparability graphs, but not all finite comparability graphs are inversion graphs.

Problems Plus

1. Prove that the polynomials $I_n(z)$ have log-concave coefficients without using generating functions.

2. Let

$$\mathrm{Exp}_q(z) = \sum_{n \geq 0} q^{\binom{n}{2}} \frac{z^n}{[n]!}.$$

Note that for $q = 1$, the power series $\mathrm{Exp}_q(z)$ reduces to the power series $\sum_{n \geq 0} \frac{z^n}{n!} = e^z$.

Express the *bivariate* generating function

$$1 + \sum_{n \geq 1} \sum_{p \in S_n} t^{d(p)} q^{i(p)} \frac{u^n}{[n]!}$$

in terms of Exp_q.

3. Let $\exp_q(z) = \sum_{n \geq 0} \frac{z^n}{[n]!}$. Express the *bivariate* generating function

$$1 + \sum_{n \geq 1} \sum_{p \in S_n} t^{exc(p)} q^{maj(p)} \frac{u^n}{[n]!}$$

   in terms of $\exp_q$.

4. Let $a(k, \ell)$ be the number of n-permutations having k descents and major index ℓ. Let $d(k, \ell)$ be the number of n-permutations q having k excedances and satisfying $den(q) = \ell$. Prove that $a(k, \ell) = d(k, \ell)$. This fact can be referred to by saying that the (den, exc) statistic is *Euler-Mahonian*.

5. Find a permutation statistic $s : S_n \to \mathbf{N}$ so that the number $c(k, \ell)$ of n-permutations p for which $s(p) = k$ and $i(p) = \ell$ is equal to the number $a(k, \ell)$ defined in Problem Plus . In other words, find a statistic s so that the joint statistic (s, i) is Euler-Mahonian.

6. Prove that the joint statistic (dmc, maj) is Euler-Mahonian. See Exercise 10 of Chapter 1 for the definition of the statistic dmc.

7. Let $1 \leq k \leq n$. Prove, using the polynomial $I_n(z)$, that the number of n-permutations p for which

$$maj(p) \equiv j \ (\mod k)$$

   does not depend on j.

8. Define the (q, r)-Eulerian polynomials by

$$A[n, k, r] = \sum_{p \in S_n} q^{rmaj(p)},$$

   where $rmaj(p)$ is the r-major index of p. Prove that

$$A[n, k, r] = [\mathbf{r} + \mathbf{k}]A[n-1, k, r] + q^{k+r-1}[\mathbf{n} + \mathbf{1} - \mathbf{k} - \mathbf{r}]A[n-1, k-1, r].$$

9. A *parking function* is a function $f : [n] \to [n]$ so that for all $i \in [n]$, there are at least i elements $j \in [n]$ for which $f(j) \leq i$. Prove that the number of parking functions on $[n]$ satisfying $\sum_{j=1}^{n} f(j) = \binom{n}{2} - k$ is equal to the number of rooted trees with root 0 and non-root vertex set $[n]$ that have k inversions. (See Exercise 7 for a definition of an inversion in a tree.)

10. Prove that the Gaussian polynomial $\begin{bmatrix} n \\ k \end{bmatrix}$ has unimodal coefficients.

11. Let $A = \{1^{a_1}, 2^{a_2}\}$, and let d be the largest common divisor of a_1 and a_2. Now let $J(A, k)$ be the number of permutations p of A that have first entry 1, and for which

$$i(p) \equiv k \mod a_1$$

holds. Let $0 \le t \le \frac{a_1}{d} - 1$. Prove that

$$J(A, k) = \frac{d}{a_1} \binom{n-1}{a_1 - 1}.$$

Note the difference from Exercise 18. Here we are looking at residue classes modulo a_1, not modulo n.

12. Log-concavity is a concept for sequences of *numbers*, but it can be extended to a concept for sequences of *polynomials* as follows.

Let $p_0(q), p_1(q), \cdots, p_m(q)$ be a sequence of polynomials with nonnegative coefficients. We say that this sequence is q-*log-concave* if the polynomial $p_k^2(q) - p_{k-1}(q)p_{k+1}(q)$ has non-negative coefficients for all k. Prove that for any fixed n, the sequence of polynomials $\begin{bmatrix} n \\ 0 \end{bmatrix}, \begin{bmatrix} n \\ 1 \end{bmatrix}, \cdots, \begin{bmatrix} n \\ n \end{bmatrix}$ is q-log-concave.

13. Let us extend the notion of unimodality to polynomials as follows. Let $p_0(q), p_1(q), \cdots, p_m(q)$ be a sequence of polynomials with non-negative coefficients. We say that this sequence is q-*unimodal* if there exists an index j so that $0 \le j \le m$ and for all i, the polynomial $p_j(q) - p_i(q)$ has non-negative coefficients. Note that a q-log-concave sequence does not have to be q-unimodal. Prove that for any fixed n, the sequence of polynomials $\begin{bmatrix} n \\ 0 \end{bmatrix}, \begin{bmatrix} n \\ 1 \end{bmatrix}, \cdots, \begin{bmatrix} n \\ n \end{bmatrix}$ is q-unimodal.

14. Generalize the result of Exercise 27 to the multiset

$$A_{m,n,r} = \{1^{r+1}, 2^{r+1}, \cdots, m^{r+1}, (m+1)^r, \cdots, n^r\}$$

as follows. Let $a_{m,r}(n)$ be the number of permutations of $A_{m,n,r}$ that contain a subword consisting of r copies of 1, then r copies of 2, and so on, ending with r copies of n. Again, the letters of the subword do not have to be consecutive entries in the permutation. Prove that

$$a_{m+1,r}(n) = (rn + 2m - r + 1)a_{m,r}(n) - m(rn + m)a_{m-1,r}(n),$$

where $a_{0,r}(n) = 1$, and $a_{1,r}(n) = rn + 1 - r$.

15. Let us call a $2n$-permutation $p = p_1 p_2 \cdots p_{2n}$ 2-ordered if $p_1 < p_3 < \cdots < p_{2n-1}$ and $p_2 < p_4 < \cdots < p_{2n}$. Prove that

$$\sum_p i(p) = n4^{n-1},$$

where the sum is taken over all 2-ordered $2n$-permutations p.

Solutions to Problems Plus

1. This result is proved by a recursively built injection in [58].

2. Richard Stanley [290] proved that

$$1 + \sum_{n \geq 1} \sum_{p \in S_n} t^{d(p)} q^{i(p)} \frac{u^n}{[n]!} = \frac{1 - t}{\text{Exp}_q(u(t-1) - t)}.$$

   Note that in this problem and the next, the exponent of t is the number of descents (respectively, excedances), and not that parameter plus one. Therefore, setting $q = 1$, we do not get the exponential generating function of the Eulerian polynomials $A_n(z)$, but of the polynomials $A_n(z)/z$.

3. John Shareshian and Michelle Wachs [283] proved that

$$1 + \sum_{n \geq 1} \sum_{p \in S_n} t^{exc(p)} q^{maj(p)} \frac{u^n}{[n]!} = \frac{(1 - tq) \exp_q(z)}{exp_q(ztq) - tq \exp_q(z)}.$$

   We point out that generating functions in which z^n is divided by $[n]!$ are sometimes called *q-exponential generating functions*.

4. This result is due to Dominique Foata and Doron Zeilberger [174], who proved it by providing alternative interpretations for the Denert statistic. In particular, they showed that

$$den(p) = i_1 + i_2 + \cdots + i_m + i(Exc\, p) + i(Nexc\, p),$$

   where $i_1, i_2, \cdots, i_m$ are the excedances of p, while $Exc\, p$ is the *substring* $p_{i_1} p_{i_2} \cdots p_{i_m}$, and $Nexc\, p$ is the substring obtained from p by removing $Exc\, p$. For our example in the text, the permutation 132, we get that

$$den(132) = 2 + 0 + 0 = 2,$$

   as we should.

5. Such a statistic was given by Mark Skandera in [285].

6. This result was obtained by Dominique Foata and Guo-Niu Han [175], who first proved that the joint statistics (d, i) and (d, maj) had the same distribution.

7. For $k = n$, the statement means that there are $(n-1)!$ permutations in S_n so that $maj(p) \equiv j(\mod n)$. This result was first proved in [29], using a heavy algebraic machinery. In that same paper, the authors

provided a bijective proof as well, but that still used Standard Young Tableaux and the Robinson–Schensted correspondence, which we will cover in Chapter 7. The general statement for $k \in [n]$ was given the following simple and beautiful proof in [30]. We know from Theorems 2.3 and 2.18 that

$$\sum_{p \in S_n} z^{maj(p)} = I_n(z) = (1+z)(1+z+z^2) \cdots (1+z+\cdots+z^{n-1}).$$

If we count our permutations according to the remainder of the major index modulo k, then we have to consider the above equation modulo the polynomial $z^k - 1$. If z is any kth root of unity other than 1, then the left-hand side vanishes as there is at least one factor $(1+z+\cdots+z^{k-1}) = \frac{z^k-1}{z-1}$ on the right-hand side, which vanishes. Therefore,

$$I_n(z) = \sum_{p \in S_n} z^{maj(p)} = c(1+z+\cdots+z^{k-1}) \qquad \mod (z^k - 1) \quad (2.20)$$

as the left-hand side and the right-hand side have $k - 1$ common roots, namely those $k - 1$ roots of unity. Setting $z = 1$, we get that $c = n!/k$. Therefore (2.20) implies that

$$I_n(z) = \frac{n!}{k} \cdot (1+z+\cdots z^{k-1}) \qquad \mod (z^k - 1).$$

That proves that the number of n-permutations p satisfying

$$maj(p) \equiv j \ (\mod k)$$

is $n!/k$, proving our claim.

8. This result is due to Don Rawlings [271].

9. This result, in a slightly different form, was found by Germain Kreweras [236].

10. There are several proofs of this fact that had first been noticed by Cayley at the end of the nineteenth century. Some of these proofs are reasonably short, but use sophisticated machinery. See [279], or [269] for such proofs. An elementary proof was given by Kathy O'Hara [258], who used the subset sum interpretation of Gaussian coefficients in her proof. Her argument was later explained in an expository article by Zeilberger [336].

11. This result was proved in [102]. The authors showed that if $p = p_1 p_2 \cdots p_n$, then exactly d of the n cyclic translates $p_1 p_2 \cdots p_n$, $p_2 \cdots p_n p_1$, $\cdots$, $p_n p_1 \cdots p_{n-1}$ have first entry 1 and inversion number k modulo a_1. As $\frac{d}{a_1}\binom{n-1}{a_1-1} = \frac{d}{n}\binom{n}{a_1}$, this proves the result. Note that the result is identical to the result of Exercise 18 (b), even if in that exercise we counted different permutations.

12. This result was proved in [105].

13. This is a special case of a more general result of Lynne Butler [106], which is of group-theoretical flavor. In her proof, Butler uses the interesting fact that the number of subgroups of order q^k of the Abelian group Z_q^n is the Gaussian polynomial $\begin{bmatrix} n \\ k \end{bmatrix}$, where q is a prime.

14. It is proved in [191] that the exponential generating function of the sequence $a_{m,r}(n)$ is

$$(1-z)^{-rn+1} \exp\left(\frac{-rz}{1-z}\right),$$

from which the proof of our statement follows. In [333], Lilly Yen sketches a bijective proof.

15. A bijective proof of this result is due to Volker Strehl, and can be found in [306].

3

In Many Circles. Permutations as Products of Cycles.

3.1 Decomposing a Permutation into Cycles

So far we have looked at permutations as *linear orders*, that is, ways of listing n objects so that each object gets listed exactly once. In this section we will discuss permutations from a different viewpoint. We will consider them as *functions*. Let us redefine permutations along these lines.

DEFINITION 3.1 *Let $f : [n] \to [n]$ be a bijection. Then we say that f is a* permutation *of the set $[n]$.*

This definition certainly does not contradict our former definition of permutations. Formerly, we said that 34152 was a permutation of length five. Now we can reformulate that sentence by saying that the function $f : [5] \to [5]$ defined by $f(1) = 3$, $f(2) = 4$, $f(3) = 1$, $f(4) = 5$, and $f(5) = 2$ is a permutation of $[5]$. Going backwards, the one-line notation simply involved writing $f(1)f(2) \cdots f(n)$ in one line.

This new look at permutations makes another way of writing them plausible. We write

$$f = \frac{1\,2\,3\,4\,5}{3\,4\,1\,5\,2},$$

expressing that f maps 1 to 3, 2 to 4, 3 to 1, 4 to 5, and 5 to 2.

This notation is called the *two-line notation* of permutations. It is more cumbersome than the one-line notation, which consists of writing the second line only, but it has its own advantages as the reader will see shortly.

Let f and g be two permutations of $[n]$. Then we can define their product $f \cdot g$ by $(g \cdot f)(i) = g(f(i))$ for $i \in [n]$. It is straightforward to verify that the set of all permutations of $[n]$ forms a *group* when endowed with this operation. Therefore, the set of all permutations of $[n]$ is often denoted by S_n and is called the *symmetric group* of degree n. We note that for $n \geq 3$, the group S_n is not commutative, so in general, $fg \neq gf$.

The symmetric group is a quintessential ingredient of group theory. It is well known, for instance, that every finite group of n elements is a subgroup of

DOI: 10.1201/9780429274107-3

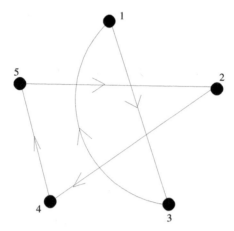

FIGURE 3.1
The cycles of $f = 34152$.

S_n. Extensive research of the symmetric group is therefore certainly justified. In this chapter, we will concentrate on the enumerative combinatorics of the symmetric group, that is, we are going to count permutations according to statistics that are relevant to this second way of looking at them.

A closer look at our running example, the permutation $f = 34152$, reveals that f permutes the elements 3 and 1 among themselves, and the elements 2, 4, 5 among themselves. That is, no matter how many times we apply f, we will always have $f^m(3) = 1$ or $f^m(3) = 3$, and $f^m(3)$ will never have any other values. In other words, f cyclically permutes 1 and 3, and f cyclically permutes 2, 4 and 5. This phenomenon is illustrated in Figure 3.1.

The facts that we have just described are highlighted by a third way of writing the permutation f. We write $f = (13)(245)$, and call it a *cycle notation* of f. When reading a permutation in this notation, we map each element to the one on its right, except for the elements that are last within their parentheses. Those are mapped to the first element within their parentheses.

Example 3.2
The permutation $g = (12)(356)(4)$ is the permutation

$$g = \frac{1\ 2\ 3\ 4\ 5\ 6}{2\ 1\ 5\ 4\ 6\ 3}.$$

❑

Several comments are in order. First, if the cycles are disjoint, the order *among* the cycles clearly does not matter, that is, $(12)(356)(4) = (356)(4)(12) = (4)(12)(356)$, and so on. Indeed, the image of each $i \in [n]$ only depends on its position within its cycle, and some other elements in its cycle. Second, if a cycle has at least three elements, then the order of elements *within* that cycle matters, up to a certain point. Indeed, the cycles (356), (563), and (635) describe the same action of g on elements 3, 5, and 6, but the cycle (365) describes a different action. This action maps 3 to 6, not to 5 as the previous actions did. A little further consideration shows that each cyclical action on k elements can be described by writing k different cycles as one can start with any of the k elements.

We would like to have a unique way of writing our permutations using the cycle notation. Therefore, we will write the largest element of each cycle first, then we will arrange the cycles in increasing order of their first elements. This way of writing permutations will be called their *canonical cycle notation*.

Example 3.3
The permutation $(312)(45)(8)(976)$ is in canonical cycle notation. ▯

3.1.1 Application: Sign and Determinants

The cycle decomposition of a permutation f contains some crucial information about f. For instance, if we know the cycle lengths of f, we can compute the smallest positive integer m for which f^m is the identity permutation. This number m is called the *order* of f in group theory. Indeed, m is obtained as the smallest common multiples of all cycle lengths of f.

A common way of fitting permutations into algebraic frameworks is by defining *permutation matrices*, that is, square matrices whose entries are all equal to 0 or 1, and that contain exactly one 1 in each row and each column. There are two equally useful ways to do this. Let $p = p_1 p_2 \cdots p_n$ be an n-permutation, and let A_p be the matrix $n \times n$ matrix in which

$$A_p(j, i) = \begin{cases} 1 \text{ if } p_i = j, \\ 0 \text{ otherwise.} \end{cases}$$

It is then straightforward to check that the map $f : S_n \to \mathbf{R}^{n \times n}$ defined by $f(p) = A_p$ is a homomorphism, that is, $A_{pq} = A_p \cdot A_q$.

Example 3.4
If $p = 2413 = (4312)$, then we have

$$A_p = \begin{pmatrix} 0 & 0 & 1 & 0 \\ 1 & 0 & 0 & 0 \\ 0 & 0 & 0 & 1 \\ 0 & 1 & 0 & 0 \end{pmatrix}.$$

▯

Now let B_p be the $n \times n$ matrix defined by

$$B_p(i,j) = \begin{cases} 1 \text{ if } p_i = j, \\ 0 \text{ otherwise.} \end{cases}$$

The following example illustrates how closely the matrices A_p and B_p are connected.

Example 3.5
If $p = 2413 = (4312)$, then we have

$$B_p = \begin{pmatrix} 0 & 1 & 0 & 0 \\ 0 & 0 & 0 & 1 \\ 1 & 0 & 0 & 0 \\ 0 & 0 & 1 & 0 \end{pmatrix}.$$

☐

Clearly, for any p, the matrix B_p is the *transpose* of A_p. Furthermore, it is straightforward to check that $A_p B_p = B_p A_p = I$, so the matrices A_p and B_p are inverses of each other. In other words, the inverse of a permutation matrix is its own transpose.

Comparing the definitions of A_p and B_p, one could ask when is the first one easier to use, and when is the second one easier to use. The advantage of the first definition is that as we said, the map $f : S_n \to \mathbf{R}^{n \times n}$ defined by $f(p) = A_p$ is a homomorphism. This is not true for the map $g : S_n \to \mathbf{R}^{n \times n}$ defined by $g(p) = B_p$. That map is an *anti-homomorphism*, that is, $B_{pq} = B_q \cdot B_p$. This can be useful when we use our matrices to permute vectors of size n.

For instance, let

$$\mathbf{x} = \begin{pmatrix} x_1 \\ x_2 \\ x_3 \\ x_4 \end{pmatrix}.$$

If the permutation $p = 2413 = (4312)$ acts on the coordinates of this vector, it takes the vector $\mathbf{x}$ into $p(\mathbf{x}) = \begin{pmatrix} x_2 \\ x_4 \\ x_1 \\ x_3 \end{pmatrix}$. We could have obtained this vector by simply taking the product $B_p \mathbf{x}$. If now another permutation q acts on the vector $p(\mathbf{x})$, then we can obtain the image by computing $B_q(B_p \mathbf{x}) = B_{pq}(\mathbf{x})$. If, instead of the column-vector $\mathbf{x}$, we had worked with the row-vector $\mathbf{y} = (y_1, y_2, y_3, y_4)$, then we would have used the matrices A_p and A_q, to compute the images yA_p and $yA_{pq} = yA_pA_q$.

DEFINITION 3.6   *A permutation is called* odd *(resp.* even*) if it has an odd (resp. even) number of inversions.*

The following proposition shows that we have to be careful throwing around the words "odd" and "even" when describing permutations. Let us call a cycle *even* (resp. *odd*) if it consists of an even (resp. odd) number of elements.

PROPOSITION 3.7

Let p be a permutation that consists of one cycle, and let that cycle be of even length. Then p is an odd permutation. If p is a permutation that consist of one cycle, and that cycle is of odd length, then p is an even permutation.

PROOF   We prove these claims simultaneously, by induction on the length n of the only cycle of our permutation p. For $n = 1$ and $n = 2$, the statement is trivially true. Now let $n \geq 3$, and consider the cycle $(p_1 p_2 \cdots p_n)$. It is straightforward to verify that $(p_1 p_2 \cdots p_n) = (p_1 p_2 \cdots p_{n-1})(p_{n-1} p_n)$. The multiplication by $(p_{n-1} p_n)$ at the end simply swaps the last two entries of $(p_1 p_2 \cdots p_{n-1})$, and therefore, either increases the number of inversions by one, or decreases it by one. So in either case, it changes the parity of the number of inversions. The proof is then immediate by the induction hypothesis.   ∎

The omnipresence of this notion is illustrated by the following Lemma.

LEMMA 3.8

Let p be a permutation. Then the following are equivalent.

(i) p is even,

(ii) $\det A_p = \det B_p = 1$,

(iii) the number of even cycles of p is even.

PROOF

- $(i) \Leftrightarrow (ii)$. This follows from the definition of determinants, as given in Theorem 2.21 because the determinant of a permutation matrix B_p has only one nonzero term, and that is $(-1)^{i(p)} \prod_{i=1}^{n} b_{ip(i)}$. The claim follows as each factor after the product sign is equal to 1.

- $(ii) \Leftrightarrow (iii)$. This is true as p is the product of its cycles. By Proposition 3.7, even cycles correspond to odd permutations, so the determinant of their permutation matrices is -1. Therefore, there has to be an even number of them for the determinant of their product to be 1.

    ∎

In particular, the product of two even permutations is even, and the inverse of an even permutation is even. As the identity permutation is even, this proves the following.

PROPOSITION 3.9
The set of all even permutations in S_n forms a subgroup.

This subgroup is called the *alternating group* of degree n, and is denoted by Alt_n. The reader should prove at this point that Alt_n has $n!/2$ elements if $n \geq 2$, then she should check her answer in Exercise 1. Like the symmetric group, the alternating group has been vigorously investigated throughout the last century. For instance, it is known that Alt_n is a *simple group* if $n \geq 5$. In fact, among all finite simple groups, Alt_n is the easiest to define, except for the cyclic groups Z_p, where p is a prime. (See any introductory book on group theory for the definition of a simple group or Z_p.) It is also interesting that Alt_n is by far larger than any other proper subgroup of S_n. Indeed, other proper subgroups of S_n are of size at most $(n-1)!$. Can you find such a subgroup?

3.1.2   An Application: Geometric Transformations

Some crucial properties of geometric transformations are easy to read off if we decompose the corresponding permutation into the product of simpler permutations. To start, consider a regular hexagon H. Apply various symmetries to this hexagon. It is a well-known fact in group theory, and is not difficult to prove, that there are 12 such symmetries, and that they can all be obtained by repeated applications of r, which is a reflection through a fixed axis a, and t, which is a rotation by 60 degrees counterclockwise. See Figure 3.2 for an illustration.

We point out that both transformations r and t correspond to permutations of the set $\{A, B, C, D, E, F\}$, or, after relabeling, [6]. In particular, $r = (43)(52)(61)$, and $t = (123456)$.

It is natural to ask the following question. Given a series of symmetries, such as $s = trrtrtrrtr$, how can we decide whether the composite transformation can be realized just by moving our hexagon in its original 2-dimensional plane?

This question is easy to answer if we have the transformation given in the above form. Clearly, transformation r changes the orientation of the hexagon, and therefore cannot be realized by a 2-dimensional movement. On the other hand, t does not change the orientation of our hexagon. Consequently, if and only if our composite transformation contains an odd number of reflections r, then it changes the orientation of H, and therefore cannot be realized by a 2-dimensional movement. The example s of the previous paragraph contains six reflections, and is therefore realizable in the plane.

Similar considerations are helpful in deciding whether a certain symmetry of a 3-dimensional solid, such as a cube, can be realized by moving the cube in the 3-dimensional space.

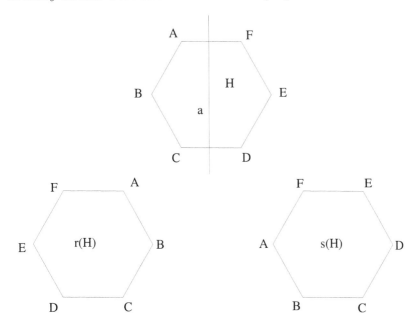

FIGURE 3.2
Two symmetries of a regular hexagon.

3.2 Type and Stirling Numbers

We start with two well-known and basic enumeration problems. It is natural to ask how many n-permutations have a given cycle structure – for instance, how many 12-permutations consist of a 4-cycle, two 3-cycles, one 2-cycle, and zero 1-cycles. It is also natural to ask the more inclusive question of how many n-permutations have exactly k cycles.

In order to facilitate the answer of the first question, we make the following definition.

3.2.1 Cycle Type of a Permutation

DEFINITION 3.10 *Let p be an n-permutation that has exactly a_i cycles of length i, for all positive integers $i \in [n]$. Then we say that p is of* cycle type $(a_1, a_2, \cdots, a_n)$.

Example 3.11
The permutation $p = (21)(534)(6)(987)$ is of cycle type $(1,1,2,0,0,0,0,0,0)$.
◻

The number of permutations with a given cycle type is fairly easy to obtain.

PROPOSITION 3.12
Let $(a_1, a_2, \cdots, a_n)$ be an n-tuple of nonnegative integers satisfying the equation $\sum_{i=1}^{n} a_i \cdot i = n$. Then the number of n-permutations of cycle type $(a_1, a_2, \cdots, a_n)$ is

$$\frac{n!}{a_1! a_2! \cdots a_n! 1^{a_1} 2^{a_2} \cdots n^{a_n}}.$$

PROOF   Let us write the elements of $[n]$ in a linear order, in one of $n!$
possible ways. Then let us place parentheses between the numbers so that the
first a_1 entries form the a_1 cycles of length one, the next a_2 entries form the
a_2 cycles of length two, and so on. The permutations we obtain this way will
all be of cycle type $(a_1, a_2, \cdots, a_n)$, but they will not all be different. Indeed,
the sets of entries forming cycles of the same length can be permuted among
each other without changing the resulting permutation. Therefore, we obtain
each permutation from $a_1! a_2! \cdots a_n!$ linear orders, due to permuting cycles of
the same length. Finally, each i-cycle can be obtained in i different ways as
any of its i entries can be at the first position. So, even if we keep the *sets* of
entries in each cycle fixed, there are $1^{a_1} 2^{a_2} \cdots n^{a_n}$ different linear orders that
could lead to any given permutation of cycle type $(a_1, a_2, \cdots, a_n)$. This shows
that on the whole, each permutation of the desired cycle type is obtained from
$a_1! a_2! \cdots a_n! 1^{a_1} 2^{a_2} \cdots n^{a_n}$ linear orders, and the statement is proved. ∎

For instance, there are $(n-1)!$ permutations of length n that consist of one
n-cycle, and there are $\frac{(2n)!}{n! 2^n}$ permutations of length $2n$ that consist of n cycles
of length two.

3.2.2   Application: Conjugate Permutations

In the symmetric group S_n, two permutations g and h are called *conjugates*
of each other if there exists an element $f \in S_n$ so that $fgf^{-1} = h$ holds. In
group theory, it is often very useful to know that two elements are conjugates
as they share many basic properties. Fortunately, the cycle decomposition of
g and h reveals whether they are conjugates or not. This is the content of the
next lemma.

LEMMA 3.13
*Elements g and h of S_n are conjugates in S_n if and only if they are of the
same cycle type.*

PROOF   Recall that if g and h are two n-permutations, then the action of gh on $[n]$ is obtained by first applying g to $[n]$, and then by applying h to the output.

First let us assume that g and h are conjugates, that is, $fgf^{-1} = h$ for some $f \in S_n$. Let $(b_1 b_2 \cdots b_k)$ be a cycle of g. Then $g^i(b_1) = b_{i+1}$ holds for all indices $i \in [k-1]$, and $g^k(b_1) = b_1$. As $h = fgf^{-1}$, we have $h^i = fgf^{-1} fgf^{-1} \cdots fgf^{-1} = fg^i f^{-1}$. Therefore, $h^i(x) = (fg^i f^{-1})(x)$.

Now choose x so that $f(x) = b_1$. Then $(fg^i f^{-1})(x) = f^{-1}(g^i(b_1)) = f^{-1}(b_{i+1})$ if $i \in [k-1]$, and $(fg^k f^{-1})(x) = f^{-1}(b_1)$. Thus multiplying by f from the left and f^{-1} by the right turns the cycle $(b_1 b_2 \cdots b_k)$ into the cycle $(f^{-1}(b_1) f^{-1}(b_2) \cdots f^{-1}(b_k))$. Therefore, the k-cycles of g are in bijection with the k-cycles of h for all k, and the "only if" part of our lemma is proved.

Now let us assume that g and h have the same cycle type. We construct a permutation f so that $fgf^{-1} = h$. If $(b_1 b_2 \cdots b_k)$ is a cycle of g and $(c_1 c_2 \cdots c_k)$ is a cycle of h, then the argument of the previous paragraph shows that we must choose f so that $f^{-1}(b_i) = c_i$ for $i \in [k]$. This defines f^{-1} for k entries. To find f^{-1} for the remaining $n - k$ entries, proceed similarly for all the remaining cycles.   ∎

Remark We point out that in the second paragraph of this proof we showed that conjugating by f turned the cycle $(b_1 b_2 \cdots b_k)$ into the cycle $(f^{-1}(b_1) f^{-1}(b_2) \cdots f^{-1}(b_k))$. This simple fact is often used in similar arguments. We will see one application in the next subsection.

It is straightforward to prove that the relation "g and h are conjugates" is an equivalence relation. The equivalence classes created by this relation are called *conjugacy* classes. The following simple consequence of the previous lemma is of fundamental importance in representation theory.

COROLLARY 3.14
The number of conjugacy classes of S_n is $p(n)$.

3.2.3   Application: Trees and Transpositions

We have seen that the decomposition of a permutation into *disjoint* cycles is unique, up to the transformations discussed above. We have also seen that the multiplication of disjoint cycles (viewed as permutations) is a commutative operation. However, if we drop the condition that the cycles be disjoint, everything falls apart. Indeed, we have $(12)(13) = 312$ and $(12)(13) = 312 \neq (13)(12) = 231$ as counterexamples. In fact, it is not hard to see that any permutation in S_n can be obtained as a product of (not necessarily disjoint) 2-cycles, or, in other words, *transpositions*. See Exercise 11 for a stronger version of this fact.

While the topic of generating permutations from given sets of not necessarily disjoint cycles belongs more to group theory than to combinatorics, the following results of János Dénes [140] certainly have a combinatorial flavor. Let us call a permutation *cyclic* if it consists of one cycle only.

LEMMA 3.15
Let $s_2, s_3, \cdots, s_n$ be transpositions of S_n. Then the product $s_n s_{n-1} \cdots s_2$ is equal to a cyclic permutation if and only if the graph $G(s_2, s_3, \cdots, s_n)$ with vertices $1, 2, \cdots, n$ and edges $s_2, s_3, \cdots, s_n$ is a tree.

In the unlikely event that the reader is not familiar with basic graph theory, the relevant definitions can be found in any introductory combinatorics book, such as [74].

PROOF   Let us assume that $s_n s_{n-1} \cdots s_2$ is cyclic. Clearly, for any $i, m \in [n]$, there is a path between i and $s_m s_{m-1} \cdots s_2(i)$ in our graph. As all entries $j \in [n]$ are in the cycle $s_n s_{n-1} \cdots s_2$, choosing the right m gives us a path between i and any j. This means that $G(s_2, s_3, \cdots, s_n)$ is connected. On the other hand, by definition, our graph has $n - 1$ edges; therefore it has to be a tree.

Now let us assume that T is a tree with vertex set $[n]$ and with edges $s_2, s_3, \cdots, s_n$. Remove the edge s_n, then T splits into two smaller trees, say T' and T''. By induction, each of these trees corresponds to a cycle, say C_1 and C_2. As these cycles are disjoint, we have

$$s_n s_{n-1} s_{n-2} \cdots s_2 = (a_1 a_2) \cdot C_1 \cdot C_2,$$

where $s_n = (a_1 a_2)$, with $a_1 \in C_1$ and $a_2 \in C_2$. These conditions imply that $s_n s_{n-1} \cdots s_2$ is a cyclic permutation. ∎

One of the best-known theorems of graph theory is Cayley's formula, stating that the number of elements of the set $\mathbf{T_n}$ of all trees on vertex set $[n]$ is n^{n-2}. See [74] for some proofs. The previous lemma enables us to link this result to cyclic permutations.

THEOREM 3.16
Let p be a given cyclic permutation of length n. Then the number of ways to decompose p into the product of $n - 1$ transpositions is n^{n-2}.

This result was first proved by Dénes [140]. The proof we are presenting is due to Moszkowski [253].

PROOF   Clearly, the choice of p is irrelevant as one can always relabel the entries of p, so we may assume that $p = (123 \cdots n)$. We are going to

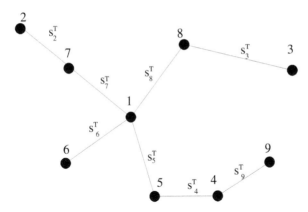

FIGURE 3.3
Labeling the edges of T.

construct a bijection $h : \mathbf{T_n} \to \mathbf{C_n}$, where $\mathbf{C_n}$ is the set of all $(n-1)$-tuples $(s_n, s_{n-1}, \cdots, s_2)$ of transpositions satisfying $s_n s_{n-1} s_{n-2} \cdots s_2 = (12 \cdots n)$.

Let $T \in \mathbf{T_n}$. Then for every vertex $i \in \{2, 3, \cdots, n\}$, there is exactly one path from 1 to i in T. If (a, i) is the last edge in this path, then we label the edge (a, i) by s_i^T. See Figure 3.3 for an example.

Let $\mathbf{s_i}$ denote the transposition interchanging the two endpoints of s_i. Set $C_T = \mathbf{s_n}^T \mathbf{s_{n-1}}^T \cdots \mathbf{s_2}^T$; then C_T is a cyclic permutation because of the previous lemma. As an intermediate step in constructing our bijection h, we define the permutation f_T by the formula

$$f_T(k) = C_T^{k-1}(1),$$

for $1 \le k \le n$. The reader is invited to verify that

$$f_T \cdot C_T \cdot f_T^{-1} = (123 \cdots n). \tag{3.1}$$

An explanation is given in the solution of Exercise 15. Moreover, we set

$$u_i^T = f_T \mathbf{s_i}^T f_T^{-1} \tag{3.2}$$

for $2 \le i \le n$. By Lemma 3.13, we see that u_i^T is also a transposition as it is the conjugate of transposition $\mathbf{s_i}^T$. Furthermore,

$$u_n^T \cdot u_{n-1}^T \cdots u_2^T = f_T \cdot \mathbf{s_n}^T \cdot f_T^{-1} \cdot f_T \mathbf{s_{n-1}}^T \cdot f_T^{-1} \cdots f_T \cdot \mathbf{s_2}^T \cdot f_T^{-1}$$
$$= f_T \mathbf{s_n}^T \mathbf{s_{n-1}}^T \cdots \mathbf{s_2}^T f_T^{-1} = (12 \cdots n),$$

where the last line holds because of (3.1). So $(u_n^T, u_{n-1}^T, \cdots, u_2^T) \in \mathbf{C_n}$. Now we define $h(T) = (u_n^T, u_{n-1}^T, \cdots, u_2^T)$.

We claim that $h : \mathbf{T_n} \to \mathbf{C_n}$ is a bijection. To see this, we will show that h has an inverse; in other words, that every element of $\mathbf{C_n}$ has exactly one preimage under h.

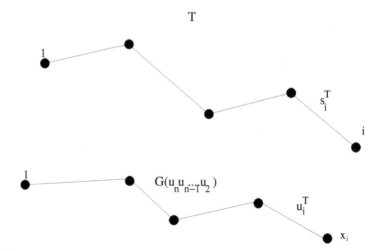

FIGURE 3.4

The connection between the labels of the two trees.

(A) First we show that $C \in \mathbf{C_n}$ cannot have more than one preimage. Let us assume that $(u_n, u_{n-1}, \cdots, u_2) \in \mathbf{C_n}$. Then $u_n \cdot u_{n-1} \cdots u_2 = (12 \cdots n)$, and by the previous lemma, the graph $G(u_n, u_{n-1}, \cdots, u_2)$ with vertex set $[n]$ and edges $u_n, u_{n-1}, \cdots, u_2$ is a tree.

(i) If T is a tree so that $h(T) = u_n \cdot u_{n-1} \cdots u_2$, then the transpositions s_i are uniquely determined by the u_i and formula (3.2). In fact, it is straightforward to check that the trees T and $G(u_n, u_{n-1}, \cdots, u_2)$ are isomorphic as unlabeled trees because T is obtained from the tree $G(u_n, u_{n-1}, \cdots, u_2)$ by applying the permutation f_T to the vertex set $[n]$. Indeed, we get

$$\mathbf{s_i}^T = f_T^{-1} u_i f_T,$$

and as we pointed out in the Remark after the proof of Lemma 3.13, the effect of conjugating a cyclic permutation (in this case, a transposition) by f_T^{-1} is precisely the same as applying f_T to the underlying set.

(ii) On the other hand, we have $f_T(1)$ by definition, and we also know that for $i > 1$, the last edge of the path from 1 to i in T is s_i^T. Therefore, if in $G(u_n, u_{n-1}, \cdots, u_2)$, the endpoint of the path from 1 to the edge u_i is x_i, then we must have $f_T(x_i) = i$.

See Figure 3.4 for an illustration of these labels.

We have seen in part (i) that T as an unlabeled tree is uniquely determined by $(u_n, u_{n-1}, \cdots, u_2)$, and then we have seen in part (ii) that the labels of T are also uniquely determined by $(u_n, u_{n-1}, \cdots, u_2)$. So if a T satisfying $h(T) = (u_n, u_{n-1}, \cdots, u_2)$ exists, it must be unique.

(B) Now we show that each $(u_n, u_{n-1}, \cdots, u_2) \in \mathbf{C_n}$ has a preimage under h. Let $T(u_n, u_{n-1}, \cdots, u_2)$ be the tree obtained from $G(u_n, u_{n-1}, \cdots, u_2)$ by applying the permutation f defined by $f(x_i) = i$ to the vertices, for $2 \le i \le n$. (We know from part (A) that this is the only tree that has a chance to be the preimage of $(u_n, u_{n-1}, \cdots, u_2)$.) Here x_i is defined as in (ii) of part (A). We will show that $h(T) = (u_n, u_{n-1}, \cdots, u_2)$, so T is the preimage we are looking for. By construction, the edges of T are the $f^{-1}u_i f$. For brevity, let us write $\mathbf{s_i} = f^{-1}u_i f$. Then we can compute C_T for our candidate T as follows.

$$C_T = \mathbf{s_n s_{n-1}} \cdots \mathbf{s_2} = f^{-1} \cdot (12 \cdots n) \cdot f = (f(1)f(2) \cdots f(n)),$$

as $u_n \cdot u_{n-1} \cdots u_2 = (12 \cdots n)$. As the next step, we compute $f_T(k)$ for T.

$$f_T(k) = C_T^{k-1}(1) = f(k).$$

Therefore, $f_T = f$, so indeed, $h(T) = (u_n, u_{n-1}, \cdots, u_2)$ as claimed.

∎

3.2.4 Permutations with a Given Number of Cycles

Let us return to the enumeration of permutations with conditions on the number of their cycles.

DEFINITION 3.17 *The number of n-permutations with k cycles is called the signless Stirling number of the first kind, and is denoted by $c(n, k)$.*

Example 3.18
For all positive integers $n > 1$, the equality $c(n, n-2) = 2\binom{n}{3} + \frac{1}{2}\binom{n}{2}\binom{n-2}{2}$ holds. ☐

SOLUTION There are two ways an n-permutation can have $n-2$ cycles. It can have a 3-cycle and $n-3$ fixed points, or it can have two 2-cycles, and $n-4$ fixed points. ∎

All we need to explain about the name "signless Stirling numbers" is the part "signless" and the part "Stirling numbers". That is, what sign are we missing, and what is the connection between these numbers and the Stirling numbers of the second kind, defined in Chapter 1? We will answer these

questions shortly. Let us start with the most basic property of the signless
Stirling numbers of the first kind.

LEMMA 3.19

*Let us set $c(n, 0) = 0$ if $n \geq 1$, and $c(0,0) = 1$. Then the numbers $c(n, k)$
satisfy*

$$c(n, k) = c(n - 1, k - 1) + (n - 1)c(n - 1, k).$$

PROOF   Take an n-permutation p with k cycles. Then the entry n of
p either forms its own 1-cycle, and then there are $c(n - 1, k - 1)$ possibilities
for the rest of the permutation, or is part of a larger cycle, and then the
permutation maps it into one of the other $n - 1$ elements. In this case, there
are $c(n - 1, k)$ possibilities for the permutation that we obtain if we omit n
from p, and the result follows.   ∎

The Stirling numbers $c(n, 0), c(n, 1), \cdots, c(n, n)$ can be easily generated as
the coefficients of a certain polynomial. This is the content of the next theorem.

THEOREM 3.20

For all positive integers n, the equality

$$z(z + 1) \cdots (z + n - 1) = \sum_{k=0}^{n} c(n, k)z^k \qquad (3.3)$$

holds. In other words, $c(n, k)$ is the coefficient of z^k in $z(z+1) \cdots (z+n-1)$.

PROOF   Let $b(n, k)$ be the coefficient of z^k in the polynomial $F_n(z) =
z(z + 1) \cdots (z + n - 1)$. It is then clear that $b(n, 0) = 0$ for $n > 0$, and we set
$b(0, 0) = 1$. We claim that the numbers $b(n, k)$ satisfy the same recurrence
relation as the numbers $c(n, k)$, that is,

$$b(n, k) = b(n - 1, k - 1) + (n - 1)b(n - 1, k).$$

Indeed, $F_n(z) = (z + n - 1)F_{n-1}(z)$, so we have

$$\sum_{k=0}^{n} b(n, k)z^k = \sum_{k=1}^{n} b(n, k)z^{k+1} + (n - 1)\sum_{k=0}^{n-1} b(n - 1, k)z^k$$

by definition. Taking the coefficient of z^k on both sides of the last equation
we get what was to be proved.   ∎

Theorem 3.20 has a plethora of applications which we will keep encountering
in this book. For now, just note that setting $z = -1$ leads to the identity (for

$n \geq 2$)

$$0 = \sum_{k} c(n,k)(-1)^k.$$

That is, there are as many permutations of length n with an odd number of cycles as there are with an even number of cycles. See Exercise 74 for a similar application.

Note that Theorem 3.20 immediately implies the following property of the signless Stirling numbers of the first kind (a property that is often difficult to prove for other sequences).

COROLLARY 3.21
For any fixed positive integer n, the sequence $\{c(n,k)\}_{0 \leq k \leq n}$ has real zeros only. In particular, this sequence is log-concave, and therefore, unimodal.

We point out that a far-reaching generalization of this result has been proved by Francesco Brenti. See Problem Plus 10 for that result.

By now, you probably know that our next question will be whether there is a combinatorial proof of the log-concavity of the sequence $\{c(n,k)\}_k$. The answer is in the affirmative, as shown by the following construction due to Bruce Sagan [278]. In this proof, we modify our cycle notation a little bit by writing our cycles with their *smallest entry first*. This will be called *reverse cycle notation*.

In the construction of Sagan, and in solving some of the exercises of this chapter, the notion of *gap positions* turns out to be useful. An n-permutation written in cycle notation has $n+1$ gap positions, one after each entry p_i (this gap position is in the same cycle as p_i), and one at the end of the permutation, in a separate cycle. For instance, if $p = (1)(23)$, then inserting 4 into the first, second, third, and fourth gap position of p, we get the permutations $(14)(23)$, $(1)(243)$, $(1)(234)$, and $(1)(23)(4)$. Note that the reverse canonical form of p is preserved, no matter where we insert 4. This is why we chose this kind of cycle notation.

Let $P(n,k)$ denote the set of n-permutations with k cycles. Define a map $\Phi : P(n,k-1) \times P(n,k+1) \to P(n,k) \times P(n,k)$ as follows.

Let $p \in P(n,m)$. Let $p_{\langle i \rangle}$ be the permutation obtained from p by removing all entries larger than i from the cycle decomposition of p.

Example 3.22
Let $p = (124)(35)$, then $p_{\langle 1 \rangle} = (1)$, $p_{\langle 2 \rangle} = (12)$, $p_{\langle 3 \rangle} = (12)(3)$, $p_{\langle 4 \rangle} = (124)(3)$, and $p_{\langle 5 \rangle} = (124)(35)$.   ☐

Note that if $p_{\langle i \rangle}$ has t cycles, then $p_{\langle i+1 \rangle}$ has either t or $t+1$ cycles. Now let $(p,q) \in P(n,k-1) \times P(n,k+1)$, and look at the sequences $\{p_{\langle i \rangle}\}_{1 \leq i \leq n}$ and $\{q_{\langle i \rangle}\}_{1 \leq i \leq n}$. Let j be the *largest index* for which $p_{\langle j \rangle}$ has one cycle less than $q_{\langle j \rangle}$. Such an index must exist as the difference between the number of cycles of $q_{\langle i \rangle}$ and $p_{\langle i \rangle}$ changes by at most one at each step, starts at 0 (when $i = 1$), and ends in 2 (when $i = n$).

Swap the elements of the pair $(p_{\langle j \rangle}, q_{\langle j \rangle})$, to get the pair $(q_{\langle j \rangle}, p_{\langle j \rangle})$. Now insert $j + 1$ into the same gap position of $q_{\langle j \rangle}$ in which it would have to be inserted if we wanted to expand $p_{\langle j \rangle}$ into $p_{\langle j+1 \rangle}$. Similarly, insert $j + 1$ into the same gap position of $p_{\langle j \rangle}$ in which it would have to be inserted if we wanted to expand $q_{\langle j \rangle}$ into $q_{\langle j+1 \rangle}$. Call the new pair of permutations obtained this way $(q'_{\langle j+1 \rangle}, p'_{\langle j+1 \rangle})$. Then continue with analogous steps. That is, construct $(q'_{\langle j+a+1 \rangle}, p'_{\langle j+a+1 \rangle})$ from $(q'_{\langle j+a \rangle}, p'_{\langle j+a \rangle})$ by inserting $j + a + 1$ into the same gap position of $q'_{\langle j+a \rangle}$ (resp. $p'_{\langle j+a \rangle}$) that we would use if we had to expand $p_{\langle j+a \rangle}$ (resp. $q_{\langle j+a \rangle}$) into $p_{\langle j+a+1 \rangle}$ (resp. $q_{\langle j+a+1 \rangle}$). At the end of the procedure, we define $\Phi(p, q) = (q'_{\langle n \rangle}, p'_{\langle n \rangle})$.

Example 3.23
Let $n = 6$ and $k = 2$, and let $p = (125463)$, and $q = (13)(24)(56)$. It is then clear that $j = 4$, and $p_{\langle 4 \rangle} = (1243)$ and $q_{\langle 4 \rangle} = (13)(24)$. After swapping, we get the pair

$$(q_{\langle 4 \rangle}, p_{\langle 4 \rangle}) = ((13)(24), (1243)).$$

To make further computations easier, we note that $p_{\langle 5 \rangle} = (12543)$ and $q_{\langle 5 \rangle} = (13)(24)(5)$.

The entry 5 would have to be inserted into the second gap position of $p_{\langle 4 \rangle}$ to get $p_{\langle 5 \rangle}$, and into the fifth gap position of $q_{\langle 4 \rangle}$ to get $q_{\langle 5 \rangle}$. So we insert 5 into the second gap position of $q'_{\langle 4 \rangle}$ and into the fifth gap position of $p'_{\langle 4 \rangle}$. We obtain

$$(q'_{\langle 5 \rangle}, p'_{\langle 5 \rangle}) = ((135)(24), (1243)(5)).$$

Finally, the entry 6 would have to be inserted into the fourth gap position of $p_{\langle 5 \rangle}$ to get $p_{\langle 6 \rangle} = p$, and into the fifth gap position of $q_{\langle 5 \rangle}$ to get $q_{\langle 6 \rangle} = q$. So we insert 6 into the fourth gap position of $q'_{\langle 5 \rangle}$ and into the fifth gap position of $p'_{\langle 5 \rangle}$. We end up with

$$\Phi(p, q) = (q'_{\langle 6 \rangle}, p'_{\langle 6 \rangle}) = ((135)(264), (1243)(56)).$$

⬚

THEOREM 3.24
The map Φ defined above is an injection that maps $P(n, k - 1) \times P(n, k + 1)$ into $P(n, k) \times P(n, k)$.

PROOF   First, Φ indeed maps into $P(n, k) \times P(n, k)$. This is because passing from $p_{\langle i \rangle}$ to $p_{\langle i+1 \rangle}$ a new cycle is formed if and only if $i + 1$ is inserted into the last gap position. By the definition of j, this happens one more time

	$k=0$	$k=1$	$k=2$	$k=3$	$k=4$	$k=5$
n=0	1					
n=1	0	1				
n=2	0	1	1			
n=3	0	2	3	1		
n=4	0	6	11	6	1	
n=5	0	24	50	35	10	1

FIGURE 3.5
The values of $c(n,k)$ for $n \leq 5$. The NE–SW diagonals contain the values for fixed k. Row n starts with $c(n,0)$.

from $q_{\langle j \rangle}$ to $q_{\langle n \rangle}$ than from $p_{\langle j \rangle}$ to $p_{\langle n \rangle}$. After swapping, this will precisely compensate for the extra cycle that we created when passing from $q_{\langle 1 \rangle}$ to $q_{\langle j \rangle}$, in comparison to the segment from $p_{\langle 1 \rangle}$ to $p_{\langle j \rangle}$.

To see that Φ is an injection, note that if $(q'_{\langle n \rangle}, p'_{\langle n \rangle})$ is in the image of Φ, its only preimage is easy to reconstruct. Indeed, remove n from both permutations, then remove $n-1$ from both permutations, and so on, and stop as soon as the first permutation has exactly one more cycle than the second. This provides the index j, and the preimage of $(q'_{\langle n \rangle}, p'_{\langle n \rangle})$ is then obtained by reversing the algorithm used in constructing Φ. ∎

See Figure 3.5 for the values of $c(n,k)$ for $0 \leq n \leq 5$.

As there seems to be no symmetry in any row of this diagram, it is natural to ask where the maximum of each row is. That is, for fixed n, what is the value (or values) of k for which $c(n,k)$ is maximal. Recall that by Corollary 3.21 and Proposition 1.33, there are either one or two such values of k. (In the latter case, the two values of k must be consecutive.) We will call these values of k the *peak* of the sequence.

Surprisingly, our main tool in answering this question comes from the real zeros property of the numbers $c(n,k)$ as proved in Corollary 3.21. This enables us to use the following powerful, and not quite well-known, theorem of Darroch.

THEOREM 3.25
Let $A(z) = \sum_{k=0}^{n} a_k z^k$ be a polynomial that has real roots only that satisfies $A(1) > 0$. Let m be an index so that $a_k \leq a_m$ for all k, that is, a_m is maximal among the coefficients of $A(z)$. Let $\mu = A'(1)/A(1)$. Then

$$|\mu - m| < 1.$$

See [130] for the proof of this spectacular result, as well as for further details. For certain values of n, an even more precise result is known.

Theorem 3.25 is remarkable for the following reason. Let us assume that $A(z)$ is the generating polynomial of certain objects according to some parameter r. For example, $A(z)$ could be the generating polynomial of n-permutations according to their number of cycles, or the generating function of all houses in a given town according to their prices. Then

$$\mu = A'(1)/A(1) = \frac{\sum_k k a_k}{\sum_k a_k}$$

is simply the *average value* of r over the set of all eligible objects. Indeed, there are a_k objects for which the value of r is k. On the other hand, m is the value that r takes *most frequently*.

In general, μ and m may not be directly connected. For example, it could certainly happen that the value of r is never taken by any one object. It could even happen that the average value of r is far away from its specific values. However, Theorem 3.25 shows that if the generating polynomial $A(z)$ has real roots only, then the average value and the most frequent value are less than 1 apart.

Let us return to the task of determining the index k for which $c(n,k)$ is maximal, while n is kept fixed. By Theorem 3.25, it suffices to compute μ, and we need to check at most two values of $c(n,k)$ to find the peak of the sequence $\{c(n,k)\}_k$. We say peak, not peaks, because Erdős [165] proved that this sequence has a unique peak if $n \geq 3$.

Let $A(z) = \sum_{k=0}^{n} c(n,k) z^k$. All we need in order to apply Theorem 3.25 is to compute $A'(1)/A(1)$, that is, the average number of cycles in n-permutations. This is the content of the next lemma.

LEMMA 3.26
The average number of cycles of a randomly chosen n-permutation is

$$H(n) = \sum_{i=1}^{n} \frac{1}{i}.$$

Lemma 3.26 has several proofs, some of which have far-reaching generalizations. For now, we provide an elementary proof. By the end of this chapter, the reader will have the techniques to prove the lemma in two additional ways, and she will be asked to do so in Exercises 65 and 66.

PROOF   We prove the lemma by induction on n, the initial case of $n = 1$ being trivial. Let us assume that the statement is true for n, and take a permutation p on $[n]$. Now insert $n+1$ into any of the $n+1$ gap positions of p. This will create a new cycle if and only if $n+1$ was inserted into the last gap position, that is, in $1/(n+1)$ of all cases. Thus the average number of

n=0				1				
n=1			0		1			
n=2			0	-1		1		
n=3		0	2		- 3		1	
n=4		0	-6	11	- 6		1	
n=5	0	24	- 50	35	-10		1	

FIGURE 3.6
The values of $s(n,k)$ for $n \leq 5$. The NE–SW diagonals contain the values for fixed k. Row n starts with $s(n,0)$.

cycles in a randomly selected $(n+1)$-permutation is

$$\frac{n}{n+1} \cdot H(n) + \frac{1}{n+1}(H(n)+1) = H(n) + \frac{1}{n+1} = \sum_{i=1}^{n+1} \frac{1}{i}$$

as claimed. ∎

Therefore, we have just shown that for any fixed n. we can locate (up to distance 1) the index k for which the Stirling number $c(n,k)$ is maximal.

THEOREM 3.27
Let $n \geq 3$ be a fixed positive integer. Then the unique value of k for which $c(n,k)$ is maximal is within distance 1 of $H(n)$.

The time has come to explain the adjective "signless" in Definition 3.17.

DEFINITION 3.28 *The Stirling numbers of the first kind are defined by $s(n,k) = (-1)^{n-k}c(n,k)$.*

The values of these numbers for $n \leq 5$ are shown in Figure 3.6.

COROLLARY 3.29
For all positive integers n, we have

$$z(z-1)\cdots(z-n+1) = \sum_{k=0}^{n} s(n,k)z^k.$$

In other words, $s(n,k)$ is the coefficient of z^k in $z(z-1)\cdots(z-n+1) = (z)_n.$

PROOF Substitute $-z$ for z in formula (3.3), proved in Theorem 3.20. Then multiply both sides by $(-1)^n$. ∎

The motivation of assigning signs to Stirling numbers of the first kind is explained by the following theorem, which also shows the close connection between the numbers $S(n, k)$ and $s(n, k)$, justifying their similar names.

THEOREM 3.30

Let S be the infinite lower triangular matrix whose rows and columns are indexed by $\mathbf{N}$, and whose entries are given by $S_{i,j} = S(i,j)$. Let s be defined similarly, with $s_{i,j} = s(i,j)$. Then $Ss = sS = I$.

PROOF   Both $A = \{1, x, x^2, x^3, \cdots\}$ and $B = \{1, x, (x)_2, (x)_3, \cdots\}$ are bases of the vector space $\mathbf{R}[x]$ of all polynomials with real coefficients. Here $(x)_m = x(x-1)\cdots(x-m+1)$. Corollary 3.29 shows that the numbers $s(n, k)$ are the coordinates of the elements of B in basis A, while Exercise 21 shows that the numbers $S(n, k)$ are the coordinates of the elements of A in basis B. In other words, s is the transition matrix from A to B, and S is the transition matrix from B to A. Therefore, they must be inverses of each other.   ∎

We mention that an explicit formula for the numbers $s(n, k)$ has been known since 1852 [281]. It is significantly more complicated than the corresponding Lemma (1.17) for the Stirling numbers of the second kind. It states that

$$
s(n, k) = \sum_{0 \leq h \leq n-k} (-1)^h \binom{n-1+h}{n-k+h} \binom{2n-k}{n-k-h} S(n-k+h, h)
$$

$$
= \sum_{0 \leq i \leq h \leq n-k} (-1)^{j+h} \binom{h}{j} \binom{n-1+h}{n-k+h} \binom{2n-k}{n-k-h} \frac{(h-j)^{n-k+h}}{h!}.
$$

This formula is called Schlömilch's formula, and we postpone its proof until the end of the chapter. The reason for this is that in the proof of that formula, we will use the generating functions of the numbers $s(n, k)$, which we are going to compute shortly.

3.2.5   Generating Functions for Stirling Numbers

Proving recurrence relations for Stirling numbers of the first kind combinatorially is not quite as easy as it is for Stirling numbers of the second kind. The additional degree of difficulty comes from the fact that the numbers $s(n, k)$ are not always positive. Therefore, in many cases, an argument using generating functions turns out to be simpler. This is the subject of the present section.

Define the following double generating function for Stirling numbers of the first kind.

$$f(z, u) = \sum_{n \geq 0} \sum_{k=0}^{n} s(n, k) z^k \frac{u^n}{n!}.$$

PROPOSITION 3.31
The equality

$$f(z, u) = (1 + u)^z$$

holds.

PROOF   The coefficient of u^n on the right-hand side is $\binom{z}{n}$ by the Binomial theorem. On the left-hand side, it is

$$\frac{\sum_{k=0}^{n} s(n, k) z^k}{n!} = \frac{(z)_n}{n!} = \binom{z}{n},$$

and the statement is proved. We have used Corollary 3.29 in the above equality. ∎

Sometimes an alternative form of $f(z, u)$ is easier to use.

PROPOSITION 3.32
The equality

$$f(z, u) = \sum_{n \geq 0} (z)_n \frac{u^n}{n!}$$

holds.

PROOF   Immediate from Corollary 3.29. ∎

The nice, compact form of the double generating function $f(z, u)$ results in a plethora of recurrence relations on the numbers $s(n, k)$. These recurrence relations are sometimes classified into three sets, *triangular*, *vertical*, and *horizontal* recurrences. These names are based on the arrays of the numbers $s(n, k)$ involved.

A triangular recurrence is readily obtained from Lemma 3.19. It is

$$s(n, k) = s(n - 1, k - 1) - (n - 1)s(n - 1, k), \tag{3.4}$$

where $n \geq 1$.

Now we are going to use our generating functions to obtain a vertical recurrence.

LEMMA 3.33

For all fixed positive integers k and n, the equality

$$ks(n, k) = \sum_{\ell=k-1}^{n-1} (-1)^{n-\ell-1} \frac{(n)_{n-\ell}}{n-\ell} \cdot s(\ell, k-1)$$

holds.

PROOF　　It is clear from Proposition 3.31 that

$$\frac{\partial f(z, u)}{\partial z} = f(z, u) \ln(1 + z).$$

The coefficient of $z^{k-1} u^n / n!$ on the left-hand side is $ks(n, k)$. The right-hand side equals

$$f(z, u) \sum_{\ell \geq 1} (-1)^{\ell+1} \frac{u^\ell}{\ell} = \sum_{i \geq 0} \sum_{j=0}^{i} s(i, j) z^j \frac{u^i}{i!} \cdot \sum_{\ell \geq 1} (-1)^{\ell+1} \frac{u^\ell}{\ell}.$$

Therefore, the coefficient of $z^{k-1} u^n / n!$ on the right-hand side is

$$n! \sum_{\ell=k-1}^{n-1} (-1)^{n-\ell-1} \cdot s(\ell, k-1) \cdot \frac{1}{n-\ell} \cdot \frac{1}{\ell!},$$

proving our claim.　■

Sometimes a univariate generating function is sufficient. The "horizontal" generating function of the numbers $s(n, k)$ was given in Corollary 3.29. Their "vertical" generating function is computed below.

LEMMA 3.34

For any fixed k, the exponential generating function of the Stirling numbers of the first kind is given by

$$f_k(u) = \sum_{n=k}^{\infty} s(n, k) \frac{u^n}{n!} = \frac{[\ln(1+u)]^k}{k!}.$$

PROOF　　Changing the order of summation in $f(z, u)$, we get

$$f(z, u) = \sum_{k=0}^{n} \sum_{n=k}^{\infty} s(n, k) \frac{u^n}{n!} z^k = \sum_{k=0}^{n} f_k(u) z^k.$$

The proof now follows by equating the coefficients of z^k in two forms of $f(z, u)$, in the one above, and in

$$f(z, u) = (1 + u)^z = \exp[z \ln(1 + u)] = \sum_{k=0}^{\infty} \frac{[\ln(1 + u)]^k}{k!} z^k.$$

▮

From here, the analogous result for $c(n, k)$ is a breeze. Note that in the end of this chapter, we will see how to deduce this corollary by a more direct and more combinatorial argument.

COROLLARY 3.35
For any fixed k, the exponential generating function of the signless Stirling numbers of the first kind is

$$h_k(u) = \sum_{n=k}^{\infty} c(n, k) \frac{u^n}{n!} = \frac{[-\ln(1 - u)]^k}{k!}.$$

Finally, we are going to prove a horizontal recurrence relation. Interestingly, the vertical generating function $f_k(u)$ will be used in our proof.

LEMMA 3.36
For all positive integers k and n, the equality

$$s(n + 1, k + 1) = \sum_{m=k}^{n} (-1)^{m-k} \binom{m}{k} s(n, m)$$

holds.

PROOF   Differentiating our generating function $f_k(u)$ (after shifting indices by one), we get

$$\sum_{n=k}^{\infty} s(n + 1, k + 1) \frac{u^n}{n!} = \frac{(1 + u)^{-1} [\ln(1 + u)]^k}{k!}$$

$$= \exp(-\ln(1 + u)) \frac{[\ln(1 + u)]^k}{k!}$$

$$= \sum_{m=k}^{\infty} \frac{(-1)^m [\ln(1 + u)]^m}{m!} \cdot \frac{[\ln(1 + u)]^k}{k!}$$

$$= \sum_{m=k}^{\infty} \frac{(-1)^{m-k} [\ln(1 + u)]^m}{(m - k)! k!}$$

Note that $f_k(u)$ itself is very similar to the last member of the last expression. This suggests that we further rearrange the last member as follows.

$$\sum_{n=k}^{\infty} s(n+1,k+1)\frac{u^n}{n!} = \sum_{m=k}^{\infty}(-1)^{m-k}\binom{m}{k}\sum_{n=m}^{\infty} s(n,m)\frac{u^n}{n!}$$

$$= \sum_{n=k}^{\infty}\left[\sum_{m=k}^{n}(-1)^{m-k}\binom{m}{k}s(n,m)\right]\frac{u^n}{n!},$$

and the result follows by equating the coefficients of $\frac{u^n}{n!}$. ∎

3.2.6   Application: Real Zeros and Probability

There is a surprisingly strong connection between polynomials having real zeros and random independent trials. It is described by the following theorem of P. Lévy.

THEOREM 3.37
[Lévy's theorem] Let $(a_0, a_1, \cdots, a_n)$ be a sequence of non-negative real numbers, let $A(z) = \sum_{k=0}^{n} a_k z^k$, and let us assume that $A(1) > 0$. Then the following are equivalent.

(i) The polynomial $A(z)$ is either constant or has real zeros only;

(ii) The sequence $(a_0/A(1), a_1/A(1), \cdots, a_n/A(1))$ is the distribution of the number d_n of successes in n independent trials with probability p_i of success on the ith trial, for some sequence of probabilities $0 \leq p_i \leq 1$. The roots of $A(z)$ are given by $-(1-p_i)/p_i$ for i with $p_i < 0$.

Note that the probability of success at trial i has to be independent of the result of all other trials, for all i.

Example 3.38
Let $a_i = \binom{n}{i}$. Then $A(z) = (1+z)^n$, so $A(z)$ indeed has real zeros only. For all i, let the ith trial be the flipping of a fair coin, and let heads be considered success. As the coin is fair, $p_i = 1/2$ for all i. Then $a_k/A(1) = \binom{n}{k}/2^n$ is indeed the probability of having exactly k successes. The roots of $A(z)$ are all equal to $-(1-p_i)/p_i = -1$. ☐

This theorem shows that we can deepen our understanding of a given sequence having real zeros by finding its probabilistic interpretation. We would like to find a probabilistic interpretation for the signless Stirling numbers of the first kind. That is, we want to find a sequence of n independent trials in which the probability of having k successes is $c(n,k)/n!$.

Consider the following sequence of trials. We have one orange ball in a box. Let us pick a ball at random, (at the first trial, we have only one choice), then let us put this ball back in the box, along with a blue ball. Keep repeating this procedure n times. A success is when we draw an orange ball. Let p_k be the probability of k successes in our n trials.

Let A_j be the event that we choose an orange ball at trial j. Then $P[A_1] = 1$, and we are going to compute $P[A_j]$. At the beginning of trial j, there is one orange ball and $(j-1)$ blue balls in the box, showing that

$$P[A_j] = \frac{1}{1+(j-1)} = \frac{1}{j}.$$

Consequently, the probability of failure at trial j is $P[\bar{A}_j] = \frac{j-1}{j}$. As our n trials are independent, the probability of k successes in n trials is computed as

$$d_k = \sum P[A_{j_1}]P[A_{j_2}]\cdots P[A_{j_k}]P[\bar{A}_{j_{k+1}}]\cdots P[\bar{A}_{j_n}],$$

where the sum is taken over the family F of all possible $(n-k)$-element subsets $\{j_{k+1}, j_{k+2}, \cdots, j_n\}$ of the set $\{2, 3, \cdots, n\}$. (There is no chance of failure on the first trial.) Using the probabilities computed above, this yields

$$d_k = \frac{1}{n!} \sum_F (j_{k+1} - 1)(j_{k+2} - 1)\cdots (j_n - 1).$$

Exercise 35 shows that the value of the sum above is $c(n, k)$, implying that

$$p_k = \frac{c(n, k)}{n!}.$$

Note that while Lévy's theorem is often a useful tool to prove the real zeros property, it is relatively easy to make a mistake while doing so. See Exercise 27 for a correct application of this method, and see Exercise 63 for a caveat.

3.3 Cycle Decomposition versus Linear Order

3.3.1 Transition Lemma

We have seen two different ways of looking at permutations. One considered permutations as linear orders of $[n]$ and denoted them by specifying the order $p_1 p_2 \cdots p_n$ of the n elements. The other considered permutations as elements of the symmetric group S_n and denoted them by parenthesized words that described the cycles of the permutations. We have not seen, however, too many connections between the two different lines of thinking. For instance, we have not analyzed the connection between the "visually" similar permutations

2417635 and (2)(41)(7635). Fortunately, such a connection exists, and it is a powerful tool in several enumeration problems. This is the content of the following well-known lemma that is due to Dominique Foata.

LEMMA 3.39

[Transition Lemma] Let p be an n-permutation written in canonical cycle notation, and let $f(p)$ be the n-permutation written in the one-line notation that is obtained from p by omitting all parentheses. Then the map $f : S_n \to S_n$ is a bijection.

Applying this bijection to the example of the paragraph preceding the lemma, we have $f((2)(41)(7635)) = 2417635$.

PROOF   It is clear that f indeed maps into S_n. What we have to show is that it has an inverse, that is, for each n-permutation q, there exists exactly one p so that $f(p) = q$.

Let $q = q_1 q_2 \cdots q_n$ be any n-permutation, and let us look for its possible preimages under f. The first cycle of any such permutation p must obviously start with q_1. Where could this cycle end? As p is in canonical notation, each cycle starts with its largest entry. Therefore, if $q_i > q_1$, then q_i cannot be part of the first cycle of p. So if j is the smallest index so that $q_j > q_1$, then the first cycle of p must end in q_{j-1} or earlier. On the other hand, we claim that this cycle cannot end earlier than in q_{j-1}. Indeed, by the definition of j, we have $q_k < q_1$ if $1 < k < j$, so if the second cycle started somewhere between q_1 and q_j, it would start with an entry less than q_1. That would contradict the canonical property of q.

Thus the second cycle of p *must* start with the *leftmost* entry q_j of q that is larger than q_1. By the very same argument, the third cycle of q *must* start with the leftmost entry of q larger than q_j, and so on. The procedure will stop with the cycle that starts with the entry n. This yields a deterministic, and always executable, algorithm to find the unique preimage of q under p.

∎

In our running example, we have $q = 2417635$. This yields $q_1 = 2$. The smallest j so that $q_j > q_1$ is $j = 2$, so the second cycle will start in the second position. Then, the leftmost entry larger than 4 is 7, so the third cycle starts with 7, and we get the permutation (2)(41)(7635).

Taking a second look at our proof, we note that we can explicitly describe the entries of q at which a new cycle is started. These are precisely the elements that are *larger than everything on their left*. This is an important notion.

DEFINITION 3.40   *Let $q = q_1 q_2 \cdots q_n$ be a permutation. We say that q_i is a* left-to-right maximum *if, for all $k < i$, the inequality $q_k < q_i$ holds.*

Example 3.41
The permutation 21546837 has four left-to-right maxima. These are the entries 2, 5, 6, and 8.   ▯

   Left-to-right minima, right-to-left maxima, and right-to-left minima are defined accordingly.

COROLLARY 3.42
The number of n-permutations with exactly k left-to-right maxima is $c(n,k)$.

PROOF   Bijection f of Lemma 3.39 maps the set of n-permutations having exactly k cycles onto this set of permutations.   ∎

3.3.2   Applications of the Transition Lemma

Lemma 3.39 has several interesting applications to random permutations. We will discuss these in Chapter 4. For now, we fulfill an old promise by proving Theorem 1.36. For easy reference, that theorem stated that the number of n-permutations having exactly $k - 1$ excedances is $A(n,k)$. Also recall that i is an excedance of the permutation $p = p_1 p_2 \cdots p_n$ if $p_i > i$. Similarly, i is a weak excedance of p if $p(i) \geq i$.

PROOF   (Of Theorem 1.36). We show that the bijection f of Lemma 3.39 maps the set of n-permutations with k *weak excedances* onto the set of n-permutations with $k - 1$ ascents.
    The proof will be somewhat surprising. The definition of excedances is given in terms of permutations as linear orders, so one would not expect to count excedances on permutations that are written in cycle notation. However, this is precisely what we will do.
    Let $\pi = (p_1 \cdots p_{i_1})(p_{i_1+1} \cdots p_{i_2}) \cdots (p_{i_{j-1}+1} \cdots p_{i_j})$ be written in canonical cycle notation.
    We now apply f to π, and count the ascents. We claim that i is an ascent of $f(\pi)$, that is, $p_i < p_{i+1}$ if and only if $i \neq n$ and $p_i \leq \pi(p_i)$.
    To see the "only if" part, note the following. Unless i is a position at the end of a cycle, we have $\pi(p_i) = p_{i+1}$ as each element of a cycle is mapped into the one immediately on its right, except for the last one. Therefore, $p_i < p_{i+1}$ implies $p_i < \pi(p_i)$. When i is a position at the end of a cycle, then $p_i < p_{i+1}$ always holds because of the canonical property, and so does $p_i \leq \pi(p_i)$. Indeed, p_i is mapped to the leftmost entry (in the canonical cycle

notation) in its cycle, and that is always larger than p_i (or equal to p_i, if the cycle containing p_i is a 1-cycle).

The proof of the "if" part is similar. If we have $p_i \leq \pi(p_i)$, then either i is a position not at the end of a cycle, or it is at the end of a cycle. In the former case, $\pi(p_i) = p_{i+1}$, and therefore the condition $p_i \leq \pi(p_i)$ implies $p_i < p_{i+1}$. In the latter case, $\pi(p_i) = p_i$ if the cycle containing p_i is a 1-cycle, and $p_i < p_{i+1}$ if the cycle containing p_i is not a 1-cycle, as in that case, $p_i \leq \pi(p_i) < p_{i+1}$. Indeed, p_i is less than the first entry of its cycle, and that entry is in turn less than the first entry of the next cycle by the canonical property.

We have shown that the bijection f of Lemma 3.39 turns the weak excedances of π other than n into ascents of $f(\pi)$. Finally, Exercise 13 of Chapter 1 shows that the number of n-permutations with k weak excedances is the same as the number of n-permutations with $k-1$ excedances, completing the proof of our theorem. ∎

Example 3.43
Let $\pi = (32)(514)(76)$. Then in the one-line notation, we have $\pi = 4325176$. This permutation has four weak excedances, namely at 1, 2, 4, and 6. On the other hand, $f(\pi) = 3251476$ has indeed three ascents, at 2, 4, and 5. ☐

Note that $f(\pi)$ had its ascents at 2, 4, and 5. This is where π has its weak excedances (besides the entry mapping into n), if we use the notations of the proof, that is, $\pi = (32)(514)(76) = (p_1 p_2)(p_3 p_4 p_5)(p_6 p_7)$. Indeed, entries 2, 1, and 4, that is, the second, fourth and fifth entries, are mapped into a larger entry.

Lemma 3.39 can also be used to prove the following surprising fact. We have computed in Lemma 3.26 that the *total* number of cycles in all $n!$ permutations of length n, is $n!H(n) = n!(1 + \frac{1}{2} + \cdots + \frac{1}{n})$. On the other hand, it is not hard to prove, either as a special case of Exercise 18 or directly, that similarly, $c(n+1, 2) = n!H(n)$. In other words, *there are as many* cycles *in all n-permutations as there are* $(n+1)$-*permutations with exactly two cycles*, if $n \geq 1$. The fact that these two sets are equinumerous certainly asks for a bijective proof, and we are going to provide one. The crucial idea will be that omitting $n+1$ from its cycle can split that cycle up into many smaller cycles.

Recall the map $f : S_n \rightarrow S_n$ of the Transition Lemma. This map was defined on permutations of length n. With a slight abuse of notation, we will also use the notation "f" for the analogous map defined on any S_m. This is because we will use f on a permutation, and then on some of its parts.

Let

$$A = \{p \in S_{n+1} : p \text{ has two cycles}\}$$

and let

$$B = \{(q, C) : q \in S_n \text{ and } C \text{ is a cycle of } q.\}$$

Define $g : A \to B$ as follows. Set E to be the cycle of p that does *not* contain the entry $n+1$. Let D be the other cycle of p, the one that contains $n+1$. Then we can consider D as a permutation, except it is not a permutation of $[n]$, but of some other set S. Apply f to D (that is, omit the parentheses), then omit $n + 1$ from the resulting permutation of S, to get the new permutation D'. Now apply f^{-1} back to D' to get a permutation q' of $S - \{n+1\}$ that is written in canonical cycle notation. Then we define $q = Eq'$, that is, q is the product of the disjoint cycles E and q'. Finally, we set $g(p) = (q, E)$.

Example 3.44
Let $n = 8$, and let $p = (4231)(97586)$. Then $E = (4231)$, and $D = (97586)$. Therefore, $D' = 7586$, so $q' = (75)(86)$. This yields $q = (4231)(75)(86)$, so we have $g(p) = (q, E) = ((4231)(75)(86), (4231))$. ⬚

THEOREM 3.45
The map $g : A \to B$ is a bijection.

PROOF   We prove that g has an inverse. That is, we show that for any $(q, C) \in B$, there is exactly one $p \in A$ so that $g(p) = (q, C)$ holds.

The crucial idea is that we can use $n + 1$ to "glue" together all cycles except for C into one cycle. Let $C_1, C_2, \cdots, C_k$ be the cycles of q in canonical order, with $C_i = C$. To get the preimage of the pair (q, C) under g, omit all parentheses from the string $C_1 C_2 \cdots C_{i-1} C_{i+1} \cdots C_k$, then prepend the obtained permutation with the entry $n + 1$, to get the single cycle H. Then define p to be product of the two disjoint cycles C and H. This shows that (q, C) has a preimage. (Note that if q had only one cycle, then the list $C_1 C_2 \cdots C_{i-1} C_{i+1} \cdots C_k$ is empty, and we just create a 1-cycle consisting of the entry $n + 1$ only.)

To see that (q, C) cannot have more than one preimage is straightforward. Indeed, a change in either cycle of p clearly results in a change of $f(p)$. ∎

Example 3.46
To find the preimage of the pair $(q, C) = ((4231)(75)(86), (4231))$ under g, we prepend the permutation 7586 by the maximum element 9. This leads to the cycle (97586), yielding $g^{-1}(q, C) = (4231)(97586)$ as we expected because of Example 3.44. ⬚

3.4   Permutations with Restricted Cycle Structure

3.4.1   Exponential Formula

We start with a very brief introduction to the main enumerative tool of this section. Many of the readers may have seen this material in a general combinatorics textbook. Other readers may consult Chapter 8 of [74] or Chapter 3 of [75] for a longer treatment on an introductory level, or Chapter 5 of [297] for a discussion at an advanced level. The book [169] is the most comprehensive treatment of the tools we will use.

Recall that the exponential generating function of the sequence $a_0, a_1, a_2, \cdots$ of complex numbers is the formal power series

$$A(z) = \sum_{n \geq 0} a_n \frac{z^n}{n!}.$$

It is not always necessary to require that the a_i be complex numbers. In general, it suffices to assume that they are elements of a field of characteristic zero.

LEMMA 3.47 Product Formula

Let $f(n)$ be the number of ways to build a certain structure on an n-element set, and let $g(n)$ be the number of ways to build another kind of structure on an n-element set. Let $F(z)$ and $G(z)$ be the exponential generating functions of the sequences $f(n)$ and $g(n)$, for $n = 0, 1, 2, \cdots$. Finally, let $h(n)$ be the number of ways to partition an n-element set into two subsets (which are allowed to be empty), then to put a structure of the first kind on the first subset, and to put a structure of the second kind on the second subset. Let $H(z)$ be the exponential generating function of the sequence $h(n)$, for $n = 0, 1, 2, \cdots$. Then the identity

$$F(z)G(z) = H(z) \tag{3.5}$$

holds.

PROOF   The number of ways to split an n-element set into a k-element set and its complement, then build the above structures on these sets is obviously $\binom{n}{k} f(k) g(n - k)$. Summing over all k, we get that

$$h(n) = \sum_{k=0}^{n} \binom{n}{k} f(k) g(n - k). \tag{3.6}$$

So our lemma will be proved if we can show that the expression on the right-hand side of (3.6) is in fact the coefficient of $z^n/n!$ in $F(z)G(z)$. This is simply a question of computation as

$$F(z)G(z) = \sum_{n \geq 0} f(n) \frac{z^n}{n!} \cdot \sum_{n \geq 0} g(n) \frac{z^n}{n!}$$

$$= \sum_{n \geq 0} z^n \sum_{k=0}^{n} \frac{1}{k!(n-k)!} f(k)g(n-k)$$

$$= \sum_{n \geq 0} \frac{z^n}{n!} \sum_{k=0}^{n} \binom{n}{k} f(k)g(n-k).$$

∎

We have two remarks about the two subsets into which we have partitioned our n-element set. First, the partition is *ordered*, that is $\{1,3\}, \{2,4,5\}$ is not the same partition as $\{2,4,5\}, \{1,3\}$. Indeed, in the first case, we build a structure of the first kind on the block $\{1,3\}$, and in the second case we build a structure of the second kind on that block. Second, empty blocks are *permitted*. This is why the generating functions $F(z)$ and $G(z)$ start with a constant term that is not necessarily 0. In other words, our two subsets form a *weak ordered partition* of our n-element set.

Iterated applications of Lemma 3.47 lead to the following.

COROLLARY 3.48
Let $k \leq n$ be positive integers. For $i \in [k]$, let $f_i(n)$ be the number of ways to build a structure of a certain kind i on an n-element set, and let $F_i(z)$ be the exponential generating function of the sequence $f_i(n)$. Finally, let $h(n)$ be the number of ways to partition an n-element set into an ordered list of blocks $(B_1, B_2, \cdots, B_k)$ where the B_i are allowed to be empty, then build a structure of kind i on B_i, for each $i \in [k]$. If $H(z)$ is the exponential generating function of the sequence $h(n)$, then the identity

$$H(z) = F_1(z)F_2(z) \cdots F_k(z)$$

holds.

Finally, note that it was not particularly important in our proof that the numbers $f(n)$ were integers, or that they counted the number of certain structures. What was important was the relation between $f(n)$ and $h(n)$. This leads to the following generalization.

THEOREM 3.49
Let K be a field of characteristic zero, and let $f_i : \mathbf{N} \to K$ be functions, $1 \leq i \leq k$. Define $h : \mathbf{N} \to K$ by

$$h(n) = \sum f_1(|A_1|)f_2(|A_2|) \cdots f_k(|A_k|),$$

where the sum ranges over all weak ordered partitions $(A_1, A_2, \cdots, A_k)$ of $[n]$ into k parts. Let $F_i(z)$ and $H(z)$ be the exponential generating functions of the sequences $f_i(n)$ and $h(n)$. Then the identity

$$H(z) = F_1(z)F_2(z) \cdots F_n(z)$$

holds.

You could ask what the advantage of letting f and h map into any field of characteristic zero is, as opposed to just the field of real or complex numbers. The answer will become obvious when we discuss cycle indices later in this section. For the time being, we just mention that K will often be chosen to be a field of rational functions over a field.

Theorem 3.49 yields an immediate second proof of Corollary 3.35. For easy reference, that corollary claimed that the exponential generating function of the signless Stirling numbers of the first kind is

$$h_k(u) = \sum_{n=k}^{\infty} c(n,k) \frac{u^n}{n!} = \frac{[-\ln(1-u)]^k}{k!}.$$

Indeed, $h_k(u)$ is just the exponential generating function of the numbers $c(n,k)$. Note that $k!c(n,k)$ is the number of ways to choose an n-permutation with k-cycles, then order its set of cycles. However, this is the same as partitioning $[n]$ into k non-empty subsets $(A_1, A_2, \cdots, A_k)$, then taking a 1-cycle on each of the A_i. Therefore, with the notations of Theorem 3.49, we have $f_i(k) = (k-1)!$ for each i, with $f(0) = 0$. This yields $F_i(u) = \sum_{k=1}^{\infty} \frac{u^k}{k} = -\ln(1-u)$ for each i, and our claim is proved.

Our main tool in this section, the Exponential formula, will be stated and proved shortly, using Theorem 3.49. First, however, we have to take some precautions. We want to work with power series of the type $\exp(F(x))$, where $F(x)$ is some formal power series.

A little thought shows that $\exp(F(z)) = 1 + F(z) + F(z)^2/2 + \cdots$ is defined if and only if $F(0) = 0$, or in other words, when F has no constant term. Indeed, it is precisely in this case that the coefficient of any z^m in $1 + F(z) + F(z)^2/2 + \cdots = \sum_{n=0}^{\infty} \frac{F(z)^n}{n!}$ will be a finite sum. Therefore, we will have to make sure we only talk about $\exp(F(z))$ when we know that F has no constant term.

Having settled that, we are ready for the exponential formula. Let $\mathbf{P}$ denote the set of all *positive* integers.

THEOREM 3.50

[The Exponential Formula] Let K be a field of characteristic zero, and let $f : \mathbf{P} \to K$ be a function. Define $h : \mathbf{N} \to K$ by $h(0) = 1$, and by

$$h(n) = \sum f(|A_1|) \cdot f(|A_2|) \cdots \cdot f(|A_m|), \qquad (3.7)$$

for $n > 0$, where the sum ranges over all partitions $(A_1, A_2, \cdots, A_m)$ *of* $[n]$ *into* any *number of parts. Let $F(z)$ and $H(z)$ be the exponential generating functions of the sequences $f(n)$ and $h(n)$. Then the identity*

$$H(z) = \exp(F(z))$$

holds.

In particular, if $f(n)$ is the number of ways one can build a structure of a certain kind on an n-element set, then $h(n)$ is the number of ways to take a set partition of an n-element set, and then to build a structure of that same kind on each of the blocks.

Note that in contrast with the product formula, here we take a *partition* of our n-element set. This means that the set of blocks is *not ordered*, that is, $\{1, 3\}, \{2, 4, 5\}$, and $\{2, 4, 5\}, \{1, 3\}$ are considered identical, and empty blocks are *not* permitted. There is no restriction on the number of the blocks, unlike in Lemma 3.47 and in Corollary 3.48. This is why we have to exclude empty blocks; otherwise the number of our partitions would be infinite.

PROOF   (of Theorem 3.50) Let

$$h_k(n) = \sum f(|A_1|) f(|A_2|) \cdots f(|A_k|)$$

where the sum ranges over all *partitions* of $[n]$ into k parts, for a fixed k.

Now that the number of blocks is fixed, we can use Corollary 3.48 (with $f(0) = 0$), keeping in mind that in that corollary, the set of blocks is ordered. Recall that $[z^n] t(z)$ denotes the coefficient of z^n in $t(z)$. We get

$$h_k(n) = \frac{1}{k!} [z^n] \left(F(z)^k \right).$$

Summing over all k, we get that

$$h(n) = \sum_k h_k(n) = \sum_k [z^n] \left(F(z)^k \right) = [z^n] \exp(F(z)),$$

which was to be proved.   ∎

We point out that there is a stronger version of this theorem, the Compositional formula, but we will not need that here. Interested readers should check Exercise 48.

Example 3.51
Let $h(0) = 1$, and let $h(n)$ be the number of ways to take a set partition of $[n]$ and then take a subset of each block. Then the Exponential Formula yields $H(z) = \sum_{n \geq 0} h(n) \frac{z^n}{n!} = \exp(\exp(2z) - 1)$.   ▯

SOLUTION Let $f(n)$ be the number of ways to choose a subset of a nonempty set. Then $f(n) = 2^n$ for $n > 0$, and $f(0) = 0$. Therefore, the exponential generating function of the sequence $f(n)$ is $F(z) = \exp(2z) - 1$, and the proof follows from the Exponential formula. ∎

COROLLARY 3.52

[Exponential formula, permutation version] Let K be a field of characteristic 0, and let $f : \mathbf{P} \to K$ be any function. Define a new function h by

$$h(n) = \sum_{p \in S_n} f(|C_1|)f(|C_2|) \cdots f(|C_k|), \tag{3.8}$$

where the C_i are the cycles of p, and $|C_i|$ denotes the length of C_i. Let $F(z)$ and $H(z)$ be the exponential generating functions of f and h. Then the equality

$$H(z) = \exp\left(\sum_{n \geq 1} f(n) \frac{z^n}{n}\right)$$

holds.

PROOF Note the subtle difference between the summation on the right-hand side of (3.8) and the summation in (3.7). The former is a summation over all $n!$ elements of S_n, the latter is a summation over all set partitions of $[n]$. As a k-element block of a partition can form $(k-1)!$ different k-cycles, a partition with blocks $B_1, B_2, \cdots, B_k$ gives rise to $\prod_{i=1}^{k}(|B_i|-1)!$ permutations whose cycles are the blocks B_i. Therefore, (3.8) can be rearranged as

$$h(n) = \sum_{B \in \Pi_n} f(|B_1|)(|B_1| - 1)!f(|B_2|)(|B_2| - 1)! \cdots f(|B_k|)(|B_k| - 1)!,$$

where Π_n denotes the set of all partitions of $[n]$.

Then the Exponential formula (applied to $(n-1)!f(n)$ instead of $f(n)$) implies

$$H(z) = \exp((n-1)! \cdot F(z)) = \exp\left(\sum_{n \geq 1} f(n) \frac{z^n}{n}\right),$$

which was to be proved. ∎

The following application of Corollary 3.52 enables us to count permutations that consist of cycles of certain lengths only.

THEOREM 3.53
Let C be any set of positive integers, and let $g_C(n)$ be the number of permutations of length n whose cycle lengths are elements of C. Then the identity

$$G_C(z) = \sum_{n \geq 0} g_C(n) \frac{z^n}{n!} = \exp\left(\sum_{n \in C} \frac{z^n}{n}\right)$$

holds.

PROOF Let $f(n) = (n-1)!$ if $n \in C$, and let $f(n) = 0$ otherwise. Then f gives the number of ways we can cover a set of n elements by a cycle of an acceptable length. Then $F(z) = \sum_{n=1}^{\infty} f(n) \frac{z^n}{n!} = \sum_{n \in C} \frac{z^n}{n}$. Our claim then follows from the Exponential formula, permutation version. ∎

An *involution* is a permutation p so that $p^{-1} = p$; in other words, p^2 is the identity permutation. It is easy to see that this happens if and only if all cycles of p have length 1 or 2. Note that 1-cycles are often called *fixed points*.

Example 3.54
Let $G_2(z)$ be the exponential generating function for fixed point-free involutions. Then the identity $G_2(z) = \exp(z^2/2)$ holds. ⧠

COROLLARY 3.55
The number of fixed point-free involutions of length $2n$ is $h(n) = \frac{(2n)!}{2^n \cdot n!} = 1 \cdot 3 \cdots (2n-1)$.

PROOF By Example 3.54, we only have to compute the coefficient $g_2(n)$ of $\frac{z^{2n}}{(2n)!}$ in $G_2(z) = \exp(z^2/2)$. We have

$$\exp(z^2/2) = \sum_{n \geq 0} \frac{z^{2n}}{2^n \cdot n!},$$

so

$$g_2(n) = \frac{(2n)!}{2^n \cdot n!} = 1 \cdot 3 \cdots (2n-1) = (2n-1)!!.$$

∎

The symbol $(2n-1)!!$ reads "$(2n-1)$ semifactorial", referring to the fact that we take the product of *every other* integer from 1 to $2n-1$.

Note that $G_2(z)$ does not contain terms with odd exponents. This makes perfect sense as a fixed point-free involution cannot be of odd length.

In what follows, we are going to look at some interesting enumerative results for some particular choices of the set C. Many of these results could be obtained by some clever combinatorial arguments and without the use of generating functions. We, however, prefer to present the general technique of generating functions, and leave the bijective proofs for the exercises.

Recall that integrating the identity $1/(1-z) = \sum_{n\geq 0} z^n$ we get the identity

$$-\ln(1-z) = \ln\left((1-z)\right)^{-1}\right) = \sum_{n\geq 0} \frac{z^{n+1}}{n+1} = \sum_{n\geq 1} \frac{z^n}{n} \qquad (3.9)$$

The following example introduces a widely researched class of permutations.

Example 3.56
Let D_n be the number of fixed-point free n-permutations (or *derangements*). Then the identity $D(z) = \sum_{n\geq 0} D_n \frac{z^n}{n!} = \frac{\exp(-z)}{1-z}$ holds. $\Box$

PROOF   In this case, we choose C to be the set of all positive integers larger than one. Then we have

$$D(z) = G_C(z) = \exp\left(\sum_{n\geq 2} \frac{z^n}{n}\right)$$

$$= \exp(\ln\left((1-z)^{-1}\right) - z) = \frac{\exp(-z)}{1-z}.$$

∎

Routine expansion of the last term yields the following formula for the number of derangements of $[n]$.

COROLLARY 3.57
Let D_n be the number of derangements of length n. Then

$$D_n = n! \sum_{i=0}^{n} (-1)^i \frac{1}{i!}.$$

Note that this formula implies that

$$\lim_{n\to\infty} \frac{D(n)}{n!} = \frac{1}{e},$$

so for n large enough (meaning in this case, $n > 1$), more than one-third of all permutations are derangements.

The number of all elements of the symmetric group S_n whose order is odd turns out to have an interesting form. These are permutations with odd cycles only. This formula is the content of our next example.

Example 3.58

Let C be the set of all odd positive integers. Then the identity $G_C(z) = \sqrt{\frac{1+z}{1-z}}$ holds. ☐

PROOF   By Theorem 3.53, we need to compute

$$G_C(z) = \exp\left(z + \frac{z^3}{3} + \frac{z^5}{5} + \cdots\right) = \exp\left(\sum_{n\geq 0} \frac{z^{2n+1}}{2n+1}\right).$$

Now note that taking derivatives yields

$$\left(\sum_{n\geq 0} \frac{z^{2n+1}}{2n+1}\right)' = \sum_{n\geq 0} z^{2n} = \frac{1}{1-z^2} = \frac{1}{2(1-z)} + \frac{1}{2(1+z)}.$$

Therefore, by r integration we obtain

$$\sum_{n\geq 0} \frac{z^{2n+1}}{2n+1} = \frac{1}{2}\ln\left((1-z)^{-1}\right) + \ln(1+z)),$$

and so

$$G_C(z) = \exp\left(\frac{1}{2}\ln\left((1-z))^{-1}\right) + \ln(1+z)\right) = \sqrt{\frac{1+z}{1-z}},$$

which was to be proved.   ∎

COROLLARY 3.59
For all positive integers n, the number of permutations of length $2n$ that have odd cycles only is $ODD(2n) = (1 \cdot 3 \cdot 5 \cdots (2n-1))^2 = (2n-1)!!^2$. Similarly, the number of permutations of length $2n+1$ that have odd cycles only is $ODD(2n+1) = (1 \cdot 3 \cdot 5 \cdots (2n-1))^2(2n+1) = (2n-1)!!^2(2n+1)$.

This remarkable result has several proofs. A combinatorial one for a more general version is given in Exercise 57.

PROOF   Using the result of Example 3.58, all we have to do is find the coefficients of $z^m/m!$ in $G_C(z) = \sqrt{\frac{1+z}{1-z}}$. Multiplying both the numerator and the denominator by $\sqrt{1+z}$, we get

$$G_C(z) = \frac{1+z}{\sqrt{1-z^2}}. \tag{3.10}$$

By the Binomial theorem, we obtain

$$(1-x^2)^{-1/2} = \sum_{m\geq 0} (-1)^m \binom{-1/2}{m} z^{2m} \tag{3.11}$$

$$= \sum_{m\geq 0} (-1)^m \frac{(-1/2)\cdot(-3/2)\cdots(-(2m-1)/2)}{m!} z^{2m} \tag{3.12}$$

$$= \sum_{m\geq 0} \frac{(2m-1)!!}{m!\cdot 2^m} z^{2m}. \tag{3.13}$$

Note that $(1-z^2)^{-1/2}$ has no terms of odd degree.

This shows that the coefficient $g_C(2m)$ of $z^{2m}/(2m)!$ in our generating function $G_C(z) = (1+z)\sum_{m\geq 0} \frac{(2m-1)!!}{m!\cdot 2^m} z^{2m}$ is

$$(2m)! \cdot \frac{(2m-1)!!}{m!\cdot 2^m} = (2m-1)!!^2,$$

while the coefficient $g_C(2m+1)$ of $z^{2m+1}/(2m+1)!$ is

$$\frac{(2m+1)!}{m!}\cdot\frac{(2m-1)!!}{2^m} = (2m-1)!!^2(2m+1),$$

proving our claims. ∎

As a natural continuation of our last subject, we will now enumerate permutations that have cycles of *even* length only.

Example 3.60
Let C be the set of all even positive integers. Then the identity $G_C(x) = \sqrt{\frac{1}{1-z^2}}$ holds. ⬚

PROOF   Theorem 3.53 implies

$$G_C(z) = \exp\left(\sum_{n\geq 1} \frac{z^{2n}}{2n}\right). \tag{3.14}$$

Note that the argument of the exponential function on the right hand side is very similar to (3.9), with z^2 playing the role of z. Therefore, (3.14) implies

$$G_C(z) = \exp\left(\frac{1}{2}(\ln(1-z^2))^{-1}\right) = \sqrt{\frac{1}{1-z^2}},$$

as claimed.   ∎

COROLLARY 3.61
For all positive integers m, the number of $(2m)$-permutations that have even cycles only is $EVEN(2m) = (2m-1)!!^2$.

PROOF Using the result of Example 3.60, all we have to do is find the coefficients of $z^{2m}/(2m)!$ in $\sqrt{\frac{1}{1-x^2}}$. We have computed this power series in (3.11), and saw that the coefficient of $z^{2m}/(2m)!$ is indeed $(2m-1)!!^2$. ∎

This leaves us with the intriguing observation that $ODD(2m) = EVEN(2m)$ for all positive integers m. This fact asks for a bijective proof. One is given in Exercise 55.

In the above examples, we have had conditions on the cycle lengths based on their parity. One could just as well compute the generating function for the number of n-permutations with no cycle lengths divisible by k, or all cycle lengths divisible by k. See Exercises 52 and 57 for generating function proofs, and Problem Plus 4 for combinatorial proofs.

3.4.2 Cycle Index and Its Applications

We would like to refine our permutation counting techniques. Theorem 3.53 allows us to count permutations with a given set of cycle lengths. However, if we want to count permutations that have, say, an even number of fixed points, or an odd number of 2-cycles, then we cannot apply Theorem 3.53 directly.

In order to be able to handle complex situations like these, we introduce extra variables. Let p be an n-permutation that has a_i cycles of length i. We then associate the monomial $\Pi_{i=1}^n t_i^{a_i}$ to p. For example, if $p = (312)(54)(6)$, then we associate the monomial $t_1 t_2 t_3$ to p, whereas if $p = (21)(3)(54)$, then we associate the monomial $t_1 t_2^2$ to p.

Summing all these monomials over all n-permutations, we get the *augmented cycle index* $\tilde{Z}(S_n)$ of the symmetric group S_n.
That is

$$\tilde{Z}(S_n) = \tilde{Z}(S_n)(t_1, t_2, \cdots, t_n) = \sum_{p \in S_n} t_1^{a_1} t_2^{a_2} \cdots t_n^{a_n},$$

where a_i is the number of i-cycles of p. We note that the *cycle index* $Z(S_n)$ of S_n is defined by $Z(S_n) = \frac{1}{n!}\tilde{Z}(S_n)$.

Example 3.62
For $n = 3$, the equality

$$\tilde{Z}(S_n) = t_1^3 + 3t_1 t_2 + 2t_3$$

holds. ◻

The reason the augmented cycle index is a useful tool for the enumeration of permutations is the following observation.

PROPOSITION 3.63
Let $\tilde{Z}(S_n)$ be defined as above. Then the equality

$$\sum_{n \geq 0} \tilde{Z}(S_n)\frac{z^n}{n!} = \exp \sum_{i \geq 1} t_i \frac{z^i}{i}$$

holds.

PROOF Let K be the field of all rational functions in variables $t_1, t_2, \cdots, t_n$ over $\mathbf{Q}$. Define $f : \mathbf{P} \to K$ by $f(n) = t_n$. Then

$$\tilde{Z}(S_n) = \sum_{p \in S_n} f(|C_1|)f(|C_2|) \cdots f(|C_k|),$$

where the C_i are the cycles of p. Our claim is then immediate from Corollary 3.52 (the permutation version of the Exponential formula). ∎

You may wonder how introducing n new variables and obtaining one new equation will help us. As you will see from the upcoming examples, the strength of Proposition 3.63 is its generality; we can choose any values for the t_i, and the equation will still hold.

Example 3.64
Let $g(n)$ be the number of n-permutations that have an even number of fixed points, and let $G(z) = \sum_{n \geq 0} g(n)z^n/n!$. Then the identity

$$G(z) = \cosh z \cdot \frac{\exp(-z)}{1 - z}$$

holds. ▯

Before we start our proof, we remind the reader of the identity $\cosh(z) = (\exp z + \exp(-z))/2 = \sum_{n \geq 0} \frac{z^{2n}}{(2n)!}$. In other words, $\cosh z$ is obtained by omitting the odd terms from $\exp z$.

PROOF The crucial observation is that

$$\frac{\tilde{Z}(S_n)(1, 1, \cdots, 1) + \tilde{Z}(S_n)(-1, 1, \cdots, 1)}{2} = g(n).$$

Applying Proposition 3.63 we get

$$\sum_{n\geq 0} g(n)t\frac{z^n}{n!} = \frac{\exp(z) + \exp(-z)}{2} \cdot \exp\left(\sum_{i\geq 2} \frac{z^i}{i}\right) = \cosh(z) \cdot \frac{\exp(-z)}{1-z}.$$

In the last step, we used the result of Example 3.56.   ∎

Example 3.65
Let $h(n)$ be the number of n-permutations that have an odd number of 2-cycles and no fixed points. Then

$$H(z) = \sum_{n\geq 0} h(n)\frac{z^n}{n!} = \sinh\left(\frac{z^2}{2}\right) \cdot \exp\left(\sum_{i\geq 3} \frac{z^i}{i}\right).$$

◻

PROOF   Note that

$$\frac{\tilde{Z}(S_n)(0, 1, 1, \cdots, 1) - \tilde{Z}(S_n)(0, -1, 1, \cdots, 1)}{2} = h(n).$$

The claim then follows just as in the previous example.   ∎

Let p be an n-permutation. We say that p has a *square root* if there exists another n-permutation q for which $q^2 = p$. Note that p can have many square roots. For instance, all involutions are square roots of the identity permutation. Whether p has a square root or not is easy to tell from its cycle decomposition.

LEMMA 3.66
The permutation p has a square root if and only if its unique decomposition into the product of distinct cycles contains an even number of cycles of each even length.

Example 3.67
The permutation $(21)(64)(753)$ has a square root, as it has two 2-cycles, and zero cycles of any other even length. The permutation (4231) does not have a square root, as it has an odd number (one) of 4-cycles.   ◻

PROOF   (of Lemma 3.66) Let us assume that $p = r^2$. When we take the square of r, the odd cycles of r remain odd cycles. The even cycles of r will split into two cycles of the same length. Therefore, the even cycles of r^2 come in pairs, proving the "only if" part of our assertion.

In order to prove that the condition is sufficient, let us assume that p has even cycles $(a_1 \cdots a_h)$ and $(b_1 \cdots b_h)$. These even cycles can be obtained by taking the square of the $(2h)$-cycle $(a_1 b_1 a_2 b_2 a_3 \cdots a_h b_h)$. Odd cycles of p, such as $(d_1 d_3 d_5 \cdots d_j d_2 d_4 \cdots d_{j-1})$ can be obtained as the square of $(d_1 d_2 \cdots d_j)$. So we can find roots for all cycles of p this way, then we can take the product of the obtained cycles to be the permutation r. ∎

As we characterized permutations having square roots by their cycle lengths, we can use Theorem 3.53 to find the exponential generating function for their numbers.

THEOREM 3.68

Let SQ_n be the number of n-permutations that have a square root, and let $SQ(z) = \sum_{n=0}^{\infty} \frac{SQ_n}{n!} z^n$. Then we have

$$SQ(z) = \sqrt{\frac{1+z}{1-z}} \prod_{i \geq 1} \cosh \frac{z^{2i}}{2i}.$$

PROOF Recall that our permutations must have an even number of cycles of each even length. Therefore, repeated application of the method seen in the proof of Example 3.64 yields

$$SQ(z) = (\exp z) \cdot (\cosh(z^2/2)) \cdot (\exp z^3/3) \cdot (\cosh(z^4/4)) \cdots . \qquad (3.15)$$

Our claim is then proved recalling that we have already computed the power series $\exp\left(z + \frac{z^3}{3} + \cdots\right)$ in Example 3.58. ∎

COROLLARY 3.69

For all positive integers n, the equality $SQ_{2n} \cdot (2n+1) = SQ_{2n+1}$ holds.

This is an interesting result. It means that when we pass from $2n$ to $2n+1$, the number of permutations with square roots grows just as fast as the number of all permutations.

PROOF It suffices to show that the coefficients of z^{2n} and z^{2n+1} in $SQ(z)$ are identical. As $\prod_{i \geq 1} \cosh(z^{2i}/(2i))$ does not contain terms with odd exponents, it suffices to show this for $\sqrt{\frac{1+z}{1-z}}$. Recall again that we have seen in Example 3.58 that

$$\sqrt{\frac{1+z}{1-z}} = \sum_{n \geq 1} ODD(n) \frac{z^n}{n!}.$$

Corollary 3.59 then shows that $ODD(2n + 1) = (2n + 1)ODD(2n)$, which is just what we needed. ∎

So the probability that a randomly chosen $2n$-permutation has a square root is equal to the probability that a randomly chosen $(2n + 1)$-permutation has a square root. This nice identity certainly asks for a combinatorial proof. One is given in Exercise 61. Not surprisingly, it builds on the combinatorial proof of the equality $ODD(2n + 1) = (2n + 1)ODD(2n)$.

3.4.2.1 Multivariate Generating Functions and the Exponential Formula

So far, our applications of the Exponential Formula involved univariate generating functions, or the cycle index. Sometimes, it is simpler to use a bivariate generating function directly.

Example 3.70
Let $h(n, k)$ be the number of n-permutations with k fixed points, and let $H(z, u) = \sum_{n \geq 0} \sum_{k=0}^{n} h(n, k) \frac{z^n}{n!} u^k$. Find an explicit formula for $H(z, u)$.
◻

SOLUTION   Set $f(n) = (n - 1)!$ for $n \geq 2$, and $f(1) = y$. Then note that $F(z) = zu + \sum_{n \geq 2} z^n/n$, and use the exponential formula to get

$$H(z, u) = \exp(F(z))$$

$$= \exp\left(zu + \sum_{n \geq 2} \frac{z^n}{n} \right)$$

$$= \exp(\ln(1/(1 - z)) + zu - z)$$

$$= \frac{e^{z(u-1)}}{1 - z}.$$

∎

If we want to know the *total* number of fixed points in all permutations of length n, then it helps if we notice that that number is just $\sum_k kh(n, k)$, which is in turn the coefficient of $z^n/n!$ in

$$\frac{\partial H(z, u)}{\partial u}\Big|_{u=1} = \sum_{n \geq 0} \sum_{k=1}^{n} h(n, k) \frac{z^n}{n!} ku^{k-1}\Big|_{u=1}.$$

It is easy to compute that

$$\frac{\partial H(z, u)}{\partial u} = z\frac{e^{z(u-1)}}{1 - z},$$

so setting $u = 1$, we get the univariate generating function $h(z) = \sum_{n \geq 1} z^n$. This shows that the *total number* of fixed points in all permutations of length n is $n!$; therefore, the average number of fixed points in such permutations is 1. This is a well-known fact that can be proved in many ways, and we will return to it in Chapter 6. For now, let us mention that $h(z)$ is called the *cumulative* generating function of all cycles.

If we want to know the *difference d_n* between the number of n-permutations with an even number of fixed points and the number of n-permutations with an odd number of fixed points, then it helps to note that that difference is just

$$d_n = \sum_{k=0}^{n} (-1)^k h(n, k) = [z^n / n!] H(z, -1).$$

Then Example 3.70 implies

$$d_n = [z^n / n!] \frac{e^{-2z}}{1 - z}$$

$$= n! \left(\sum_{i=0}^{n} \frac{(-2)^i}{i!} \right).$$

Similarly, if $C(z, u) = \sum_{n \geq 0} \sum_{k=0}^{n} c(n, k) \frac{z^n}{n!} u^k$ is the bivariate generating function for permutations according to their length and their number of cycles, then we can prove the following interesting-looking formula.

Example 3.71

The identity $C(z, u) = (1 - z)^{-u}$ holds.   ☐

SOLUTION   Set $f(n) = (n - 1)! y$ for all positive integers n. Then the Exponential formula yields

$$C(z, u) = \exp(F(z))$$

$$= \exp \left(\sum_{n \geq 2} \frac{z^n u}{n} \right)$$

$$= \exp(u \ln(1/(1 - z)))$$

$$= (1 - z)^{-u}.$$

∎

Note that in the special case of $u = -1$, we get the generating function $1 - z$, meaning that if $n \geq 2$, then there are as many n-permutations with an even number of cycles as there are permutations with an odd number of cycles.

The reader is invited to find proofs of Proposition 3.31 and Corollary 3.35 using the result of the preceding example.

3.4.2.2   Proving the Formula of Schlömilch

We return to the task of proving the formula of Schlömilch. This is a some-what complicated computational argument. First, and this is interesting, we will need the Lagrange Inversion Formula. It is remarkable that as simple a problem as the one at hand calls for this advanced technique.

Let f and g be two formal power series in $\mathbf{R}[[z]]$. We say that g is the (compositional) inverse of f if $f(g(z)) = z$ holds. One can prove that in that case, $g(f(z)) = z$ also holds. It is also straightforward to check that if f has an inverse, then this inverse is unique. We will denote the inverse of f by $f^{\langle -1 \rangle}$. Finally, we point out that f has an inverse if and only if $f(0) = 0$ and the coefficient of z in f is non-zero.

Recall that if $f \in \mathbf{R}[[z]]$ is a formal power series, then $[z^n]f$ denotes the coefficient of z^n in f.

LEMMA 3.72
[Lagrange Inversion Formula] Let $f \in \mathbf{R}[[z]]$, and let k and n be positive integers satisfying $1 \le k \le n$. Then the equality

$$[z^n](f^{\langle -1 \rangle})^k = \frac{k}{n}[z^{n-k}]\left(\frac{f(z)}{z}\right)^{-n} \tag{3.16}$$

holds.

The Lagrange Inversion Formula has several interesting proofs, among which we can find combinatorial, algebraic, and analytic arguments. See [297] for one proof of each kind. The original argument of Lagrange actually proves a stronger statement in that k and n are only required to be integers. Gilbert Labelle *et al.* found combinatorial proofs for various versions of the Lagrange Inversion formula, including multivariate generalizations. See [86], [192] and [241] for these results.

Now we are in a position to state and prove the formula of Schlömilch.

THEOREM 3.73
[281] For all positive integers k and n satisfying $k < n$, the identity

$$s(n,k) = \sum_{1 \le j \le n-k} (-1)^j \binom{n-1+j}{n-k+j}\binom{2n-k}{n-k-j}S(n-k+j,j)$$

$$= \sum_{1 \le i \le j \le n-k} (-1)^{j+i} \binom{j}{i}\binom{n-1+j}{n-k+j}\binom{2n-k}{n-k-j}\frac{(j-i)^{n-k+j}}{j!}$$

holds.

PROOF   First note that the result of Lemma 3.34 provides a formal power series whose coefficients are very close to the numbers $s(n,k)$. Indeed,

$$\frac{k!}{n!}s(n,k) = [u^n](\ln(1+u))^k.$$

Then we can apply the Lagrange Inversion formula to the power series $f(u) = (\exp u) - 1$ and $f^{\langle -1 \rangle}(u) = \ln(1+u)$. We get

$$\frac{k!}{n!}s(n,k) = \frac{k}{n}[u^{n-k}]\left(\frac{(\exp u) - 1}{u}\right)^{-n}. \tag{3.17}$$

In order to proceed further, we need the following technical lemma.

LEMMA 3.74
Let r be any complex number, and let $G(u) = \sum_{n\geq 0} g_n \frac{u^n}{n!}$ be a formal power series with $g_0 = 1$. Define the polynomials $P_m^{(r)}(u)$ by the equation

$$(G(u))^r = 1 + \sum_{m\geq 1} P_m^{(r)} \frac{u^m}{m!}.$$

In other words, $P_m^{(r)}$ is the coefficient of u^m/m in $(G(u))^r$. Then the identities

$$P_m^{(-r)} = r\binom{m+r}{m}\sum_{j=1}^{m}(-1)^j \frac{1}{r+j}\binom{m}{j}P_m^{(j)} \tag{3.18}$$

hold.

See page 142 of [123] for a proof of this lemma.

Now let us apply the lemma with $G(u) = \frac{(\exp u) - 1}{u}$, $r = n$, and $m = n - k$. Using these values, (3.17) implies

$$\frac{k}{n}\cdot s(n,k) = \frac{k}{n}\cdot n\binom{2n-k}{n}\sum_{j=1}^{n-k}(-1)^j \frac{1}{n+j}\binom{n-k}{j}\left([u^{n-k}]\left(\frac{e^u - 1}{u}\right)u^j\right).$$

The last step is certainly counterintuitive. With significant effort, we have made a simple expression much more complicated. The reason for this is that now we can recognize $S(n - k + j, j)$ in the last term. Indeed, a simple application of the Exponential Formula (see Exercise 29) shows that

$$\sum_{n=k}^{\infty} S(n,k)\frac{u^n}{n!} = \frac{1}{k!}\cdot(e^u - 1)^k.$$

Therefore,

$$[u^{n-k}]\left(\frac{e^u - 1}{u}\right)x^j = \frac{j!S(n - k + j, j)}{(n - k + j)!},$$

and the first equality of the formula of Schlömilch is proved by routine cancellations. One then obtains the second equality by substituting the exact formula for $S(n - k + j, j)$ given in Lemma 1.17. ∎

Exercises

1. (−) Prove that Alt_n has $n!/2$ elements for $n \geq 2$.

2. (−) Find a subgroup of S_n that has $(n - 1)!$ elements.

3. (−) For $p \in S_n$, prove that p and p^{-1} have the same number of inversions.

4. For $p, q \in S_n$, prove that the number of fixed points of pq and qp is the same. Try to find a solution using linear algebra.

5. (−) Find a surjective homomorphism $f : S_n \to S_2$.

6. (Basic knowledge of Linear Algebra required.) It is routine to check that the map $f(p) = A_p$ mapping each element of S_n to a permutation matrix is a homomorphism. Is this homomorphism irreducible? That is, is there a nonzero n-dimensional vector $\mathbf{v}$ (say, with real coefficients) for which $A_p(\mathbf{v}) = \mathbf{v}$ for all $p \in S_n$?

7. (−) Let $n \geq 3$. Let us assume that we know that $f \in S_n$ is such that $fg = gf$ for all $g \in S_n$. What can f be?

8. (−) Let n be an even positive integer. Prove that at least half of all permutations of length n contain a cycle of length at least $n/2$.

9. Let $L(n, k)$ be the number of n-permutations whose longest cycle is of length k. Let us assume that $\frac{n}{2} < k \leq n$. Find a formula for $L(n, k)$.

10. Let $L(n, k)$ be defined as in the previous exercise.

    (a) Let $\frac{n}{3} < k \leq \frac{n}{2}$. Find a formula for $L(n, k)$.
    (b) (+) Now let $1 < k < n$. Find a formula for $L(n, k)$.

11. An *adjacent transposition* is a transposition interchanging two consecutive entries in a permutation. Prove that any element of S_n can be obtained as a product of (not necessarily distinct) adjacent transpositions.

12. (−) Let $p \in S_n$ be a permutation that can be obtained as a product of m transpositions and also as a product of k transpositions. What can be said about $m + k$?

13. Let $n > 1$. Find all values of n for which S_n does not have two conjugacy classes of the same size.

14. Prove that any even permutation of S_n can be obtained as a product of (not necessarily distinct) 3-cycles.

15. Prove formula (3.1).

16. Let T be a tree on vertex set $[n]$, with degree sequence $d_1, d_2, \cdots, d_n$. Prove that the number of all cyclic permutations of $[n]$ that are generated by the transpositions defined by the edges of T is $D(T) = d_1! d_2! \cdots d_n!$.

17. (–) Find a simple closed formula for $c(n, 1)$ and $c(n, n-1)$.

18. Prove, preferably combinatorially, that

$$c(n, k) = \frac{n!}{k!} \sum_{r_1 + r_2 + \cdots + r_k = n} \frac{1}{r_1 r_2 \cdots r_k},$$

where the r_i are positive integers.

19. Prove a formula for $S(n, k)$ that is similar to that of the previous exercise.

20. (+) Give a generating function proof of the identity

$$s(n+1, k+1) = \sum_{m=k}^{n} (-1)^{n-m} (n)_{n-m} \cdot s(m, k). \qquad (3.19)$$

21. Define $a_k(n) = S(n+k, n)$ and $b_k(n) = c(n, n-k)$. Prove that for any fixed positive integer k, both $a_k(n)$ and $b_k(n)$ are polynomial functions of n. What is their degree and leading coefficient?

22. Let $a_k(n)$ and $b_k(n)$ be defined as in the previous exercise. We have just proved that these functions are polynomials of n. We can therefore substitute any real number into these polynomials, not just positive integers. Prove that for all positive integers k and all integers n, we have

(a)
$$a_k(0) = a_k(-1) = \cdots = a_k(-k) = 0,$$

(b)
$$a_k(-n) = b_k(n).$$

23. Give a combinatorial proof of the identity

$$c(n+1, k+1) = \sum_{m=0}^{n} (n)_m c(n, m),$$

where we set $(n)_0 = 1$, and where $0 \le k \le n$.

24. Let $c_2(n, k)$ be the number of n-permutations with k cycles in which *each cycle length is at least two*. Find a recurrence relation for the numbers $c_2(n, k)$.

25. Let $c_3(n, k)$ be defined in a way analogous to the previous exercise. Find a recurrence relation satisfied by the numbers $c_3(n, k)$.

26. Find a closed form for the generating function

$$g_k(u) = \sum_{n \ge 0} \sum_{k=0}^{n} S(n, k) \frac{u^n}{n!}.$$

27. A box originally contains m white balls. We run a multiple step experiment as follows. At each step, we draw one ball from the box at random. If the drawn ball is white, we put a black ball in the box instead of that white ball. If the drawn ball is black, we put it back to the box. For $k \le n$, compute the probability of drawing k white balls in n trials. What is the connection between these probabilities $p(k, n)$, and the Stirling numbers of the second kind?

28. Prove that $S(n, k) = \sum_{k \le m \le n} S(m-1, k-1)k^{n-m}$.

29. Find a closed form of the exponential generating function $g_k(u) = \sum_{n=k}^{\infty} S(n, k)u^n/n!$.

30. Find a simple combinatorial definition for the numbers $R(n, k)$ so that we have

$$c(n, k) = \sum_{k \le m \le n} c(m-1, k-1)R(n-m, k).$$

Then find a formula for the numbers $R(n, k)$.

31. Let p be a prime number. Prove that $c(p, k)$ is divisible by p unless $k = p$, or $k = 1$.

32. (Wilson's theorem.) Let p be a prime. Prove that $1+(p-1)!$ is divisible by p.

33. Let p be a prime. Prove that $S(p, k)$ is divisible by p, unless $k = p$ or $k = 1$.

34. (a) Prove that

$$\sum_{1 \le i_1 \le i_2 \le \cdots \le i_{n-k} \le k} i_1 i_2 \cdots i_{n-k} = S(n, k).$$

(b) Prove that

$$\sum_{1 \le i_1 \le i_2 \le \cdots \le i_{n-k} \le k} (i_1 i_2 \cdots i_{n-k})^\ell = S(n, k, \ell),$$

where $S(n, k, \ell)$ is the ℓ-Stirling number of the second kind as defined in Problem Plus 3 of Chapter 1.

35. Prove that

$$\sum_{1 \le i_1 < i_2 < \cdots < i_{n-k} \le n-1} i_1 i_2 \cdots i_{n-k} = c(n, k).$$

36. Prove (preferably with a combinatorial argument) that for fixed positive integers n and k satisfying $k \le n$, the equality

$$\sum_{m=k}^{n} c(n, m) S(m, k) = \frac{n!}{k!} \binom{n-1}{k-1}$$

holds.

37. Give a proof of Corollary 3.57 without the use of generating functions.

38. Prove that D_n (the number of derangements of length n) is the integer closest to $n!/e$.

39. Which set is larger, the set of all derangements of length n, or the set of all n-permutations with exactly one fixed point? By how much?

40. Give a combinatorial proof (no generating functions, or recurrence relations) for your answer to the previous exercise.

41. Let p be an n-permutation, and let $aa(p)$ be *the smallest* of all the ascents of p, if p has any ascents, and let $aa(p) = n$ otherwise. A *desarrangement* is a permutation for which $aa(n)$ is even. For example, 213 and 54312 are desarrangements. Let $J(n)$ be the number of desarrangements of length n.

    (a) Prove that $J(n) = nJ(n-1) + (-1)^n$.
    (b) Prove that $J(n) = D_n$.

42. Find a combinatorial proof for the identity $J(n) = D_n$, where $J(n)$ is defined in the previous exercise.

43. Prove combinatorially that $D_n = (n-1)(D_{n-1} + D_{n-2})$.

44. Prove that for any fixed n, the sequence $\{S(n,k)\}_{1 \le k \le n}$ is log-concave. (Be careful.)

45. Prove that

$$\det C_k = \det \begin{pmatrix} c(n+1,1) & c(n+1,2) & \cdots & c(n+1,k) \\ c(n+2,1) & c(n+2,2) & \cdots & c(n+2,k) \\ \cdots & \cdots & \cdots & \cdots \\ c(n+k,1) & c(n+k,2) & \cdots & c(n+k,k) \end{pmatrix} = (n!)^k.$$

46. How many n-permutations have exactly three left-to-right maxima and exactly one left-to-right minimum?

47. Give a combinatorial proof for the result of Corollary 3.55.

48. [The Compositional Formula] Let K be a field of characteristic zero, let $f : \mathbf{P} \to K$ be a function, and let $g : \mathbf{N} \to K$ be a function satisfying $g(0) = 1$. Define $h : \mathbf{N} \to K$ by

$$h(n) = \sum f(|A_1|)f(|A_2|) \cdots f(|A_m|)g(m),$$

where the sum ranges over all *partitions* $(A_1, A_2, \cdots, A_m)$ of $[n]$ into *any* number of parts. Let $F(z)$ (resp. $G(z), H(z)$) be the exponential generating function of the sequence $f(n)$ (resp. $g(n), h(n)$). Prove that

$$\exp(H(z)) = G(F(z)).$$

49. Let $h(n)$ be the number of ways in which n people can sit down around an unspecified number of *numbered* tables, without leaving any tables empty. Two seating arrangements are considered identical if each person has the same left neighbor in both of them. Find the exponential generating function of the numbers $h(n)$.

50. Let $h(n)$ be the number of ways in which n people can sit down around an unspecified number of tables *arranged in a circle*. Two seating arrangements are considered identical if each person has the same left neighbor in both of them, and, the same holds for the *tables*.

In order to further explain this second condition, let $A_1, A_2, \cdots, A_k$ be the tables of one circular arrangement, listed counterclockwise. Here, each A_i is a cycle that is formed by people. Let $B_1, B_2, \cdots, B_k$ be another circular arrangement of tables listed counterclockwise. We say that these two arrangments are identical if they are circular translates of each other, that is, if there is an index j so that $A_j = B_{i+j}$ for all i, where $i + j$ is understood modulo k.

Find the exponential generating function of the numbers $h(n)$.

51. (–) Find a formula for the number of n-permutations whose third power is the identity permutation.

52. Find a formula for the number of permutations of length n in which each cycle length is divisible by k.

53. Find the exponential generating function $A(z)$ for the numbers $a(n)$ of n-permutations that have an even number of cycles of each odd length.

54. Find the exponential generating function $B(z)$ for the numbers $b(n)$ of partitions of $[n]$ that have an even number of blocks of each even size.

55. Prove combinatorially that $ODD(2m) = EVEN(2m)$.

56. Find a combinatorial proof for Exercise 52.

57. Find a formula for the number of permutations of length n in which no cycle length is divisible by k.

58. Find a bijective proof of the equality $ODD(2n) \cdot (2n+1) = ODD(2n+1)$.

59. Find a bijective proof of the equality

$$ODD(2n + 1) \cdot (2n + 1) = ODD(2n + 2).$$

60. (+)

    (a) Prove that the number of all fixed point-free n-permutations with descent set $\{i\}$ is equal to $\binom{n-2}{i-1}$.
    (b) What can we say about the number of all n-permutations having exactly one fixed point whose descent set is equal to $\{i\}$?

61. (+) Find a bijective proof of the equality $SQ_{2n} \cdot (2n + 1) = SQ_{2n+1}$.

62. Prove, without using generating functions, that $SQ_{2n+1} \cdot (2n + 2) \geq SQ_{2n+2}$.

63. Exercise 6 of Chapter 1 provided a probabilistic interpretation for the Eulerian numbers. Why cannot we use that interpretation to deduce by Lévy's theorem (Theorem 3.37) that the Eulerian polynomials have real roots only? Note that the Eulerian polynomials have irrational roots in general, while the interpretation of the mentioned exercise would yield rational roots.

64. Find a combinatorial proof for the fact that

$$D_n^2 \leq D_{n-1}D_{n+1},$$

if $n \geq 3$. Here D_n denotes the number of derangements of length n.

65. Prove Lemma 3.26 (that is, the statement that the average number of cycles in an n-permutation is the harmonic number $H(n)$), by analyzing the *derivative* of both sides of the equation

$$\sum_{k=0}^{n} c(n,k)z^k = z(z+1)\cdots(z+n-1),$$

which we proved in Theorem 3.20.

66. Prove Lemma 3.26 using the Product formula. (Hint: compute the total number of cycles in all n-permutations.)

67. Let $h(n,k)$ be the number of n-permutations with exactly k fixed points. Find a closed formula for the exponential generating function $H(z,u) = \sum_{n\geq 0} \sum_{k=0}^{n} h(n,k)u^k \frac{z^n}{n!}$.

68. Let C be a subset of the set of all positive integers. Consider the power series

$$W_C(z) = \prod_{i \in C} \left(\exp\left(\frac{z^i}{i} - 1 \right) \right).$$

Find the class of permutations whose exponential generating function is $W_C(z)$.

69. Let $f(n,i)$ be the number of ways to choose an i-element subset S of $[n]$, and then to choose an involution q on S. Set $f(n,0) = 1$. Prove, by a combinatorial argument, that $\sum_{i=0}^{n} f(n,i)(-1)^{n-i}$ is equal to the number of fixed point-free involutions on $[n]$.

70. Solve the previous exercise using generating functions.

71. Let p be a permutation of length n. Let us define the *increasing tree* $IT(p)$ of p as follows. The tree $IT(p)$ is a nonplane tree that has vertex set $\{0, 1, \cdots, n\}$, and root 0. The parent of vertex p_i is the vertex p_j that is the closest vertex to p_i that precedes p_i for which $p_j < p_i$ holds. If there is no such vertex, that is, when p_i is a left-to-right minimum, then the parent of p_i is the root of $IT(p)$.

Prove that the map $p \to IT(p)$ is a bijection from the set of all permutations of length n to the set of all non-plane trees on vertex set $\{0, 1, \cdots, n\}$ in which each vertex has a larger label than its parent.

72. Consider all $n!$ increasing trees on vertex set $\{0, 1, \cdots, n\}$. What is the total number of all *leaves* of these trees?

73. Consider all $n!$ increasing trees on vertex set $\{0, 1, \cdots, n\}$. What is the total number of all vertices in these trees that have at least one leaf as a neighbor?

74. A group of n tourists enters a restaurant. They sit down around an unspecified number of identical tables, then each table orders one of two kinds of drinks. In how many ways is this possible?

75. What is the total number of 2-cycles in all derangements of length n?

76. Let $d(n, k)$ be the number of derangements of length n that consist of k cycles. Find a closed form for the bivariate generating function $D(z, u) = \sum_{n \geq 0} \sum_{k=0}^{n} d(n, k) \frac{z^n}{n!} u^k$.

77. Find an explicit formula for

$$f(n) = \sum_{p} i(p),$$

where the sum is taken over all permutations of length n that consist of a single n-cycle.

Problems Plus

1. Prove combinatorially that $D_n = nD_{n-1} + (-1)^n$.

2. (a) Prove that $SQ_{2n+2} \leq (2n + 2)SQ_{2n+1}$.

   (b) For what values of n does equality hold?

3. Let $p_2(n)$ be the probability that a randomly selected n-permutation has a square root. Exercises 61 and 2 show that $p_2(n)$ is a decreasing sequence. Show that $p_2(n) \to 0$.

4. Find a combinatorial proof for Exercise 57.

5. (a) Let $H_n = \sum_{i=1}^{n} \frac{1}{i}$. Prove that

$$c(n + 1, 3) = \frac{n!}{2} \left(H_n^2 - \sum_{i=1}^{n} \frac{1}{i^2} \right).$$

   (b) Show that

$$c(n + 1, 4) = \frac{n!}{6} \left(H_n^3 - 3 \sum_{i=1}^{n} \frac{1}{i^2} + 2 \sum_{i=1}^{n} \frac{1}{i^3} \right).$$

6. Let $q > 2$ be a prime, and let $p_q(n)$ be the probability that a randomly selected n-permutation has a qth root. Prove that $p_q(n) = p_q(n + 1)$, except when $n + 1$ is divisible by q.

7. Show that
$$c(n+1, k+1) \simeq n![\ln(n+1) + C]^k/k!.$$

   What can we say about the constant C above?

8. Show that
$$S(n, k) \simeq \frac{k^n}{k!}.$$

9. In this problem, we are investigating how many inversions an n-permutation can have if it has a fixed number k of cycles. Recall that $i(p)$ denotes the number of inversions of p, and that $c(p)$ denotes the number of cycles of p. Let
$$b(n, p) = \min\{i(p)|p \in S_n, \; c(p) = k\},$$

   and
$$B(n, p) = \max\{i(p)|p \in S_n, \; c(p) = k\}.$$

   (a) Find a formula for $b(n, k)$.

   (b) Find a formula for $B(n, k)$.

   (c) Find a formula for the number $m(n, k)$ of n-permutations with k-cycles for which the minimum is attained, that is, that have $b(n, k)$ inversions.

   (d) Find a formula for the number $M(n, k)$ of n-permutations with k-cycles for which the maximum is attained, that is, that have $B(n, k)$ inversions.

10. An r-derangement is a permutation in which each cycle length is larger than r. So a 0-derangement is a permutation, while a 1-derangement is a derangement. Let
$$G_{n,r}(z) = \sum_p z^{c(p)},$$

    where the sum is taken over the set $D_r(n)$ of all r-derangements of length n. Prove that for any fixed n and r, the polynomial $G_{n,r}(z)$ has real zeros only.

11. (a) Let D be a conjugacy class of S_n. Prove that the polynomial
$$E_D(z) = \sum_{p \in D} z^{exc(p)},$$

    where $exc(p)$ denotes the number of excedances of p, has real zeros only.

    (b) It follows from part [(a)] that the polynomial $E_D(z) = \sum_{p \in D} z^{exc(p)}$ is unimodal. Where does it take its maximal value?

(c) Now let $D_{k,r}(n)$ be the set of all n-permutations that have exactly k cycles, and each of these cycles is longer than r. Prove that the polynomial

$$E_{D,k,r,n}(z) = \sum_{p \in D_{k,r}(n)} z^{exc(p)}$$

is symmetric and unimodal.

(d) Prove that the polynomial

$$E_{D_r(n)}(z) = \sum_{p \in D_r(n)} z^{exc(p)}$$

is symmetric and unimodal. Here $D_r(n)$ denotes the set of all r-derangements as defined in the previous Problem Plus.

12. Exercises 41 and 42 defined desarrangements, and discussed some of their close connections to derangements. This problem takes that approach further.

Let $T \in [n-1]$. Prove the following statement: The number of derangements of length n having descent set T is equal to the number of desarrangements of length n whose *inverse* has descent set T.

13. It follows from the solution of Exercise 22 that there exist integers $B_{k,i}$, with $i \in [k]$, such that

$$\sum_{k=0}^{\infty} a_k(n)z^n = \frac{\sum_{i=1}^{k} B_{k,i} z^i}{(1-z)^{2k+1}},$$

and

$$\sum_{k=0}^{\infty} b_k(n)z^n = \frac{\sum_{i=k+1}^{2k} B_{2k-i+1,i} z^i}{(1-z)^{2k+1}},$$

where $\sum_{i=1}^{k} B_{k,i} = (2k-1)!!$. Prove that the numbers $B_{k,i}$ are always non-negative by finding a set of permutations that these numbers enumerate.

14. Let S be any proper subset of $[n-1]$. Prove that the number of fixed point-free n-permutations with descent set S is equal to the number of n-permutations with one fixed point having descent set S.

15. Let S be any subset of $[n-1]$. Prove that there are as many involutions on $[n]$ with descent set S as involutions on $[n]$ with descent set $[n-1]-S$.

16. Let $p_{n,k}$ be the number of n-permutations in which every cycle has less than k elements. Prove that for any fixed k, the sequence $(p_{n,k})_{k\geq 2}$ is *log-convex*, that is,

$$p_{n,k}^2 \leq p_{n-1,k}p_{n+1,k}.$$

17. Generalize Exercise 41 in the following way. Let $F_{n,k}$ be the number of n-permutations with exactly k fixed points. Let $E_{n,k}$ be the number of n-permutations $p_1 p_2 \cdots p_n$ in which $p_1 p_2 \cdots p_k$ is an increasing sequence, and $p_{k+1} p_{k+2} \cdots p_n$ is a desarrangement. A little thought shows that for each n-permutation, there is exactly one non-negative k with this property. Prove that $F_{n,k} = E_{n,k}$.

18. Let $fix(p)$ be the number of fixed points of the permutations p. Let $p^{\lfloor 0 \rfloor}$ be the multiset-permutation obtained from p by replacing each fixed point of p by 0, and let $dez(p)$ be the number of descents of $p^{\lfloor 0 \rfloor}$. Prove that the three-variable statistics (fix, exc, d) and (fix, exc, dez) are equidistributed. In other words, prove that there is a bijection ϕ on the set of all n-permutations so that p and $\phi(p)$ have the same number of fixed points, the same number of excedances, and $d(p) = dez(\phi(p))$.

19. Let J be a subset of $[n-1]$.

   (a) Let p be an n-permutation, and let us assume that $D(p) = J$. At most how many fixed points can p have?

   (b) Let $F_n(J)$ be the set of n-permutations with descent set J and exactly $n - |J|$ fixed points. Let $G(J)$ be the set of derangements q on J so that whenever both i and $i+1$ are elements of J, the inequality $q(i) > q(i+1)$ holds. Prove that then

   $$\sum_{p \in F_n(J)} z^{exc(p)} = \sum_{q \in G(J)} z^{exc(q)}.$$

20. (a) What is the largest possible number of fixed points in an alternating n-permutation? Answer the same question for reverse alternating permutations.

   (b) Let $d_k(n)$ be the number of alternating n-permutations with exactly k fixed points, and let $d_k^*(n)$ denote the number of reverse alternating n-permutations with exactly k fixed points. Prove that if $n \geq 4$, then
   $$d_{\lceil n/2 \rceil}(n) = D_{\lfloor n/2 \rfloor},$$
   and if $n \geq 5$, then
   $$d_{\lceil (n+1)/2 \rceil}^*(n) = D_{\lfloor (n-1)/2 \rfloor}.$$

   (c) Prove that the number of alternating permutations of length $2n$ having k excedances and n fixed points is equal to the number of derangements of length n having k excedances.

   (d) Prove that $d_0 n = d_1(n)$, and $d_0^*(n) = d_1^*(n)$.

21. (a) Find a formula for the number of derangements of length n that are even permutations minus the number of derangements of length n that are odd permutations.

   (b) Let $m \geq 2$ be a fixed positive integer, and let $0 \leq r \leq m - 1$ be a fixed integer. Let $p(n, m, r)$ be the probability that the number of cycles of a randomly selected n-permutation is of the form $mq + r$. Prove that

$$\lim_{n \to \infty} p(n, m, r) = \frac{1}{m}.$$

22. Let n be a fixed positive integer, and let $G_{n,r}(z)$ be defined as in Problem Plus 10, that is, as the generating polynomial of all r-derangements of length n with respect to their number of cycles.

   (a) Prove that for all *negative integers* m, and for all real numbers $\epsilon > 0$, there exists a threshold N so that if $n > N$, then the polynomial $G_{n,r}(z)$ has a root in the interval $(m - \epsilon, m + \epsilon)$.

   (b) Deduce a statement on r-derangements that is analogous with the claim of part (b) of Problem Plus 22.

23. Recall that if $S \subseteq [n - 1]$, then $\beta(S)$ denotes the number of all permutations of length n whose descent set is equal to S. We have computed $\beta(S)$ in Theorem 1.4. Let $\beta^{cyc}(S)$ be the number of permutations of length n that have descent set β and consist of a single cycle of length n.

   Prove that for each n, there exists a collection F_n of subsets of $[n - 1]$ so that $\lim_{n \to \infty}(|F_n|/2^{n-1}) = 1$, and

$$\lim_{n \to \infty} \max_{S \in F_n} \left| \frac{n\beta^{cyc}(S)}{\beta(S)} - 1 \right| = 0.$$

   In other words, for most subsets S, the ratio of the number of *cyclic* permutations with descent set S to the number of *all* permutations with descent set is close to $1/n$. Note that if we drop the requirement that these permutations have descent set S, then the corresponding ratio is exactly $1/n$. So, that ratio is fairly independent of the required descent set.

Solutions to Problems Plus

1. This result was first proved by Jeffrey Remmel in [275] as a corollary to a more general argument involving q-analogues. Other proofs can be found in [141] and (in a slightly different context), in [328].

2. (a) The proofs are different for the cases of odd n and even n. The case of even n can be proved very similarly to Exercise 61. For odd n, however, we cannot solely rely on our previous methods, because $2n + 2$ is divisible by 4, and so there are permutations counted by SQ_{2n+2} whose odd part is empty. (And therefore, their even part is of length $2n + 2$, and as such, cannot be found in shorter permutations). One can show, however, that the number of this kind of permutations is very small if $n > 8$.

    (b) Equality holds if and only if $n = 1$. As part [(a)] shows that $SQ_{2n+1} < SQ_{2n+2}$ for $n > 8$, this can be seen by checking the cases of odd n for $n \leq 7$.

3. A more general version of this result is proved in [53].

4. The first combinatorial proof was given in [43]. A slightly simpler proof appears in [34].

5. (a) Formula (3.3) is clearly equivalent to

$$\sum_{k=0}^{n} c(n+1, k+1)z^k = (z+1)(z+2)\cdots(z+n)$$

$$= n!(1+z)\left(1 + \frac{z}{2}\right)\cdots\left(1 + \frac{z}{n}\right).$$

    Then $c(n+1, 3)$ is the coefficient of z^2 on the right-hand side. The proof then follows by routine computations.

    (b) Similar to part (a).

6. This result is proved in [53].

7. This can be proved like Problem Plus 5, with more complex computations involving symmetric functions. See [123] or [109] for those details. Here C is the Euler constant, that is,

$$C = \lim_{n \to \infty} \left(\sum_{k=1}^{n} \frac{1}{k} - \ln n\right).$$

8. We have seen in Lemma 1.17 that $k!S(n, k) = \sum_{i=0}^{k}(-1)^i \binom{k}{i}(k - i)^n$. We claim that all summands of the right-hand side are negligibly small compared to the last one. That is, we claim that for fixed k and fixed $i \in [k]$, we have that

$$\frac{(k - i)^n \binom{k}{i}}{k^n} \to 0$$

as $n \to \infty$. This is clearly true as $\binom{k}{i} \leq 2^k$, and $(k - i)/k < 1$.

9. This line of research was essentially initiated by Paul Edelman [150].

(a) We have $b(n, k) = n - k$. On one hand, $b(n, k) \leq n - k$ is easy to see for instance by induction. On the other hand, the permutation $(12 \cdots n - k + 1)(n - k + 2)(n - k + 3) \cdots (n)$ shows that this is indeed possible.

(b) We have

$$B(n, k) = \begin{cases} \binom{n}{2} - \lceil \frac{n}{2} \rceil + k, & \text{if } k \leq \lceil \frac{n}{2} \rceil, \\ \binom{n}{2} - \binom{2k-n}{2}, & \text{if } k > \lceil \frac{n}{2} \rceil. \end{cases}$$

(c) Let us call a cycle $(c_1 c_2 \cdots c_m)$ unimodal if the sequence of entries $c_1, c_2, \cdots, c_m$ is unimodal. Let us call a permutation p unimodal if all its cycles (when started with their smallest elements) are unimodal, and when the underlying sets of all cycles of p consist of *intervals* of positive integers. For instance, $(1342)(58976)$ is a unimodal permutation. Then it can be proved that $b(n, k)$ is attained by p if and only if p is a unimodal permutation. This implies that for $1 \leq k \leq n - 1$, the number $m(n, k)$ we are looking for is given by

$$m(n, k) = \sum_{r=0}^{k} \binom{n - k - 1}{r - 1} \binom{k}{r} 2^{n-k-r}.$$

(d) If $n - 1 \geq k > \lceil \frac{n}{2} \rceil$, then $B(n, k)$ is attained by only one permutation. This permutation p is given by

$$p(i) = \begin{cases} n + 1 - i, & \text{if } i \in [1, n - k] \cup [k + 1, n], \\ i, & \text{if } i \in [n - k + 1, k] \end{cases}$$

In other words, p is an involution that leaves the middle of the set $[n]$ element-wise fixed.

If $k \leq \lceil \frac{n}{2} \rceil$, then the numbers $M(n, k)$ we are looking for are given by

$$M(n, k) = \sum_{r=0}^{k} \binom{\lceil \frac{n}{2} \rceil - k - 1}{r - 1} \binom{k}{r} 3^{\lceil \frac{n}{2} \rceil - k - r} 2^r.$$

See [150] for further details, proofs and generalizations.

10. This result is due to Francesco Brenti, and can be found in [96].

11. (a) This result (and the rest of the results in this exercise) is due to Francesco Brenti, and can be found in [97]. In fact, it is proved in that paper that

$$E_D(t) = \frac{n!}{z_D} \prod_{i=1}^{k} \frac{A_{d_i-1}(z)}{(d_i - 1)!},$$

where $d = (d_1, d_2, \cdots, d_k)$ is the partition defining the conjugacy class D, the $A_j(z)$ are the Eulerian polynomials, and finally, $z_D = \prod_{i \geq 1} i^{m_i(d)} m_i(d)!$. In this last formula, $m_i(d)$ is the multiplicity of i as a part of d.

(b) It is proved in [97] that $E_D(z)$ is symmetric. Therefore, it has to take its maximal value in the middle. This is then proved to be at $(n - m)/2$, where m is the number of fixed points in the partition defining D. (If this is not an integer, then there are two maximal values, and they are bracketing $(n - m)/2$. In other words, $(n - m)/2$ is always the *center of symmetry* of the sequence of coefficients of $E_D(z)$.) One main tool of the proof is the result of Exercise 24 of Chapter 1.

(c) Let D be a conjugacy class of S_n, and let the partition defining D be $(d_1, d_2, \cdots, d_k)$, where $d_i > r$ for $i \in [k]$. Then it follows from part [(b)] that $E_D(z)$ is a symmetric and unimodal polynomial with center of symmetry $n/2$ as permutations in D have no fixed points. That is, the center of symmetry does *not* depend on D. Our definitions imply

$$E_{D,k,r,n}(z) = \sum_D E_D(z),$$

where D ranges over all conjugacy classes that are given by partitions with all parts larger than r. The claim then follows as the right-hand side is the sum of symmetric and unimodal polynomials with center of symmetry $n/2$.

(d) As we have

$$E_{D_r(n)}(z) = \sum_{k \geq 1} E_{D,k,r,n}(z),$$

part [(c)] implies our claim.

12. The first proof of this result was given in [143] by the use of symmetric functions. Two further, and more bijective, proofs were given in [144], one of which is the proof of a special case of a more general theorem.

13. This result is due to Richard Stanley and Ira Gessel [194]. Let Q_k be the set of Stirling permutations, that is, the set of all permutations $p_1 p_2 \cdots p_{2n}$ of the multiset $\{1, 1, 2, 2, \cdots, k, k\}$ in which $u < v < w$ and $p_u = p_w$ imply $p_v > p_u$. Then the number $B_{k,i}$ is equal to the number of permutations $p_1 p_2 \cdots p_{2n} \in Q_k$ that have exactly i descents in the following sense: $p_j > p_{j+1}$ or $j = 2k$ for exactly i values of $j \in [2k]$. The above reference contains two proofs of this fact, a combinatorial one and an inductive one.

14. A proof using symmetric functions can be found in [193]. It would still be interesting to find a bijective proof.

15. The first proof is due to Volker Strehl [305] who used the classic Robinson–Schensted bijection in his proof. We will discuss this bijection in Chapter 7. A very short proof using quasi-symmetric functions was given in [193].

16. This follows from the following general result of Ed Bender and Rodney Canfield [33]. Let $Z_1, Z_2, \cdots$ be a log-concave sequence of non-negative real numbers, and define the sequences A_n and P_n by

$$\sum_{n \geq 0} A_n z^n = \sum_{n \geq 0} \frac{P_n}{n!} z^n = \exp\left(\sum_{j \geq 1} \frac{Z_j z^j}{j}\right).$$

Then the A_n are log-concave and the P_n are log-convex.

In order to get our result from this theorem, set $Z_j = 1$ for $j < k$, and set $Z_j = 0$ otherwise. Then the far right-side of the above formula simplifies to $\exp\left(\sum_{j \geq 1}^{k-1} \frac{Z_j z^j}{j}\right)$, which is precisely the exponential generating function for the numbers $p_{n,k}$. In other words, in this special case, we get $P_n = p_{n,k}$.

17. This result is proved in [143] in two different ways. That paper also contains further generalizations, such as refinements of the proved equality with respect to descent sets.

18. This result is due to Dominique Foata and Guo-Niu Han [175].

19. (a) We claim that p cannot have more than $n - |J|$ fixed points. Indeed, if $i \in D(p)$, then it is impossible for both i and $i + 1$ to be fixed points. On the other hand, $n - |J|$ fixed points are possible as shown by the increasing permutation, or the decreasing permutation if n is odd.

   (b) This result is due to Guo-Niu Han and Guoce Xin [212]. They first rewrite the statement by using the result of Problem Plus 18 to replace the number of descents by $dez(p)$. After that, they give a one-paragraph proof for the new statement. This is remarkable, because several of the upcoming Problems Plus will be consequences of this statement.

20. (a) If p is alternating, then it is impossible for both $2i - 1$ and $2i$ to be fixed points of p at the same time. So p has at most $\lceil n/2 \rceil$ fixed points. This maximum is reached, for instance by permutations 623154 and 6235417. If p is reverse alternating, then it is impossible for both $2i$ and $2i + 1$ to be a fixed point of p. This means that p cannot have more than $\lfloor n/2 \rfloor + 1$ fixed points. This maximum is achieved by permutations like 15342 and 153426.

   (b) This result is due to Robin Chapman and Lauren Williams [113].

(c) This result is due to Guo-Niu Han and Guoce Xin [212]. Let p be an alternating $(2n)$-permutation with n fixed points. It is then not difficult to prove that for each $i \in [n]$, either $2i - 1$ is a fixed point of p, and $p_{2i} < 2i$, or $2i$ is a fixed point of p, and $2i - 1$ is an excedance of p. Conversely, if every $i \in [n]$ has this property, then p is alternating. Now remove all fixed points from p, to get a derangement of an n-element subset of $[2n]$. If we replace the rth smallest entry of this derangement by r, we get a derangement $\phi(p)$ of $[n]$ that has the same number of excedances as p. It can then be proved that ϕ is bijective.

(d) The first proofs of these facts were given by Richard Stanley in [298], who asked for combinatorial proofs. Such proofs were then given in [212].

21. (a) A short combinatorial proof is given by Robin Chapman [112].

(b) Let us first assume that $r = 0$. Let α be an mth root of unity with the smallest non-zero argument, and set $\alpha = z$ in (3.3). Then sum the obtained formula for all mth roots of unity, that is, for all α^t, with $t = 0, 1, \cdots m - 1$. Using the summation formula of a geometric progression, we get that

$$\sum_{t=0}^{m-1} (\alpha^k)^t = \begin{cases} 0 \text{ if } \alpha^k \neq 1, \text{ that is, if } m \text{ is a divisor of } k, \\ \\ m \text{ if } \alpha^k = 1, \text{ that is, if } m \text{ is not a divisor of } k. \end{cases}$$

Therefore on the right-hand side, all summands $c(n, k)(\alpha^t)^k$ will cancel except for those in which k is a multiple of m; if k is a multiple of m, then the summands containing $c(n, k)$ will add up to $m \cdot c(n, k)$. On the left-hand side, we will have a real number that is the sum of m terms, $m - 1$ of which have a negligibly small absolute value compared to the one in which $z = 1$, and has absolute value $n!$. Dividing both sides by $n!$ and taking limits, our claim is proved.

If $r \neq 0$, then the proof is very similar, but instead of simply adding the m equations obtained by setting $z = \alpha^t$ in (3.3), we multiply each such equation by α^{-rt}, and then sum them over $t = 0, 1, \cdots, m - 1$. The only change is that on the left-hand side, all summands $c(n, k)(\alpha^t)^k$ will cancel, except for those in which k is congruent to r modulo m.

22. (a) See *On a balanced property of derangements* [59]. The key idea is that the polynomial $G_{n,r}(z)/n!$ (as defined in Problem Plus 10) takes very small values for $z = -1, -2, \cdots, -r$ if r is small compared to n. Hence this polynomial must have a root close to such negative integers.

(b) The probability that the number k of cycles of a randomly selected r-derangement of length n is congruent to r modulo m converges to $1/m$. The proof is analogous to that of part (b) of Problem Plus 21.

23. This result can be found in Section 5.2 of [162].

4

In Any Way but This. Pattern Avoidance. The Basics.

4.1 Notion of Pattern Avoidance

In earlier chapters, we have studied inversions of permutations. These were *pairs* of elements that could be anywhere in the permutation, but always related to each other the same way, that is, the one on the left was always larger.

There is a far-fetching generalization of this notion from pairs of entries to k-tuples of entries. Consider a "long" permutation, such as $p = 25641387$, and a shorter one, say $q = 132$. We then say that the 3-tuple of entries (2,6,4) in p forms a *pattern* or *subsequence* of type 132 because the entries (2,6,4) of p relate to each other as the entries 132 of q do. That is, the first one is the smallest, the middle one is the largest, and the last one is of medium size. The reader is invited to find a pattern of type 321 in p. On the other hand, there is no pattern of type 12345 in p; therefore, we will say that p *avoids* 12345.

The notion of pattern avoidance is at the center of this entire chapter, so we will make it formal.

DEFINITION 4.1   *Let $q = (q_1, q_2, \cdots, q_k) \in S_k$ be a permutation, and let $k \leq n$. We say that the permutation $p = (p_1, p_2, \cdots, p_n) \in S_n$ contains q as a pattern if there are k entries $p_{i_1}, p_{i_2}, \cdots, p_{i_k}$ in p so that $i_1 < i_2 < \cdots < i_k$, and $p_{i_a} < p_{i_b}$ if and only if $q_a < q_b$. If p does not contain q, then we say that p avoids q.*

Example 4.2
The permutation $p = 57821346$ contains the pattern $q = 132$ as shown in Figure 4.1.
▯

In other words, p contains q as a pattern if p has a subsequence of elements that "relate" to one another the same way the elements of q do.

DOI: 10.1201/9780429274107-4

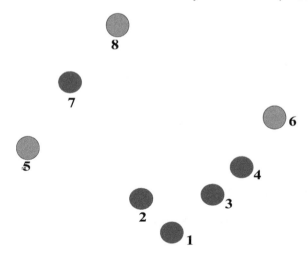

FIGURE 4.1
Containing the pattern 132.

4.1.1   Permutation classes

Let P be the infinite partially ordered set of all finite permutations ordered
by pattern containment. That is, in this poset we have $p \leq_P q$ if q contains
p as a pattern. This P is an interesting poset and we will learn more about
it in Section 7.2. Recall that a subset C of a partially ordered set R is called
an *ideal* if for all elements $x \in C$ and all elements $y \in R$ satisfying $y \leq x$, the
inclusion $y \in C$ also holds. In other words, ideals are "closed downward".

DEFINITION 4.3   *If C is an ideal of the poset of P of all finite permutations ordered by pattern containment, then we will call C a* permutation
class.

The set of all permutations that avoid a given pattern, the set of permutations that avoid a given set of patterns, or the set of all permutations that
are shorter than a given bound are all examples of permutation classes. If the
permutation class C consists of all permutations that avoid one given pattern
q, then we call $C = C(q)$ a *principal* permutation class.

4.2   Patterns of Length Three

We have enumerated n-permutations with a given number of inversions and
obtained rather precise results. The corresponding question, that is, a formula

for the number of n-permutations with a given number of occurrences of a pattern q, is in general too difficult. We will therefore first concentrate on the special case when this given number is zero. That is, we will try to find the number $\mathrm{Av}_n(q)$ of n-permutations that avoid the pattern q.

Obviously, we have $\mathrm{Av}_n(12) = \mathrm{Av}_n(21) = 1$, so the first nontrivial case is that of patterns of length three. There are six such patterns, but as we will shortly see, there are many symmetries between them.

Recall that for the permutation $p = p_1 p_2 \cdots p_n$, we define the *reverse* of p as the permutation $p^r = p_n p_{n-1} \cdots p_1$, and the *complement* of p as the permutation p^c whose ith entry is $n + 1 - p_i$.

It is clear that if a permutation avoids 123, then its reverse avoids 321, thus $\mathrm{Av}_n(123) = \mathrm{Av}_n(321)$. Similarly, if a permutation avoids 132, then its reverse avoids 231, its complement avoids 312, and the reverse of its complement avoids 213. Therefore, we also have $\mathrm{Av}_n(132) = \mathrm{Av}_n(231) = \mathrm{Av}_n(312) = \mathrm{Av}_n(213)$.

If we can prove that $\mathrm{Av}_n(123) = \mathrm{Av}_n(132)$, then we will have the remarkable result that all patterns of length three are avoided by the same number of n-permutations.

LEMMA 4.4
For all positive integers n, the equality $\mathrm{Av}_n(123) = \mathrm{Av}_n(132)$ holds.

PROOF The are several ways to prove this first nontrivial result of the subject. The one we present here is due to Rodica Simion and Frank Schmidt [284]. Recall that an entry of a permutation that is smaller than all the entries that precede it is called a *left-to-right minimum*. Note that the left-to-right minima form a decreasing subsequence.

We are going to construct a bijection f from the set of all 132-avoiding n-permutations to the set of all 123-avoiding n-permutations that leaves all left-to-right minima fixed.

The map f is defined as follows: Keep the left-to-right minima of p fixed, and write all the other entries in decreasing order in the positions between the left-to-right minima. The obtained permutation $f(p)$ is always 123-avoiding as it is the union of two decreasing subsequences, one of which is the sequence of all left-to-right minima, and the other is the decreasing sequence into which we arranged the remaining entries.

Example 4.5
If $p = 67341258$, then the left-to-right minima of p are 6, 3, and 1; therefore, $f(p) = 68371542$. □

We would like to point out that the left-to-right minima of p and $f(p)$ are the same, even if some other entries of p have moved. Indeed, we can say that

f simply rearranges the m entries that are not left-to-right minima pair by pair. That is, whenever we (impersonating the function f) see a pair of these entries that are not in decreasing order, we swap them. This algorithm stops in at most $\binom{m}{2}$ steps. Moreover, each step of this algorithm moves a smaller entry to the right and a larger one to the left and therefore never creates a new left-to-right minimum.

We note that this is the only 123-avoiding permutation with the given set and position of left-to-right minima. Indeed, if there were two entries x and y that are not left-to-right minima and form a 12-pattern, then the left-to-right minimum z that is closest to x on the left, and the entries x and y, would form an increasing sequence.

Now we prove that f is a bijection by showing that it has an inverse. Let q be an n-permutation that avoids 123. Keep the left-to-right minima of q fixed, and fill in the remaining positions with the remaining entries, moving left-to-right, as follows. At each step, place the smallest element not yet placed that is larger than the closest left-to-right minima on the left of the given position. Call the obtained permutation $g(q)$.

Example 4.6
If $q = 68371542$, then the left-to-right minima of q are 6, 3, and 1. To the empty slot between 6 and 3, we put the smallest of the two entries that are larger than 6, that is, 7. In the empty slot between 3 and 1, we put the smallest entry not used yet that is larger than 3, that is, 4. Immediately on the right of 1, we place the smallest entry not used yet that is larger than 1, that is, 2. We finish this way, by placing 5 and 8 in the remaining slots, to get $g(q) = 67341258$. $\quad\Box$

The obtained permutation is always 132-avoiding. Indeed, if there were a 132-pattern in it, then there would be one that starts with a left-to-right minimum, but that is impossible as entries larger than any given left-to-right minimum are written in increasing order.

Note again that $g(q)$ is the only 132-avoiding permutation that has the same set and position of left-to-right minima as q. Indeed, if at any given instance, two entries $u < v$ that are larger than a left-to-right minimum a were in decreasing order, then avu would be a 132-pattern.

This proves that $g(f(p)) = p$, implying that f is a bijection, and proving our theorem. ∎

The techniques and notions used in the above proof will be used so often in the coming sections that it is worth visualizing them. Figure 4.2 shows our running example, the permutation $p = 67341258$, and its image, $f(p) = 68371542$.

Note again that the left-to-right minima (the decreasing sequence of gray circles) are left fixed. In $f(p)$, all the remaining entries form a decreasing

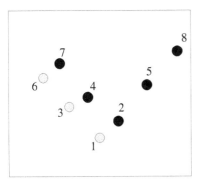 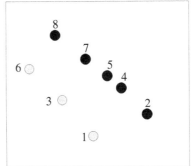

FIGURE 4.2
A 132-avoiding permutation and its image.

sequence, while in p, all entries larger than a given left-to-right minimum form an increasing sequence.

Now that we know that $\mathrm{Av}_n(q)$ does not depend on q as long as the length of q is three, it suffices to determine $\mathrm{Av}_n(q)$ for one possible choice of q of length three.

THEOREM 4.7
For all positive integers n, the equality

$$\mathrm{Av}_n(132) = C_n = \frac{\binom{2n}{n}}{n+1}$$

holds.

PROOF   Set $C_n = \mathrm{Av}_n(132)$. Suppose we have a 132-avoiding n-permutation in which the entry n is in the ith position. Then any entry to the left of n must be larger than any entry to the right of n. Indeed, if x and y violate this condition, then xny is a 132-pattern. Therefore, the set of entries on the left of n must be $\{n-i+1, n-i+2, \cdots, n-1\}$, and the set of entries on the right of n must be $[n-i]$. Moreover, there are C_{i-1} possibilities for the order of entries to the left of n and C_{n-i} possibilities for the order of entries on the right of n. Summing over all i we get the recurrence relation

$$C_n = \sum_{i=0}^{n-1} C_{i-1} C_{n-i}. \tag{4.1}$$

Therefore, if $C(z) = \sum_{n\geq 0} C_n z^n$ is the ordinary generating function of the numbers C_n, then (4.1) implies $C^2(z)z + 1 = C(z)$. That functional equation in turn yields

$$C(z) = \frac{1 - \sqrt{1-4z}}{2z}. \tag{4.2}$$

By standard methods this yields

$$C(z) = \sum_{n=1}^{\infty} \frac{1}{n} \binom{2n-2}{n-1} z^{n-1} = \sum_{n=0}^{\infty} \frac{1}{n+1} \binom{2n}{n} z^n \qquad (4.3)$$

and the statement of the theorem is proved. ∎

Because of the trivial identities $\mathrm{Av}_n(132) = \mathrm{Av}_n(231) = \mathrm{Av}_n(312) = \mathrm{Av}_n(213)$, and $\mathrm{Av}_n(123) = \mathrm{Av}_n(321)$, we have proved the following.

COROLLARY 4.8
Let q be a pattern of length three. Then

$$\mathrm{Av}_n(q) = C_n = \frac{\binom{2n}{n}}{n+1}.$$

The numbers C_n are called the *Catalan numbers*, and are perhaps the most famous counting sequence in combinatorics. Richard Stanley [297] has collected over 150 different classes of objects that are enumerated by these numbers. We will shortly see that the condition that q is of length three is of crucial importance in several aspects.

The following definition will make future discussion of pattern avoiding permutations simpler.

DEFINITION 4.9   *We say that the permutation patterns q and q' are Wilf-equivalent if $\mathrm{Av}_n(q) = \mathrm{Av}_n(q')$ for all n.*

So Corollary 4.8 implies that all patterns of length three are Wilf-equivalent to each other.

We end this section by a simple consequence of Lemma 4.4 that we will need later in this chapter. Recall that we have seen in the proof of Lemma 4.4 that a 132-avoiding permutation is completely determined by the set and position of its left-to-right minima. The following proposition makes this observation explicit.

PROPOSITION 4.10
Let $p = p_1 p_2 \cdots p_n$ be a permutation of length n that avoids 132. Let the ordered pairs of words $(u(p), v(p))$ of length n be defined as follows. The ith letter of $u(p)$ is A if p_i is a left-to-right minimum in p, and B otherwise. The ith letter of $v(p)$ is A if the entry of value i is a left-to-right minimum in p, and B otherwise.

Then the map $r(p) = (u(p), v(p))$ is injective.

Example 4.11
If $p = 43512$, then $u(p) = AABAB$ and $v(p) = ABAAB$.   ⬚

Taking reverse permutations, we get a dual version of Proposition 4.10 that we will use soon as well. Therefore, we announce it on its own.

PROPOSITION 4.12
Let $p = p_1 p_2 \cdots p_n$ be a permutation of length n that avoids 213. Let the ordered pairs of words $(x(p), y(p))$ of length n be defined as follows. The ith letter of $x(p)$ is C if p_i is not a right-to-left maximum in p, and D otherwise. The ith letter of $y(p)$ is C if the entry of value i is not a right-to-left maximum in p, and D otherwise.
Then the map $s(p) = (x(p), y(p))$ is injective.

Example 4.13
If $p = 35412$, then $x(p) = CDDCD$, and $y(p) = CDCDD$.   ⬚

4.3   Monotone Patterns

After obtaining satisfying results for the case of patterns of length three, the reader may think that we will now turn to patterns of length four, and then to the general case, and obtain similarly strong results. Unfortunately, that is quite a mountainous task. As we will shortly see, finding an exact formula is difficult even for patterns of length four, and is an open problem for all longer patterns. However, a long-standing conjecture (proved in 2003 [250] by Adam Marcus and Gábor Tardos) connected all patterns by claiming that no matter what q is, the number of n-permutations avoiding q is very small compared to $n!$.

CONJECTURE 4.14 *[Stanley–Wilf conjecture, 1980] Let q be any pattern. Then there exists a constant c_q so that for all positive integers n,*

$$\mathrm{Av}_n(q) \le c_q^n. \tag{4.4}$$

It is worth pointing out that Vaughan Pratt [267] *almost* made this conjecture explicit in 1973!

Note that the conjectured number c_q^n is very small compared to the number of all n-permutations, which is $n!$. In other words, this is a quite ambitious conjecture.

The following conjecture may look even more ambitious, but that is a false appearance.

CONJECTURE 4.15 *[Stanley–Wilf conjecture, alternative version] Let q be any pattern. Then the limit*

$$L(q) = \lim_{n \to \infty} \sqrt[n]{\operatorname{Av}_n(q)}$$

exists.

The number $L(q)$ is a crucial piece of information about the pattern q, and the principal permutation class generated by it. We will usually call it the *growth rate* of the class of q-avoiding permutations, though in the past it was sometimes called the *Stanley-Wilf limit* of q. The advantage of the terminology *growth rate* is the following. As we will see, for *principal* permutation classes, the limit $\lim_{n \to \infty} \sqrt[n]{\operatorname{Av}_n(q)}$ always exists. However, for permutation classes in general, it may not. This warrants the following definition.

DEFINITION 4.16 *If C is a permutation class, and $|C_n|$ is the number of elements of C that are of length n, then*

$$\limsup_{n} \sqrt[n]{|C_n|}$$

is called the growth rate *of C.*

The first published proof of the fact that the above two versions of the conjecture are equivalent is given in [17]. On one hand, it is obvious that Conjecture 4.15 implies Conjecture 4.14. On the other hand, to prove the implication in the other direction, we claim that $\operatorname{Av}_m(q)\operatorname{Av}_n(q) \le \operatorname{Av}_{m+n}(q)$, for all patterns q, and all positive integers m and n.

DEFINITION 4.17 *Let p be a permutation of length m, and let p' be a permutation of length n. Then their* sum *is the permutation $p \oplus p'$ of length $m + n$ given by*

$$(p \oplus p')_i = \begin{cases} p_i \ if \ i \le m, \\ \\ m + p'_{i-m}, \ if \ m + 1 \le i \le m+n. \end{cases}$$

Example 4.18
If $p = 213$ and $p' = 3142$, then $p \oplus p' = 2136475$.   ⬚

Sometimes, the dual concept of *skew sum* is useful.

DEFINITION 4.19 *Let p be a permutation of length m, and let p' be a permutation of length n. Then their* skew sum *is the permutation $p \ominus p'$ of length $m + n$ given by*

$$(p \ominus p')_i = \begin{cases} p_i + n \ \text{if } i \leq m, \\ p'_{i-m}, \ \text{if } m + 1 \leq i \leq m + n. \end{cases}$$

Example 4.20
If $p = 213$ and $p' = 3142$, then $p \ominus p' = 6573142$. ▯

Let us assume, without loss of generality, that the first entry of q is larger than the last entry of q. Let p and p' be permutations of length m and n, respectively, that avoid q.

Then the permutation $p \oplus p' \in \operatorname{Av}_{m+n}$ clearly avoids q, proving (injectively) that $\operatorname{Av}_m(q)\operatorname{Av}_n(q) \leq \operatorname{Av}_{n+m}(q)$. Therefore the sequence $\sqrt[n]{\operatorname{Av}_n(q)}$ is bounded and monotone, and must thus be convergent.

As we have mentioned, the Stanley–Wilf conjecture was proved to be true [250] in 2003. We will present that spectacular proof in Section 4.5. For now, we will consider some special cases in which more precise results are available.

We have seen in the previous section that the Stanley–Wilf conjecture is true if q is of length three, with $c_q = 4$. Indeed, $C_n < \binom{2n}{n} < 4^n$. There is only one other class of patterns for which it is similarly easy to prove the Stanley–Wilf conjecture, namely *monotone* patterns.

THEOREM 4.21
For all positive integers n and $k > 2$, we have

$$\operatorname{Av}_n(123 \cdots k) \leq (k - 1)^{2n}.$$

PROOF Let us say that an entry x of a permutation is of rank i if it is the end of an increasing subsequence of length i, but there is no increasing subsequence of length $i + 1$ that ends in x. Then for all i, elements of rank i must form a decreasing subsequence. Therefore, a q-avoiding permutation can be decomposed into the union of $k - 1$ decreasing subsequences.

This means that a permutation p avoiding $12 \cdots k$ is completely determined by the following two ordered partitions of $[n]$ into $k - 1$ blocks. (An ordered partition is a partition whose set of blocks is linearly ordered.)

1. The partition P_1 of the set of n *entries* of p, into the blocks of entries of rank 1, the block of entries of rank 2, and so on.

2. The partition P_2 of the n *positions* of p, into the block of positions containing entries of rank 1, the block of positions containing entries of rank 2, and so on.

Indeed, once these partitions are known, the entries of rank i must occupy the positions assigned to them, in decreasing order.

There are at most $(k-1)^n$ possibilities for each of P_1 and P_2, proving our theorem. ∎

Note that for $k = 3$, we get that $\mathrm{Av}_n(123) < 4^n$, agreeing with our earlier results. As we have seen that $\mathrm{Av}_n(123) = C_n$, we see that the constant 4 provided by Theorem 4.21 is actually the best possible constant. This is not an accident.

Indeed, Theorem 4.21 has a stronger version, which needs heavy analytic machinery, and therefore will not be proved here. The result is significant as it shows that no matter what k is, the constant $(k-1)^2$ cannot be replaced by a smaller number. Recall that functions $f(n)$ and $g(n)$ are said to be *asymptotically equal* if $\lim_{n \to \infty} \frac{f(n)}{g(n)} = 1$.

THEOREM 4.22

[272] *For all $k \geq 2$, the asymptotic equality*

$$\mathrm{Av}_n(1234 \cdots k) \simeq \lambda_k \frac{(k-1)^{2n}}{n^{(k^2-2k)/2}}$$

holds. Here

$$\lambda_k = \gamma_k^2 \int \int_{x_1 \geq x_2 \geq \cdots \geq x_k} \cdots \int [D(x_1, x_2, \cdots, x_k) \cdot e^{-(k/2)x^2}]^2 dx_1 dx_2 \cdots dx_k,$$

where $D(x_1, x_2, \cdots, x_k) = \prod_{i<j}(x_i - x_j)$, and $\gamma_k = (1/\sqrt{2\pi})^{k-1} \cdot k^{k^2/2}$.

Note that the multiple integral in the above formula evaluates to a constant.

4.4 Patterns of Length Four

In this section we will study the number $\mathrm{Av}_n(q)$ for patterns of length four. There are 24 patterns of length four, but there are many (trivial and nontrivial) Wilf-equivalences between them. First, we will use these equivalences to significantly decrease the number of patterns that need individual attention.

By taking reverses, complements, and reverse complements, we can restrict our attention to those patterns of length four in which

(a) the first element is smaller than the last one and

(b) the first element is 1 or 2.

This still leaves us 10 patterns, 1234, 1243, 1324, 1342, 1423, 1432, 2134, 2143, 2314, and 2413. Note that if p contains q, then the inverse of p clearly contains the inverse of q (as the inverse of a permutation matrix is its transpose), so $\mathrm{Av}_n(q) = \mathrm{Av}_n(q^{-1})$. Therefore, we can drop 1423, too, as its inverse 1342 remains on the list. Similarly, we can drop 2314 as its complement is 3241 and the reverse of that is again 1423.

We cannot proceed further without new tools. The following general theorem of Backelin, West, and Xin has a quite involved proof, but some of its special cases are easier.

THEOREM 4.23

[26] Let k be any positive integer, and let q be a permutation of length r. Then for all positive integers n, the equality

$$\mathrm{Av}_n((12\cdots k) \oplus q) = \mathrm{Av}_n((k(k-1)\cdots 1) \oplus q)$$

holds.

For instance, if $r = 2$, $k = 2$, and $q = 34$, then Theorem 4.23 says that $\mathrm{Av}_n(1234) = \mathrm{Av}_n(2134)$. If $q = 43$ and the other parameters do not change, then Theorem 4.23 says that $\mathrm{Av}_n(1243) = \mathrm{Av}_n(2143)$. As 2134 and 1243 are reverse complements of each other, the patterns 2134, 2143, and 2134 can all be removed from our list. Now let $k = 3$, and $r = 1$. Then clearly $q = 4$, and Theorem 4.23 shows that $\mathrm{Av}_n(1234) = \mathrm{Av}_n(3214)$. Therefore, by taking complements, we can remove the complement of the reverse of 3214, that is, the pattern 1432, from our list.

This leaves us with the patterns 1234, 1324, 1342, and 2413. The following result of Zvezdelina Stankova [288] eliminates one more pattern.

LEMMA 4.24

For all positive integers n, the equality

$$\mathrm{Av}_n(1342) = \mathrm{Av}_n(2413)$$

holds.

We note that Stankova proved her result in the equivalent form of $\mathrm{Av}_n(4132) = \mathrm{Av}_n(3142)$.

So we are left with three patterns, 1234, 1342, and 1324. It is high time that we took a look at numerical evidence computed by Julian West [325]. This evidence shows that as n grows starting at $n = 1$,

- for $\mathrm{Av}_n(1342)$ we have 1, 2, 6, 23, 103, 512, 2740, 15485
- for $\mathrm{Av}_n(1234)$ we have 1, 2, 6, 23, 103, 513, 2761, 15767
- for $\mathrm{Av}_n(1324)$ we have 1, 2, 6, 23, 103, 513, 2762, 15793.

These data are startling for at least two reasons. First, the numbers $\mathrm{Av}_n(q)$ are no longer independent of q. That is, there are some patterns of length four that are easier to avoid than others. Second, the monotone pattern 1234 does not provide either extremity. That is, it seems that the monotone pattern is neither the easiest nor the hardest one to avoid.

The numerical evidence shown above also raises some intriguing questions.

QUESTION 4.25 *Is it true that for all $n \geq 7$, the inequality*

$$\mathrm{Av}_n(1234) < \mathrm{Av}_n(1324)$$

holds?

QUESTION 4.26 *Is it true that for all $n \geq 6$, the inequality*

$$\mathrm{Av}_n(1342) < \mathrm{Av}_n(1234)$$

holds?

QUESTION 4.27 *In general, what makes a pattern easier to avoid than another pattern?*

QUESTION 4.28 *Is it true that if $\mathrm{Av}_n(q_1) < \mathrm{Av}_n(q_2)$ for some n, then for all $N > n$, we have $\mathrm{Av}_n(q_1) < \mathrm{Av}_n(q_2)$?*

In this section, we will answer Questions 4.25 and 4.26 in the affirmative. There is no known answer for Question 4.27, but we will mention some related interesting facts. The answer for Question 4.28 is, in general, negative. The first counterexample [289] is for two patterns q_1 and q_2 of size five, and for $n = 12$. So whether a pattern is easy or hard to avoid depends in some cases not just on the pattern itself, but also on n.

4.4.1 Pattern 1324

The following result is the earliest example [44] in which a pattern was shown to be more restrictive than another pattern of the same size.

THEOREM 4.29
For all $n \geq 7$, we have

$$\mathrm{Av}_n(1234) < \mathrm{Av}_n(1324).$$

The first step in our proof is the following classification of all n-permutations.

DEFINITION 4.30 *We will say that two permutations x and y have the same profile if*

- *the left-to-right minima of x are the same as those of y, and*

- *the left-to-right minima of x are in the same positions as the left-to-right minima of y, and*

- *the above two conditions hold for the right-to-left maxima of x and y as well.*

Note that profiles are equivalence classes.

Example 4.31
Permutations $x = 51234$ and $y = 51324$, have the same profile, but $z = 24315$ and $v = 24135$ do not, as the third entry of z is not a left-to-right minimum, whereas that of v is. ☐

The outline of our proof is going to be as follows: we show that for each (nonempty) profile T, there is *exactly* one 1234-avoiding permutation and *at least* one 1324-avoiding permutation having profile T. Then we show that "at least one" means "more than one" at least once, completing the proof.

The first half of our argument is simple.

LEMMA 4.32
For each non-empty profile T, there is exactly one 1234-avoiding permutation having profile T.

PROOF Let us assume that we have chosen a profile T, that is, we fixed the positions and values of all the left-to-right minima and right-to-left maxima. It is clear that if we put all the remaining entries into the remaining slots in decreasing order, then we get a 1234-avoiding permutation. Indeed, the permutation obtained this way consists of three decreasing subsequences, that is, the left-to-right minima, the right-to-left maxima, and the remaining entries. If the permutation just constructed contained a 1234-pattern, then by the pigeon-hole principle two of the entries of that 1234-pattern would be in the same decreasing subsequence, which would be a contradiction. Note that if T non-empty, then we can indeed write the remaining entries in decreasing order without conflicting with the existing constraints — otherwise T would be empty. This can be seen as in the proof of Lemma 4.4. Therefore, at least one 1234-avoiding permutation has profile T.

On the other hand, we claim that the decreasing order of the remaining entries is the only one that will result in a 1234-avoiding permutation. Indeed, if we put two of the remaining entries, say a and b, in increasing order, then together with the rightmost left-to-right minimum on the left of a and the leftmost right-to-left maximum on the right of b they would form a 1234-pattern. ∎

The second half of our argument is somewhat subtler.

LEMMA 4.33
Each nonempty profile contains at least one 1324-avoiding permutation.

PROOF   First note that if a permutation contains a 1324-pattern, then we can choose such a pattern so that its first element is a left-to-right minimum and its last element is a right-to-left maximum. Indeed, we can just take any existing pattern and replace its first (last) element by its closest left (right) neighbor that is a left-to-right minimum (right-to-left maximum). Therefore, to show that a permutation avoids 1324, it is sufficient to show that it does not contain a 1324-pattern having a left-to-right minimum for its first element and a right-to-left maximum for its last element. Such a pattern will be called a *bad pattern*. Also note that a left-to-right minimum (right-to-left maximum) can only be the first (last) element of a 1324-pattern.

Now take any 1324-containing permutation. By the above argument, it has a bad pattern. Interchange the second and third entries of this bad pattern. Observe that we can do this without violating the existing constraints, that is, no element x goes on the left of a left-to-right minimum that is larger than x, and no element y goes on the right of a right-to-left maximum that is smaller than y. The resulting permutation has the same profile as the original one because the left-to-right minima and right-to-left maxima have not been changed. Repeat this procedure as many times as possible, that is, as long as 1324-patterns can be found. Note that crucially, *each step of the procedure decreases the number of inversions of our permutation by at least one.* Therefore, we will have to stop after at most $\binom{n}{2}$ steps. Then, the resulting permutation will have the same profile as the original one, but it will have no bad pattern and therefore no 1324-pattern, as we claimed. ∎

We are only one step away from proving Theorem 4.29. We have to show that if $n \geq 7$, then classes that contain more than one 1324-avoiding permutations indeed exist. Let $n = 7$, then the class $3*1*7*5$ (with the stars denoting the positions of the remaining entries) contains two such permutations, namely 3612745 and 3416725. If $n > 7$, then put the entries $n, n-1, \cdots, 8$ in front of $3*1*7*5$, with no additional stars. The obtained class will again contain two 1324-avoiding permutations, coming from the subwords 624 and 462 of the remaining entries. This completes the proof of Theorem 4.29.

At this point, readers with a penchant for asymptotic results will surely say something like "OK, $\text{Av}_n(1324)$ is larger than $\text{Av}_n(1234)$, but is it really significantly larger?". That is, could it be that $\text{Av}_n(1234)$ and $\text{Av}_n(1324)$ are in fact asymptotically equal, meaning that $(\text{Av}_n(1234)/\text{Av}_n(1324)) \to 1$? The following proposition answers this question in the negative.

PROPOSITION 4.34
The sequences $\text{Av}_n(1234)$ and $\text{Av}_n(1324)$ are not asymptotically equal.

PROOF   See the solution of Exercise 29.   ∎

What can we say about $L(1324)$? In [5], it is proved that $L(1324) \geq 9.47$. In [38], David Bevan improved that lower bound by proving the inequality $L(1324) \geq 9.81$. Note that it follows from Theorem 4.22 that $L(1234) = 9$. Therefore, the sequences $\text{Av}_n(1234)$ and $\text{Av}_n(1324)$ have different growth rates, and the sequence $\text{Av}_n(1234)/\text{Av}_n(1324)$ converges to 0 at an exponential rate.

4.4.1.1   Exponential Upper Bound

In the previous subsection, we have seen that it is easier to avoid 1324 than 1234, but how much easier? We will now discuss three results concerning upper bounds for $L(1324)$, in the order they were discovered.

We start by presenting a result of Anders Claesson, Vit Jelinek, and Einar Steingrímsson [119] showing that $\text{Av}_n(1324) < 16^n$ for all positive integers n. The following simple definition will be crucial for the proof.

DEFINITION 4.35   *Let p' be a permutation of length k and let p'' be a permutation of length m, with $k + m = n$. We say that the n-permutation $p = p_1 p_2 \cdots p_n$ is a merge of p' and p'' if there is a partition of the set $[n]$ into two blocks, $I = \{i_1, i_2, \cdots, i_k\}$ and $J = \{j_1, j_2, \cdots, j_m\}$, so that the subsequence $p_{i_1} p_{i_2} \cdots p_{i_k}$ is a p'-pattern, and the subsequence $p_{j_1} p_{j_2} \cdots p_{j_m}$ is a p''-pattern.*

In other words, p is a merge of p' and p'' if we can color each entry of p red or blue so that the red entries form a p'-pattern, and the blue entries form a p''-pattern.

Example 4.36
The permutation $p = 4751326$ is a merge of $p' = 123$ and $p'' = 4132$, as can be seen by choosing $I = \{1, 3, 7\}$ and $J = \{2, 4, 5, 6\}$, leading to the subsequences 456 and 7132.   ⧠

The following lemma proves an upper bound for the number of permutations avoiding a pattern that is a merge of two shorter patterns. An earlier, less general version of this lemma was first used in [60]. The version we present is slightly less general than that of [119], but will suffice for our purposes.

LEMMA 4.37

Let q, q', and q'' be three permutation patterns so that every q-avoiding permutation can be obtained as a merge of a q'-avoiding permutation and a q''-avoiding permutation.

Let a and b be positive constants so that for all positive integers n, the inequalities $\mathrm{Av}_n(q') < a^n$ and $\mathrm{Av}_n(q'') < b^n$ hold. Then for all positive integers n, the inequality

$$\mathrm{Av}_n(q) < \left(\sqrt{a} + \sqrt{b}\right)^{2n}$$

holds.

PROOF   Let p be a q-avoiding permutation of length n. Then, we can color each entry of p red or blue so that the subsequence of all red entries avoids q_1 and the subsequence of all blue entries avoids q_2. If there are exactly k red entries, then this leaves less than $\binom{n}{k}^2 a^k b^{n-k}$ possibilities for p. Indeed, there are $\binom{n}{k}$ choices for the set of red entries, and there are $\binom{n}{k}$ choices for the set of positions of the red entries.

Summing over all possible values of k, this shows that

$$\mathrm{Av}_n(q) < \sum_{k=0}^{n} \binom{n}{k}^2 a^k b^{n-k}$$

$$\leq \sum_{k=0}^{n} \left(\binom{n}{k}\sqrt{a}^k \sqrt{b}^{n-k}\right)^2$$

$$\leq \left(\sum_{k=0}^{n} \binom{n}{k}\sqrt{a}^k \sqrt{b}^{n-k}\right)^2$$

$$\leq \left(\sqrt{a} + \sqrt{b}\right)^{2n}.$$

∎

In order to be able to use Lemma 4.37 to prove an upper bound for $\mathrm{Av}_n(1324)$, we need to show that each 1324-avoiding permutation is a merge of two permutations avoiding shorter patterns. The following lemma, which is a special case of a lemma in [119], accomplishes that.

LEMMA 4.38

Let p be a 1324-avoiding permutation. Then p is a merge of a 132-avoiding permutation and a 213-avoiding permutation.

PROOF Let $p = p_1 p_2 \cdots p_n$ be a 1324-avoiding permutation. Let us color the entries of p one-by-one, from left to right, according to the following rules.

1. If coloring p_i red would create a 132-pattern with all red entries, then color p_i blue, and

2. if there already is a blue entry smaller than p_i, then color p_i blue;

3. otherwise, color p_i red.

It is obvious from the first rule that the subsequence of red entries will indeed avoid 132. We claim that the subsequence of blue entries will avoid 213.

Let us assume that this is not the case, that is, there exist blue entries b_2, b_1, b_3 (located in this order from left to right) that form a 213-pattern. Let $b_1 = p_t$. Then let p_s be a blue entry such that $s \leq t$ and $p_s \leq p_t$, and p_s is minimal among entries satisfying these conditions. That means that p_s was colored blue because of the first rule, that is, coloring it red would have completed a red 132-pattern consisting of the entries r_1 and r_3, which are red, and p_s. Note that in particular, $r_1 < p_s \leq p_t = b_1$.

Now there are two possibilities. If r_1 is on the left of b_2, then $r_1 b_2 b_1 b_3$ forms a 1324-pattern in p, which is a contradiction. If r_1 is on the right of b_2, then of course, r_3 is also on the right of b_2. That implies that $r_3 < b_2$; otherwise, the second rule would force r_3 to be blue. In particular, $r_3 < b_3$. This means that $r_1 r_3 p_s b_3$ is a 1324-pattern in p, which is again a contradiction. ∎

We can now easily prove the result promised at the beginning of this subsection.

THEOREM 4.39

[119] For all positive integers n, the inequality

$$\mathrm{Av}_n(1324) \leq 16^n$$

holds.

PROOF Use Lemma 4.37 with $q_1 = 132$, $q_2 = 213$, and note that $\mathrm{Av}_n(132) = \mathrm{Av}_n(213) = C_n < 4^n$ as we proved in Corollary 4.8. ∎

Reading the proof of Theorem 4.39, it is tempting to think that there is a tremendous waste in our methods. In particular, the 2^n-upper bound for the set and position of the red entries, coming from Lemma 4.37, seems to be wasteful. The following result refines the techniques used in proving Theorem 4.39, and significantly improves its exponential upper bound.

THEOREM 4.40
For all positive integers n, the inequality

$$\mathrm{Av}_n(1324) \leq (7 + 4\sqrt{3})^n$$

holds.

Let us color each entry of the 1324-avoiding permutation $p = p_1 p_2 \cdots p_n$ red or blue using the same rules as in the proof of Theorem 4.39. Furthermore, let us mark each entry of p with one of the letters A, B, C, or D as follows.

1. Mark each red entry that is a left-to-right minimum in the partial permutation of red entries by A,

2. Mark each red entry that is not a left-to-right minimum in the partial permutation of red entries by B,

3. Mark each blue entry that is not a right-to-left maximum in the partial permutation of blue entries by C, and

4. Mark each blue entry that is a right-to-left maximum in the partial permutation of blue entries by D.

Call entries marked by the letter X entries of *type X*. Let $w(p)$ be the n-letter word over the alphabet $\{A, B, C, D\}$ defined above. In other words, the ith letter of $w(p)$ is the type of p_i in p. Let $z(p)$ be the n-letter word over the alphabet $\{A, B, C, D\}$ whose ith letter is the type of the entry of value i in p.

REMARK 4.41   Note that the function $f(p) = (w(p), z(p))$ in fact applies the map r of Proposition 4.10 to the string p_{red} of red entries of p, and the map s of Proposition 4.12 to the string p_{blue} of blue entries of p. So given $f(p) = (w(p), z(p))$, we can immediately recover $r(p_{red})$ and $s(p_{blue})$. Indeed, $r(p_{red})$ is the pair of subwords of $w(p)$ and $z(p)$ that consist of letters A and B, whereas $s(p_{blue})$ is the pair of subwords of $w(p)$ and $z(p)$ that consist of letters C and D.

Conversely, if we are given $r(p_{red}) = (u(p_{red}), v(p_{red}))$ and $s(p_{blue}) = (x(p_{blue}), y(p_{blue}))$, *and* we know in which positions of p the red entries are, and entries of which value of p are red, we can recover $f(p)$ as follows. Shuffle the words $u(p_{red})$ and $x(p_{blue})$ so that letters A and B are in positions that belong to red entries in p, and shuffle the words $v(p_{red})$ and $y(p_{blue})$ so that letters A and B are in positions j for which the entry of value j is red in p.

∎

Example 4.42
Let $p = 3612745$. Then the subsequence of red entries of p is 36127, the subsequence of blue entries of p is 45, so $w(p) = ABABBCD$, while $z(p) = ABACDBB$. ⬜

The following lemma shows a property of $w(p)$ that will enable us to improve the upper bound on $\mathrm{Av}_n(1324)$. Let us say that a word w has a *CB-factor* if somewhere in w, a letter C is immediately followed by a letter B.

LEMMA 4.43
If p is 1324-avoiding, then $w(p)$ has no CB-factor.

PROOF Let us assume that in $p = p_1 p_2 \cdots p_n$, the entry p_i is of type C, while the entry p_{i+1} is of type B. That means that $p_i > p_{i+1}$, otherwise the fact that p_i is blue would force p_{i+1} to be blue. Furthermore, since p_i is not a right-to-left maximum, there is an entry d on the right of p_i (and on the right of p_{i+1}) so that $p_i < d$. Similarly, since p_{i+1} is not a left-to-right minimum, there is an entry a on its left so that $a < p_{i+1}$. However, then $a p_i p_{i+1} d$ is a 1324-pattern, which is a contradiction. ∎

LEMMA 4.44
If p is 1324-avoiding, then there is no entry i in p so that i is of type C and $i + 1$ is of type B.

PROOF Analogous to the proof of Lemma 4.43. If such a pair existed, i would have to be on the right of $i + 1$, since i is blue and $i + 1$ is red. As i is not a right-to-left maximum, there would be a larger entry d on its right. As $i + 1$ is not a left-to-right minimum, there would be a smaller entry a on its left. However, then $a(i + 1)id$ would be a 1324-pattern. ∎

LEMMA 4.45
Let h_n be the number of words of length n that consist of letters A, B, C and D that have no CB-factors. Then we have

$$H(z) = \sum_{n \geq 0} h_n z^n = \frac{1}{1 - 4z + z^2}.$$

This implies

$$h_n = \frac{3 + 2\sqrt{3}}{6} \cdot \left(2 + \sqrt{3}\right)^n + \frac{3 - 2\sqrt{3}}{6} \cdot \left(2 - \sqrt{3}\right)^n. \tag{4.5}$$

PROOF   This is straightforward, using the recurrence relation $h_{n+2} = 4h_{n+1} - h_n$ for $n \geq 0$, with $h_0 = 1$ and $h_1 = 4$. ∎

The following, simple, but crucial, lemma tells us that the ordered pair $(w(p), z(p))$ completely determines the 1324-avoiding permutation p.

LEMMA 4.46
Let $Av_n(1324)$ be the set of all 1324-avoiding n-permutations. Then the map $f : Av_n(1324) \to \mathcal{H}_n \times \mathcal{H}_n$, given by $f(p) = (w(p), z(p))$ is injective.

PROOF   Let $(w, z) \in \mathcal{H}_n \times \mathcal{H}_n$, and let us assume that $f(p) = (w, z)$, that is, that $w(p) = w$, and $z(p) = z$ for some $p \in Av_n(1324)$.

Then w tells us for which indices i the entry p_i will be of type A, namely for the indices i for which the ith letter of w is A. Similary, w tells us the indices j for which the entry p_i is of type B, type C, or type D.

After this, we can use z to figure out which *entries* of p are of type A, type B, type C or type D.

Now let w_{AB} (resp. z_{AB}) be the subword of w (resp. z) that consists of all the letters A and B in w (resp. z). In other words, the pair (w_{AB}, z_{AB}) contains all information about the *red* entries of p. It then follows from Proposition 4.10 that there exists at most one 132-avoiding permutation p' for which $r(p') = (w_{AB}, z_{AB})$.

Define w_{CD} and z_{CD} in an analogous manner. Then Proposition 4.12 shows that there exists at most one 213-avoiding permutation p'' for which $s(p'') = (w_{CD}, z_{CD})$.

It is now immediate from Remark 4.41 that f is injective. Indeed, if $f(p) = (w, z)$, then the red entries of p must form the unique permutation p' for which $r(p') = (w_{AB}, z_{AB})$, and the blue entries of p must form the unique permutation p'' for which $s(p'') = (w_{CD}, z_{CD})$. Finally, as we said in the second and third paragraphs of this proof, the pair (w, z) uniquely determines the set and positions of red entries of p, and the set and positions of blue entries of p. ∎

We are now ready to prove Theorem 4.40.

PROOF   (of Theorem 4.40) The fact that $Av_n(1324) < h_n^2$ is immediate from the injective property of f that we have just proved in Lemma 4.46. In order to complete the proof of the first inequality, note that the image of f

consists of ordered pairs $(w(p), z(p))$ in which both $w(p)$ and $z(p)$ starts with an A, since both p_1 and 1 are always red, and left-to-right minima within the string of red entries (and even in all of p). The rest follows from formula (4.5), since in that formula, the second summand is negative, and in the first summand, the coefficient $(3 + 2\sqrt{3})/6$ is smaller than the base $(2 + \sqrt{3})$. ∎

Theorem 4.40 shows that $L(1324) = 7 + 4\sqrt{3} < 13.93$. Exploring some dependencies between the words $w(p)$ and $z(p)$, this upper bound can be improved to $L(1324) \leq 13.73718$. See [72] for details.

An even sharper upper bound was found by David Bevan, Robert Brignall, Andrew Elvey Price and Jay Pantone [39], who proved that $L(1324) \leq 13.5$ holds. The first step in their proof is the notion of a *domino*.

DEFINITION 4.47 *A domino of size n is a 1324-avoiding permutation p of length n and an integer $k \in [0, n]$ so that the entries of p that are larger than k form a 213-avoiding permutation (the* top part *of the domino), and the entries of p that are at most k (the* bottom part *of the domino) form a 132-avoiding permutation.*

LEMMA 4.48
The number of dominoes of size n is

$$d_n = \frac{2(3n + 3)!}{(n + 2)!(2n + 3)!}.$$

Example 4.49
There are 22 dominoes of size three. Indeed, there are six permutations of length three, and they all avoid 1324. There are four choices for k, and almost all of them can be made with almost all permutations of length three. There are only two exceptions. When $p = 213$, then k cannot be 0, and when $p = 132$, then k cannot be three. ∎

See [39] for a proof of Lemma 4.48.

We will now show how to decompose a 1324-avoiding permutation into a sequence of dominoes. Let $\mathcal{A}v_n(q)$ denote the set of all permutations of length n that avoid the pattern q. Let $p \in \mathcal{A}v_n(1324)$. We will define a subsequence $a_1, a_2, \cdots, a_r$ of p as follows. Let $a_1 = p_1$, the leftmost entry of p.

- If i is even, then let a_i be the largest entry of p that plays the role of 1 in a 213-pattern consisting only of entries weakly on the right of a_{i-1}.

- If $i > 1$ is odd, then let a_i be the leftmost entry of p that plays the role of 2 in a 132-pattern consisting only of entries weakly below a_{i-1}.

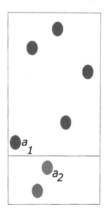

FIGURE 4.3

The domino decomposition of $p = 3612745$.

The sequence of the a_i ends when no entry satisfies the requirements to be the next a_i. The *domino decomposition* of p is obtained by inserting horizontal lines immediately above a_i for i even, and vertical lines immediately to the left of a_i for i odd, if $i > 1$. See Figures 4.3 and 4.4 for illustrations.

Note that it follows from the 1324-avoiding property of p that all entries in the top part of a domino in the domino decomposition of p are larger than all entries in subsequent dominoes in that decomposition. Similarly, all entries in the bottom part of a domino in the domino decomposition of p are larger than all entries in the bottom part of subsequent dominoes of that decomposition.

Also note that the top part of a domino is always 213-avoiding, and the bottom part of a domino is always 132-avoiding.

Example 4.50

Let $p = 3612745$. Then $a_1 = 3$, $a_2 = 2$, and the sequence of the a_i ends here.
□

Example 4.51

Let $p = 76285314$. Then $a_1 = 7$, $a_2 = 6$, $a_3 = 3$, $a_4 = 1$ and the sequence of the a_i ends here.   □

Let $\mathcal{D}_n$ be the set of dominoes of size n, and let $\mathcal{B}_n$ be the set of words of length n over the alphabet $\{u, d\}$.

LEMMA 4.52

There exists an injection $f : \mathcal{A}v_n(1324) \rightarrow \mathcal{D}_n \times \mathcal{B}_n$.

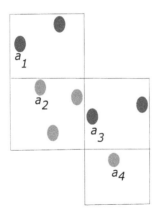

FIGURE 4.4
The domino decomposition of $p = 76285314$.

PROOF   Let $p \in \mathcal{A}v_n(1324)$. Take the domino decomposition of p defined above. At the same time, create a word $w = w_1 w_2 \cdots w_n$ over $\{u, d\}$ as follows. Go through all entries of p, in *decreasing order*, from n to 1. Consider the domino decomposition of p. If, in that decomposition, $n + 1 - i$ goes to the top part of a domino, then let $w_i = u$; otherwise, let $w_i = d$. At the end, concatenate the top part of all dominoes in the domino decomposition of p into a 213-avoiding permutation p' on the largest entries of $[n]$ preserving their pattern, and the bottom parts into a 132-avoiding permutation on the smallest entries of $[n]$ preserving their pattern. Finally, create the domino D whose top part is p', whose bottom part is p'', and the horizontal location of the entries that belong to p' is the same as it was in p.

The obtained domino D and the obtained word over $\{u, d\}$ form $f(p)$.

This map f is injective, since p can be recovered from its image (D, w). The horizontal position of each entry of p can be recovered from the position of the corresponding entry in D, and the vertical position of each entry, starting with n and moving in decreasing order, can be recovered from w. ∎

Example 4.53
Let $p = 76285314$ as in Example 4.51, illustrated in Figure 4.4. The entries 8, 7, 4, and 3 are in top parts of dominoes, so it is the first, second, fifth and sixth letters of the word w that will be letters u, implying that $w = uudduudd$.

The top entries determine the partial permutation 7834, and the bottom entries determine the partial permutation 6251. That leads to a domino D whose top part is $p' = 7856$, and whose bottom part is $p'' = 4231$. The entries that belong to the top part are in the first, fourth, sixth, and eighth horizontal position of D, just as they were in those positions of p. See Figure 4.5 for an illustration.

□

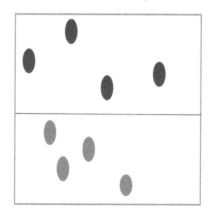

FIGURE 4.5
The domino D, where $p = 76285314$, and $f(p) = (D, uudduudd)$.

Example 4.54
We will show how to recover p from its image (D, w), where D is the domino
shown in Figure 4.5, and $w = uudduudd$. The word w shows that entries
8, 7, 4, and 3 of p are in the top part of their dominoes in the domino
decomposition of p, while entries 6, 5, 1, and 2 are in the bottom part. In
order to recover p, the vertical positions of the four entries in the top part
of D have to be adjusted so that their pattern does not change, but their
set becomes $\{3, 4, 7, 8\}$, that is, their sequence becomes the subsequence 7834.
The horizontal positions of these four entries do not change, they are still in
the first, fourth, sixth, and eighth positions. Similarly, the vertical positions
of the remaining four entries must be rearranged so that their pattern does
not change, but their set becomes $\{1, 2, 5, 6\}$, that is, their sequence becomes
the subsequence 6251, and they are still in the second, third, fifth and seventh
horizontal positions. This leads to the permutation $p = 76285314$. $\quad\square$

THEOREM 4.55
For all positive integers n, the inequality

$$\mathrm{Av}_n(1324) \leq 13.5^n$$

holds.

PROOF   A routine application of Stirling's formula shows that Lemma
4.48 implies that $d_n \leq 6.75^n$, and then our claim follows from Lemma 4.52,
since $|\mathcal{B}_n| = 2^n$. ∎

4.4.2 Pattern 1342

In this subsection, we turn our attention to the pattern 1342. Interestingly, we will be able to provide an *exact formula* for $\text{Av}_n(1342)$. This is exceptional; the only other pattern longer than three for which an exact formula is known is 1234. The formula is given by the following theorem.

THEOREM 4.56

For all positive integers n, the equality

$$\text{Av}_n(1342) = (-1)^{n-1} \cdot \frac{(7n^2 - 3n - 2)}{2}$$
$$+ 3 \sum_{i=2}^{n} (-1)^{n-i} \cdot 2^{i+1} \cdot \frac{(2i-4)!}{i!(i-2)!} \cdot \binom{n-i+2}{2}$$

holds.

This is a very surprising result. It is straightforward to prove from this formula that $\text{Av}_n(1342) < 8^n$ for all n, and that $\lim_{n \to \infty} \sqrt[n]{\text{Av}_n(1342)} = 8$.

The result itself is not the only interesting aspect of the facts surrounding the pattern 1342. We will see that permutations avoiding 1342 are in bijection with two different kinds of objects that at first look totally unrelated. The first, and for our purposes, more important, type of objects is a specific kind of labeled trees.

DEFINITION 4.57 *[127] A rooted plane tree with non-negative integer labels $\ell(v)$ on each of its vertices v is called a $\beta(0,1)$-tree if it satisfies the following conditions:*

- *if v is a leaf, then $\ell(v) = 0$ (this explains the number 0 in the name of $\beta(0,1)$-trees),*

- *if v is the root and $v_1, v_2, \cdots, v_k$ are its children, then $\ell(v) = \sum_{i=1}^{k} \ell(v_k)$,*

- *if v is an internal node and $v_1, v_2, \cdots, v_k$ are its children, then the inequality $\ell(v) \leq 1 + \sum_{i=1}^{k} \ell(v_k)$ holds (this explains the number 1 in the name of $\beta(0,1)$-trees).*

Example 4.58

Figure 4.6 shows a $\beta(0,1)$-tree on 12 vertices. ⬚

Let us call a permutation $p = p_1 p_2 \cdots p_n$ *skew indecomposable* if there exists no $k \in [n-1]$ so that for all $i \leq k < j$, we have $p_i > p_j$. In other words, p is skew indecomposable if it cannot be cut into two parts so that everything

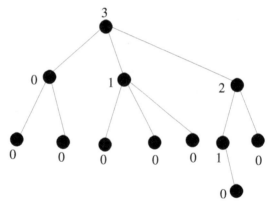

FIGURE 4.6
A $\beta(0,1)$-tree.

before the cut is larger than everything after the cut. For instance, 3142 is skew indecomposable, but 43512 is not as we could choose $k = 3$, that is, we could cut between the third and fourth entries. If a permutation is not skew indecomposable, then we will call it *skew decomposable*.

The importance of $\beta(0,1)$-trees for us is explained by the following theorem.

THEOREM 4.59

For all positive integers n, there is a bijection F from the set of skew in-decomposable 1342-avoiding n-permutations to the set $D_n^{\beta(0,1)}$ of $\beta(0,1)$-trees on n vertices.

Example 4.60

Let $n = 3$. Then there are three skew indecomposable n-permutations, 123, 132, and 213, and they all avoid 1342. Correspondingly, there are indeed three $\beta(0,1)$-trees on three vertices, as can be seen in Figure 4.7. □

Let $t_n = \left| D_n^{\beta(0,1)} \right|$. If we can prove Theorem 4.59, then we have made a crucial step forward as it is known [127] that

$$t_n = 3 \cdot 2^{n-2} \cdot \frac{(2n-2)!}{(n+1)!(n-1)!}. \tag{4.6}$$

We start by treating two special types of $\beta(0,1)$-trees on n vertices. There are two things that contribute to the structure of a $\beta(0,1)$-tree, namely its (un-labeled) tree structure, and its labels. We will therefore first look at $\beta(0,1)$-trees in which one of these two ingredients is trivial, that is, $\beta(0,1)$-trees that consist of a single path only, and $\beta(0,1)$-trees in which all labels are zero.

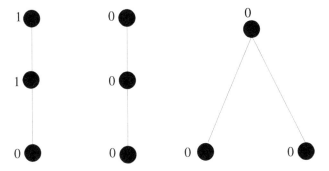

FIGURE 4.7
The three $\beta(0,1)$-trees on three vertices.

LEMMA 4.61
There is a bijection f from the set of 1342-avoiding n-permutations starting with the entry 1 and the set of $\beta(0,1)$-trees on n vertices consisting of one single path.

Note that a simpler description of the domain of f is that it is the set of 231-avoiding permutations of the set $\{2, 3, 4 \cdots, n\}$.

PROOF   Let $p = p_1 p_2 \cdots p_n$ be an 1342-avoiding n-permutation so that $p_1 = 1$. Take an unlabeled tree on n nodes consisting of a single path and give the label $\ell(i)$ to its ith node ($1 \leq i \leq n-1$) by the following rule:

$$\ell(i) = \begin{cases} |\ \{j \leq i \text{ so that } p_j > p_s \text{ for at least one } s > i,\}\ | & \text{if } i < n \\ \ell(n-1) \text{ if } i = n. \end{cases}$$

That is, $\ell(i)$ is the number of entries weakly on the left of p_i which are larger than at least one entry on the right of p_i. Note that this way we could define f on the set of *all* n-permutations starting with the entry 1, but in that case, f would not be a bijection. (For example, the images of 1342 and 1432 would be identical.)

Example 4.62
If $p = 143265$, then the labels of the nodes of $f(p)$ are, from the leaf to the root, 0,1,2,0,1,1. See Figure 4.8. For easy reference, we wrote p_i to the ith node of the path $f(p)$. To avoid confusion, in this figure, and for the rest of this subsection, the entries of p will be written in small, Roman letters, and the labels of the nodes will be written in large italic letters.   ⬚

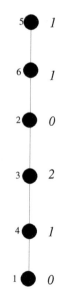

FIGURE 4.8
The $\beta(0,1)$-tree of $p = 143265$.

It is easy to see that f indeed maps into the set of $\beta(0,1)$-trees: $\ell(i+1) \leq \ell(i) + 1$ for all i because there can be at most one entry counted by $\ell(i+1)$ and not counted by $\ell(i)$, namely the entry p_{i+1}. All labels are certainly non-negative and $\ell(1) = 0$.

To prove that f is a bijection, it suffices to show that it has an inverse, that is, for any $\beta(0,1)$-tree T consisting of a single path, we can find the only permutation p so that $f(p) = T$. We claim that given T, we can recover the entry n of the preimage p. First note that p is 1342-avoiding and starts with 1, so any entry on the left of n must be smaller than any entry on the right of n. In particular, the node preceding n must have label 0. Moreover, as n is larger than any entry following it in p, the entry n is the leftmost entry p_i of p so that $\ell(j) > 0$ for all $j \geq i$ if there is such an entry at all, and $n = p_n$ if there is none. That is, n corresponds to the node that starts the uninterrupted sequence of strictly positive labels that ends in the last node as long as there is such a sequence. Otherwise, n corresponds to the last node.

Once we locate where n is in p, we can simply delete the node corresponding to it from T and decrement all labels after it by 1. (If this means deleting the last node, we just change $\ell(n-1)$ so it is equal to $\ell(n-2)$ to satisfy the root-condition.) We can indeed do this because the node preceding n had label 0 and the node after n had a positive label (1 or 2), by our algorithm to locate n. Then we can proceed recursively, by finding the position of the entries $n-1, n-2, \cdots, 1$ in p. This clearly defines the inverse of f, so we have proved that f is a bijection. ∎

As we promised, we continue by explaining which skew indecomposable 1342-avoiding permutations correspond to $\beta(0,1)$-trees in which all labels are equal to zero.

LEMMA 4.63
There is a bijection g from the set of 132-avoiding n-permutations ending with n to the set of $\beta(0,1)$-trees on n vertices with all labels equal to zero.

Note that we could describe the domain of g as the set of *skew indecomposable* 132-avoiding n-permutations, or as the set of $(n-1)$-permutations that avoid 132.

PROOF   In this proof, we can obviously think of our $\beta(0,1)$-trees as unlabeled rooted plane trees. A *branch* of a rooted tree is a subtree whose root is one of the root's children. Some rooted trees may have only one branch, which does not necessarily mean they consist of a single path.

We will construct g inductively. There is only one unlabeled $\beta(0,1)$-tree on 2 vertices and it is the image of the only 1-permutation; $p = 1$. Using induction, suppose we have already constructed g for all positive integers $k < n$. Let p be a 132-avoiding permutation of length n. Let $p' = p_1 p_2 \cdots p_{n-1}$. Then there are two possibilities.

(a) The first case is when p' is skew decomposable, that is, we can cut p' into several (at least two) strings $p_{\langle 1 \rangle}, p_{\langle 2 \rangle}, \cdots, p_{\langle h \rangle}$ so that all entries of $p_{\langle i \rangle}$ are larger than all entries of $p_{\langle j \rangle}$ if $i\langle j$. In this case, $g(p)$ will have h branches, the branch b_i satisfying $g(p_{\langle i \rangle}) = b_i$. We then obtain $g(p)$ by connecting all branches b_i to a common root. Given that we are in a $\beta(0,1)$-tree, the label of the root is determined by the labels of its children.

(b) The second case is when p' is skew indecomposable. As p avoids 132, this is equivalent to saying that p' ends with its maximal entry $n - 1$. In this case, $g(p)$ will have just one branch b_1, that is, the root of $g(p)$ will have only one child. We define $b_1 = g(p')$.

Again, we prove that g is a bijection by showing that it has an inverse. Let T be an unlabeled plane tree on n vertices with root q. Let q have t children, and say that, going left-to-right, they are roots of the branches $b_1, b_2, \cdots, b_t$, which have $n_1, n_2, \cdots, n_t$ nodes. Then by induction, for each i, the branch b_i corresponds to a 132-avoiding n_i-permutation ending with n_i. Now add $\sum_{j=i+1}^{t} n_j$ to all entries of the permutation p_i associated with b_i, then concatenate all these strings and add n to the end to get the permutation p associated with T.

It is straightforward to check that this procedure always returns the original permutation, proving our claim.   ∎

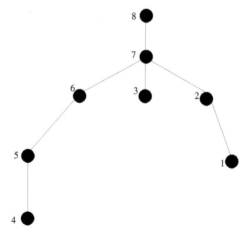

FIGURE 4.9
The $\beta(0,1)$-tree of $p = 45631278$.

Example 4.64
   The permutation 45631278 corresponds to the $\beta(0,1)$-tree with all labels equal to 0 shown in Figure 4.9. For easy reference, we write p_n to the root of $g(p)$, and proceed analogously for the other entries in the recursively defined subtrees.
   ☐

   An easy way to read off the corresponding permutation once we have its entries written to the corresponding nodes is the well-known *postorder* reading: for every node, first write down the entries associated with its children from left to right, then the entry associated with the node itself, and do this recursively for all the children of the node.
   The reader is invited to prove that if p is as in Lemma 4.63 and the entries of p are associated to the vertices of $g(p)$ as in Example 4.64, vertex i will be a descendant of vertex j if and only if i is on the left of j in p and $i < j$.
   Our plan is to bring Lemmas 4.61 and 4.63 together to prove Theorem 4.59. This needs some preparation. Optimally, we would take a 1342-avoiding skew indecomposable n-permutation p, associate its entries to the nodes of an unlabeled plane tree T, then define the labels of this tree so that it becomes a $\beta(0,1)$-tree. The question is, however, how do we know what T we should use, and if T is given, how do we know in what order we should write the entries of p to the nodes of T? In what follows, we develop the notions to decide these questions.

DEFINITION 4.65   *Two n-permutations x and y are said to have the same weak type if the left-to-right minima of x are the same as those of y,* *and they are in the same positions.*

Example 4.66
Permutations 456312 and 465312 have the same weak type since their left-to-right minima are 4, 3, and 1, and they are located at the same positions. Permutations 31524 and 34152 do not have the same weak type. □

PROPOSITION 4.67
Each nonempty weak type has exactly one 132-avoiding permutation.

PROOF   Take all entries that are not left-to-right minima and fill all empty positions between the left-to-right minima with them as follows: In each step place the smallest entry that has not been placed yet that is larger than the previous left-to-right minimum. (This is just what we did in the proof of the Simion-Schmidt bijection in Lemma 4.4.)

On the other hand, the resulting permutation will be the only 132-avoiding permutation of this weak type because any time we deviate from this procedure, that is, we place something else, not the smallest such entry, we create a 132-pattern. ∎

DEFINITION 4.68   *The normalization $N(p)$ of an n-permutation p is the only 132-avoiding permutation having the same week type as p.*

Example 4.69
If $p = 356214$, then $N(p) = 345216$.   □

DEFINITION 4.70   *The normalization $N(T)$ of a $\beta(0,1)$-tree T is the $\beta(0,1)$-tree which is isomorphic to T as a plane tree, with all labels equal to zero.*

It turns out that normalization preserves the skew indecomposable property.

PROPOSITION 4.71
A permutation p is skew indecomposable if and only if $N(p)$ is skew indecomposable.

PROOF   (The author is grateful to Aaron Robertson, who found and corrected a mistake in his original argument.) We will show that whether p is skew decomposable or not depends only on the set and position of its left-to-right minima, which is obviously equivalent to the claim to be proved. Let C

be set of permutations having the same weak type as p, given by the set and position of its left-to-right minima. It is clear that if $p \in C$ is decomposable, then the only way to cut it into two parts (so that everything before the cut is larger than everything after the cut) is to cut it immediately on the left of a left-to-right minimum $a < n$. Now if there is a left-to-right minimum in position $n - a + 2$, then the entries $1, 2, \cdots, a - 1$ must occupy positions $n - a + 2, n - a + 1, \cdots, n$. Therefore, we can cut immediately on the left of position $n - a + 2$, and p is decomposable.

If there is no such a, then for all left-to-right minima m, all the entries $1, 2, \cdots, m - 1$ must be to the right of m. However, in one of the positions $n - m + 2, n - m + 1, \cdots, n$, there exists an element $y > m$, implying that our permutation p is not decomposable. ∎

COROLLARY 4.72

If p is a skew indecomposable n-permutation, then $N(p)$ always ends in the entry n.

PROOF Note that the only way for a 132-avoiding n-permutation to be skew indecomposable is for it to end with n. If p is a 132-avoiding n-permutation and n is not the last entry, then we may cut p immediately after the entry n. Then the statement follows from Proposition 4.71. ∎

Now we are in a position to prove Theorem 4.59.

PROOF (of Theorem 4.59.) Let p be a skew indecomposable 1342-avoiding n-permutation. Take $N(p) = r$. By Corollary 4.72 its last element is n. Define $F(r)$ to be the unlabeled plane tree S associated with r by the bijection g of Lemma 4.63. So g is just the restriction of F to the set of skew indecomposable 132-avoiding permutations.

This unlabeled tree S is the tree we are going to work with. First, we will write the entries of p to the nodes of S. (The reader should recall that we did this in the proof of Lemma 4.61, and that the *entries* of p written to the nodes of S are not to be confused with the *labels* of the nodes.) We will do this in the order specified by p and $N(p)$. That is, $N(p)$ is a 132-avoiding permutation, so its entries are in natural bijection with the nodes of S as we saw in Lemma 4.63. We then let the permutation $p(N(p))^{-1}$ act on the entries of $N(p)$ (written to the nodes of S) to get the order in which we write the entries of p to S. Note in particular that the left-to-right minima are kept fixed.

Example 4.73

Let $p = 4621357$. Then $N(p) = 4521367$, and the unlabeled plane tree S

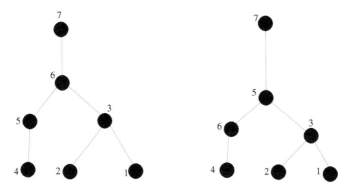

FIGURE 4.10
The unlabeled trees of $N(p) = 4521367$ and $p = 4621357$.

associated with these permutations is shown in Figure 4.10, together with the order in which the entries of p are written to the nodes. Note that p and $N(p)$ only differ in the transposition (56). This is why it is these two entries whose positions have been swapped. □

Now we are going to define the label $\ell(v)$ of each node v for the new $\beta(0, 1)$-tree $T = F(p)$ that we are constructing from S. As an unlabeled tree, T will be isomorphic to S, but its labels will be different. Let v_i be the ith node of T in the postorder reading, the node to which we associate p_i. We say that the entry p_i *beats* the entry p_j if there is an entry p_h so that p_h, p_i, p_j are written in this order (so $h < i < j$) and they form a 132-pattern. Moreover, we say that p_i *reaches* p_k if there is a subsequence $p_i, p_{i+a_1}, \cdots p_{i+a_t}, p_k$ of entries so that $i < i + a_1 < i + a_2 < \cdots < i + a_t < k$ and that each entry in this subsequence beats the next one. In particular, if x beats y, then x also reaches y.

Example 4.74
Let $p = 3716254$. Then 7 beats 6, 6 beats 2; therefore, 7 reaches 2. □

Finally, we define the label $\ell(v_i)$ of the non-root vertex v_i of $T = F(p)$ to which the entry p_i is associated as follows. The label $\ell(v_i)$ is the number of descendants v_j of v_i (including v_i itself) for which there exists at least one index $k > i$ so that the entry p_j reaches the entry p_k.

Let $F(p)$ be the $\beta(0, 1)$-tree defined by these labels. Note that a descendant of i is an element of the subtree rooted at i. Note that this rule is an extension of the labeling rule we have in Lemma 4.61.

First, it is easy to see that F indeed maps into the set $D_n^{\beta(0,1)}$. Indeed, let v be an internal node and let $v_1, v_2, \cdots, v_k$ be its children. Then $\ell(v) \leq 1 + \sum_{i=1}^{k} \ell(v_k)$ because there can be at most one entry counted by $\ell(v)$ and not

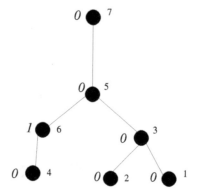

FIGURE 4.11

The image $F(p)$ of $p = 4621357$.

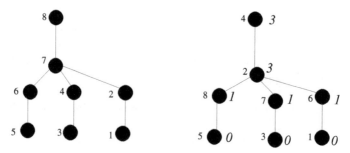

FIGURE 4.12

The tree S and $F(p)$ for $p = 58371624$.

counted by any of its children's labels, namely v itself. Second, all labels are certainly non-negative and all leaves, that is, the left-to-right minima, have label 0.

Example 4.75

In Example 4.73 we have created the unlabeled tree S for $p = 4621357$. Application of the above rule shows that $F(p)$ is the $\beta(0, 1)$-tree shown in Figure 4.11. Indeed, the only 132-pattern of p is 465, and that is counted only once, at the entry 6. ▯

Example 4.76

Let $p = 58371624$, then we have $N(p) = 56341278$, giving rise to the unlabeled tree shown on the left of Figure 4.12. We then compute the labels of $F(p)$ to obtain the tree shown on the right of Figure 4.12. ▯

To prove that F is a bijection, it suffices to show that it has an inverse. That is, it suffices to show that for any $\beta(0,1)$-tree $T \in D_n^{\beta(0,1)}$, we can find a unique permutation p so that $F(p) = T$.

Before we explain why F has an inverse, we prove two propositions that show a simple way to find the position of the entry n in p if $F(p)$ is given.

PROPOSITION 4.77

Suppose $p_n \neq n$, that is, n is not associated with the root vertex. Then each ancestor of n, including n itself, has a positive label. If $p_n = n$, then $\ell(v_n) = 0$ and thus there is no vertex with the above property.

PROOF   If $p_n = n$, then there is nothing on the right of n to reach, thus $\ell(v_n)$ enumerates an empty set, yielding $\ell(v_n) = 0$. Suppose now that p_n is not the root vertex. We will show that the entry n always reaches p_n, which implies the statement of the proposition.

Note that $n \neq p_1$, because then p would be skew decomposable. Let a_1 be the smallest entry on the left of n. Let n_1 be the rightmost entry that n beats; in other words, the rightmost entry that is still larger than a_1.

If $n_1 = p_n$, we are done; otherwise, let a_2 be the smallest entry on the left of n_1, and let n_2 be the rightmost entry that n_1 beats; in other words, the rightmost entry that is still larger than a_2. If $n_2 = p_n$, we are done; otherwise, continue this same process. We claim that this process eventually reaches the rightmost entry p_n.

Indeed, how could the process get stuck? In the first step, if n_1 cannot be defined, that means that all entries on the right of n are smaller than all entries on the left of n, which means that p is skew decomposable, contradicting our assumptions. Then, in the second step, a_2 is always well-defined, and if n_2 does not exist, that means that all entries on the right of n_1 are smaller than a_2, (they cannot be larger than a_1 because of the definition of n_1), while all entries on the left of n_1 are, by the definition of a_2, at least as large as a_2. This again implies that p is skew decomposable, which is a contradiction.

Continuing in this way, we see that the process will end only when the rightmost entry p_n is reached and $p_n = n_k$ for some k. As n beats n_1, then n_1 beats n_2, and so on, n reaches all the n_i, including $p_n = n_k$, and our proof is complete.   ∎

The only problem is that there could be many vertices with the property that all their ancestors have a positive label. If that happens, we resort to the following proposition to locate the vertex associated with n.

PROPOSITION 4.78

Suppose $p_n \neq n$. Then n is the leftmost entry of p that has the property that each of its ancestors (including itself) has a positive label.

PROOF   Suppose p_k and n both have this property and that p_k is on the left of n. (If there are several candidates for the role of p_k, choose the rightmost one). If p_k beats an element y on the right of n by participating in the 132-pattern $x p_k y$, then $x p_k n y$ is a 1342-pattern, which is a contradiction. So p_k does not beat such an element y. In other words, all elements after n are smaller than all elements before p_k. Still, p_k must reach elements on the right of n, thus it beats some element v between p_k and n. This element v in turn beats some element w on the right of n by participating in some 132-pattern tvw. However, this would imply that $tvnw$ is a 1342-pattern, a contradiction, which proves our claim. ∎

Now we are ready to prove that F is a bijection. The author is grateful to Julika Mehrgardt for valuable discussions on this proof. We prove our statement by induction on n, the case of $n = 1$ being obvious. Let us assume that we have already proved the statement for all positive integers less than n. With a slight abuse of notation, we denote the bijections belonging to these smaller integers also by F. Let T be a $\beta(0, 1)$-tree on n vertices. Note that T, even without its labels, specifies the weak class of any purported permutation p for which $F(p) = T$. Let p_1 be the leftmost entry (and so, the leftmost left-to-right minimum) of p, determined from T. Let v_1 be the leftmost leaf of T. We consider the following four cases.

1. Let us assume that v_1 is not the only child of its parent x. In this case, we can simply remove v_1 from T. The remaining tree T' is a valid $\beta(0, 1)$-tree that is smaller than T, so by our induction hypothesis, it has a unique preimage $F^{-1}(T') = p'$. Now prepend p' by the entry p_1 (and add 1 to all entries in p' that are larger than or equal to p_1) to get the unique preimage $F^{-1}(T) = p$. Note that the addition of p_1 to the front of p' does not create a 1342-pattern since p_1 is larger than the first entry of p' in this case.

2. Let us assume that v_1 is the only child of its parent x, but $\ell(x) = 0$. In this case, we can still remove v_1 from T and proceed as in the previous case. Note that in this case, the entry associated to x will be $p_1 + 1$.

3. Now let us assume that v_1 is the only child of x, and while $\ell(x) \neq 0$, there exists an ancestor of x that has label 0. Let v be the closest such ancestor of x. That means that v is not the root of T, since the only way for the root to have label 0 is by all of its children to have label 0.

    Now temporarily remove the subtree T_v rooted at v from T except for its root. In other words, shrink T_v to its root v. Let T' be the remaining tree. Note that since $\ell(v) = 0$, the tree T' is a valid $\beta(0, 1)$-tree . As T' is smaller than T, it has a unique preimage p' under F by our induction hypothesis. We claim that there is exactly one possibility for the set of entries of p that are associated to the vertices of T_v, and there is exactly one possibility for the partial permutation p_v of these entries.

For the first claim, note that the set S of entries associated to the vertices of T_v is equal to the set Z of entries associated to the vertices of T_v by $N(p)$. Recall from the proof of Proposition 4.67 that the normalization procedure keeps the left-to-right minima of a permutation fixed, and then puts the smallest eligible entry into each available slot, proceeding from left to right. (This is the only way to get a 132-avoiding permutation in the given weak class.) If $S \neq Z$ held, that would mean that this procedure was not followed, and hence, a 132-pattern starting in T_v but ending outside T_v was formed, contradicting the assumption $\ell(v) = 0$.

For the second claim, note that T_v becomes a $\beta(0,1)$-tree if we change the label of its root v from 0 to the sum of the label of its children (which does not change anything since neither the label 0 nor the new label carries any new information for T_v). As T_v is smaller than T, it has a unique preimage p_v under F by our induction hypothesis.

The partial permutations p_v, p', and the set S on which p_v is taken together determine the unique permutation $F^{-1}(p)$.

4. Finally, let us assume that v_1 is the only child of x, and no ancestor of x has zero for its label, that is, all vertices on the path from x to the root have positive labels. By Proposition 4.78, that means that x is the vertex to which the entry n is associated. In other words, n is the second entry of our purported permutation p. Now remove x from T (that is, connect v_1 directly to the parent of x, and decrease the labels of each ancestor of x by 1). This can be done since all those labels were positive. Let T' be the obtained tree. Construct the unique permutation $p' = f^{-1}(T)$ whose existence is guaranteed by our induction hypothesis. Insert n to its second position to get $F^{-1} = p$. This certainly does not create a 1342-pattern, since n could only play the role of 4 in such a pattern, but there is only one entry preceding n in p.

This proves that F is a bijection, completing the proof of Theorem 4.59. ∎

COROLLARY 4.79
The number of skew indecomposable 1342-avoiding n-permutations is

$$|D_n^{\beta(0,1)}| = t_n = 3 \cdot 2^{n-2} \cdot \frac{(2n-2)!}{(n+1)!(n-1)!}. \tag{4.7}$$

PROOF   Follows from (4.6) and Theorem 4.59. ∎

Computing the numbers $\text{Av}_n(1342)$ is now a straightforward task using generating functions.

LEMMA 4.80
Let $H(z) = \sum_{n=0}^{\infty} \text{Av}_n(1342)z^n$. Furthermore, let $F(z) = \sum_{n=1}^{\infty} t_n z^n$, that is, let $F(z)$ be the generating function of the numbers of skew indecomposable 1342-avoiding permutations. Then

$$H(z) = \sum_{i \geq 0} F^i(z) = \frac{1}{1 - F(z)} = \frac{32z}{-8z^2 + 20z + 1 - (1 - 8z)^{3/2}}. \qquad (4.8)$$

PROOF   Tutte [313] has computed the ordinary generating function of the numbers t_n and obtained

$$F(z) = \sum_{n=1}^{\infty} t_n z^n = \sum_{n=1}^{\infty} 3 \cdot 2^{n-1} \cdot \frac{(2n-2)!}{(n+1)!(n-1)!} z^n \qquad (4.9)$$

$$= \frac{z^2 + 12z - 1 + (1 - 8z)^{3/2}}{32z}. \qquad (4.10)$$

The coefficients of $F(z)$ are the numbers of skew indecomposable 1342-avoiding n-permutations. Any 1342-avoiding permutation has a unique decomposition into skew indecomposable permutations. This can consist of $1, 2, \cdots, n$ blocks, implying that $\text{Av}_n(1342) = \sum_{i=1}^{n} t_i s_{n-i}$. Therefore, $H(z) = 1/(1 - F(z))$ as claimed.   ∎

It is time that we mentioned the other kind of objects that are in bijection with these permutations. These are *rooted bicubic maps*, that is, planar maps in which each vertex has degree three, there is a distinguished half-edge (the root), and the underlying graph is bipartite. Tutte was enumerating these maps (according to the number $2(n+1)$ of vertices) when he obtained formula (4.9), and Cori, Jacquard, and Schaeffer [127] then used the $\beta(0,1)$-trees to find a more combinatorial proof of Tutte's result.

In Chapter 8, we will mention another surprising occurrence of the numbers $\text{Av}_n(1342)$.

Now that we have the generating function of the numbers $\text{Av}_n(1342)$, we are in a position to obtain an explicit formula for their number. That formula will prove Theorem 4.56.

PROOF   (of Theorem 4.56). Multiply both the numerator and the denominator of $H(z)$ by $(-8z^2 + 20z + 1) + (1 - 8z)^{3/2}$. After simplifying, we get

$$H(z) = \frac{(1 - 8z)^{3/2} - 8z^2 + 20z + 1}{2(z+1)^3}. \qquad (4.11)$$

As $(1 - 8z)^{3/2} = 1 - 12z + \sum_{n \geq 2} 3 \cdot 2^{n+2} z^n \frac{(2n-4)!}{n!(n-2)!}$, formula (4.11) implies our claim.   ∎

So the first few values of $\mathrm{Av}_n(1342)$ are 1, 2, 6, 23, 103, 512, 2740, 15485, 91245, 555662. We mention that Jonathan Bloom and Sergi Elizalde [41] found a different proof for Theorem 4.56 in 2013.

In particular, one sees easily that the formula for $\mathrm{Av}_n(1342)$ given by Theorem (4.56) is dominated by the last summand; in fact, the alternation in sign assures that this last summand is larger than the whole right hand side if $n \geq 8$. As $\frac{(2n-4)!}{n!(n-2)!} < \frac{8^{n-2}}{n^{2.5}}$ by Stirling's formula, we have proved the following exponential upper bound for $\mathrm{Av}_n(1342)$.

COROLLARY 4.81
For all n, we have $\mathrm{Av}_n(1342) < 8^n$.

On the other hand, it is routine to verify that the numbers t_n satisfy the recurrence $t_n = (8n - 12)t_{n-1}/(n + 1)$. As we explained immediately after Conjecture 4.15, the fact that $\mathrm{Av}_n(1342) < 8^n$ implies that the limit $L(1342) = \lim_{n \to \infty} \sqrt[n]{\mathrm{Av}_n(1342)}$ exists. Therefore, by the Squeeze Principle, we obtain the following Corollary.

COROLLARY 4.82
The equality

$$L(1342) = 8$$

holds.

This result was striking for two different reasons. On one hand, this is the third time that the exact value of $L(q) = \lim_{n \to \infty} \sqrt[n]{\mathrm{Av}_n(q)}$ for some pattern q was computed. Indeed, we have seen that

$$\lim_{n \to \infty} L(q) = \begin{cases} 4 \text{ if } q \text{ is of length 3}, \\ (k-1)^2 \text{ if } q = 123 \cdots k, \\ 8 \text{ if } q = 1342, \end{cases}$$

In other words, in every case when we saw an exact answer, the exact answer was an integer. In general, however, that does not hold. It is proved in [57] that $L(12453) = 9 + 4\sqrt{2}$. So these limits are not even always *rational*. Articles [57] and [60] contain other examples for patterns q for which $L(q)$ is irrational. However, in all cases when $L(q)$ is known, it is a real number of the form $a + b\sqrt{2}$, where a and b are integers.

The other surprise provided by Corollary 4.82 is that

$$L(1342) = 8 \neq L(1234) = 9.$$

That is, even in the logarithmic sense, the sequences $\mathrm{Av}_n(1342)$ and $\mathrm{Av}_n(1234)$ are different. This phenomenon is not well understood. If we could understand why the pattern 1342 is *really* so difficult to avoid, then maybe we could use that information to find other, longer patterns that are difficult to avoid.

As far as the fact that $L(1324) > L(1234)$ goes, that is not an accident in the following sense. A pattern q is called *layered* if q consists of a union of decreasing subsequences (the layers) so that the entries increase from one subsequence to the next. For instance, $q = 3215476$ is a layered pattern. It is proved in [57] that if q is a layered pattern of length k, then

$$L(q) \geq L(12 \cdots k) = (k - 1)^2.$$

We will formally define layered patterns in the next section.

4.4.3   Pattern 1234

The pattern 1234 is a monotone pattern; therefore Theorem 4.22, which provides an asymptotic formula and a very good upper bound for the numbers $\text{Av}_n(123 \cdots k)$, applies to it. However, we would like to point out that using certain techniques beyond the scope of this book, Ira Gessel [191] proved the following *exact formula* for these numbers

$$\text{Av}_n(1234) = 2 \cdot \sum_{k=0}^{n} \binom{2k}{k} \binom{n}{k}^2 \frac{3k^2 + 2k + 1 - n - 2nk}{(k+1)^2(k+2)(n-k+1)}. \qquad (4.12)$$

The alert reader has probably noticed that the summands on the right-hand side are not always non-negative, which decreases the hopes for a combinatorial proof. However, a few years later Gessel found the following alternative form for his formula [190]

$$\text{Av}_n(1234) = \frac{1}{(n+1)^2(n+2)} \sum_{k=0}^{n} \binom{2k}{k} \binom{n+1}{k+1} \binom{n+2}{k+1}. \qquad (4.13)$$

In this new form, all terms are non-negative, but there is still a division, suggesting that a direct combinatorial proof is probably difficult to find.

We will return to the surprising complexity of Gessel's formulae in the next chapter.

4.5   Proof of the Stanley-Wilf Conjecture

The goal of this section is to present a proof of the Stanley–Wilf conjecture. In order to achieve this goal, we will discuss one more conjecture involving 0-1 matrices, show why it implies the Stanley–Wilf conjecture, and then prove the conjecture on the 0-1 matrices. The proof we present was announced by Adam Marcus and Gábor Tardos in the Fall of 2003. Before that, Noga Alon and Ehud Friedgut [12] proved a somewhat weaker result.

4.5.1 Füredi–Hajnal Conjecture

Let us extend the notion of pattern avoidance to *0-1 matrices* as follows.

DEFINITION 4.83 *Let A and P be matrices whose entries are either equal to 0 or to 1, and let P be of size $k \times \ell$. We say that A contains P if A has a $k \times \ell$ submatrix Q so that if $P_{i,j} = 1$, then $Q_{i,j} = 1$, for all i and j. If A does not contain P, then we say that A avoids P.*

In other words, A contains P if we can delete some rows and some columns from A and obtain a matrix Q that has the same shape as P, and has a 1 in each position P does. Note that Q can have more 1 entries than P, but not less. Note that if all entries of A are equal to 1, then A contains all matrices P that have shorter side lengths than A.

Example 4.84

Let $A = \begin{pmatrix} 1 0 0 1 \\ 0 1 1 0 \\ 0 0 1 1 \\ 1 0 0 0 \end{pmatrix}$, and let $P = \begin{pmatrix} 0 0 \\ 1 1 \end{pmatrix}$. Then A contains P as can be seen by deleting everything from A except the intersection of the first and third rows with the third and fourth columns. If $Q = \begin{pmatrix} 1 1 \\ 1 1 \end{pmatrix}$, then A avoids Q.

☐

All matrices in this section will be 0-1 matrices, so we will not repeat that condition any more. The famous Füredi–Hajnal conjecture, which was originally stated as a question, sounds similar to the Stanley–Wilf conjecture.

CONJECTURE 4.85 *[186] [Füredi–Hajnal conjecture] Let P be any permutation matrix, and let $f(n, P)$ be the maximum number of 1 entries that a P-avoiding $n \times n$ matrix A can have. Then there exists a constant c_P so that*

$$f(n, P) \leq c_P n. \tag{4.14}$$

We point out that the condition that P is a *permutation matrix* is important. See Exercise 43 for a non-permutation matrix for which (4.14) does not hold.

4.5.2 Avoiding Matrices versus Avoiding Permutations

In this ssection, we present an argument proving that the Füredi–Hajnal conjecture implies the Stanley–Wilf conjecture. The original form of the argument is due to Martin Klazár. The simplified form of the argument that we present is due to Doron Zeilberger and Vincent Vatter.

Let q be a pattern of length k, and let p be an n-permutation that avoids q. Then the permutation matrix of p avoids the permutation matrix B_q of q. Now let $M_n(q)$ be the number of $n \times n$ matrices with 0 and 1 entries that avoid B_q. Then $\text{Av}_n(q) \leq M_n(q)$. We will now find an upper bound for $M_n(q)$.

If Conjecture 4.85 is true, then we know that there exists a constant b so that $f(n, B_q) \leq bn$. Let A be an $n \times n$ matrix that avoids B_q, and let us cut A up into 2×2 blocks. (There will be smaller blocks at the end if n is odd.) If a block contains no 1s, then replace that block with a 0; if it contains at least one 1, then replace that block with a 1.

This results in a matrix of size $\lceil n/2 \rceil \times \lceil n/2 \rceil$ that is also B_q-avoiding, and therefore has at most $f(\lceil n/2 \rceil, B_q) \leq b\lceil n/2 \rceil$ entries equal to 1. Furthermore, each of these ones comes from a nonzero 2×2 block of A. There are $2^4 - 1 = 15$ possibilities for each such block. Therefore, the number $M_n(q)$ of all possible matrices A we could have started with satisfies

$$M_n(q) \leq 15^{b\lceil n/2 \rceil} \cdot M_{\lceil n/2 \rceil}(q).$$

Repeat this argument until the right-hand side becomes $M_1(q) = 2$. Then we get

$$\text{Av}_n(q) \leq M_n(q) \leq 15^{2bn}. \qquad (4.15)$$

Therefore, we have proved the following.

PROPOSITION 4.86

[226] If the Füredi–Hajnal conjecture is true, then the Stanley–Wilf conjecture is also true.

4.5.3   Proof of the Füredi–Hajnal Conjecture

We close this chapter by presenting the spectacular proof of Conjecture 4.85, given by Adam Marcus and Gábor Tardos in 2003 [250].

Let P be a $k \times k$ permutation matrix, and let A be an $n \times n$ matrix that avoids P and contains exactly $f(n, P)$ entries 1 (as defined in Conjecture 4.85). Let us assume for simplicity that n is divisible by k^2. The crucial idea of the proof is a decomposition of A into blocks. While the simple idea of decomposing a matrix into smaller matrices is not new, the novelty of the Marcus–Tardos method is that it decomposes A into $\frac{n}{k^2} \cdot \frac{n}{k^2}$ blocks, which are each of size $k^2 \times k^2$.

For $(i, j) \in \left[\frac{n}{k^2}\right] \times \left[\frac{n}{k^2}\right]$, let $S_{i,j}$ denote the submatrix (block) of A that consists of the intersection of rows $(i-1)k^2 + 1, (i-1)k^2 + 2, \cdots, ik^2$ and columns $(j-1)k^2 + 1, (j-1)k^2 + 2, \cdots, jk^2$. We will now contract A into a much smaller matrix B as follows. Each entry of B will contain some information about a block of A. The matrix $B = (b_{i,j})$ is of size $\frac{n}{k^2} \times \frac{n}{k^2}$, and

$$b_{i,j} = \begin{cases} 0 \text{ if all entries of } S_{i,j} \text{ are zero,} \\ \\ 1 \text{ if not all entries of } S_{i,j} \text{ are zero.} \end{cases}$$

PROPOSITION 4.87
The matrix B avoids P.

PROOF Let us assume that B contains P, and let us take a copy P_c of P in B. Then considering A, and using the fact that P is a permutation matrix, we can take a 1 from each block of A that defined an entry of P_c, and get a copy of P in A. ∎

The next crucial step is the following definition.

DEFINITION 4.88 A block $S_{i,j}$ of A is called wide *if it contains a 1 in at least k different columns. Similarly, a block is called* tall *if it contains a 1 in at least k rows.*

Note that a block has k^2 columns, but it is called wide if at least k of these columns contain a 1.

LEMMA 4.89
For any fixed j, the set of blocks $C_j = \{S_{i,j} | 1 \le i \le \frac{n}{k^2}\}$ of the matrix A contains less than $(k-1)\binom{k^2}{k} + 1$ wide blocks.

PROOF We show that if the statement of the lemma were false, then A would contain a copy of P. Indeed, assuming that the number of wide blocks in C_j is at least $(k-1)\binom{k^2}{k} + 1$, by the pigeon-hole principle there would be a k-tuple of integers $1 \le c_1 < c_2 < \cdots < c_k \le k^2$ so that there are k blocks $S_{a_1,j}, S_{a_2,j}, \cdots, S_{a_k,j}$ that all contain a 1 entry in column c_i, for $1 \le i \le k$. In that case, it is easy to find a copy of P in A, which is a contradiction. Indeed, if the single 1 in column i of P is in row $p(i)$, then choose a 1 from column c_i of $S_{a_{p(i)},j}$. As the blocks $S_{a_1,j}, S_{a_2,j}, \cdots, S_{a_k,j}$ are positioned in a column, the n entries 1 chosen this way will prove that A contains P, which is a contradiction. ∎

It goes without saying that the same argument can be made for the array of blocks $R_i = \{S_{i,j} | 1 \le j \le \frac{n}{k^2}\}$, thereby giving the following lemma.

LEMMA 4.90

For any fixed i, the set of blocks $R_i = \{S_{i,j}|1 \leq j \leq \frac{n}{k^2}\}$ of the matrix A contains less than $(k-1)\binom{k^2}{k} + 1$ tall blocks.

We have seen that A cannot have too many wide or tall blocks, and Proposition 4.87 seems to suggest that A cannot have too many nonzero blocks either. Putting together these observations, we get the following recursive estimate.

LEMMA 4.91

For any $k \times k$ permutation matrix P, and any positive multiples n of k^2, we have

$$f(n, P) \leq (k-1)^2 f\left(\frac{n}{k^2}, P\right) + 2k^3 \binom{k^2}{k} n.$$

PROOF By Proposition 4.87, the number of nonzero blocks is at most $f\left(\frac{n}{k^2}, P\right)$. By Lemmas 4.89 and 4.90, there are at most $\frac{n}{k^2}(k-1)\binom{k^2}{k}$ wide blocks, and at most $\frac{n}{k^2}(k-1)\binom{k^2}{k}$ tall blocks.

Let us count how many 1 entries the various blocks of A can contribute to $f(n, P)$. A wide (or tall) block can contribute at most k^4 entries 1, so the total contribution of these blocks is at most

$$2\frac{n}{k^2}(k-1)\binom{k^2}{k} \cdot k^4 < 2k^3\binom{k^2}{k}n.$$

If a block is neither wide nor tall, then by the pigeon-hole principle, it can contain at most $(k-1)^2$ entries 1. Multiplying this bound by the number of nonzero blocks yields that the contribution of all blocks that are not tall or wide is at most

$$(k-1)^2 f\left(\frac{n}{k^2}, P\right).$$

Adding all the contributions of all these blocks, the lemma is proved. ∎

We now have all necessary tools to prove the Füredi–Hajnal conjecture.

THEOREM 4.92

For all permutation matrices P of size $k \times k$, we have

$$f(n, P) \leq 2k^4\binom{k^2}{k}n.$$

PROOF We prove the statement by induction on n, the initial case of $n = 1$ being obvious. (In fact, the statement is obvious when $n \leq k^2$, because then A has at most k^4 entries.)

Now assume the statement is true for all positive integers less than n, and prove it for n. Let $n' = \lceil n/k^2 \rceil k^2$. Then by Lemma 4.91, we have

$$f(n, P) \le f(n', P) + 2k^2 n$$
$$\le (k-1)^2 f\left(\frac{n'}{k^2}, P\right) + 2k^3 \binom{k^2}{k} n' + 2k^2 n$$

as in the worst case, we fill the part of A that cannot be partitioned into $k^2 \times k^2$ blocks by entries 1. Applying the induction hypothesis to $f\left(\frac{n'}{k^2}, P\right)$, we get

$$f(n, P) \le (k-1)^2 \left[2k^4 \binom{k^2}{k} \frac{n'}{k^2} \right] + 2k^3 \binom{k^2}{k} n' + 2k^2 n$$
$$\le 2k^2 ((k-1)^2 + k + 1) \binom{k^2}{k} n$$
$$\le 2k^4 \binom{k^2}{k} n,$$

since $k^2 > (k-1)^2 + k + 1$ whenever $k \ge 2$. ∎

The proof of the Stanley–Wilf conjecture is now obvious. We include it because we want to show the numerical result obtained.

COROLLARY 4.93

For any permutation pattern q of length k, we have

$$\mathrm{Av}_n(q) \le c_q^n,$$

where $c_q = 15^{2k^4 \binom{k^2}{k}}$.

PROOF   Immediate from Theorem 4.92 and (4.15). ∎

The value $c_q = 15^{2k^4 \binom{k^2}{k}}$ is certainly very high. For instance, if $k = 3$, then we get $c_q = 15^{27216}$, while we have seen in Corollary 4.8 that $c_q = 4$ is the best possible result. Josef Cibulka [118] improved the upper bound of Corollary 4.93 as follows. Let c_k be the smallest constant satisfying $\mathrm{Av}_n(q) < c_k^n$ for all patterns q of length k, and all positive integers n. Then there is a constant C so that the inequality

$$c_k \le 2^{Ck \log k} \tag{4.16}$$

holds. Though this is a significant improvement, the upper bound obtained for c_k is still an *exponential* function of k.

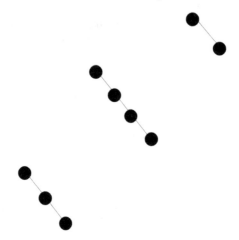

FIGURE 4.13
The diagram of a generic layered permutation.

The following definition enables us to discuss a disproved conjecture that suggested a fairly low (polynomial) value for c_k.

DEFINITION 4.94   *A permutation q is called* layered *if q can be written as the concatenation $q_1 q_2 \cdots q_m$ of the strings q_i, for $1 \leq i \leq m$, where*

  (a) each q_i is a decreasing sequence of consecutive integers, and

  (b) the leading entry of q_i is smaller than the leading entry of q_{i+1}, for $1 \leq i \leq m - 1$.

In other words, a permutation is layered if it consists of decreasing subsequences (layers) so that the entries decrease within each layer but even the largest entry of a layer is smaller than the smaller entry of the next layer. In Exercise 13, of the next chapter you are asked to characterize layered permutations by pattern avoidance.

Example 4.95
The patterns 1432765 and 4321576 are both layered.   ⫿

See Figure 4.13 for an illustration.
In [119], the authors use generalized versions of the methods we presented in the proof of Theorem 4.39 to show the following.

THEOREM 4.96

Let q be a layered permutation pattern of length k. Then the inequality

$$\text{Av}_n(q) \le (4k^2)^n = (2k)^{2n}$$

holds.

We will give the proof of Theorem 4.96 in Problems Plus 21 and 22.

Note that numerical evidence suggests that for small values of k, among all permutation patterns of length k, layered patterns are the easiest to avoid. Therefore, Theorem 4.96 suggested the following conjecture, published in [119].

CONJECTURE 4.97 *Let q be any permutation pattern of length k. Then $\text{Av}_n(q) \le (2k)^{2n}$.*

In 2013, Jacob Fox [178] disproved Conjecture 4.97. He proved the following result.

THEOREM 4.98

There exists an infinite sequence $\pi_1, \pi_2, \cdots$ of patterns so that π_k is a pattern of length k, and if $f(k) = L(\pi_k)$, then

$$f(k) = 2^{\Omega(k^{1/4})}.$$

In other words, c_k is an *exponential* function of k, not a polynomial function as Conjecture 4.97 claimed. We note that Fox's proof of Theorem 4.98 is nonconstructive, so we still do not have an example of an infinite sequence of patterns that exhibit the behavior described in the theorem.

We mention that Jacob Fox has also improved Cibulka's upper bound for c_k that we stated in (4.16). See Problem Plus 28 for that result.

Exercises

1. Prove that $\text{Av}_n(123 \cdots k) = \text{Av}_n(123 \cdots k(k-1))$. Here the pattern on the right is the monotone pattern of length k, with its last two entries reversed. Do not use Theorem 4.23.

2. (a) Find a formula for $\text{Av}_{k+1}(q)$ where q is any pattern of length k.

   (b) Find a formula for $\text{Av}_{k+2}(q)$ where q is any pattern of length k.

3. For patterns p and q, let $\text{Av}_n(p, q)$ denote the number of permutations of length n that avoid both p and q. Find a formula for $\text{Av}_n(132, 312)$.

4. Find a formula for $\text{Av}_n(p, q)$, where p and q are patterns of length three. Go through all possible choices of p and q. How many different sequences are there?

5. Find a formula for $\text{Av}_n(132, 1234)$.

6. Prove that $f(n) = \text{Av}_n(132, 3421) = 1 + (n - 1)2^{n-2}$.

7. Prove that $g(n) = \text{Av}_n(132, 4231) = 1 + (n - 1)2^{n-2}$.

8. Find a formula for $\text{Av}_n(132, 4321)$.

9. (a) Find the ordinary generating function $\sum_{n \geq 0} \text{Av}_n(3142, 2413)z^n$.

   (b) Find other pairs of patterns (p, q) so that $\text{Av}_n(p, q) = \text{Av}_n(3142, 2413)$. (Do not look for the easy way out. We are interested in pairs that cannot be obtained from $(3142, 2413)$ by iterating the trivial equivalences.)

10. The result of Exercise 6 shows that $\text{Av}_n(132, 3421)$ is always odd if $n \geq 3$. Prove this fact by finding an involution on the set enumerated by $\text{Av}_n(132, 3421)$ that has exactly one fixed point.

11. (a) Let $h(n) = \text{Av}_n(1324, 2413)$. Compute the ordinary generating function $H(z) = \sum_{n=0}^{\infty} h(n)z^n$.

    (b) Find other pairs of patterns (p, q) so that $\text{Av}_n(p, q) = \text{Av}_n(1324, 2413)$. Again, we are looking for nontrivial examples.

12. In Exercises 6 and 7, we have seen examples for pairs of patterns p and q so that $\text{Av}_n(p, q)$ is odd for sufficiently large n. Are there pairs of patterns p and q so that $\text{Av}_n(p, q)$ is *even* for sufficiently large n?

13. Let $I_n(q)$ be the number of *involutions* of length n that avoid the pattern q. Find the ordinary generating function for the numbers $I_n(2143)$. Note that involutions enumerated by $I_n(2143)$ are called *vexillary involutions*.

14. Are two patterns less restrictive than three patterns? More precisely, let a, b, c, d, and e be five patterns of the same length. Is it then true that
$$\text{Av}_n(a, b) \geq \text{Av}_n(c, d, e)$$
as long as $(a, b) \neq (123 \cdots k, k(k - 1) \cdots 1)$?

15. Find an explicit formula for the number of 132-avoiding n-permutations that are decomposable into three skew blocks, each block consisting of a skew indecomposable permutation.

16. (a) Find a formula for the number of skew indecomposable 321-avoiding permutations.

    (b) Find a bijective argument proving the result of part (a).

17. (−) Find a bijection f from the set of 231-avoiding n-permutations to the set of northeastern lattice paths from $(0,0)$ to (n,n) that do not go above the main diagonal. What parameter of these paths will correspond to the number of ascents of the corresponding permutations?

18. Give an example of two patterns p and q so that $Av_n(p,q)$ is a polynomial.

19. (−) Are there any patterns p and q so that $Av_n(p,q) = 0$ if n is sufficiently large?

20. Let $E_n(q)$ (resp. $O_n(q)$) denote the number of *even* (resp. odd) permutations of length n that avoid a given pattern q. Prove that

$$E_n(132) = \frac{C_n + C_{(n-1)/2}}{2}.$$

21. Prove that

$$E_n(231) - O_n(231) = E_n(312) - O_n(312) = (-1)^{\lfloor n/2 \rfloor} C_{(n-1)/2}.$$

22. Prove that

$$E_n(213) - O_n(213) = C_{(n-1)/2}.$$

23. Give an example of three patterns m, q, and r so that

$$\lim_{n \to \infty} \sqrt[n]{Av_n(m,q,r)}$$

is not an integer.

24. Let us assume that we know the numbers $Av_n^{(k)}(q)$ of permutations of length n having exactly k fixed points that avoid a certain pattern q. For what other patterns q' can we obtain the numbers $Av_n^{(k)}(q')$ directly from these data?

25. Prove that if $k > 3$, then there exists a pattern of length k so that for n sufficiently large, the inequality $Av_n(123 \cdots k) < Av_n(q)$ holds.

26. Find a formula for $I_n(132)$.

27. (a) A *binary plane tree* is a rooted plane tree with unlabeled vertices in which each vertex that is not a leaf has one or two children, and each child is either a left child or a right child of its parent. So the two binary plane trees shown in Figure 4.14 are different. Prove (preferably by a bijection from the set of 132-avoiding permutations of length n to the set of binary plane trees on n vertices) that the number of binary plane trees on n vertices is C_n.

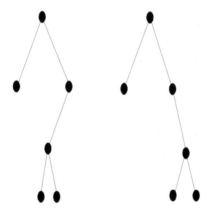

FIGURE 4.14

Two distinct binary plane trees.

  (b) What parameter of these plane trees will correspond to the number
       of descents of the corresponding 132-avoiding permutation?

  (c) How can we decide from a binary plane tree whether the corre-
       sponding 132-avoiding permutation has a descent in i?

28. Let $D_n(k)$ denote the number of 321-avoiding n-permutations that start
    in an ascending run consisting of exactly k elements. Prove that

$$D_n(k) = \binom{2n-k-1}{n-1} - \binom{2n-k-1}{n}.$$

29. Prove by a simple direct argument that $\mathrm{Av}_n(1234)$ and $\mathrm{Av}_n(1324)$ are
    not asymptotically equal.

30. Let r be a pattern that starts with the entry 1, and assume that $\mathrm{Av}_n(r) <
    c^n$ for some positive constant c. Let $w = 1 \oplus q$. Prove that $\mathrm{Av}_n(w) <
    (c + 1 + 2\sqrt{c})^n$.

31. Let q be a pattern that has an entry x that is a left-to-right maxi-
    mum and a right-to-left minimum at the same time, and assume that
    $\mathrm{Av}_n(q) < c^n$ for all n. In other words, we have $q = L \oplus 1 \oplus R$ for some
    strings L and R. Let us *replace* x in q by a pattern p so that the ob-
    tained pattern q' has form $L \oplus p \oplus R$. Let us assume that $\mathrm{Av}_n(p) < C^n$
    for all n.

    State certain conditions for p, then prove that if those conditions hold,
    then there exists a constant K so that $\mathrm{Av}_n(q') < (4cC)^n$ for all n.

32. Prove that the number of 1324-avoiding n-permutations in which the
    entry 1 precedes the entry n is less than 8^n.

33. Let q be a pattern of length k. Prove that $\mathrm{Av}_n(q) \geq c(k-1)^n$, for some positive constant c.

34. For what positive integers n will $\mathrm{Av}_n(123)$ be an odd integer?

35. The bijection F that we constructed in the proof of Theorem 4.59 maps the set of indecomposable 1342-avoiding n-permutations to the set $D_n^{\beta(0,1)}$ of $\beta(0,1)$-trees on n vertices. Let A be the set of 123-avoiding n-permutations that are in the domain of F. Describe $F(A)$.

36. Find a formula for $I_n(231)$.

37. Let f be the bijection defined in Exercise 17. Let B_m be the set of permutations in the domain of f that have exactly m inversions. Give a simple description of $f(B_m)$.

38. Find three patterns p, q, and r so that for all positive integers n, the equality $\mathrm{Av}_n(p, q, r) = n$ holds.

39. Are there infinitely many nontrivial pairs of patterns p and q so that $\mathrm{Av}_n(p, q) < h_{(p,q)}(n)$ for all n, where $h_{(p,q)}(n)$ is a polynomial function of n? By nontrivial pairs we mean pairs in which at least one of the two patterns has at least two alternating runs.

40. Find a formula for the number of 132-avoiding n-permutations whose longest decreasing subsequence is of length exactly $k + 1$.

41. Find a formula for the number $B_1(a_1, a_2, a_3)$ of all permutations of the *multiset* $\{1^{a_1} 2^{a_2} 3^{a_3}\}$ that avoid both 123 and 132.

42. Let us extend the notion of permutation pattern avoidance for *words over a finite alphabet* $\{1, 2, \cdots, m\}$ as follows. We say that a word w contains the permutation pattern $q = q_1 q_2 \cdots q_d$ if we can find d distinct entries in w, denoted $a_1, a_2, \cdots, a_d$ from left to right so that $a_i < a_j$ if and only if $d_i < d_j$.

   We say that a word t on the alphabet $M = \{1, 2, \cdots, m\}$ is k-*regular* if the distance between two identical letters in t is at least k. Let $l_k(q, m)$ be the maximum length that a k-regular word over M can have if it avoids the permutation pattern q. Let us assume that we know that there exists a constant $c = c_{k,q,m}$ so that

$$l_k(q, m) \leq cm. \tag{4.17}$$

   Prove the Stanley–Wilf conjecture from this assumption.

43. Let

$$B = \begin{pmatrix} 1 & 1 & 0 \\ 1 & 0 & 0 \\ 0 & 0 & 1 \end{pmatrix}.$$

Let the sequence of matrices A_i be defined by $A_0 = 1$, $A_1 = \begin{pmatrix} 1 & 1 \\ 1 & 0 \end{pmatrix}$, and $A_{n+1} = \begin{pmatrix} I_{2^n} & A_n \\ A_n & 0 \end{pmatrix}$, where I_{2^n} is the identity matrix of size $2^n \times 2^n$. Set $m = 2^n$, and prove that $f(m, B) = \Omega(m \log m)$, so inequality (4.14) of the Füredi–Hajnal conjecture does not hold for B.

44. Let us say that the permutation $p = p_1 p_2 \cdots p_n$ *tightly* contains the permutation $q = q_1 q_2 \cdots q_k$ if there exists an index $0 \le i \le n - k$ so that $q_j < q_r$ if and only if $p_{i+j} < p_{i+r}$. In other words, p tightly contains q if there is a string of k entries in p in *consecutive* positions that relate to each other as the entries of q do.

    If p does not tightly contain q, then we say that p *tightly avoids* q. Let $T_n(q)$ denote the number of n-permutations that tightly avoid q. For instance, 1436725 tightly contains 123 (consider the third, fourth, and fifth entries), but tightly avoids 321 or 4231.

    Prove that if $n \ge 4$, then $T_n(123) > T_n(132)$.

45. Prove that if $n \ge 7$, then $T_n(1342) > T_n(1243)$.

46. Find a *direct* bijection between 321-avoiding permutations of length n and lattice paths using steps $(0, 1)$ and $(1, 0)$ from $(0, 0)$ to (n, n) that never go above the main diagonal.

47. Alternating permutations are defined in Definition 1.54. Find a formula for the number of 312-avoiding alternating permutations of length n.

48. Find all pairs of patterns (p, q) of length three or more so that $\mathrm{Av}_n(p, q) = I_n(p, q)$ for all n. That is, all permutations that avoid p and q are involutions.

49. Let q be a skew indecomposable involution. Prove that for all n, the inequality
$$\mathrm{Av}_n(q) \le I_{2n}(q)$$
holds.

50. Use the framework of Theorem 4.40 to prove an upper bound for
$$\limsup_n \sqrt[n]{I_n(1324)}.$$

51. Let $k \ge 3$, and let H_k be the set of $k - 1$ patterns of length k that start with an increasing subsequence of length $k - 1$, and end with a descent. For instance, $H_4 = \{1243, 1342, 2341\}$. Prove that $\mathrm{Av}_n(H_k) \le ((k-1)^2 + 1)^n$.

52. Let $u_n(q)$ denote the number of permutations that avoid q and have a *unique longest increasing subsequence*, or ULIS. Find the closed form of the generating function $\sum_{n \geq 0} u_n(231) z^n$.

53. A *Dyck path* of semilength n is a lattice path from $(0,0)$ to $(2n,0)$ using steps $(1,1)$ and $(1,-1)$ that never goes below the horizontal axis.

   (a) Find a natural bijection f from the set $Av_n(132)$ to the set of all Dyck paths of semilength n.

   (b) Let $UI_n(132)$ be the set of all 132-avoiding involutions of length n that have a ULIS. Describe the image of $UI_n(132)$ under the bijection f of part (a).

54. Let us say that permutation p *strongly* avoids pattern q if p and p^2 both avoid q. Let $Sav_n(q)$ denote the number of strongly q-avoiding permutations of length n. Find an explicit formula for the ordinary generating function for the sequence of numbers $Sav_n(312)$.

55. Let k be a positive integer, and let $n \geq (k-1)^3 + 1$. Prove that $Sav_n(12 \cdots k) = 0$.

56. Let q' be the reverse complement of q. Prove that for all n, the equality $Sav_n(q) = Sav_n(q')$ holds.

57. (+) Let us call permutations of length n that consist of one n-cycle *cyclic*. Let $Cyc_n(q_1, q_2)$ denote the number of cyclic permutations of length n that avoid the patterns q_1 and q_2. Find an explicit formula for $Cyc_n(123, 132)$.

58. Prove the following formulas.

   (a) For $n \geq 5$, $Cyc_n(123, 321) = 0$.
   (b) For $n \geq 3$, $Cyc_n(231, 312) = 0$.
   (c) For all positive integers n, $Cyc_n(231, 321) = 1$.
   (d) For all positive integers n, $Cyc_n(132, 321) = \phi(n)$.

59. Let $q = q_1 q_2 \cdots q_k$ be any *involution* of length $k > 2$ such that if $q_i = k$, then $i \leq k - 2$. Prove that for all $n \geq 2$, the inequality

$$2Cyc_n(q) \leq Cyc_{n+1}(q)$$

holds.

60. Prove that $Av_n(1234, 1243) \leq (3 + 2\sqrt{2})^n$.

61. Let $k \geq 2$. Find a permutation class $\mathcal{C}$ so that its number of elements of length n is equal to the number of compositions of the integer n into parts that are at most k.

62. (-) Are there any real numbers in the interval $(0, 1)$ that are growth rates of a permutation class?

Problems Plus

1. Prove that the number of n-permutations that avoid 132 and have $k+1$ left-to-right minima is $\frac{1}{n}\binom{n}{k}\binom{n}{k+1}$.

2. Prove a formula for the number $d_n(132)$ of *derangements* of length n that avoid 132.

3. (a) Let p be a permutation that is the union of an increasing and a decreasing sequence. Such permutations are called *skew-merged*. Can we characterize p in terms of pattern avoidance?

   (b) Find a formula for the number of skew-merged n-permutations.

4. (a) Prove that
$$\mathrm{Av}_n(1234, 3214) = \frac{4^{n-1} + 2}{3}.$$

   (b) Find other pairs of patterns p and q so that the equality $\mathrm{Av}_n(p,q) = \mathrm{Av}_n(1234, 3214)$ holds for all n, besides those pairs obtained from $(1234, 3214)$ by trivial symmetries.

5. Find an upper bound for $\mathrm{Av}_n(23451, 13452, 12453)$.

6. Prove that for all positive integers n, the equality $\mathrm{Av}_n(1243, 2143, 321) = n + 2\binom{n}{3}$ holds.

7. Prove that for all positive integers n, the equality $\mathrm{Av}_n(1243, 2143, 231) = (n+2)2^{n-3}$ holds.

8. Let us generalize the notion of pattern avoidance as follows. A *generalized* pattern is a pattern in which certain consecutive elements *may* be required to be consecutive entries of a permutation. For instance, a generalized $31 - 2$ pattern is a 312-pattern in which the entries playing the role of 3 and 1 must be consecutive entries of the permutation (this is why there is no dash between 3 and 1), but the entries playing the role of 1 and 2 do not have to be consecutive (this is why there is a dash between 1 and 2).

   For example, a $3 - 1 - 2$ generalized pattern is just a traditional 312-pattern, a $2-1$ generalized pattern is an inversion, while a 21 generalized pattern occurs when the permutation has a descent.

   Prove that $\mathrm{Av}_n(1-23) = B_n$, where B_n is the nth Bell number, denoting the number of all partitions of the set $[n]$.

9. Is generalized pattern avoidance always stricter than traditional pattern avoidance? That is, let q be a traditional pattern, and let q' be a generalized pattern so that q' becomes q if all the consecutiveness restrictions

are released, but there *are* some consecutiveness restrictions. Is it then true that $\text{Av}_n(q) > \text{Av}_n(q')$ if n is large enough?

10. For a permutation p, and a generalized pattern q, let $q(p)$ denote the number of occurrences of q in p. So the examples of Problem Plus 8 can be written as

$$d(p) = 21(p), \text{ and } i(p) = (2-1)(p).$$

Find a similar expression for $maj(p)$.

11. In Exercise 42, we proved the Stanley–Wilf conjecture from formula (4.17), which we assumed without proof. Prove formula (4.17).

12. Let q be any pattern of length $k-3$. Prove that for all n, the equality $\text{Av}_n((k-1)(k-2)kq) = \text{Av}_n((k-2)k(k-1)q)$ holds.

13. Prove that for any permutation patterns q of length k, the inequality

$$L(q) = \lim_{n\to\infty} \sqrt[n]{\text{Av}_n(q)} \geq \frac{k^2}{e^3}$$

holds.

14. Let q be an indecomposable pattern, let $L(q)$ be defined as in the previous Problem Plus, and let $q' = 1 \oplus q$.

(a) Prove that $L(q') \geq 1 + L(q) + 2\sqrt{L(q)}$.

(b) Prove that in the special case when q starts in the entry 1, the equality $L(q') = 1 + L(q) + 2\sqrt{L(q)}$ holds.

15. Let $f(z) = \sum_{n\geq 0} \text{Av}_n(1324, 4321)z^n$. Prove that

$$f(z) = \frac{1 - 12z + 59z^2 - 152z^3 + 218z^4 - 168z^5 + 58z^6 - 6z^7}{(1-z)(1-2z)^4(1-4z+2z^2)}.$$

Use this fact to compute $L(1324, 4321)$.

16. Let $T_n(q)$ be defined as in Exercise 44. Prove that for any given pattern q, there exists a constant c_q so that

$$\lim_{n\to\infty} \sqrt[n]{T_n(q)/n!} = c_q.$$

17. Prove that for all positive integers $k \geq 3$, the identity

$$\sum_{n\geq 0} T_n(12\cdots k)\frac{z^n}{n!} = \left(\sum_{i\geq 0} \frac{z^{ik}}{(ik)!} - \sum_{i\geq 0} \frac{z^{ik+1}}{(ik+1)!} \right)^{-1}.$$

holds.

18. Find a closed formula for $A(z) = \sum_{n \geq 0} T_n(123) \frac{z^n}{n!}$.

19. A permutation pattern $q = q_1 q_2 \cdots q_k$ is called *nonoverlapping* if there is no permutation $p = p_1 p_2 \cdots p_n$ so that $k < n < 2k - 1$ and both $p_1 p_2 \cdots p_k$ and $p_{n-k+1} p_{n-k+2} \cdots p_n$ form a q-pattern. For instance, $q = 132$ is nonoverlapping, but $q' = 2143$ is not since $p = 214365$ has the property that both its first four entries and its last four entries form a 2143-pattern. In other words, a permutation is called non-overlapping if it is impossible for two of its copies to overlap in more than one entry. Let a_n be the probability that a randomly selected n-permutation is nonoverlapping.

   (a) Prove that if n is even, then

   $$a_n = 1 - \sum_{j=2}^{n/2} \frac{a_j}{j!}. \tag{4.18}$$

   (b) What modification of (4.18) is necessary if n is odd?

   (c) Prove that $\lim_{n \to \infty} a_n$ exists, and compute this limit up to four decimal points.

20. Let q be a nonoverlapping pattern of length k, and let $V_n(q)$ be the number of permutations of length n that do not contain a *very tight* copy of q. A tight copy of q is called very tight if the set of entries of q forms an interval. For instance, in 25687314, the entries 5687 form a very tight copy of 1243, but the entries 314 do not form a very tight copy of 213. Prove that $V_n(q) \leq V_n(12 \cdots k)$.

21. For a permutation pattern r, let $r' = r \oplus 1$.

   Prove the following generalized version of Lemma 4.38. Let q_1, q_2, and q_3 be three permutation patterns. Then every permutation avoiding $q_1 \oplus q'_2 \oplus q_3$ is a merge of a permutation avoiding $q_1 \oplus q'_2$ and a permutation avoiding $q'_2 \oplus q_3$.

22. Use the result of the preceding Problem Plus to prove Theorem 4.96 in the following stronger form. Let q be the layered pattern of layer lengths $a_1, a_2, \cdots, a_t$. Then

   $$\text{Av}_n(q) \leq \left(a_1 + a_t - t + 1 + 2 \sum_{i=2}^{t-1} a_i \right)^2.$$

23. Reverse alternating permutations were introduced in Definition 1.54. Let $RA_n(q)$ denote the number of reverse alternating n-permutations

that avoid the pattern q. Prove that for all $n \geq 1$, the equality

$$RA_{2n}(2143) = RA_{2n}(1234) = \frac{2 \cdot (3n)!}{n!(n+1)!(n+2)!}$$

holds.

24. Keep the notation of the previous problem. Prove that for all $n \geq 1$, and for all k, the equality

$$RA_{2n}(12 \cdots k) = RA_{2n}(12 \cdots k(k-1))$$

holds. Explain why the identity does not hold if $2n$ is replaced by $2n+1$.

25. Keep the notation of the previous problem. Prove that for all n (odd or even), the equality

$$RA_n(12 \cdots k) = RA_n(21 \cdots (k-1)k)$$

holds.

26. Prove that for all positive integers n, the equality

$$Av_n(1324) = Av_n(1234, 3416725)$$

holds.

27. Prove the following version of Theorem 4.98. For each k, there exists a family U consisting of almost all patterns of length k so that $L(U)$ exists, and $L(U) = 2^{\Omega((k/\log k)^{1/4})}$. In other words, even if we count permutations that *simultaneously* avoid almost all patterns of length k, we get the described, relatively high number.

28. Recall the notation $c_k = \max_{q \in S_k} L(q)$. Prove that $c_k = O(2^k)$.

29. Find the generating function for the sequence $Av_n(3124, 4312)$.

30. Prove that

$$\lim_{n \to \infty} \frac{u_n(132)}{C_n} = 1/2.$$

(See Exercise 52 for the definition of $u_n(132)$.)

31. Find an asymptotic formula for the number of 132-avoiding involutions of size n that have a unique longest increasing subsequence.

32. Prove that

$$Cyc_n(132, 231) = \frac{1}{2n} \sum_{\substack{d \mid n \\ d=2k+1}} \mu(d)2^{n/d}.$$

See Exercise 57 for the definition of $Cyc_n(q_1, q_2)$.

33. Let ϕ be the Euler totient function. That is, for a positive integer z, let $\phi(z)$ be the number of positive integers less than z that are relatively prime to z. Prove that

$$\mathrm{Cyc}_n(123, 231) = \begin{cases} \phi(2k) = \phi(n/2) \text{ if } n = 4k, \\ \phi(k+1) + \phi(2k+1) = \phi\left(\frac{n+2}{4}\right) + \phi\left(\frac{n}{2}\right) \text{ if } n = 4k+2, \\ \phi(m) = \phi((n+1)/2) \text{ if } n = 2m-1. \end{cases}$$

$$(4.19)$$

34. Prove that for all positive integers n, the inequality $\mathrm{Cyc}_n(132, 213) \leq n^2 2^{n/2}$ holds.

35. Let $\mathcal{C}$ be a permutation class, and let $|\mathcal{C}_n|$ be the number of elements of $\mathcal{C}$ that are of length n. Prove that exactly one of the following two statements holds.

    (a) There is a constant K so that $|\mathcal{C}_n| \leq n^K$, for all $n \geq 1$.

    (b) For all $n \geq 1$, the inequality $F_n \leq |\mathcal{C}_n|$ holds, where F_n is the nth Fibonacci number, defined by $F_0 = F_1 = 1$, and $F_n = F_{n-1} + F_{n-2}$ for $n \geq 2$.

    Provide an example when equality holds in the statement in (b).

36. The preceding Problem Plus shows an example of a permutation class $\mathcal{C}$ whose growth rate (which is equal to $(1+\sqrt{5})/2$) is in the open interval $(1, 2)$. What other numbers in the interval $(1, 2)$ are growth rates of permutation classes?

37. What is the smallest real number $c > 2$ so that c is the growth rate of a permutation class?

38. What is the smallest real number κ so that there are uncountably many permutation classes that have growth rate κ?

39. It follows from the answer to Problem Plus 36 that the subset of the interval $(1, 2)$ that consists of growth rates of permutation classes is countable. What is the largest real number ξ so that the same statement remains true for the interval $(1, \xi)$?

40. Recall that a real number r is *algebraic* if there exists a polynomial $A(z)$ with rational coefficients so that $A(r) = 0$. Is it true that the growth rate of a permutation class is always an algebraic number?

41. Does there exist a real number λ so that if $\lambda < c$, then c is the growth rate of a permutation class?

Solutions to Problems Plus

1. First, note that in a 132-avoiding permutation, the entry p_i is a left-to-right minimum if and only if either $i = 1$, or $i - 1$ is a descent. Therefore, we are looking for the number of 132-avoiding n-permutations with k descents, or, by taking the reverse, 231-avoiding n-permutations with k ascents. We know from Exercise 17 that the number of such n-permutations is equal to the number of northeastern lattice paths from $(0,0)$ to (n, n) that have k north-to-east turns and that never go above the main diagonal. A comprehensive survey of lattice paths of this and more general kinds can be found in [233].

In order to count these paths, note that the north-to-east turns of such a path completely determine the path. It therefore suffices to count the possible positions for the k-element set of north-to-east turns of such a path. Let (a_i, b_i) be the coordinates of the ith north-to-east turn of a northeastern lattice path r; then the vector

$$a_1, a_2, \cdots, a_k, b_1, b_2, \cdots, b_k \tag{4.20}$$

completely determines r. Disregard for the time being the requirement that r does not go above the main diagonal. Then we have

$$0 \le a_1 < a_2 < \cdots < a_k \le n - 1$$

and

$$1 \le b_1 < b_2 < \cdots < b_k \le n.$$

Clearly, all vectors (4.20) satisfying these conditions define a northeastern lattice path with k turns, proving that the number of these lattice paths is $\binom{n}{k}\binom{n}{k}$.

Now we have to count the vectors (4.20) defining a lattice path that goes above a main diagonal, that is, we have to determine for how many vectors (4.20) there exists an index i so that $a_i < b_i$. Let s be such a vector, and let i be the largest index for which $a_i < b_i$. Define

$$f(s) = (b_1, b_2, \cdots, b_{i-1}, a_{i+1}, \cdots, a_k, a_1, \cdots, a_i, b_i, \cdots, b_k).$$

Then $f(s)$ satisfies the chains of inequalities

$$1 \le b_1 \le b_2 \le \cdots \le b_{i-1} \cdots \le a_{i+1} \le \cdots \le a_k \le n - 1$$

and

$$0 \le a_1 \le a_2 \le \cdots \le a_i \cdots \le b_{i+1} \le \cdots \le b_k \le n - 1.$$

Therefore, we have $\binom{n+1}{k+1}\binom{n-1}{k-1}$ possibilities for $f(s)$. As f is a bijection (this has to be shown), this means that this is also the number of possibilities for s. So the number of northeastern lattice paths with k north-to-east turns that do not go above the main diagonal is

$$\binom{n}{k}\binom{n}{k} - \binom{n+1}{k+1}\binom{n-1}{k-1} = \frac{1}{n}\binom{n}{k}\binom{n}{k+1}.$$

These numbers are often called the *Narayana numbers*, and are denoted $A(n, k)$.

2. It is proved in [276] that

$$d_n(132) = \frac{1}{2} \cdot \sum_{i=0}^{n-2} \left(\frac{-1}{2}\right)^i C_{n-i}.$$

The numbers d_n are called the *Fine numbers*. An alternative formula for these numbers is

$$d_n(132) = \sum_{1 \leq k \leq n/2} \binom{2n-2k-1}{n-1} - \binom{2n-2k-1}{n}.$$

3. (a) It is proved in [288] that the skew-merged permutations are precisely the permutations that avoid both 2143 and 3412.

   (b) Michael Atkinson [19] proved that the number of these permutations is

$$\text{Av}_n(2143, 3412) = \binom{2n}{n} - \sum_{m=0}^{n-1} 2^{n-m-1}\binom{2m}{m}.$$

4. (a) This result was proved in [235] by Darla Kremer.

   (b) It is proved in [235] that

$$\frac{4^{n-1} + 2}{3} = \text{Av}_n(4123, 3214) = \text{Av}_n(2341, 2143)$$
$$= \text{Av}_n(1234, 2143).$$

5. The number of these permutations is less than $(6 + 4\sqrt{2})^n$. Take a permutation $p \in$ of length n that is in that class. Remove its left-to-right minima. We get a permutation p' that must avoid 1342, otherwise p would have contained 12453. More interestingly, p' must also avoid 2341, because otherwise p would have contained either 13452 or 23451, and both are impossible.

So p' avoids both 1342 and 2341. Such permutations are counted by the Schröder numbers, which have exponential order $(3+2\sqrt{2}) = (1+\sqrt{2})^2$. We mentioned these numbers in the solution of Exercise 9. The proof is then analogous to that of Lemma 4.37.

6. This result is due to Astrid Reifegerste [274], who explored the connections between permutations counted by Schröder numbers and lattice paths.

7. This result is due to A. Reifegerste, *ibid.*

8. The concept of generalized patterns was introduced in [25]. This result was proved in [120] in two different ways, together with many other results concerning short generalized patterns. One way is to recall that the Bell numbers satisfy the recurrence relation $B(n+1) = \sum_{i=0}^{n} B(i)\binom{n}{i}$, and then show that the 1-23 avoiding permutations satisfy this same recurrence. The other way is a direct bijective proof, based on a way of writing each partition in a canonical form similar to what we have seen for permutations in Chapter 3. (Write each block with its smallest element first, and then in decreasing order, then order the blocks in decreasing order of their smallest elements.)

9. No, that is not true, and one can find a counterexample using any of the four non-monotone patterns of length three. For instance, a permutation contains 132 if and only if it contains 13-2. The "if" part is obvious. For the "only if" part, let acb be a 132-pattern in p, and assume that the distance between the positions of a and c is minimal among all 132-patterns in p. That implies that if there is an entry d located between a and c, then d cannot be less than a (for dcb), cannot be larger than b (for adb), and cannot be between a and b in size (for dcb). This means there cannot be any entries among a and c.

10. This result is from the paper that introduced generalized patterns [25]. In that paper, the authors showed that essentially all Mahonian statistics in the literature can be expressed by generalized patterns. For the major index, one just has to add the descents of p; in other words, count the entries that precede a descent, then add these numbers for all descents. With this in mind, it is straightforward to see that

$$maj(p) = (1-32)(p) + (2-31)(p) + (3-21)(p) + (21)(p).$$

11. This result can be found in Martin Klazar, The Füredi–Hajnal conjecture implies the Stanley–Wilf conjecture, *Formal Power Series and Algebraic Combinatorics,* Springer, Berlin, 250–255, 2000. In that paper, the author in fact proves that formula (42) and the Füredi–Hajnal conjecture are *equivalent.*

12. See [289], where the authors prove this result introducing the interesting notion of shape–Wilf–equivalence. For two patterns q and q', it is necessary for $Av_n(q) = Av_n(q')$ to hold for all n in order for the patterns to be shape-Wilf-equivalent, but it is not sufficient.

13. This result was proved by Pavel Valtr, but was first published in [223]. One can assume without loss of generality that q is indecomposable, and then build a sufficient number of decomposable permutations in which each block avoids q.

14. (a) See [57] for a proof.

    (b) This follows from part (a), and the proof of part (b) of Exercise 30.

15. This result is due to Michael Albert, Michael Atkinson, and Vincent Vatter [6]. As f is a rational function, the exponential order of its coefficients is equal to the reciprocal of its singularity of the smallest modulus, which is $2 + \sqrt{2}$.

16. This result is due to Sergi Elizalde [159]. It is worth pointing out that Elizalde and Marc Noy [158] conjectured that if q is any pattern of length k, then $T_n(q) \leq T_n(12 \cdots k)$. Recently, in [160], Elizalde proved this conjecture. Furthermore, in the same paper, he proved that if q is any pattern of length k, then $T_n(q) \geq 12 \cdots (k-2)k(k-1)$.

17. This result is due to Richard Warlimont [323, 324].

18. It has been proved by Sergi Elizalde and Marc Noy [158] that

$$A(z) = \frac{\sqrt{3}}{2} \cdot \frac{e^{z/2}}{\cos\left(\frac{\sqrt{3}}{2}z + \frac{\pi}{6}\right)}.$$

In that same paper the authors provide a (somewhat less closed) formula for the corresponding generating function for the tight pattern 132.

19. (a) If q is nonoverlapping, then there is a smallest index j so that the first and last j entries of q form identical patterns. That pattern has to be non-overlapping.

    (b) If $n = 2k + 1$ is odd, then there are overlapping patterns of length n for which the smallest index j described in the solution of part (a) is $j = k + 1$. If b_n is the number of such patterns, then $a_n = 1 - \sum_{j=2}^{n/2} \frac{a_j}{j!} - b_n$.

    (c) It can be shown that $b_{2k+1} < a_{k+1}/(k+1)!$, and so the sequence of the a_n is monotone decreasing. The details can be found in [68]. The limit L is close to 0.3641.

20. The proof of this result can be found in [61]. The main idea of the proof is that if q is nonoverlapping, then it is very difficult to pack many very tight copies of q into a permutation, so many permutations are needed in order to contain all very tight copies of q.

21. This result has been published in [119]. The proof is similar to that of Lemma 4.38.

22. This result has also been published in [119]. Let q be the layered pattern of layer lengths $a_1, a_2, \cdots, a_t$. The claim of Theorem 4.96 is proved by induction on t, the case of $t = 1$ being obvious, and the case of $t = 2$ directly following from Theorem 4.23. Let us assume that we know the statement for all positive integers less than t. We can now use Lemma 4.37 with q_1 being the layered pattern consisting of two layers, of length a_1 and a_2, and q_2 being the layered pattern consisting of $t - 1$ layers, of length $a_2, a_3, \cdots, a_t$ to get the desired upper bound.

23. This result is due to Joel Brewster Lewis [243], who proved that both sides are equal to the number of Standard Young Tableaux (which will be defined in Chapter 7) of shape (n, n, n).

24. This result is due to the present author [70], who proved that if the length of our permutations is even, then the bijection of Exercise 1 preserves the alternating property.

25. The present author proved [70] that a slight modification of the bijection used in the solution of the previous problem is possible here, and parity restrictions are not needed.

26. See the paper of Alexander Burstein and Jay Pantone [104] that contains the first examples of this kind of unbalanced Wilf equivalence.

27. See [178] for this surprising result.

28. This result is also proved in [178].

29. Jay Pantone [259] proved the exact formula

$$\frac{(8z^5 - 16z^4 + 28z^3 - 26z^2 + 9z - 1) + \sqrt{1 - 4z}(2z^4 - 8z^3 + 14z^2 - 7z + 1)}{2z^2(1 - 6z + 9z^2 - 4z^3)}$$

for this generating function. Note that the denominator has rational roots.

30. The number $u_n(132)$ is equal to the number of plane unlabeled trees on $n + 1$ vertices that have a unique leaf at a maximum distance from the root. Then our statement is the special case of the following general theorem. Let us select a rooted plane unlabeled tree on n vertices uniformly at random, and let $a_{n,k}$ be the probability that the selected tree has k leaves at maximum distance from the root. Then

$$\lim_{n \to \infty} a_{n,k} = 2^{-k}.$$

A very general theorem that contains this result is Theorem 2 in [263]. An earlier, and more specific, reference is [213].

A more direct, more elementary proof would certainly be welcome.

31. It follows [81] from the solution of Exercise 53 that such involutions are in bijection with lattice paths of length $n+1$ consisting of U steps and D steps so that each prefix and each suffix contains strictly more U steps than D steps. The main term of their asymptotic enumeration can be found in [90], while the most precise enumerative result is proved in [208], where these lattice paths are called *bidirectional ballot sequences*, and the number of such paths of length n is denoted by B_n. The result is that

$$\frac{B_n}{2^n} \sim \frac{1}{4n} + \frac{1}{6n^2} + O\left(\frac{1}{n^3}\right).$$

In particular,

$$i_n(132) = B_{n+1} \sim \frac{2^{n+1}}{4(n+1)} = \frac{2^{n-1}}{n+1}.$$

32. This was the first result of this kind, and it can be found in [15].

33. This result is proved in [80] using a case-by-case analysis.

34. This result is due to Brice Huang [218].

35. This result was proved by Tomáš Kaiser and Martin Klazar in [223]. The class $\mathrm{Av}_n(123, 132, 213) = F_n$ is an example for equality.

36. Let the generalized Fibonacci numbers $F_{n,k}$ be defined by the identity

$$\sum_{n\geq 0} F_{n,k} z^n = \frac{1}{1 - z - z^2 - z^k}.$$

Then it is proved in [223] that exactly one of the following two statements hold for any infinite class $\mathcal{C}$ of permutations.

(a) There is a unique k and a constant c so that $F_{n,k} \leq |\mathcal{C}_n| \leq F_{n,k} \cdot n^c$ for all $n \geq 1$.

(b) The inequality $|\mathcal{C}_n| \geq 2^{n-1}$ holds for all $n \geq 1$.

It follows that the only numbers in $(1, 2)$ that are growth rates are those that are growth rates of the generalized Fibonacci numbers $F_{n,k}$ for some fixed k. These are the reciprocals of roots of the smallest modulus of the polynomials $(1 - z - z^2 - \cdots - z^k)$, or, equivalently, the roots of the largest modulus of $(z^k - z^{k-1} - \cdots - 1)$, or, after multiplication by $(z - 1)$, of the polynomial $z^{k+1} - 2z^k + 1$. In other words, there are only countably many growth rates in the interval $(1, 2)$.

37. Vincent Vatter [316] has proved that the smallest such number c is the unique positive root of the polynomial $1 + 2z + z^2 + z^3 - z^4$, which is about 2.06599.

38. It is proved in [316] that the largest such value for κ is the unique real root of the polynomial $z^3 - 2z^2 - 1$, so $\kappa \approx 2.20577$.

39. The smallest such real number is the unique positive real root of the polynomial $z^5 - 2z^4 - z^2 - z - 1$, so $\xi \approx 2.30522$. See [318] for a proof of this, and [260] for a characterization of all growth rates below ξ.

40. No. In [9], Michael Albert and Steve Linton give an example of a *perfect set* of growth rates. A perfect set is a closed set S so that each element $s \in S$ is an accumulation point of $S \backslash s$. As any perfect set is uncountable, while the set of algebraic numbers is countable, non-algebraic growth rates exist. Note that all the known examples of non-algebraic growth rates involve permutation classes with an infinite base. Even more strongly, every known example of a growth rate of a finitely based permutation class is an *algebraic integer* (the root of a polynomial with rational coefficients and leading coefficient 1).

41. Yes. The existence of such a real number λ was first proved by Vincent Vatter [317], who also showed that the smallest possible value of λ is less than 2.49. This bound was later improved to 2.36 by David Bevan [37].

5

In This Way, but Nicely. Pattern Avoidance.
Follow-Up.

5.1 Polynomial Recurrences

5.1.1 Polynomially Recursive Functions

In the previous chapter, we spent considerable time and effort to find out how large the numbers $\mathrm{Av}_n(q)$ are. In this chapter, we will mostly concentrate on *how nice* they are, or rather, how nice the sequence $\{\mathrm{Av}_n(q)\}_{1 \leq n}$ is. By abuse of language, we will often refer to this sequence as the sequence $\mathrm{Av}_n(q)$.

First, of course, we should define what we mean by "nice." We have already made one important definition, that of P-recursive (or polynomially recursive) sequences in Exercise 29 of Chapter 1, but for easy reference we repeat that definition here.

DEFINITION 5.1 *A sequence $f : \mathbf{N} \to \mathbf{C}$ is called P-recursive if there exist polynomials $P_0, P_1, \cdots, P_k \in \mathbf{C}[n]$, with $P_k \neq 0$ so that*

$$P_k(n+k)f(n+k) + P_{k-1}(n+k-1)f(n+k-1) + \cdots + P_0(n)f(n) = 0 \quad (5.1)$$

for all natural numbers n.

This definition, and some of the most important theorems in the theory of P-recursive sequences, can be found in [292], which is the earliest paper completely devoted to this subject. Chapter 6 of [297] is a comprehensive source for results on P-recursiveness. We do not want to duplicate existing literature on the topic, so we will not prove theorems that belong to the general theory of P-recursiveness. There will be one exception to this, Theorem 5.8.

See the exercises following Exercise 29 of Chapter 1 for some examples of P-recursive sequences.

Example 5.2
The sequence of Catalan numbers is P-recursive. Therefore, the sequence $\mathrm{Av}_n(q)$ is P-recursive for all patterns q of length 3. ☐

DOI: 10.1201/9780429274107-5

PROOF   We have

$$\frac{C_n}{C_{n-1}} = \frac{\binom{2n}{n}}{n+1} \cdot \frac{n}{\binom{2n-2}{n-1}} = \frac{(2n)!(n-1)!(n-1)!n}{(2n-2)!n!n!(n+1)} = \frac{4n-2}{n+1};$$

therefore, $(n+1)C_n - (4n-2)C_{n-1} = 0$, providing a polynomial recurrence relation. ∎

At this point, the reader should try to prove that the sum of two P-recursive sequences is also P-recursive.

Example 5.2, and some other examples of P-recursive sequences $\mathrm{Av}_n(q)$, such as $\mathrm{Av}_n(1234)$, or in general, $\mathrm{Av}_n(123\cdots k)$, suggested the following conjecture.

CONJECTURE 5.3   *[191] [Gessel, 1990] For any permutation pattern q, the sequence $\mathrm{Av}_n(q)$ is P-recursive.*

Six years later, John Noonan and Doron Zeilberger looked at a larger family of sequences. Instead of trying to count permutations that *avoid* a pattern, they wanted to count permutations that contain it exactly r times, where r was a fixed non-negative integer. Numerical data and some theorems that we will cover in Chapter 7 led them to the following conjecture.

CONJECTURE 5.4   *[256] [Noonan and Zeilberger, 1996] Let $S_{n,r}(q)$ be the number of n-permutations that contain the pattern q exactly r times. Then for any fixed r and q, the function $S_{n,r}(q)$ is polynomially recursive.*

5.1.2   Permutation classes again

It seems that Conjecture 5.4 is much more ambitious than Conjecture 5.3. Indeed, the latter seems to be a special case of the former, namely the special case in which $r = 0$. However, certain versions of the two conjectures are actually *equivalent* as has been proved by Atkinson [20].

Recall the definition of permutation classes that we made in Section 4.1.1. It is straightforward to check that the intersection of permutation classes, as well as the union of permutation classes, is always a permutation class. So, for instance, the class $\mathcal{A}v(p_1, p_2)$ of permutations avoiding both p_1 and p_2 is a permutation class.

Recall that in a partially ordered set, an element x is called a *minimum* if it is smaller than all other elements of the poset, while the element y is called *minimal* if no element of the poset is smaller than y. So a poset can have any number of minimal elements, but only one or zero minimum elements.

Just as any ideal in any poset, a permutation class C is determined by the minimal elements of the complement of C. These minimal elements obviously form an antichain, that is, a set in which no two elements are comparable. This antichain is called the *basis* of C.

Example 5.5
The basis of the permutation class $C(p)$ is the one-element antichain consisting of p. ⬚

Let C be a class that has basis $p_1, p_2, \cdots, p_k$. The reader is invited to prove the simple fact that $C = C(p_1, p_2, \cdots, p_k)$ must hold, that is, C must be the class of all finite permutations avoiding all p_i, for $i \in [k]$.

Taking a second look at Conjecture 5.3, we see that it in fact claims that if $f(n)$ is the number of n-permutations of a class C that has a one-element basis, then $f(n)$ must be a P-recursive function. A stronger version of this conjecture was the following.

CONJECTURE 5.6 *For any permutation patterns $q_1, q_2, \cdots, q_k$, the sequence $\mathrm{Av}_n(q_1, q_2, \cdots, q_k)$ is P-recursive. In other words, if $f(n)$ counts the elements of a finitely based class of permutations that have length n, then $f(n)$ is P-recursive.*

In other words, we replaced the condition of having a one-element basis by the condition of having a finite basis.

Conjecture 5.6 was disproved in 2015 by Scott Garrabrant and Igor Pak [187]. They proved that there exists a set S of patterns of length 80 so that the sequence $\mathrm{Av}_n(S)$ of numbers counting permutations of length n that avoid all elements of S is not P-recursive.

5.1.3 Algebraic and Rational Power Series

Sometimes the generating function of a sequence has a much simpler form than the sequence itself. The following notion of d-finiteness, and the theorem after that, show how we can use these simple generating functions to prove P-recursiveness.

DEFINITION 5.7 *We say that the power series $u(z) \in \mathbf{C}[[z]]$ is d-finite if there exists a positive integer d and polynomials $p_0(n), p_1(n), \cdots, p_d(n)$ so that $p_d \neq 0$ and*

$$p_d(z)u^{(d)}(z) + p_{d-1}(z)u^{(d-1)}(z) + \cdots + p_1(z)u'(z) + p_0(z)u(z) = 0, \quad (5.2)$$

Here $u^{(j)} = \frac{d^j u}{dz^j}$.

We promised that we will prove one theorem from the theory of P-recursive functions. We are going to fulfill that promise now. The theorem in its explicit form is due to Richard Stanley [292].

THEOREM 5.8

The sequence $f(0), f(1), \cdots$ is P-recursive if and only if its ordinary generating function

$$u(z) = \sum_{n=0}^{\infty} f(n)z^n \tag{5.3}$$

is d-finite.

PROOF

- First suppose u is d-finite, then (5.2) holds with $p_d \neq 0$. Fix $i \leq d$. We start by finding an expression for $p_i(z)$ involving linear combinations of the $f(n+t)$ with polynomial coefficients. Differentiate both sides of (5.3) i times and then multiply both sides by z^j to get

$$z^j u^{(i)} = \sum_{n \geq 0} (n+i-j)_i f(n+i-j)z^n. \tag{5.4}$$

  Here $(m)_i = m(m-1)\cdots(m-i+1)$. Now let a_j be the coefficient of z^j in $p_i(z)$. Multiply both sides by a_j, then repeat the entire procedure for each nonzero coefficient of $p_i(z)$, and add the obtained equations. We get that

$$p_i(z)u^{(i)} = \sum_{n \geq 0} \left(\sum_t f(n+t)q_t(n) \right) z^n,$$

  where the $q_t(n)$ are polynomials in n, and the sum in the parentheses is finite.

  Repeating the above procedure for all $i \leq d$, and then adding all the obtained equations, we get an equation that is similar to the last one, except that on the left-hand side, we will have $\sum_{i=0}^{d} p_i(z)u^{(i)}$, which is, by (5.2), equal to zero. If we compute the coefficient of z^{n+k} on both sides of the last equation and equate the two expressions, we get an equation that involves only linear combinations of some $f(n+t)$, $0 \leq t \leq k$, with coefficients that are polynomials in n. Therefore, this equation can be rearranged to yield a polynomial recurrence for f. This recurrence will not be $0=0$ as $p_d \neq 0$.

- Now suppose $g(n)$ is P-recursive in n, so (5.1) holds. Note that for any fixed natural number i, the polynomials $(n+i)_j$, $j \geq 0$, form a **C**-basis for the vector space $\mathbf{C}[n]$. In particular, $P_i(n)$ is a linear combination of polynomials of the form $(n+i)_j$. Therefore, using generating functions,

$\sum_{n \geq 0} P_i(n) f(n+i) z^n$ is a linear combination of generating functions of the form

$$\sum_{n \geq 0} (n+i)_j f(n+i) z^n \tag{5.5}$$

with complex coefficients.

Compare formulae (5.4) and (5.5). We see that the left-hand side of (5.5) almost agrees with $z^{j-i} u^{(j)}$, that is, they can only differ in finitely many terms with all negative coefficients. Let the sum of these terms be $R_i(z) \in z^{-1} K[z^{-1}]$, a Laurent-polynomial. If we multiply (5.1) by z^n and sum over all non-negative n, we get

$$0 = \left(\sum_i a_{ij} z^{j-i} u^{(j)} \right) + R(z). \tag{5.6}$$

Here the sum is finite by the definition of P-recursiveness and $R(z)$ is a Laurent-polynomial. If we multiply both sides by z^q where q is sufficiently large, the terms with negative exponents will disappear and we get an equation of the form (5.2).

∎

LEMMA 5.9
The product of two d-finite power series is d-finite.

PROOF   See [297], page 192.   ∎

A useful consequence of Theorem 5.8 and Lemma 5.9 is that the *convolution* of two P-recursive sequences is P-recursive.

LEMMA 5.10
Let $\{f(k)\}_k$ and $\{g(m)\}_m$ be two polynomially recursive sequences, and let

$$h(n) = \sum_{k=0}^{n} f(k) g(n-k). \tag{5.7}$$

Then $\{h(n)\}_n$ is a P-recursive sequence in n.

PROOF   Let $F(z)$, $G(z)$, and $H(z)$ denote the ordinary generating functions of the sequences $\{f(k)\}_k$, $\{g(m)\}_m$, and $\{h(n)\}_n$. Then $F(z)$ and $G(z)$ are both d-finite by Theorem 5.8. It is well-known that $F(z)G(z) = H(z)$, so $H(z)$ is d-finite by Lemma 5.9; therefore $\{h(n)\}_n$ is P-recursive.   ∎

Note that by repeatedly applying Lemma 5.10, we get the statement that the convolution of several P-recursive sequences is always P-recursive in the *sum* of the variables.

Knowing that a function is P-recursive is helpful, but often not sufficient to determine the function itself by interpolation. This is because we often do not know the degree of the recursion, or the degrees of the polynomials that appear in the recursion. Therefore, it is certainly useful to look for other, stronger properties that the sequence $Av_n(q)$ has, at least for some q.

DEFINITION 5.11 *The formal power series $f \in \mathbf{C}[[z]]$ is called algebraic if there exist polynomials $P_0(z), P_1(z), \cdots, P_d(z) \in \mathbf{C}[z]$ that are not all equal to zero so that*

$$P_0(z) + P_1(z)f(z) + \cdots + P_d(z)f^d(z) = 0. \tag{5.8}$$

The smallest positive d for which such polynomials exist is called the degree *of f.*

Example 5.12
Let $f(z) = \frac{1-\sqrt{1-4z}}{2z}$, the generating function of the Catalan numbers. Rearranging the previous equation as $1 - 2zf(z) = \sqrt{1 - 4z}$, then taking squares and rearranging again, we see that

$$z^2 f^2(z) - zf(z) + z = 0,$$

so f is algebraic of degree 2. ☐

The following lemma shows the connection between algebraic and d-finite generating functions.

LEMMA 5.13
If $f \in \mathbf{C}[[z]]$ is algebraic, then it is d-finite.

PROOF See [297], page 190. ∎

Note that the converse of Lemma 5.13 is not true. The reader should try to find an example of a power series that is d-finite but not algebraic, then check Exercise 9.

Even if having an algebraic generating function is a stronger property than being P-recursive, sometimes it is easier to prove the latter by proving the former.

Example 5.14
The sequence $Av_n(1342)$ is P-recursive. ☐

PROOF   We have seen in (4.8) that the ordinary generating function $H(z)$ of this sequence satisfies

$$H(z) = \frac{32z}{-8z^2 + 20z + 1 - (1 - 8z)^{3/2}}.$$

We claim that $H(z)$ is algebraic. To see this, it suffices to show that $Z(z) = \frac{1}{H(z)}$ is algebraic, because we can multiply both sides of the polynomial equation satisfied by $Z(z) = \frac{1}{H(z)}$ by a power of $H(z)$, and obtain a polynomial equation satisfied by $H(z)$. On the other hand, routine transformations show that

$$\left(32zZ(z) + 8z^2 - 20z - 1\right)^2 = (1 - 8z)^3,$$

so $Z(z)$ does satisfy a polynomial equation.   ∎

We would like to point out that the ordinary generating function of the sequence $\mathrm{Av}_n(1234)$ is *not algebraic* (see Problem Plus 2). So yet again, patterns of length four turn out to be quite surprising. The monotone pattern 1234 is not only not the easiest or the hardest to avoid, it is also not the "nicest"!

It is known that $\mathrm{Av}_n(123 \cdots k)$ is *P*-recursive for any k. We will prove this result in Chapter 7, when we learn about the interesting connections between permutations and *Standard Young Tableaux*.

An even smaller class of power series is that of *rational functions*. A rational function is just the ratio of two polynomials, such as $\frac{2z-3}{z^2-3z+5}$. A rational function is always algebraic of degree 1 (as multiplying it by its denominator we get a polynomial), and therefore, *d*-finite.

5.1.4   The Generating Function Of Most Principal Classes Is Nonrational

DEFINITION 5.15   *Let F and G be two combinatorial generating functions with nonnegative real coefficients that are analytic at 0, and let us assume that G(0) = 0. Then the relation*

$$F(z) = \frac{1}{1 - G(z)} \tag{5.9}$$

is called supercritical *if $G(R_G) > 1$, where R_G is the radius of convergence of G.*

The interested reader should consult [169] for a detailed treatment of supercritical relations.

Note that in the above definition, we allow for $G(R_G) = \infty > 1$, which indeed happens when R_G is a pole of G.

COROLLARY 5.16

If the relation between F and G described in Definition 5.15 is supercritical, then the radius of convergence of F is less than that of G, and so the exponential growth rate of the coefficients of F is larger than that of G.

PROOF   Note that as the coefficients of $G(z)$ are nonnegative, $G(R_G) > 1$ implies that $G(\alpha) = 1$ for some $\alpha \in (0, R_G)$. So $F(z)$ has a singular point closer to 0 than $G(z)$ does.   ∎

Example 5.17

If F and G are two *rational* functions with nonnegative coefficients so that $G(0) = 0$ and (5.9) holds, then the relation (5.9) is supercritical.   ▯

SOLUTION   Let R_G be the radius of convergence of G. It then follows from Pringsheim's theorem that the positive real number R_G is a singular point of G. As G is a rational function, that singular point R_G is necessarily a pole, so $\lim_{z \to R_G} G(z) = \infty > 1$.   ∎

What does this mean for permutation classes? Let us recall that a permutation p is *skew indecomposable* if it is not possible to cut p into two parts so that each entry before the cut is larger than each entry after the cut. For instance, $p = 2413$ is skew indecomposable, but $r = 364512$ is not as we can cut it into two parts by cutting between entries 5 and 1, to obtain $3645|12$.

If p is not skew indecomposable, then there is a unique way to cut p into nonempty skew indecomposable strings $s_1, s_2, \cdots, s_\ell$ of consecutive entries so that each entry of s_i is larger than each entry of s_j if $i < j$. We call these strings s_i the *skew blocks* of p. For instance, $p = 67|435|2|1$ has four skew blocks, while skew indecomposable permutations have one skew block.

This leads to the following fact. Let $A_q(z) = \sum_{n \ge 0} \mathrm{Av}_n(q) z^n$, and let $A_{1,q}(z) = \sum_{n \ge 1} \mathrm{Av}_{n,1}(q) z^n$ be the ordinary generating function of the sequence of the numbers $\mathrm{Av}_{n,1}(q)$ of *skew indecomposable* permutations of length n that avoid q.

PROPOSITION 5.18

Let q be a skew indecomposable pattern. Then the equality

$$A_q(z) = \frac{1}{1 - A_{1,q}(z)}, \tag{5.10}$$

holds.

PROOF   The right-hand side is equal to $\sum_{i \ge 0} (A_{1,q}(z))^i$, and in that sum, the summand indexed by i is the generating function for the number of q-avoiding permutations that have i skew blocks.   ∎

We will now show that for the generating function of most principal permutation classes, relation (5.10) is *not* supercritical, and therefore, by Proposition 5.18, these generating functions cannot be rational. Let $L(q, 1) = \lim_{n\to\infty} (\mathrm{Av}_{n,1}(q))^{1/n}$. It is easy to prove that $L(q, 1)$ exists and is finite, just as we proved that $L(q)$ exists and is finite, in Section 4.3.

LEMMA 5.19

Let $q = q_1 q_2 \cdots q_k$ be a skew indecomposable pattern so that at least one of the equalities $q_1 = 1$ and $q_k = k$ does not *hold. Then the exponential growth rates of the sequences $\mathrm{Av}_n(q)$ and $\mathrm{Av}_{n,1}(q)$ are equal, that is, $L(q) = L(q, 1)$.*

PROOF Let us assume that $q_k \neq k$. Let $p \in \mathcal{Av}_n$. Let $p' = p \oplus 1$; in other words, p' is p, with the new maximal entry $n + 1$ affixed to the end. As $q_k \neq k$, it follows that p' avoids q, and as p' ends in $n + 1$, it is skew indecomposable. This injective map proves that

$$\mathrm{Av}_n(q) \leq \mathrm{Av}_{n+1,1}(q),$$

so

$$L(q) \leq L(q, 1).$$

As the inequality $L(q, 1) \leq L(q)$ trivially holds, our claim is proved.

If $q_k = k$, but $q_1 \neq 1$, then we can repeat the above argument for the *reverse complement* of q, which will not end in k, and which will be a skew indecomposable pattern. ∎

Now the proof of our main result concerning the nonrationality of the generating functions of principal permutation classes is immediate.

THEOREM 5.20

Let $q = q_1 q_2 \cdots q_k$ be a skew indecomposable pattern so that at least one of the equalities $q_1 = 1$ and $q_k = k$ does not *hold. Then the generating function $A_q(z)$ is not rational.*

PROOF If $A_q(z)$ were rational, then by Example 5.17, the relation $A_q(z) = 1/(1 - A_{1,q}(z))$ would be supercritical, implying that $L(q) > L(q, 1)$, but we saw in Lemma 5.19 that $L(q) = L(q, 1)$. ∎

Note that if $q_1 = 1$ and $q_k = k$, but q is Wilf-equivalent to a pattern q' that does satisfy the criteria of Theorem 5.20, then of course, $A_q(z) = A'_q(z)$, so $A_q(z)$ is not rational. This happens for instance when $q = 12 \cdots k$, and $q' = 213 \cdots k$.

Note that Theorem 5.20 obviously fails for the pattern $q = 12$. Indeed, in that case, we have $\mathrm{Av}_n(q) = 1$ for all n, so $A_q(z) = 1/(1-z)$. The shortest pattern for which we cannot decide whether $A_q(z)$ is rational is $q = 1324$.

The following corollary will be useful in the Problems Plus.

COROLLARY 5.21

Let q be a pattern that is either equal or Wilf-equivalent to a pattern for which Theorem 5.20 applies. Let R be the radius of convergence of $A_q(z)$. Then $A_q(z) < \infty$.

PROOF   Lemma 5.19 shows that the convergence radius of $A_{1,q}(z)$ is also R. As the relation between $A_{1,q}(z)$ and $A_q(z)$ is not supercritical, it follows that $A_{1,q}(R) < 1$, and so

$$A_q(R) = \frac{1}{1 - A_{1,q}(R)} < \infty.$$

∎

5.1.5   Polynomial Recursiveness of $S_{n,r}(132)$

In this subsection, we present a fairly general result by proving the following theorem.

THEOREM 5.22

Let r be any fixed non-negative integer. Then the sequence $S_{n,r}(132)$ is P-recursive. Moreover, the ordinary generating function of this sequence is algebraic of degree 2.

The proof of this theorem will have a quite complicated, but elegant structure. So that we do not get lost in the details, it is important to get an overview of our goals first.

Note that no matter how large the number r is, it is a *fixed* number. On the other hand, n is changing, and will eventually be much larger than r. So the requirement that our permutations have only r copies of 132 really means that they have these copies in *exceptional cases* only.

In particular, every time there is an entry on the left of the maximal entry n that is smaller than an entry on the right of n, a 132-pattern is formed. This simple observation is crucial for us. Therefore, we introduce special terminology to discuss it.

Entries of an n-permutation p on the left of the entry n will be called *front entries*, whereas those on the right of n will be called *back entries*. Front entries of p that are smaller than the largest back entry of p will be called *black entries*, whereas back entries of p that are larger than the smallest front

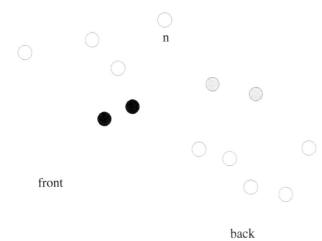

FIGURE 5.1

A generic permutation and its colored entries.

entry of p will be called *gray entries*. Note that not all entries will have colors. See Figure 5.1 for an illustration.

Why are we coloring our entries? First, any black entry is smaller than any front entry which is not black, while any gray entry is larger than any back entry which is not gray. In other words, black entries are the smallest front entries, while gray entries are the largest back entries. Moreover, any black and any gray entry is part of at least one 132-subsequence. Indeed, take any black entry x, the entry n, and any back entry larger than x. A dual argument applies for gray entries. Finally, if a 132-subsequence spans over the entry n, that is, it starts with a front entry and ends with a back entry, then it must start with a black one and end with a gray one.

Starting now, we are going to partition, and partition, and partition. That is, we will partition the set of n-permutations that contain r copies of 132 into many equivalence classes, which we will call *blocks*, in order to avoid confusion with the notion of permutation classes. How many blocks? A lot, but the number of these blocks will not depend on n, just on r, so it will be a *fixed* number. Then we are going to show that within each block C, the number of permutations $S_{132,r,C}(n)$ that are in C and have r copies of 132 is a polynomially recursive function of n. With that, we will be done by summing over all blocks C. Indeed, the sum of a fixed number of P-recursive functions is P-recursive.

Let us start this partitioning.

(a) As we said above, any colored entry is part of at least one subsequence of type 132 that spans over n. Therefore if p has exactly r subsequences of type 132, and B (resp. G) denotes the number of black (resp. gray) entries, then $\max(B, G) \leq r$. This implies that we have at most r^2 choices for the values of B and G.

(b) Once the values of B and G are given and we know in which position the entry n is, then we only have a bounded number of choices for the set of black and gray entries. Indeed, if x is the smallest black entry, then x is larger than all but G back entries. Thus, if n is in the ith position, then $x \geq n - i - G$. On the other hand, x is the smallest front entry, thus $x \leq n - i + 1$. A similar argument applies for the largest gray entry.

(c) Finally, there is only a bounded number of positions where a black entry can be. Indeed, if x is black and $y > x$ is a back entry, then $x\,z\,y$ is a 132-pattern for any front entry z on the right of x which is not black (and thus, is larger than any back entry). Recall that $x\,n\,y$ is such a pattern as well. Thus, if t is the number of such (x, z) pairs, then we have $t + G \leq r$. In particular, the distance between any black entry and the entry n cannot be larger than r.

The following definition makes use of the observations we have just made:

DEFINITION 5.23 *We say that the n-permutations p_1 and p_2 are in the same* strong equivalence class *if they agree in all of the following:*

- *position of the entry n*
- *set of black entries*
- *set of gray entries*
- *pattern formed by the gray entries*
- *position of the black entries*
- *pattern of consecutive entries starting with the leftmost black entry and ending with the entry n.*

In other words, permutations of the same block agree in everything that can be part of a 132-pattern spanning through the entry n. Note that once the position of n is given, there is only a bounded number of possibilities for the blocks of those permutations.

DEFINITION 5.24 *Let p be an n-permutation. The subsequence of p consisting of*

- *all black and gray entries and*
- *all front entries that are preceded by at least one black entry and*
- *the entry n*

is called the fundamental subsequence *of p.*

This means that permutations of the same block have identical fundamental subsequences, and the part of these subsequences that is weakly on the left of n is in the same position in each permutation belonging to the class.

DEFINITION 5.25 *The blocks C and C' are called* similar *if their permutations have fundamental subsequences that are identical as permutation patterns.*

Thus in this case there is no requirement that the left part of the fundamental subsequences must be in identical positions.

Example 5.26
The blocks containing the permutations 3 4 1 5 2 and 4 2 5 1 3 are similar. Indeed, permutations in the first block have fundamental subsequence 152, and permutations in the second block have fundamental subsequence 253.
□

5.1.5.1 Inductive Proof

Our proof of Theorem 5.22 will be an induction on r.

5.1.5.2 Initial Step

To make our argument easier to follow, we introduce some new notions:

DEFINITION 5.27 *Let q be a pattern. Then to insert the entry y to the jth position of q is to put y between the $j-1$st and the jth entry of q and to increase the value of any entry v for which originally $v \geq y$ held by 1.*
  The deletion of an entry is obtained similarly: erase the entry and decrease all entries larger than it by 1.

LEMMA 5.28
Let q be any pattern of length k. Then the number $C_q(n)$ of 132-avoiding n-permutations that end with a subsequence of type q is a P-recursive function of n.

PROOF The proof is by induction on k. If $k = 0$, then $C_q(n) = C_n = \binom{2n}{n}/(n+1)$, the nth Catalan-number. We say in Example 5.2 that C_n is a P-recursive sequence, so the initial case is proved.

Let us now assume that we know that the statement is true for all subsequences of length $k - 1$. We will prove it for the subsequence q, which has length k.

If q is not 132-avoiding, then clearly $C_q(n) = 0$. So we can suppose that q is 132-avoiding. Now we consider two separate cases.

(a) If the first element q_1 of q is *not* the smallest one, then the entry 1 of our n-permutation p cannot be on the left of q_1, thus in particular, the entry 1 of p is one of the last k entries, which form a subsequence of type q. Then it must be the smallest of these last k entries, thus we know exactly where the entry 1 of p is located. Let us delete the smallest entry of q to get the subsequence q'. Apply the induction hypothesis to q' to get that $C_{q'}(n-1)$ is P-recursive. Then insert 1 to its original place to see that $C_q(n)$ is P-recursive.

(b) If q_1 is the smallest element of q, then it is easy to apply what we have just shown in the previous case. Let q'' be the subsequence obtained from q by deleting q_1. Moreover, let $r_2, r_3, \cdots, r_k$ (resp.) be the subsequences whose last $k - 1$ elements determine a subsequence of type q'' and whose first elements (resp.) are $2, 3, \cdots, k$. Then it is obvious that

$$C_q(n) = C_{q''}(n) - \sum_{i=2}^{k} C_{q_i}(n) \qquad (5.11)$$

The first term of the right-hand side is P-recursive by induction and the second one is P-recursive by the previous case, and the lemma is proved.

This completes the proof. We will refer to the method applied in case (b) as the *complementing* method. ∎

COROLLARY 5.29
Let q be a pattern of length k. Then the number $K_q(n)$ of 132-avoiding n-permutations in which the k largest entries form a subsequence of type q is a P-recursive function of n.

PROOF   This is true as $K_q(n) = C_{q-1}(n)$. Indeed, taking inverses turns the last k entries of a permutation p into the k largest entries of the permutation p^{-1}. ∎

5.1.5.3   Induction Step

Now would be the time to carry out the induction step of our inductive proof. It turns out that this is easier to do if we generalize our statement even further.

THEOREM 5.30

Let q be any pattern of length k. Then the number $C_{q,r}(n)$ of n-permutations that contain exactly r subsequences of type 132 and end with a subsequence of type q is a P-recursive function of n.

Note that Theorem 5.22 is a very special case of Theorem 5.30. Also note that equivalently we can state the theorem for the numbers $K_{q,r}(n)$ of n-permutations that contain exactly r subsequences of type 132 and in which the largest k entries form a subsequence of type q.

PROOF   Induction on r. If $r = 0$, then Theorem 5.30 reduces to Lemma 5.28 and Corollary 5.29.

The inductive step will be carried out in a *divide and conquer* fashion. For brevity, we call the statement of Theorem 5.22 the *weak statement* and we call the statement of Theorem 5.30 the *strong statement*.

For our induction proof, we have to show that the strong statement for $r - 1$ implies the strong statement for r. We do this in two parts.

(a) We show that if the strong statement holds for all non-negative integers at most as large as $(r - 1)$, then the weak statement holds for r.

(b) Then we prove that if the weak statement holds for r, then the strong statement holds for r.

Let us follow up on these promises.

(a) Let us assume now that the strong statement holds for all non-negative integers at most as large as $r-1$. Choose any block C of n-permutations. Suppose the fundamental subsequence type of C contains exactly s subsequences of type 132, where $s \leq r$.

- Let $s \geq 1$. How can a permutation in C contain 132-patterns that are *not* contained in the fundamental subsequence? Clearly, they must be either entirely before the entry n or entirely after it. If there are i such subsequences before n and j such subsequences after n, then $i + j + s = r$ must hold. Denote by q_1 the pattern of all front entries in the fundamental subsequence and by q_2 the pattern of all back entries there. Then with the previous notation we have

$$f(n_1, n_2, q_1, q_2, i, j, s) = C_{q_1,i}(n_1) \cdot K_{q_2,j}(n_2)$$

such permutations, where n_1 (resp. n_2) denotes the number of front (resp. back) entries which are not in the fundamental subsequence. Indeed, entries of the fundamental subsequence are either the rightmost front entries, or the largest back entries. We know by induction that $C_{q_1,i}(n_1)$ is P-recursive in n_1 and $K_{q_2,j}(n_2)$ is P-recursive in n_2. Therefore, their convolution

$$f(n, q_1, q_2, i, j, s) = \sum_{n_1+n_2=n} f(n_1, n_2, q_1, q_2, i, j, s) \quad (5.12)$$

$$= \sum_{n_1+n_2=n} C_{q_1,i}(n_1) \cdot K_{q_2,j}(n_2) \quad (5.13)$$

is P-recursive in n. Clearly this convolution expresses the number of n-permutations with exactly r subsequences of type 132 in all blocks similar to C.

It is clear now that we have only a bounded number of choices for i, j, and s so that $i + j + s = r$, thus we can sum (5.12) over all these choices and still get that

$$f(n, q_1, q_2) = \sum_{i,j,s} f(n, q_1, q_2, i, j, s) \quad (5.14)$$

is P-recursive in n. (Recall that $s > 0$, thus we can always use the induction hypothesis.) Summing (5.14) over all q_1 and q_2 we get that

$$f(n) = \sum_{q_1,q_2} f(n, q_1, q_2) \quad (5.15)$$

As in this section, the only pattern we study is 132, we simplify our notations a little bit by writing $S_r(n)$ for $S_{n,r}(132)$.

- Now suppose $s = 0$. Then any 132-subsequence must be either entirely on the left of the entry n or entirely on the right of n. Moreover, the position of n completely determines the set of the front and back entries. If n is in the i-th position, and we have j 132-subsequences in the front and $r - j$ in the back, then this gives us

$$g(i, j) = S_j(i - 1)S_{r-j}(n - i) \quad (5.16)$$

permutations of the desired kind. If $1 \le j \le r - 1$, then the induction hypothesis applies for S_j and S_{r-j}, therefore, after summing (5.16) over all i,

$$g(n) = \sum_i S_j(i - 1)S_{r-j}(n - i) \quad (5.17)$$

is P-recursive in n. If $j = 0$ or $j = r$, then we cannot apply the induction hypothesis. By a similar argument as above we get nevertheless that in this case we have

$$2 \cdot \sum_i S_r(i - 1)C_{n-i} \quad (5.18)$$

n-permutations with exactly r 132-subsequences. (We remind the reader that $S_0(n-i) = Av_{n-i}(132) = C_{n-i}$, the $(n-i)$th Catalan-number.)

Summing (5.15), (5.17), and (5.18), we get

$$S_r(n) = f(n) + g(n) + 2 \cdot \sum_i S_r(i-1) \cdot C_{n-i}. \qquad (5.19)$$

Let F, G, C, and S_r denote the ordinary generating functions of $f(n)$, $g(n)$, C_n, and $S_r(n)$. Then the previous equation yields

$$S_r(z) = F(z) + G(z) + 2z \cdot C(z)S(z),$$

that is,

$$S_r(z) = \frac{F(z) + G(z)}{1 - 2z \cdot C(z)}. \qquad (5.20)$$

Therefore, $S_r(z)$ is d-finite as it is the product of two d-finite power series. Indeed, the numerator is d-finite and $1/(1-2zC(z)) = 1/\sqrt{1-4z}$ is d-finite as it is algebraic. Thus, $S_r(n)$ is P-recursive and we are done with the first part of the proof.

(b) Now we prove that the weak statement for r implies the strong statement for r. Let q be any subsequence of length k. We must prove that the number $C_{q,r}(n)$ of n-permutations that end with a subsequence of type q and contain exactly r subsequences of type 132 is a P-recursive function of n. Clearly, if q contains more than r 132-subsequences, then $C_{q,r}(n) = 0$ and we are done. Otherwise, we will do induction on k, the case of $k = 1$ being obvious. There are three different cases to consider.

- If q has more than r inversions, then it is obvious that no such permutation can have its entry 1 on the left of the last k elements. Therefore, this entry 1 must be a part of the q-subsequence formed by the last k elements. Now, deleting this entry 1 we may or may not lose some 132-patterns, as there may or may not be inversions on its right, but we can read off this information from q. (See the next example.) Again, let q' be the pattern obtained from q by deleting its entry 1. If we do not lose any 132-patterns by this deletion, then we are left with an $(n-1)$-permutation ending with the pattern q' and having r subsequences of type 132. If we lose t such patterns, then we are left with an $(n-1)$-permutation ending with q' and having $r - t$ such subsequences. Both cases give rise to a P-recursive function of n by our induction hypothesis as q' is shorter than q.

Example 5.31

If $q = 3\,41\,6\,5\,2$, then $q' = 2\,3\,5\,4\,1$; thus we lose three subsequences of type 132 when deleting 1. Therefore, we can apply our inductive hypothesis for $r - 3$, then reinsert the entry 1 to its place. If $q = 3124$, then we do not lose any 132-patterns when deleting the entry 1 and getting $q' = 2\,1\,3$. Thus we still need to count permutations with r 132-patterns, but they must end with q', not with q. The pattern q' is shorter than q, thus the induction hypothesis on k can be applied. □

- If q has at most r inversions, but q is not the monotonic pattern $1\,2\cdots k$, then it can also happen that the entry 1 is not among the last k entries of our permutation. However, we claim that it cannot be too far away from them. Indeed, let y be an element from the last k elements of the permutation (thus one of the elements of the set L of the last k entries that form the ending q) that is smaller than some other element $x \in L$ on its left. Then clearly, if n is large enough, then y must be smaller than $r+k+1$; otherwise, we would have too many 132-patterns of the form $w\,x\,y$. So y is bounded. If the entry 1 of the permutation were more than $2r + k + 1$ to the left of y, then there would necessarily be more than r elements between 1 and y that are larger than y, a contradiction. Thus the distance between 1 and y is bounded. Therefore, we can consider all possibilities for the position of the entry 1 of the permutation and for the subsequence on its right. In each case we can delete the entry 1 and reduce the enumeration to one with a smaller value of r (as q has at least one inversion), then use the inductive hypothesis on r. Thus this case contributes a bounded number of P-recursive functions, too.

- Finally, if q is the monotonic pattern $1\,2\cdots k$, then use the complementing method of Lemma 5.28.

Now note that speaking in terms of ordinary generating functions, all operations we made throughout the induction step were either adding or multiplying a finite number of power series together. In particular, the ordinary generating function $C(z)$ of our initial C_n-sequence (that is, when $r = 0$ and $k = 0$) is $C(z) = \frac{1-\sqrt{1-4z}}{2z}$, thus an algebraic power series. Therefore, the ordinary generating function of $S_r(n)$, the power series $S_r(z)$, is algebraic, too.

Now note a bit more precisely that throughout our proof we have either added formal power series together, or, as in (5.16), multiplied two functions of type $S_j(i-1)$ together, or, as in (5.20), multiplied a power series by the power series $1/(1 - 2z \cdot C(z)) = 1/\sqrt{1 - 4z}$. Therefore, the following proposition is immediate:

PROPOSITION 5.32

Let $r \geq 1$, and recall that $S_r(z)$ denotes the ordinary generating function of the numbers $S_r(n)$. Then $S_r(z) \in \mathbf{C}[[z, \sqrt{1-4z}]]$. Moreover, when written in smallest terms, the denominator of $S_r(z)$ is a power of $(\sqrt{1-4z})$.

It is convenient to work in this setting as the square of $\sqrt{1-4z}$ is an element of $\mathbf{C}[z]$, which makes computations much easier.

We are going to determine the exponent $f(r)$ that $\sqrt{1-4z}$ has in the denominator of $G_r(z)$. Equations (5.16) and (5.20) show that

$$f(r) = max_{1 \leq i < r}(f(i) + f(r-i)) + 1. \tag{5.21}$$

We now claim that $f(r) = 2r - 1$. It is easy to compute (see [47]) that $f(1) = 1$. Now suppose by induction that we know our claim for all positive integers smaller than r. Then (5.21) and the induction hypothesis yield that for some i, we have

$$f(r) = (f(i) + f(r-i)) + 1 = (2i-1) + (2r-2i-1) + 1 = 2r-1,$$

which was to be proved.

Recall now that $1/\sqrt{1-4z} = \sum_{n \geq 0} \binom{2n}{n} z^n$ and that the sequence $\binom{2n}{n}$ satisfies a linear recursion. Differentiate both sides of this equation several times. On the left-hand side, each differentiation will add two to the exponent of $\sqrt{1-4z}$ in the denominator. On the right-handvside, it will add one to the degree of the highest-degree polynomials appearing in the recursive formula for the coefficients. Thus, differentiating $r-1$ times we get that the denominator of $S_r(z)$ gives rise to a polynomial recursion of degree r. The numerator of $S_r(z)$ cannot increase this degree.

We collect our observations in our last lemma:

LEMMA 5.33

Let $r \geq 1$ and let $S_r(x)$ be the generating function for the sequence $\{S_{n,r}(132)\}_n$. Write $S_r(z)$ in lowest terms. Then the denominator of $S_r(z)$ is equal to

$$(\sqrt{1-4z})^{2r-1} = (1-z)^{r-1} \cdot \sqrt{1-4z}.$$

Therefore, the sequence $\{S_{n,r}(132)\}_n$ satisfies a polynomial recurrence relation with degree at most r.

∎

5.2 Containing a Pattern Many Times

5.2.1 Packing Densities

In this section we will turn around and instead of trying to avoid a pattern, or containing a pattern just a few times, we will try to contain a given pattern q as many times as possible. More precisely, for fixed q and n, we want to find the largest integer $M_{n,q}$ so that there exists an n-permutation that contains exactly $M_{n,q}$ copies of q. However, for divisibility or other reasons, it can happen that not all integers n behave in the same way as far as containment of many copies of q is concerned. Therefore, we would like to have a more global way of measuring how many copies of q can be packed into an n-permutation. The following definition establishes this measurement.

DEFINITION 5.34 *Let q be a pattern of k elements. Then the* packing density *of q, denoted $g(q)$, is given by*

$$g(q) = \lim_{n \to \infty} \frac{M_{n,q}}{\binom{n}{k}}.$$

In other words, the packing density of q tells us how large a portion of all k-element subwords of an n-permutation can be copies of q as n goes to infinity.

The alert reader probably caught us red-handed in committing the sin of using $\lim_{n \to \infty} \frac{M_{n,q}}{\binom{n}{k}}$ to define packing density, without proving first that that limit exists. The limit does exist, however, as can be seen from the following result of Galvin, which appeared in [268]. In what follows, if an n-permutation p contains $M(n, q)$ copies of q, then we will say that p is a q-*optimal* permutation.

LEMMA 5.35
Let q be a pattern of length k. Then for $n \geq k$, the sequence

$$\frac{M_{n,q}}{\binom{n}{k}}$$

is nonincreasing, therefore $\lim_{n \to \infty} \frac{M_{n,q}}{\binom{n}{k}}$ exists.

PROOF Let $g(n, q) = M_{n,q} / \binom{n}{k}$. We show that

$$g(n, q) \leq g(n - 1, q),$$

which will clearly imply our claim.

Let p be an n-permutation that is q-optimal, that is, contains a maximal number of copies of q. Let $p_{\langle 1 \rangle}, p_{\langle 2 \rangle}, \cdots, p_{\langle n \rangle}$ denote the n permutations of length $n - 1$ that we obtain from p by omitting the first, second, and so on, entry of p. It is clear that the proportion $g(n, q)$ of q-copies among all k-element subwords of p is the average of the proportions $prop(p, q, i)$ of q-copies among all k-element subwords of the $p_{\langle i \rangle}$. (The average is taken over all $i \in [n]$.) There has to be at least one i for which $prop(p, q, i)$ is not below that average, that is, for which

$$g(n, q) = \frac{\sum_{j=1}^{n} prop(p, q, j)}{n} \leq prop(p, q, i) \leq g(n - 1, q).$$

This shows that our sequence is nonincreasing, and therefore, consisting of nonnegative elements, convergent. ∎

If q is a monotone pattern of length k, then the answer is trivial. Indeed, we have $M_{n,q} = \binom{n}{k}$, and this maximum is attained on the monotone permutation of length n. Therefore, the packing density of any monotone pattern is 1. Hence, for the remainder of this section, we will assume that q is not monotone (unless otherwise stated).

5.2.2 Layered Patterns

For a general pattern, however, the question of determining $M_{n,q}$ is still unsolved. However, for layered patterns, which we introduced in Definition 4.94, we can say significantly more. This is because of the following unpublished theorem of Walter Stromquist. While Stromquist has never published his proof, it can be found in [268].

THEOREM 5.36
Let q be a layered pattern. Then there exists a layered *permutation p so that p has exactly $M_{n,q}$ copies of q.*

That is, among the n-permutations that are q-optimal, there is at least one layered permutation. The number of layered permutations of length n is obviously 2^{n-1}, which is much less than $n!$, so the above theorem is extremely useful if we want to find a q-optimal permutation with the help of a computer program.

We point out that while it is not true in general that *all* q-optimal permutations are layered, there are special cases when they are. See Exercise 27 for a large special case.

What can we say about the structure of a layered q-optimal permutation when q is layered? Even this question is unsolved in this generality. However, there are certain precise answers in the case when q has only two layers. We will discuss the shortest of these patterns, the pattern $q = 132$. For the more general case of $q = 1k(k-1)\cdots 32$, see [55]. For a different, and also more general treatment, see [268].

5.2.2.1 Pattern 132

We want to construct an n-permutation that contains 132 as many times as possible. Our intuition might suggest that (disregarding questions of divisibility) we split the set $[n]$ into three parts, then place entries from 1 to $n/3$ into the first $n/3$ positions, place entries from $\frac{n}{3}+1$ to $\frac{2n}{3}$ into the *last* $n/3$ positions, finally place entries from $\frac{2n}{3}+1$ to n into the middle $n/3$ positions. Then we apply the same strategy recursively in each of the three tiers we have just created. This construction imitates the pattern 132 quite closely, so it seems reasonable to conjecture that it is optimal. Nevertheless, this conjecture is *false*. For instance, if $n = 9$, then this method constructs the permutation 132 798 465 that has 36 copies of 132 (27 copies containing one entry from each tier, three copies within one tier, and six copies containing one entry from the first tier and two entries in decreasing order from one of the other tiers). On the other hand, the permutation 1 32 987654 contains 46 copies of 132. Indeed, choosing one of the three leftmost entries and two of the six rightmost entries in $3 \cdot \binom{6}{2} = 45$ ways we get a copy of 132, and we also get a copy of 132 if we select the three leftmost entries. Therefore, we have to look for 132-optimal permutations with other methods.

Let p be a 132-optimal n-permutation that is layered. We know from Theorem 5.36 that such a permutation exists. We will show that layered 132-optimal permutations have a simple recursive structure. This is because of the fact, which we will use many times, that to form a 132 pattern in a layered permutation one must take a single element from some layer and a pair of elements from a subsequent layer.

PROPOSITION 5.37
Let p be a layered 132-optimal n-permutation whose last layer is of length m. Then the leftmost $k = n - m$ elements of p form a 132-optimal k-permutation.

PROOF Let D_k be the number of 132-copies of p that are disjoint from the last layer. The number of 132-copies of p is clearly $k\binom{m}{2} + D_k$. So once k is chosen, p will have the maximum number of copies if D_k is maximal. (Note that this argument works even if $m = 1$ as $\binom{1}{2} = 0$.) ∎

We point out that the proof of this proposition uses the fact that 132 has only two layers, the first of which is a singleton.

As we are now concentrating on copies of 132, we simplify our notation by setting $M_n = M_{n,132}$. That is, M_n denotes the maximum number of 132-copies an n-permutation can have. Then the previous proposition implies that

$$M_n = \max_k \left(M_k + k \binom{m}{2} \right). \tag{5.22}$$

The integer k for which the right-hand side attains its maximum will play a crucial role throughout this section and the next one. Therefore, we introduce specific notation for it.

DEFINITION 5.38 *For any positive integer n, let k_n be the positive integer for which*

$$M_n = \max_{k<n} \left(M_k + k \binom{m}{2} \right)$$

is maximal. If there are several integers with this property, then let k_n be the largest among them.

In other words, k_n is the largest possible length of the remaining permutation after removing the last layer of an optimal n-permutation p. When there is no danger of confusion, we will only write k instead of k_n, to simplify notation. We will also always use $m = n - k$ to denote the length of the last layer of p.

We continue our search for 132-optimal n-permutations. The construction at the beginning of this section, which was shown not to be optimal, leads to n-permutations with

$$\left(\frac{n}{3}\right)^3 + 3 \cdot \left(\frac{n}{9}\right)^3 + \cdots + 3^d \cdot \left(\frac{n}{3^{d+1}}\right)^3 = \frac{n^3}{27} \sum_{i=0}^{d} 9^{-i} \sim \frac{n^3}{24}$$

copies of 132. Our goal is to find a construction, for general n, that provides more copies. The proof of the following lemma provides such a construction.

LEMMA 5.39
The inequality

$$g(132) = \lim_{n \to \infty} \frac{M_n}{\binom{n}{3}} \geq 2\sqrt{3} - 3 \approx 0.464$$

holds.

PROOF Let us look for 132-optimal n-permutations in a specific form, which we will call *geometric form*. In these permutations, disregarding questions of divisibility, the last layer is of length $m = t \cdot n$ (for some $t < 1$), the penultimate layer is of length $(1 - t)tn$, the one before that is of length $(1 - t)^2 tn$, and so on. In other words, the layer lengths form a geometric progression with quotient $1 - t$.

How many 132-patterns will such a permutation $p(t)$ contain? It follows from the layered structure of $p(t)$ that in a copy of 132 in $p(t)$, the last two entries will come from the same layer. We will count the 132-copies according to the layers that contain their last two entries.

There are $\binom{m}{2}k = \frac{1}{2} \cdot t^2(1-t)n^3 - O(n^2)$ copies of 132 that have their last two entries on the last layer of $p(t)$. Similarly, there are roughly $\frac{1}{2} \cdot t^2(1-t)^4 n^3$ copies of 132 that have their last two entries on the penultimate layer. There are roughly $\frac{1}{2} \cdot t^2(1-t)^7 n^3$ copies that have their last two entries on the layer before that, and so on. Therefore, if n is large enough, the n-permutation $p(t)$ created in this way will have

$$\frac{n^3}{2}t^2(1-t)\sum_{j\geq 1}(1-t)^{3j} - O(n^2\log n) = \frac{n^3 t^2(1-t)}{2(1-(1-t)^3)} - O(n^2\log n) \quad (5.23)$$

copies of 132. Indeed, there are at most $O(\log n)$ layers, and on each layer, the error term is at most $O(n^2)$. Note that, for instance, for $t = 2/3$ (this is what we used in our counterexample above), the above formula shows that we have $n^3/13 - O(n^2\log n)$ copies of 132, which is significantly more than the $n^3/24$ achieved by our first construction. However, $t = 2/3$ is still not the best choice. To find the best choice for t, we simply use elementary calculus to find the maximum of the function

$$f(t) = \frac{t^2(1-t)}{2(1-(1-t)^3)} = \frac{t(1-t)}{2(t^2-3t+3)}.$$

We find that the maximum of f on the interval $[0,1]$ is taken at $t = \frac{3-\sqrt{3}}{2} \approx 0.634$, so this is the best choice for t.

Our construction works for all n, proving that

$$M_n \geq \frac{t(1-t)}{2(t^2-3t+3)}n^3 - O(n^2\log n).$$

It is then clear from the definition of $g(132)$ that

$$g(132) = \lim_{n\to\infty}\frac{M_n}{\binom{n}{3}} \geq \lim_{n\to\infty}\max_{t\in[0,1]}\frac{f(t)}{1/6} = 2\sqrt{3} - 3 \approx 0.464. \quad (5.24)$$

∎

We claim that in the long run, nothing beats patterns in geometric form. In other words, the above construction is asymptotically optimal among all n permutations. By this we mean that the weak inequality in (5.24) is actually an *equality*. This is the content of the following theorem.

THEOREM 5.40

[268] The above construction is asymptotically optimal, that is,

$$g(132) = 2\sqrt{3} - 3 \approx 0.464.$$

PROOF   Let

$$B = \max_{z \in [0,1]} \frac{3z(1-z)}{1+z+z^2} = \max_{z \in [0,1]} h(z). \tag{5.25}$$

We will first prove that $B = g(132)$, then we will prove that

$$B = 6 \max_{t \in [0,1]} f(t),$$

which will complete the proof of our theorem by (5.24). To prove the first claim, it suffices to prove that

$$\frac{M_n}{\binom{n}{3}} \le B + O\left(\frac{1}{n}\right), \tag{5.26}$$

or, in other words, that for all n, we have

$$M_n \le B\frac{n^3}{6}.$$

We are going to prove this claim by induction on n. For $n = 1$, the statement is true as the left-hand side is equal to 0. Now assume that we know the statement for all positive integers less than n.

Recall that we have described the structure of 132-optimal permutations in Proposition 5.37, and in the subsequent formula (5.22). Therefore, we have

$$M_n = \max_k \left(M_k + k\binom{m}{2} \right) \tag{5.27}$$

$$= \max_k \left(M_k + k\binom{m}{2} \right) \le \max_k \left(M_k + km^2 \right) \tag{5.28}$$

$$= \max_k \left(B\frac{k^3}{6} + \frac{km^2}{6} \right) \tag{5.29}$$

$$= \max_{t \in (0,1)} \left(\frac{n^3}{6}(B(1-t)^3 + 3(1-t)t^2) \right). \tag{5.30}$$

In the last step, we dropped the requirement that $tn = m$ and $(1-t)n = k$ be integers.

Remember that B was defined as a maximum in (5.25). From that definition it follows that

$$\frac{3z(1-z)}{1+z+z^2} \le B$$

$$3z(1-z)^2 \le B(1-z^3)$$

$$Bz^3 + 3z(1-z)^2 \le B,$$

or, setting $t = 1 - z$,

$$B(1-t)^3 + 3(1-t)t^2 \le B.$$

Comparing this last equation with the last term of (5.30) proves (5.26). So, we have shown that $g(132) = B$.

Therefore, our theorem will be proved if we show that $B = 6 \max_{t \in [0,1]} f(t)$. This is immediate if we observe that $h(z) = 6f(1 - z)$. ∎

As this is our first result in the theory of packing densities, it is worth pointing out the interesting fact that while our problem is about integers, the best ratio of consecutive layer lengths in order to contain many copies of 132 was provided by an *irrational* number, the number $\frac{\sqrt{3}-1}{2}$.

Most results on packing densities are about layered patterns, which is not surprising since Theorem 5.36 is a very powerful tool in studying permutations containing a maximum number of copies of such patterns. A sampling of these results is contained in the Exercises. We point out that Problem Plus 21 contains an interesting result about a *non-layered* pattern.

5.3   Containing a Pattern a Given Number of Times

In the last section, we looked for permutations that contained a given pattern as many times as possible. In this section, we will take a different approach and we will be looking for permutations that contain a given pattern a *given number of times*.

DEFINITION 5.41   *Let n be a fixed positive integer, and let q be a fixed pattern. Let $S_{n,c}(q)$ be the number of n-permutations with exactly c patterns of type q. The sequence $\{S_{n,c}(q)\}_{c \geq 0}$ is called the* frequency sequence *of the pattern q for n.*

The alert and meticulous reader will point out that we have already seen an example for a frequency sequence, and we have proved several interesting statements about it. Indeed, *inversions*, our main topic in Section 2.1, are nothing but 21-patterns. We have seen that the frequency sequence of inversions is log-concave, and is therefore unimodal.

When q is longer than 2, numerical evidence suggests that the frequency sequence of q will no longer be unimodal, let alone log-concave. In fact, disturbing objects called *internal zeros* seem to be present in most frequency sequences.

DEFINITION 5.42   *An integer c is called an* internal zero *of the sequence $\{S_{n,c}(q)\}_{c \geq 0}$ if we have $S_{n,c}(q) = 0$, but there exist c_1 and c_2 so that*

(a) $c_1 < c < c_2$, and

(b) $S_{n,c_1}(q) > 0$, *and* $S_{n,c_2}(q) > 0$.

Example 5.43

Let $q = 123$, and let $n > 3$ be any positive integer. Then $c = \binom{n}{3} - 1$ is an internal zero of the sequence $(S_{n,c}(q))_{c \geq 0}$ as we can have $c_1 = \binom{n-1}{3}$ and $c_2 = \binom{n}{3}$ for the permutations $n12 \cdots n - 1$ and $123 \cdots n$, respectively. □

5.3.1 Construction with a Given Number of Copies

In this subsection we show that there are infinitely many integers n so that the sequence $F_n = S_{n,c}(132)$ does not have internal zeros. We will call such an integer, or its corresponding sequence, *NIZ (no internal zero)*, and otherwise *IZ*. Our strategy is recursive; we will show that if k_n is NIZ, so is n, where k_n was described in Definition 5.38. As $k_n < n$, this will lead to an infinite sequence of NIZ integers. There is a problem, however. In order for this strategy to work, we must ensure that given k, there is an n such that $k = k_n$. This is the purpose of the following theorem.

THEOREM 5.44

The sequence $\{k_n\}_{n \geq 1}$ *diverges to infinity and satisfies*

$$k_n \leq k_{n+1} \leq k_n + 1$$

for all $n \geq 1$. *So, in particular, for all positive integers* k *there is a positive integer* n *so that* $k_n = k$.

The next subsection is devoted to a proof of this theorem. We suggest that the reader assume the result now and continue with this section to preserve continuity. Keep the notation M_k from the previous section. Before starting the proof of our main theorem, we need only note the useful fact that

$$M_k \geq \binom{k-1}{2} \tag{5.31}$$

which follows by considering the permutation $1k(k-1)(k-2) \cdots 32$.

THEOREM 5.45

There are infinitely many NIZ integers.

PROOF By Theorem 5.44, it suffices to show (for sufficiently large k_n) that if k_n is NIZ, then so is n. By the same theorem, we can choose n so that k_n is NIZ, and $k_n \geq 4$. To simplify notation in what follows, we will write k for k_n.

Now given c with $0 \le c \le M_n = M_k + k\binom{m}{2}$, we will construct a permutation $p \in S_n$ having $c = c(p)$ copies of 132. Because of (5.31) and $k \ge 4$, we have $M_k \ge k - 1$. So it is possible to write c (not necessarily uniquely) as $c = ks + t$ with $0 \le s \le \binom{m}{2}$ and $0 \le t \le M_k$. Since k is NIZ, there is a permutation $p' \in S_k$ with $c(p') = t$. Also, it is not difficult to prove (see Exercise 15) that there is a permutation in S_m with no copies of 132 and s copies of 21. Let p'' be the result of adding k to every element of that permutation. Then, by construction, $p = p'p'' \in S_n$ and $c(p) = ks + t = c$ as desired. ∎

One can modify the proof of the previous theorem to locate precisely where the internal zeros could be for an IZ sequence. We will need the fact (established by computer) that for $n \le 12$ the only IZ integers are 6, 8, and 9, and that they all satisfied the following result.

THEOREM 5.46

For any positive integer n, the sequence F_n does not have internal zeros, except possibly for $c = M_n - 1$ or $c = M_n - 2$, but not both.

PROOF   We prove this theorem by induction on n. Numerical evidence shows that the statement is true for $n \le 12$. Now suppose we know the statement for all integers smaller than n, and prove it for n. If n is NIZ, then we are done.

If n is IZ then, by the proof of Theorem 5.45, $k = k_n$ is IZ. So $k \ge 6$ and we have $M_k \ge k + 2$ by (5.31). Now take c with $0 \le c \le M_n - 3$ so that we can write $c = ks + t$ with $0 \le s \le \binom{m}{2}$ and $0 \le t \le M_k - 3$. Since the portion of F_k up to $S_{k,132}(M_k - 3)$ has no internal zeros by induction, we can use the same technique as in the previous theorem to construct a permutation p with $c(p) = c$ for c in the given range. Furthermore, this construction shows that if $S_{k,132}(M_k - i) \ne 0$ for $i = 1$ or 2 then $S_{n,132}(M_n - i) \ne 0$. This completes the proof. ∎

5.3.2   Sequence $\{k_n\}_{n \ge 0}$

In order to prove Theorem 5.44, we first need a lemma about the lengths of various parts of a 132-optimal permutation p. In all that follows, we use the notation

$b = $ length of the next-to-last layer of p

$a = $ length of the string created by removing the last two layers of p

$\quad = n - m - b$

$\quad = k - b.$

The following lemma summarizes some innocent-looking properties of the above layer lengths that we will need in our proof of Theorem 5.44. As the reader will see, these properties are actually not all as immediate as they look.

LEMMA 5.47

The following inequalities hold:

(i) $b \le m$,

(ii) $a \le (m-1)/2$,

(iii) $m > k$, *which implies* $m > n/2$ *and* $k < n/2$,

(iv) $m \le 2(n+1)/3$.

PROOF   The basic idea behind all four of the inequalities is as follows. Let p' be the permutation obtained from our 132-optimal permutation p by replacing its last two layers with a last layer of length m' and a next-to-last layer of length b'. Then in passing from p to p' we lose some 132-patterns and gain some. Since p was optimal, the number lost must be at least as large as the number gained. And this inequality can be manipulated to give the one desired.

For the details, the following chart gives the relevant information to describe p' for each of the four inequalities. In the second case, the last two layers of p are combined into one, so the value of b' is irrelevant.

m'	b'	number of gained 132-patterns $\le$ number of lost 132-patterns
b	m	$m\binom{b}{2} \le b\binom{m}{2}$
$b+m$	0	$abm \le b\binom{m}{2}$
$m+1$	$b-1$	$(a+b-1)m \le \binom{m}{2} + a(b-1)$
$m-1$	$b+1$	$\binom{m-1}{2} + ab \le (a+b)(m-1)$

Now (i) and (ii) follow immediately by canceling bm from the inequalities in the first two rows of the table. From these two, it follows that $a(b-1) < \binom{m}{2}$. So using the third line of the chart

$$(k-1)m = (a+b-1)m \le \binom{m}{2} + a(b-1) < 2\binom{m}{2} = m(m-1)$$

and cancelling m gives (iii). For (iv) we have

$$\binom{m-1}{2} \le \binom{m-1}{2} + ab \le (a+b)(m-1) = (n-m)(m-1).$$

Canceling $m - 1$ and solving for m completes the proof.   ∎

We now turn to the proof of Theorem 5.44. First note that, by Lemma 5.47 (iv), we have

$$k = n - m \geq \frac{n-2}{3}.$$

So $\{k_n\}_{n \geq 1}$ clearly diverges to infinity. For our next step, we prove that $\{k_n\}_{n \geq 1}$ is monotonically weakly increasing. Let $p_{n,i}$ denote an n-permutation whose last layer is of length $n - i$, and whose leftmost i entries form an optimal i-permutation, and let $c_{n,i} = c(p_{n,i})$. Clearly

$$c_{n,i} = M_i + i \binom{n-i}{2}.$$

PROPOSITION 5.48
For all $n \geq 1$, the inequality $k_n \leq k_{n+1}$ holds.

PROOF   As usual, let $k = k_n$. Then it suffices to show that $c_{n+1,k} \geq c_{n+1,i}$ for all $i < k$. This is equivalent to showing that

$$M_k + k \binom{n+1-k}{2} \geq M_i + i \binom{n+1-i}{2}. \tag{5.32}$$

However, by the definition of k, we know that for all $i < k$,

$$M_k + k \binom{n-k}{2} \geq M_i + i \binom{n-i}{2}. \tag{5.33}$$

Comparing (5.33) and (5.32), we see that all we need to do is to prove the inequality $k(n - k) \geq i(n - i)$. Rearranging terms in order to cancel $k - i$ gives the equivalent inequality $n \geq k + i$. However, $k < n/2$ by Lemma 5.47 (iii), and so the bound on i gives $k + i < 2k < n$.   ∎

The proof of the upper bound on k_{n+1} is a bit more involved but follows the same general lines as the previous demonstration. Note that this will finish the proof of Theorem 5.44.

LEMMA 5.49
For all positive integers n, the inequality $k_n \leq k_{n+1} \leq k_n + 1$ holds.

PROOF   We are going to use induction on n. The statement is easy to check for $n \leq 2$. Suppose we know that the lemma is true for integers smaller than or equal to n, and prove it for $n + 1$. For simplicity, set $k = k_n$,

$m = n - k$, and $c_i = c_{n+1,i}$. Since we have already proved the lower bound, it suffices to show that

$$c_i \geq c_{i+1} \text{ for } k + 1 \leq i < \left\lfloor \frac{n+1}{2} \right\rfloor. \tag{5.34}$$

Note that we do not have to consider $i \geq \lfloor (n+1)/2 \rfloor$ because of Lemma 5.47 (iii).

We prove (5.34) by induction on i. For the base case, $i = k + 1$, we wish to show

$$M_{k+1} + (k+1)\binom{m}{2} \geq M_{k+2} + (k+2)\binom{m-1}{2}. \tag{5.35}$$

However, since $p_{n,k}$ is optimal by assumption, we have

$$M_k + k\binom{m}{2} \geq M_{k+1} + (k+1)\binom{m-1}{2}. \tag{5.36}$$

Comparing (5.36) and (5.35) and rearranging terms, it suffices to prove

$$m - 1 \geq (M_{k+2} - M_{k+1}) - (M_{k+1} - M_k). \tag{5.37}$$

Let $p' \in S_k$, $p'' \in S_{k+1}$, and $p''' \in S_{k+2}$ be layered 132-optimal permutations having last layer lengths m', m'', and m''', respectively, as short as possible. Since $n \geq 2$ and $k < n/2$, we have $k + 2 \leq n$ and so, by induction, these three permutations satisfy the lemma. If $m'' = m' + 1$ then let x be the largest element in the last layer of p'' (namely $x = k + 1$). Otherwise, $m'' = m'$ and removing the last layer of both p' and p'' leaves permutations in $S_{k-m'}$ and $S_{k-m'+1}$, respectively. We can iterate this process until we find the single layer where p' and p'' have different lengths (those lengths must differ by 1) and let x be the largest element in that layer of p''. Similarly we can find the element y, which is largest in the unique layer where p'' and p''' have different lengths.

Now let

r = number of 132-patterns in p''' containing neither x nor y,

s = number of 132-patterns in p''' containing x but not y,

t = number of 132-patterns in p''' containing y but not x, and

u = number of 132-patterns in p''' containing both x and y.

Note that there is a bijection between the 132-patterns of p''' not containing y and the 132-patterns of p''. A similar statement holds for p'' and p'. So

$$M_k = r, \quad M_{k+1} = r + s, \quad M_{k+2} = r + s + t + u.$$

Note also that $s \geq t$ because increasing the length of the layer of x results in the greatest number of 132-patterns being added to p'. It follows that

$$(M_{k+2} - M_{k+1}) - (M_{k+1} - M_k) = t + u - s \leq u.$$

However, there are only k elements of p''' other than x and y, so $u \leq k \leq m-1$ by Lemma 5.47 (iii). This completes the proof of (5.37) and of the base case for the induction on i.

The proof of the induction step is similar. Assume that (5.34) is true for $i-1$ so that

$$M_{i-1} + (i-1)\binom{l+1}{2} \geq M_i + i\binom{l}{2}. \tag{5.38}$$

where $l = n+1-i$. We wish to prove

$$M_i + i\binom{l}{2} \geq M_{i+1} + (i+1)\binom{l-1}{2}. \tag{5.39}$$

Comparing the last two ineqaualities as usual and simplifying, we need to show

$$2l - i - 1 \geq (M_{i+1} - M_i) - (M_i - M_{i-1}).$$

Proceeding exactly as in the base case, we will be done if we can show that $2l-i-1 \geq i-1$ or equivalently $l \geq i$. However, this is straightforward because $l = n+1-i$ and $i < \lfloor (n+1)/2 \rfloor$. ∎

We have seen that there are infinitely many NIZ integers. It is natural to ask whether there are infinitely many IZ integers as well. The answer is in the affirmative, and can be found in [55].

Exercises

1. Find an explicit formula for $S_{132,1}(n)$.

2. Prove that the number of ways to dissect a convex $(n+1)$-gon into $n-2$ parts with noncrossing diagonals is $S_{132,1}(n)$.

3. (+) Generalizing the previous problem, find a formula for the number $f(n,d)$ of ways to dissect a convex $n+2$-gon by d non-intersecting diagonals.

4. Let f_n and g_n be two sequences that differ only in a finite number of terms. Prove that f_n is P-recursive if and only if g_n is P-recursive.

5. Prove that for any fixed k, the sequence $\{S(n,k)\}_n$ is P-recursive.

6. Prove that for any fixed k, the sequence $\{P(n,k)\}_n$ is P-recursive. See Exercise 33 of Chapter 1 for the definition of $P(n,k)$.

7. Let $A(z)$ and $B(z)$ be two algebraic power series. Prove that $A(z)B(z)$ and $A(z) + B(z)$ are also algebraic power series.

8. Let $p_k(n)$ be the number of partitions of the integer n into at most k parts. Is $p_k(n)$ a P-recursive sequence?

9. Prove that $f(z) = \sin z$ is d-finite, but not algebraic.

10. Prove that $Av_n(3142, 4231)$ is a P-recursive sequence.

11. Prove that $Av_n(1342, 2431)$ is a P-recursive sequence.

12. Prove that the sequence $f(n) = (n!)^n$ is not P-recursive.

13. (–) Characterize layered permutations by pattern avoidance.

14. Find the packing density of the pattern $q_k = 1 \, (k+1) \, k \cdots 2$.

15. Prove that for any nonnegative integer s with $s \le \binom{n}{2}$ there is a permutation $p \in S_n$ having s copies of the pattern 21 and no copies of 132.

16. Let N be a positive integer. Show that there exists a pattern q and a positive integer n so that the frequency sequence $(S_{n,q}(c))_{c \ge 0}$ contains N consecutive internal zeros.

17. (–) Let p be a 321-avoiding n-permutation. At most how many inversions can p contain?

18. Let p be a $k \cdots 321$-avoiding n-permutation. At most how many copies of $q = (k-1) \cdots 321$ can p contain?

19. A k-*superpattern* is a permutation that contains all $k!$ patterns of length k. Let $sp(k)$ be the length of the shortest k-superpattern. For instance, $sp(2) = 3$, as 132 is a 2-superpattern of length three, and obviously, there is no shorter 2-superpattern.

    (a) Determine $sp(3)$.
    (b) Prove that $sp(4) \le 10$.
    (c) Prove that $sp(k) \le k^2$.

20. Prove that $sp(k) \le (k-1)^2 + 1$.

21. (a) Prove that $sp(4) > 6$.
    (b) Prove that $sp(4) > 7$.

22. (a) Prove that if k is large enough, then $sp(k) \ge \frac{k^2}{e^2}$.
    (b) Prove that for *all* integers $k \ge 4$, we have $sp(k) \ge 2k$.

23. Let $p \in S_8$. Then p has $2^8 = 256$ subwords. Let S be the set of these subwords. Prove that the number of *distinct* permutations patterns that occur in S is at most 127.

24. Find $M_{n,2143}$.

25. Let q be a layered pattern consisting of k layers of length two each. Is it true that the q-optimal layered n-permutation consists of k layers of length n/k each? (We can assume that n is divisible by k.)

26. Find all 1243-optimal n-permutations.

27. (+) Let q be a layered pattern in which all layer lengths are at least two. Prove that all q-optimal permutations are layered.

28. Prove that for any fixed k, the sequence $G(n,k)$ of the numbers of n-permutations with k alternating runs is P-recursive.

29. Let M_n be the number of lattice paths from $(0,0)$ to $(n,0)$ using steps $(1,0)$, $(1,1)$, and $(1,-1)$ that never go below the line $y = 0$. (These lattice paths are called *Motzkin paths*. Prove that M_n is a P-recursive sequence.

    The numbers M_n enumerate many kinds of pattern-avoiding involutions, a fact to which we will return several times.

30. Let r_n be the number of lattice paths from $(0,0)$ to (n,n) using steps $(1,1)$, $(1,0)$, and $(0,1)$ that never go above the diagonal $x = y$. Prove that r_n is a P-recursive sequence. What permutations are enumerated by these numbers?

31. (−) Let allperm(n) be the smallest positive integer for which there exists a sequence S of length allperm(n) whose elements are elements of the set $[n]$ such that S contains each of the $n!$ permutations of length n as a subsequence (not necessarily in consecutive positions).

    For example, allperm(2) = 3, since $S = 121$ contains both 12 and 21 as a subsequence, and no sequence of length two has this property.

    Find allperm(3).

32. Keeping the notation of the previous exercise, prove that allperm(n) $\leq n^2 - 2n + 4$.

33. Prove that allperm(n) $\geq (n+1)n/2$.

34. Prove that if p contains exactly one copy of 321, then p is not a derangement.

35. Prove that for any k, the sequence $Av_n(132, 12 \cdots k)$ is P-recursive.

36. Let us call a sequence s of elements of $[n]$ an *n-superpermutation* if s contains all $n!$ permutations of length n as a factor, that is, s contains each permutation $p \in S_n$ in *consecutive* positions. For instance, 121 is a 2-superpermutation, while 123121321 is a 3-superpermutation. Let spp(n) be the length of the shortest n-superpermutation.

   (a) Prove that $\mathrm{spp}(n) \geq n! + n - 1$.

   (b) Prove that the upper bound of part (a) cannot be achieved if $n \geq 3$.

37. Prove that $\mathrm{spp}(n) \leq \sum_{k=1}^{n} k!$.

38. Prove that $\mathrm{spp}(n) \geq n! + (n-1)! + n - 2$.

Problems Plus

1. Find a formula for $S_{123,1}(n)$, and prove from that formula that $S_{123,1}(n)$ is P-recursive.

2. (a) Prove that the ordinary generating function of the sequence $\mathrm{Av}_n(1234)$ is *not algebraic*.

   (b) Let $k \geq 4$ be an *even* positive integer. Prove that the ordinary generating function of the sequence $\mathrm{Av}_n(12 \cdots k)$ is *not algebraic*.

3. Let $\mathcal{S}_{q,i}(n)$ denote the set of all permutations of length n that contain exactly i copies of the pattern q. Construct an injection from $\mathcal{S}_{321,1}(n)$ to $\mathcal{A}v_{n+2}(231)$.

4. Construct an injection from $\mathcal{S}_{321\ominus r,1}(n)$ to $\mathcal{A}v_{231\ominus r}(n)$, where r is any pattern.

5. Let r be any pattern. Prove that the ordinary generating function of the sequence $\mathcal{S}_{321\ominus r,1}(n)$ is not rational.

6. Let $k > 2$ be an *even* integer. Prove that the ordinary generating function of the sequence $\mathcal{S}_{k(k-1)\cdots21,1}(n)$ is not algebraic.

7. Prove that
$$S_{132,2}(n) = \binom{2n-6}{n-4} \frac{n^3 + 17n^2 - 80n + 80}{2n(n-1)}.$$

8. Prove that
$$S_{123,2}(n) = \binom{2n}{n-4} \frac{59n^2 + 117n + 100}{2n(2n-1)(n+5)}.$$

9. Let π be a partition of the set $[n]$. We say that the 4-tuple of elements (a, b, c, d) is a *crossing* of π if $a < b < c < d$, and a and c are in a block π_1 of π, and b and d are in a different block π_2 of π. Let r be a fixed natural number, and let $H_r(n)$ be the number of partitions of $[n]$ with exactly r crossings. Prove that $H_r(n)$ is a P-recursive function of n. If possible, strengthen this claim.

10. Let p be an n-permutation, and let $f(p)$ be the number of all distinct patterns contained in p. For instance, if $p = 1324$, then $f(p) = 7$ for the patterns 1, 12, 21, 213, 123, 132, 1324. Let $\text{pat}(n) = \max_{p \in S_n} f(p)$. Prove that

$$\lim_{n \to \infty} \frac{\text{pat}(n)}{1.61^n} \geq 1.$$

11. Let us keep the notation of the previous Problem Plus, and let us assume for simplicity that $n = k^2$. Prove that $\text{pat}(n) \geq 2^{(k-1)^2}$.

12. Improve the lower bound for $\text{pat}(n)$ given in the previous Problem Plus.

13. Prove that

$$2^n - O\left(n^2 2^{n-\sqrt{2n}}\right) \leq \text{pat}(n) \leq 2^n - \Theta\left(n 2^{n-\sqrt{2n}}\right).$$

14. Find a 4-superpattern of length nine.

15. Prove that there exists a k-superpattern of length $\frac{3}{4}k^2$.

16. We call the permutation p a *weak k-superpattern* if for all patterns q of length k, at least one of q and q^{-1} is contained in p. Prove that there exists a weak k-superpattern of length $\binom{k+1}{2}$.

17. (a) Prove that there exists a k-superpattern of length $\binom{k+1}{2}$.

    (b) Prove that there exists a k-superpattern of length $\lceil (n^2/2) \rceil$.

18. Can one pattern be Wilf-equivalent to infinitely many? That is, do there exist patterns q and $q_1, q_2, \cdots$ so that $\text{Av}_n(q) = \text{Av}_n(q_1, q_2, \cdots)$? In order to exclude trivial answers, we require that the infinite sequence $q_1, q_2, \cdots$ consist of patterns of pairwise distinct sizes.

19. Let q be a layered pattern that contains exactly two layers, and each of those layers is of length at least two. Prove that q-optimal permutations are all layered, and also have exactly two layers.

20. Prove that for all constants $c < 1$, the inequality $\text{allperm}(n) > cn^2$ holds if n is large enough. See Exercise 31 for the definition of $\text{allperm}(n)$.

21. Prove that $g(2413) \geq 0.10472$.

22. Let $n \geq 10$. Prove that $\text{allperm}(n) \leq n^2 - 2n + 3$. Note that this result is barely sharper than the result of Exercise 32.

23. Let $n \geq 7$. Prove that $\text{allperm}(n) \leq n^2 - \frac{7n}{3} + \frac{19}{3}$. This bound is sharper than that of the preceding Problem Plus if $n \geq 11$.

24. Prove that the upper bound given in Exercise 37 for $\text{spp}(n)$ is not optimal if $n \geq 6$.

25. What is the length of the shortest permutation that contains all 2^{n-1} *layered* permutations of length n?

26. (a) Let p be a permutation that contains all q-avoiding permutations of length k, and is of minimal length among all permutations with that property. Is it true that p avoids q?

(b) Consider the set S of all permutations of minimal length that contain all q-avoiding permutations of length k. Is it true that at least one element of S is q-avoiding?

27. Improve the lower bound that was given for $sp(k)$ in Exercise 22, part (a).

Solutions to Problems Plus

1. This result is due to John Noonan [255], who proved that

$$f(n) = S_{123,1}(n) = \frac{3}{n}\binom{2n}{n+3}.$$

Doron Zeilberger [339] found a simpler proof of this formula. The P-recursive property then follows as

$$\frac{f(n)}{f(n-1)} = \frac{n-1}{n} \cdot \frac{(2n)!}{(n+3)!(n-3)!} \frac{(n+2)!(n-4)!}{(2n-2)!} = \frac{2n(2n-1)}{(n+3)(n-3)}.$$

2. (a) Recall that Theorem 4.22 says that

$$\mathrm{Av}_n(1234\cdots k) \sim \lambda_k \frac{(k-1)^{2n}}{n^{(k^2-2k)/2}},$$

where λ_k is a constant given by a multiple integral. In particular, if $k > 2$ is even, then the denominator of the right-hand side is a polynomial of the form n^p, where n is a positive integer. However, it can be proved (see for instance [168], Theorem D, page 293) that the coefficients of an algebraic power series cannot be of this form.

(b) In fact, if $f(z) = \sum_{n\geq 0} f_n z^n$ is an algebraic power series and $a_n \sim cn^d a^n$, where a and c are non-zero constants, and d is a *negative* real number, then $d = s + 0.5$ for some integer s. (See the solution of Exercise 6.3 in [297].) Therefore, the generating function of the sequence $SAv_n(1234\cdots k)$ is not algebraic if k is an even number that is larger than 2.

3. This injection can be found in [83], where it is built up from an idea of Doron Zeilberger.

4. This result is proved in [83].

5. It follows from the previous Problem Plus that for all n, the inequality $S_{321\ominus r,1}(n) \leq \mathrm{Av}_{n+2}(231 \ominus r)$ holds. It is easy to see that the two sequences have the same growth rates, so their generating functions have the same convergence radius R.

   By Corollary 5.21, the generating function of the sequence $\mathrm{Av}_{n+2}(231 \ominus r)$ has a *finite* value at R; therefore, so does that of the sequence $S_{321\ominus r,1}(n)$. However, all singular points of a rational function are poles, and power series go to infinity at poles.

6. This result is proved in [83]. The main idea is that

$$\mathrm{Av}_n(k \cdots 21) \leq S_{k(k-1)\cdots 21,1}(n) \leq \mathrm{Av}_{n+2}(k \cdots 21),$$

   and the two sequences squeezing $S_{k(k-1)\cdots 21,1}(n)$ have non-algebraic generating functions, as shown in Problem Plus 2.

7. This formula was first conjectured in [256]. It was proved by Toufik Mansour and Alek Vainshtein [249], who found a general method to compute the ordinary generating function of the numbers $S_{132,r}(n)$, for $r \leq 6$.

8. This formula was also conjectured in [256]. It was proved by Marcus Fulmek [184], who used a lattice path approach.

9. This result is due to the present author and can be found in [50]. The proof is very similar to the proof of $S_{132,r}(n)$ being polynomially recursive. The generating function is algebraic, and has only one quadratic irrationality, namely $\sqrt{1 - 4z}$.

10. Let $p(n)$ be the n-permutation $1\, n\, 2\, n-1 \cdots$, with $p(0) = \emptyset$, $p(1) = 1$, $p(2) = 12$, $p(3) = 132$, $p(4) = 1423$, $p(5) = 15243$, and so on. Then the value of $f(p(n))$ for these permutations is 1, 2, 3, 5, 8, 13. This suggests that $f(p(n+2)) \geq f(p(n+1)) + f(p(n))$.

    We are now going to prove this claim. Note that all patterns contained in $p(n+2)$ start either in their minimal or in their maximal entries.

    (a) The patterns that start in their maximal entries. These patterns are clearly contained in $p(n+2)$ even if the initial entry 1 of $p(n+2)$ is omitted. Then the entry $n+2$ can play the role of the maximal entries of these patterns, showing that there are $pat(p(n))$ such patterns. (Removing 1 and $n+2$ from $p(n+2)$, we get $p(n)$.)

(b) The patterns that start in their minimal entry. The entry 1 of $p(n+2)$ can play the role of the minimal entry for these patterns. Then these patterns can continue $p(n+1)$ different ways. Indeed, removing 1 from $p(n+2)$, we get a permutation with $f(p(n+1))$ different patterns as the obtained permutation is the complement of $p(n+1)$.

Therefore, the sequence $f(p(n))$ is at least as large as the Fibonacci sequence, proving our claim. This construction is due to Herb Wilf.

11. This result is due to Micah Coleman [122]. The proof is as follows. Let p_k be the n-permutation

$$p_k = k \, 2k \, \ldots \, k^2 \quad (k-1) \, \ldots \, (k^2-1) \quad \ldots\ldots \quad 1 \, (k+1) \, \ldots \, (k^2-k+1).$$

For example,

$$p_3 = 3\,6\,9 \quad 2\,5\,8 \quad 1\,4\,7.$$

In other words, p_k consists of k segments, each of which consist of k entries that are congruent to one another modulo k.

We claim that p_k has at least $2^{(k-1)^2}$ different patterns. Indeed, let P_k be the set of subsequences of p_k that

(a) Contain all k entries divisible by k, and

(b) Contain all entries of $[k]$.

In other words, subsequences in P_k must contain the whole first segment, and the first element of each segment. This means that these patterns must contain these $2k-1$ entries, and are free to contain or not to contain the remaining $(k-1)^2$ entries. Therefore, P_k consists of $2^{(k-1)^2}$ subsequences. The reader is invited to verify that these subsequences are all different as patterns.

12. The first construction proving this fact was given by four authors in [7], who showed that for all positive integers n, the inequality

$$\mathrm{pat}(n) \geq 2^n \left(1 - 6\sqrt{n}2^{-\sqrt{n}/2}\right)$$

holds.

13. This result is due to Alison Miller [252]. This is the best currently known estimate for $\mathrm{pat}(n)$. The extra strength of Miller's method is that she obtains estimates on the number of patterns that have to be contained multiple times in a permutation.

14. We claim that 519472683 is a 4-superpattern. To verify this, note that the last five entries form a 3-superpattern. The entries 1 and 9 are on

7	⑧	9
④	⑤	⑥
1	②	3

FIGURE 5.2

The chosen squares lead to the subword 48526, then to $p(S) = 25314$.

the left of that 3-superpattern, so we certainly have all patterns that start with 1 or 4. One then verifies that the remaining 12 patterns are also present. This construction is due to Rebecca Smith. Note that computer data proves that there is no 4-superpattern of length 8.

15. This result was proved in [166]. The authors proved the upper bound constructively as follows. Let us consider an $m \times n$ chess board, in which, like in real chess, the bottom right corner square is white. Let us pick a subset S of t squares on this chess board. We will associate a t-permutation $p(S)$ to S as follows. From left to right, write the numbers 1 through n in the first (bottom) row in increasing order, the numbers $n + 1, \cdots, 2n$, in the second row in increasing order, and so on. Then read the entries that belong to S column-by-column, starting with the leftmost column, and going down in each column. See Figure 5.2 for an example.

This will result in a pattern of length t, which then defines a t-permutation by relabeling. The authors then take the $k \times \lfloor 3n/2 \rfloor$ chess board, and choose S to be the set of all white squares. They prove that the resulting permutation is in fact a k-superpattern.

16. This result was proved in [166]. Let $S = T_k$ be the set of squares weakly below the diagonal of the $k \times k$ rectangular chess board. It is then proved that the permutation $p(k)$ defined by these squares (in the sense of the solution of the previous Problem Plus) contains q or q^{-1} for each pattern q of length k.

17. (a) This result is due to Alison Miller [252]. Let Z_k be the word of length $k(k + 1)/2$ that starts with the increasing sequence $a_k = 1\,3\ldots(2\lfloor k/2 \rfloor + 1)$, continues with the decreasing sequence $(2\lfloor (k + 1)/2 \rfloor)\ldots 4\,2$, then continues with alternating copies of a_k and b_k, until $\lceil k/2 \rceil$ copies of a_k and $\lfloor k/2 \rfloor$ copies of b_k are used. For instance,

$$Z_5 = 135642135642135.$$

We call Z_k a *zigzag word*. The author then proves that either Z_k or $Z_k + 1$ (the word obtained from Z_k by adding 1 to each entry) contains all permutations of length k. She then shows that if $p = p_1 p_2 \cdots p_{k(k+1)/2}$ is an n-permutation so that $p_i < p_j$ if the

ith letter of Z_k is smaller than the jth digit of Z_k, then p is a k-superpattern.

(b) This result is due to Michael Engen and Vincent Vatter [164]. The proof is similar to that seen in part (a), but the definition of the zigzag words is more careful. If n is even, then there is an increasing run of the first $n/2$ odd integers, followed by a decreasing run of $n/2$ integers, and so on. For instance, we get

$$z_6 = 135643135642135642.$$

If n is odd, then the odd runs are of length $(n+1)/2$, and there are $(n+1)/2$ of them, while the even runs are of length $(n-1)/2$, and there are $(n-1)/2$ of them. For instance,

$$z_5 = 1354213542135.$$

Just as in part (a), breaking the ties in z_k results in a k-superpattern.

18. Yes. It is proved in [22] that $\mathrm{Av}_n(1342) = \mathrm{Av}_n(q_2, q_3, \cdots)$, where

$$q_m = 2\ 2m - 1\ 4\ 1\ 6\ 3 \cdots 2m\ 2m - 3,$$

for $m \geq 2$.

19. This result was proved in two parts. First, it was shown in [268] that if p is a *layered* q-optimal permutation, then p has only two layers. The somewhat lengthy argument shows that if there exists a layered n-permutation with $s+1$ layers containing k copies of q, then a layered n-permutation containing more than k copies but consisting of only s layers also exists, as long as $s \geq 2$.) Then, it was shown in [4] that if q is as specified in the exercise, then all q-optimal permutations are layered. Note that the latter result does not need the requirement that q has only two layers, just that all layers of q are of length at least two.

20. It is proved in [227] that for any real number $\alpha \in (1.75, 2)$, there exists a constant c_α so that $\mathrm{allperm}(n) > n^2 - c_\alpha n^\alpha$, for all $n \geq 1$. The proof is quite complicated, though it can be called an enhanced version of the argument given in the solution of Exercise 33.

21. This strong result was proved by Cathleen Battiste Presutti and Walter Stromquist [266] by a high-powered argument. The authors conjecture that their lower bound, given by an integral and computed to 20 decimals, is actually equal to $g(2413)$. In [265], a simpler argument is used to obtain the slightly weaker result that $g(2413) = 0.10425$.

22. This result is due to Eugen Zălinescu [335]. While it is numerically a small improvement, it is remarkable because it ended a 35-year long dormant period in the history of this problem.

23. This result is due to Saša Radomirović [270], who improved on Zălinescu's result just one year later.

24. For $n = 6$, Exercise 37 provides the upper bound $\text{spp}(6) \geq 873$, but Robin Houston [217] found a 6-superpermutation of length 872. It then follows from the recursive result given in the solution of Exercise 37 that $\text{spp}(n) \leq \sum_{k=2}^{n} k!$ if $n \geq 6$.

25. Let $a(n)$ be the length of the shortest such permutation. Then it is proved in [8] that

$$a(n) = n + \min_{0 \leq k \leq n-1} \{(a) + a(n - k - 1)\}.$$

It has been shown by Knuth ([230], Section 5.3.1, equation (3)) that this implies that

$$a(n) = (n + 1)\lceil \log_2(n + 1) \rceil - 2^{\lceil \log_2(n+1) \rceil} + 1.$$

26. (a) No. The following counterexample is given in [8]. Let $k = 5$, let $q = 231$, and let

$$p = 1\ 5\ 11\ 9\ 3\ 2\ 8\ 4\ 7\ 6\ 10.$$

(b) No. The counterexample of part (a) is a counterexample for this statement as well.

27. It is proved in [117] that if $n < 1.000076k^2/e^2$, then every permutation of length n contains only $o(k!)$ different patterns. So for k large enough, $sp(k) > 1.000076k^2/e^2$.

6

Mean and Insensitive. Random Permutations.

6.1 Probabilistic Viewpoint

In the previous chapters we have enumerated permutations according to various statistics. A similar line of research is to choose an n-permutation uniformly p *at random*, and compute the probability of the event that p has a given property A. Throughout this chapter, when we say that we select an n-permutation at random, we mean that each of the $n!$ permutations of length n is chosen with probability $1/n!$.

Theoretically speaking, this is not a totally new approach. Indeed, the probability of success (that is, the event that p has property A) is defined as the number of favorable outcomes of our random choice divided by the number of all outcomes. In other words, this is the number of n-permutations having property A divided by the number of all n-permutations, which is, of course, $n!$. Therefore, the task of computing the probability that p has property A is reduced to the task of enumerating n-permutations that have property p.

The formal definition of discrete probability that we are going to use in this book is as follows.

DEFINITION 6.1 *Let Ω be a finite set of outcomes of some sequence of trials, so that all of these outcomes are equally likely. Let $B \subseteq \Omega$. Then we call Ω a* sample space, *and we call B an* event. *The ratio*

$$P(B) = \frac{|B|}{|\Omega|}$$

is called the probability of B.

Methods borrowed from probability theory are often extremely useful for the enumerative combinatorialist. In this chapter, we will review some of these methods.

In the most direct examples, no specific knowledge of Probability Theory is needed. We just need to look at our object from a probabilistic angle. Let us start with some classic examples.

DOI: 10.1201/9780429274107-6

Example 6.2

Let i and j be two distinct elements of $[n]$, and let p be a randomly chosen n-permutation. Then the probability that i and j belong to the same cycle of p is $1/2$.   ▯

SOLUTION   For obvious reasons, the probability in question does not depend on the values of i and j, so we might as well assume that $i = n - 1$ and $j = n$.

The crucial idea is that we can use the Transition Lemma (Lemma 3.39). Indeed, entries $n - 1$ and n are in the same cycle of p if, when p is written in canonical cycle notation and the parentheses are omitted, n precedes $n - 1$. This obviously happens in half of all n-permutations, proving our claim.   ∎

This result can be illustrated as follows. I go to a movie theater where each customer is given a ticket to a specific seat. Careless moviegoers, however, do not respect the assigned seating, and just take a seat at random. When I go to my seat, it can be free, or it can be taken. If it is free, fine. If somebody else has already taken it, I ask her to go to her own seat. She does that, and if that seat is free, then everybody is happy; if not, then she will ask the person illegally taking her seat to go to his or her assigned seat. This procedure continues until the person just chased from his illegally taken seat finds his assigned seat empty. Then the probability that a randomly selected person will have to move during the procedure (that is, that he is in my cycle) is one half.

The previous example discussed a "yes-or-no" event, that is, two entries were either contained in the same cycle, or they were not. Our next observation is more refined.

Example 6.3

Let $i, k \in [n]$, and let p be a randomly selected n-permutation. Then the probability that the entry i is part of a k-cycle of p is $1/n$; in particular, it is independent of k.   ▯

This fact is a bit surprising. After all, one could think that it is easier to be contained in a large cycle than in a small cycle as there is "more space" in a large cycle.

SOLUTION   We would like to use the Transition Lemma again. For obvious reasons related to symmetry, the choice of the entry i is insignificant, so we might as well assume that $i = n$. The bijection f of the Transition Lemma maps the set of n-permutations in which n is part of a k-cycle into the set of n-permutations in which $p_{n+1-k} = n$. The latter is clearly a set of size $(n - 1)!$, and our claim is proved.   ∎

1	2	5	7	10
3	4	8		
6	9			

FIGURE 6.1
A Standard Young Tableau on ten boxes.

In the context of the explanation following the previous example, this means that it is just as likely that i people in the movie theater have to move as it is that j people have to move, no matter what i and j are.

6.1.1 Standard Young Tableaux

Our simple examples in the previous subsection might have given the impression that a proof using elementary probability arguments is necessarily simple. This is far from the truth. We illustrate this by presenting two classic proofs given by Curtis Greene, Albert Nijenhaus, and Herbert Wilf [201].

A *Standard Young Tableau* is a Ferrers shape on n boxes in which each box contains one of the elements of $[n]$ so that all boxes contain different numbers, and the rows and columns increase going down and going to the right. Standard Young Tableaux have been around for more than one hundred years by now, being first defined by the Reverend Young in a series of papers starting with [334] at the beginning of the twentieth century. Figure 6.1 shows an example.

Given this definition, a real enumerative combinatorialist will certainly not waste any time before asking how many Standard Young Tableaux exist on a given Ferrers shape, or more generally, on all Ferrers shapes consisting of n boxes. Fortunately, we can answer both of these questions, which is remarkable as we could not tell how many Ferrers shapes (that is, partitions of the integer n) exist on n boxes. Standard Young Tableaux, or, in what follows, SYT, are very closely linked to permutations. We will see in the next chapter that the entries of an SYT determine a certain (restricted) permutation, indeed, an involution, of $[n]$, and that there are numerous beautiful connections between SYT on n boxes, and the set of $n!$ permutations of length n. All the above motivate us to discuss SYT in this book.

6.1.1.1 Hooklength Formula

DEFINITION 6.4 *Let F be a Ferrers shape, and let b be a box of F. Then the* hook *of b is the set H_b of boxes in F that are weakly on the right of b (but in the same row) or weakly below b (but in the same column). The size of H_b is called the* hooklength *of b, and is denoted by h_b. Finally, the box b is called the* peak *of H_b.*

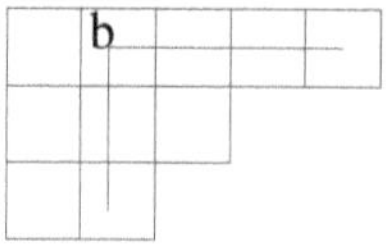

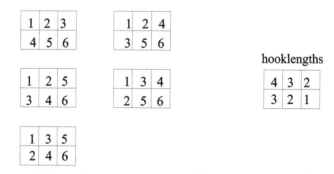

FIGURE 6.2

A hook and all the hooklengths of a Ferrers shape.

FIGURE 6.3

The five SYT of shape $F = 2 \times 3$ and the hooklengths of F.

See Figure 6.2 for the hook H_b of b and the hooklengths associated to each box of a Ferrers shape.

The number of SYT of a given shape is given by the following classic theorem.

THEOREM 6.5

[Hooklength Formula] Let F be any Ferrers shape on n boxes. Then the number of Standard Young Tableaux of shape F is equal to

$$\frac{n!}{\prod_b h_b},\tag{6.1}$$

where the product is over all n boxes b of F.

Example 6.6

Let F be a 2×3 rectangle. Then there are five Standard Young tableaux of shape F, and the hooklengths of F are, row-by-row, 4, 3, 2, 3, 2, 1. So the Hooklength formula is verified as $\frac{6!}{4\cdot3\cdot2\cdot3\cdot2\cdot1} = \frac{720}{144} = 5$. See Figure 6.3 for an illustration.

□

FIGURE 6.4
Events A_b and A_c are not independent.

This is one of those theorems that looks unbelievable at first sight, then believable, and then unbelievable again until a proof is completed. Indeed, at first sight it is not even obvious that the expression in (6.1) is an *integer*.

On second thought, we could argue like this. Let us write in the entries of $[n]$ into F in some random order. The obtained tableau will be a Standard Young Tableau if and only if each hook has its largest entry in its peak. The probability of that happening for a given hook H_b is $\frac{1}{h_b}$, so *if we can multiply these probabilities together*, we are done.

Unfortunately, this is too big of a leap in the general case. Let A_b be the event that H_b has its largest entry at its peak. While it is true that $P[A_b] = \frac{1}{h_b}$, it is in general *not true* that $P[A_b \cap A_c] = P[A_b] \cdot P[A_c]$. In other words, the events A_b and A_c are *not independent*; the occurrence of one can influence the occurrence of the other. To see this, consider the Ferrers shape shown in Figure 6.4, with the boxes b and c marked. In the unlikely case that the reader has not seen an introductory probability text on independent events, it is very easy to catch up by consulting such a text. As we only need discrete probability in this book, we recommend Discrete Probability by Hugh Gordon [198].

Clearly, $P[A_b] = P[A_c] = \frac{1}{2}$ holds, but $P[A_b \cap A_c] = \frac{1}{3}$ as that is the chance that the element x in the bottom right corner is the smallest of the three elements involved. This kind of dependence of the events A_b will always be present, unless F itself is a hook. Therefore, in the general case, this argument will not work. And this is why our theorem looks so unbelievable again. What we have to prove is that when everything is taken into account, we do have $P[\cap_{b \in F} A_b] = \prod_b \frac{1}{h_b}$ after all.

PROOF (of Theorem 6.5) We will define an algorithm that generates a random SYT on F. We will show that each Standard Young Tableau T of shape F has the same chance, namely

$$P[T] = \frac{\prod_b h_b}{n!} \tag{6.2}$$

chance to be generated by our algorithm. As our algorithm always stops by producing an SYT of shape F, these probabilities have to sum to 1, meaning that there must be $\frac{n!}{\prod_b h_b}$ of them, proving that this is the number of SYT on F.

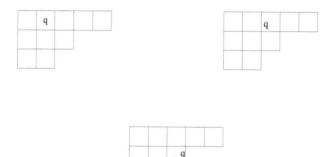

FIGURE 6.5
Random choices leading to the placement of n.

The algorithm proceeds by first placing the entry n somewhere in F, then placing the entry $n - 1$ somewhere else in F, and so on, until all entries have been placed.

The entry n is placed as follows. First, a box in F denoted by q is chosen at random. (Here, and throughout this proof, all eligible boxes have the same chance to be chosen.) Then, q is moved to any of the other $h_q - 1$ positions of the original hook H_q. Call this new box q now. Then repeat the same for the new q, that is, move it to a different position within the new H_q, and call that box q. Continue this until q becomes an *inner corner*, that is, a box whose hook consists of one box only, namely q itself. When that happens, place the entry n into q. The process of moving n to its final destination that we described above is called a *hook walk*.

Example 6.7
Figure 6.5 shows a possible hook walk to the placement of $n = 10$ in our Ferrers shape. The first choice has $1/10$ of probability, the second one has $1/5$, and the third one has probability $1/3$. Therefore, the probability that this particular hook walk will be chosen when we place n is $\frac{1}{150}$.   □

Once n has been placed, temporarily remove the box containing n from F to get the Ferrers shape F_1, and repeat the algorithm with $n - 1$ playing the role of n. Continue this until all elements of $[n]$ are placed and an SYT is obtained. It is obvious that this algorithm always produces an SYT of shape F.

We promised to prove that all SYT on F are obtained with the same probability by this algorithm. We prove this statement by induction on n, the base case of $n = 1$ being trivially true.

Let T be an SYT having shape F that contains the entry n in box q. Removing that box, we get the Standard Young Tableau T_1 that has shape

F_1. It is then clear by the structure of our algorithm that

$$P[T] = P[T_1]P[q], \tag{6.3}$$

where $P[q]$ is the probability that n gets placed into box q, while $P[T]$ and $P[T_1]$ are the probabilities that the mentioned tableaux are obtained by our algorithm.

As T_1 is an SYT on $n-1$ boxes, by the induction hypothesis, we have

$$P[T_1] = \frac{\prod_{b \in F_1} h_b}{(n-1)!}. \tag{6.4}$$

Comparing equations (6.3) and (6.4), we see that our main claim (6.2) will be proved if we can show that

$$P[q] = \frac{1}{n} \cdot \prod_{c \in L_q} \frac{h_c}{h_c - 1}, \tag{6.5}$$

where L_q is the set of boxes that are in the same line (row or column) as q, but are different from q. Indeed, if $d \notin L_q$, then the removal of q does not affect the hooklength h_d, so h_d will be part of both the numerator and denominator, and will therefore cancel. Otherwise, the removal of q will decrease h_d by one, explaining the remaining terms of the last equation. Recall that q is a corner, so for all boxes $c \in L_q$, the inequality $h_c > 1$ holds, so (6.5) will never contain a division by 0.

In order to better understand what $P[q]$ is, we decompose it as a sum of several summands corresponding to the various sequences leading to q. If we have a sequence of random choices leading to the placement of n into $q = (x, y)$, then we denote by I the set of rows (not including x) and by J the set of columns (not including y) that contained the moving box q at some point during the sequence. Then I and J are called the *horizontal* and *vertical projection* of that sequence.

Let $P_{I,J}[x, y]$ be the probability that a random sequence of choices leads to $q = (x, y)$ and has horizontal projection I and vertical projection J.

Example 6.8
Let F be as shown in Figure 6.6, and let $q = (2, 3)$. If $I = \{1\}$ and $J = \{2\}$, then there are two possible sequences ending in q; one is $(1, 2) \to (1, 3) \to (2, 3)$, and the other is $(1, 2) \to (2, 2) \to (2, 3)$. We have seen in Example 6.7 that the first sequence has $\frac{1}{150}$ probability to be selected, whereas the second one has $\frac{1}{10} \cdot \frac{1}{5} \cdot \frac{1}{2} = \frac{1}{100}$ probability to be selected. This yields

$$P_{I,J}[x, y] = \frac{1}{150} + \frac{1}{100} = \frac{1}{60}.$$

☐

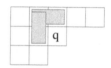

FIGURE 6.6
The projections I and J.

It is now obvious that

$$P[q] = \sum_{I,J} P_{I,J}[x,y]. \tag{6.6}$$

It is time that we proved (6.5). Note that we can rearrange (6.5) as follows.

$$P[q] = \frac{1}{n} \prod_{c \in L_q} \left(1 + \frac{1}{h_c - 1}\right) = \frac{1}{n} \prod_{1 \le i \le x-1} \left(1 + \frac{1}{a_i}\right) \prod_{1 \le j \le y-1} \left(1 + \frac{1}{b_j}\right),$$
$$\tag{6.7}$$

where we split the set L_q into two parts, those boxes that are in the same row as q and those boxes that are in the same column as q. So a_i is the hooklength of the box (i, y) decreased by one, and b_j is the hooklength of the box (x, j) decreased by one.

We are going to prove (6.7) by showing that each expansion term of the last expression is equal to a suitably selected $P_{I,J}[x, y]$. Then (6.7) will immediately follow by (6.6).

In fact, we claim that the following holds.

LEMMA 6.9
Let x and y be the coordinates of an inner corner q, and let $I \subseteq [x-1]$ and $J \subseteq [y-1]$. Then the equality

$$P_{I,J}[x,y] = \frac{1}{n} \prod_{i \in I} \frac{1}{a_i} \prod_{j \in J} \frac{1}{b_j}$$

holds.

Example 6.10
Let F be the same shape as in Example 6.8, let $x = 2$, and $y = 3$, and let $I = \{1\}$, and $J = \{2\}$. Then we need to compute the product $\frac{1}{10} \cdot \frac{1}{a_1} \cdot \frac{1}{b_2}$. As we said after equation (6.7), a_1 is the hooklength of the box $(1, 3)$ decreased by one, that is, $a_1 = 4 - 1 = 3$. Similarly, b_2 is the hooklength of the box $(2, 2)$ decreased by one, that is, $b_2 = 3 - 1 = 2$. This yields $\frac{1}{10} \cdot \frac{1}{3} \cdot \frac{1}{2} = \frac{1}{60}$. This is what we expected as in Example 6.8 we have computed that $P_{I,J}[x, y] = \frac{1}{60}$.

□

PROOF   (of Lemma 6.9) In this proof, it will be advantageous to denote a box by its coordinates (a, b) rather than just by one letter, and consequently, hooks and hooklengths will also be denoted by $H_{a,b}$ and $h_{a,b}$.

Let us assume first that J is empty. That means that the hook walk leading to q started in row y, and therefore its horizontal projection I completely determines it. So in this case, $P_{I,J}[x, y]$ is equal to the probability of that single hook walk, and that is obviously $\frac{1}{n} \prod_{i \in I} \frac{1}{a_i}$. We argue analogously if I is empty.

Let us assume now that both I and J are nonempty. In this case we prove our lemma by induction on $|I| + |J|$. The initial case is when $I = \{i\}$ and $J = \{j\}$, and this case is easy to verify.

To prove the inductive step, note that if a hook walk has $I = \{i_1, i_2, \cdots, i_k\}$ as its horizontal projection, and $J = \{j_1, j_2, \cdots, j_k\}$ for its vertical projection, then it must start at (i_1, j_1). Then it must continue either to (i_2, j_1) or to (i_1, j_2). The hook walks from either of these two points to (x, y) have a truncated horizontal or vertical projection, that is, $I' = I - \{i_1\}$ or $J' = J - \{j_1\}$; therefore, the induction hypothesis applies. If h is the hooklength of the hook H of (i_1, j_1), then it is immediate from the definitions that

$$P_{I,J}[x, y] = \frac{1}{h - 1} (P_{I',J}[x, y] + P_{I,J'}[x, y]).$$

However, by the induction hypothesis this implies

$$P_{I,J}[x, y] = \frac{1}{h - 1} \cdot \frac{1}{n} \left(\frac{a_{i_1}}{\prod_{i \in I} a_i \prod_{j \in J} b_j} + \frac{b_{j_1}}{\prod_{i \in I} a_i \prod_{j \in J} b_j} \right)$$

$$= \frac{a_{i_1} + b_{j_1}}{h - 1} \cdot \frac{1}{n \prod_{i \in I} a_i \cdot \prod_{j \in J} b_j}$$

Finally, note that the first fraction of the last row is equal to one. Indeed, by the definition of the a_i and b_j, we have

$$a_{i_1} + b_{j_1} = h_{i_1,y} - 1 + h_{x,j_1} - 1 = h - 1,$$

as $|H_{i_1,y} \cup H_{x,j_1}| = |H| = h$. Note that $|H_{i_1,y} \cap H_{x,j_1}| = 1$, because that intersection consists of the box q (see Figure 6.7). So the lemma is proved. ∎

This proves (6.7) and therefore the equivalent equality (6.5), which in turn implies our main claim (6.2). ∎

6.1.1.2   Frobenius Formula

The other spectacular result on the enumeration of Standard Young Tableaux is the Frobenius formula, which also dates back to the beginning of the twentieth century [181], [182].

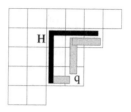

FIGURE 6.7
The gray hooks intersect in one box.

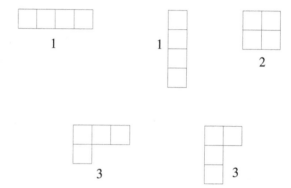

FIGURE 6.8
The Ferrers shapes on five boxes and their f^F-value.

THEOREM 6.11

[The Frobenius formula] For a Ferrers shape F, let f^F denote the number of Standard Young Tableaux that have shape F. Then for any positive integer n, we have

$$\sum_{|F|=n} (f^F)^2 = n!, \tag{6.8}$$

where the sum on the left-hand side is taken over all Ferrers shapes on n boxes.

Example 6.12

Let $n = 4$. Then there are five SYT on n boxes, and the values of f^F for these SYT are 1, 1, 2, 3, 3 as shown in Figure 6.8. So the Frobenius formula is verified as $1^2 + 1^2 + 2^2 + 3^2 + 3^2 = 24$.   □

The Frobenius formula is even more surprising than the hooklength formula. It has a nice and simple bijective proof, which we will cover in Section 7.1. It also has various algebraic proofs. Here we will show a probabilistic proof that is again due to Greene, Nijenhaus, and Wilf [202].

The outline of the proof will be as follows. We will define a random procedure that produces a random SYT on n boxes. Note that we do not know in advance what the shape of the obtained SYT will be. Then we prove that our procedure produces any given SYT of shape F with probability $f^F/(n!)$. Therefore, the total probability that we get a tableaux of shape F is $f^F \cdot \frac{f^F}{n!} = \frac{(f^F)^2}{n!}$, proving that

$$\sum_F \frac{(f^F)^2}{n!} = 1$$

as claimed.

We will need the notion of *conditional probabilities*.

DEFINITION 6.13 *Let A and B be two events so that $P[B] > 0$. Then we define*

$$P[A|B] = \frac{P[A \cap B]}{P[B]}$$

and we call $P[A|B]$ the probability of A given B.

In other words, $P[A|B]$ is the probability of the occurrence of A if we assume that B occurs. As we mentioned before, A and B are called *independent* if $P[A|B] = P[A]$, that is, if the occurrence of B does not make the occurrence of A any more likely or any less likely.

One basic and well-known application of conditional probabilities is Bayes' Theorem, which is also called the law of total probability. It can be found in any introductory probability textbook, such as [198]. It states the following.

THEOREM 6.14
Let $A \subset \Omega$ be an event, and let $X_1, X_2, \cdots, X_m$ be events in Ω so that the X_i are pairwise disjoint, and $X_1 \cup X_2 \cup \cdots \cup X_m = \Omega$. Then the equality

$$P(A) = \sum_{i=1}^{m} P(A|X_i)P(X_i)$$

holds.

In the proof of the hooklength formula, we considered various probabilities that a hook walk ends in a particular box. Now we will be looking at conditional probabilities $P[ab|xy]$ that a hook walk ends in a given box (a, b) provided that it went through another given box (x, y).

We want to prove some general facts, preferably explicit formulae, about $P[ab|xy]$. The first step is the following proposition that provides these formulae for hook walks that lie within one row or column.

PROPOSITION 6.15

Let $x \geq a$ and let $y \geq b$. Then

$$P[xy|ay] = \prod_{a \leq i < x} \frac{h_{i+1,y}}{h_{i,y} - 1}, \tag{6.9}$$

and similarly,

$$P[xy|xb] = \prod_{b \leq j < y} \frac{h_{x,j+1}}{h_{x,j} - 1}. \tag{6.10}$$

PROOF   If a hook walk passes through (a, b) and goes to (x, y), then its first stop after (a, b) can be any of the other $h_{a,b} - 1$ boxes of $H_{a,b}$. After that stop, the walk can proceed to (x, y) in many ways. Therefore, by Theorem 6.14,

$$P[xy|ab] = \frac{1}{h_{a,b} - 1} \left(\sum_{a < i \leq x} P[xy|ib] + \sum_{b < j \leq y} P[xy|aj] \right). \tag{6.11}$$

If $b = y$, then the second sum is empty, and we get, after rearrangement,

$$(h_{a,y} - 1)P[xy|ay] = \sum_{a < i \leq x} P[xy|iy]. \tag{6.12}$$

In order to get rid of the factor $h_{a,y} - 1$, we apply the following clever trick [202]. We take (6.12) for $a + 1$, instead of a. Then we get

$$(h_{a+1,y} - 1)P[xy|(a + 1)y] = \sum_{a+1 < i \leq x} P[xy|iy], \tag{6.13}$$

Subtracting (6.13) from (6.12) we get

$$(h_{a,y} - 1)P[xy|ay] - (h_{a+1,y} - 1)P[xy|(a + 1)y] = P[xy|(a + 1)y],$$

$$P[xy|ay] = \frac{h_{a+1,y}}{h_{a,y} - 1} P[xy|(a + 1)y].$$

We have obtained a formula for $P[xy|ay]$ that is the product of $P[xy|(a+1)y]$ and another term, namely $\frac{h_{a+1,y}}{h_{a,y}-1}$. Applying the same procedure again, this time for $P[xy|(a + 1)y]$, then $P[xy|(a + 2)y]$, and so on, we will be left with the product of fractions, and get (6.9).

     The proof of (6.10) is analogous.   ∎

     A direct consequence of this Proposition is the following, surprisingly compact expression for $P[xy|ab]$.

COROLLARY 6.16

Let $x \geq a$ and $y \geq b$. Then

$$P[xy|ab] = P[xy|ay] \cdot P[xy|xb]. \tag{6.14}$$

PROOF Let F be the rectangular Ferrers shape with northwestern and southeastern corners $(1,1)$ and (x,y), and let $S \subseteq F$ be the subset of boxes (a,b) in F for which (6.14) is not proved yet. So, at the beginning, $S = F - (x,y)$. Let (a,b) be an inner corner of S. We will now prove that (6.14) must hold for (a,b).

Let us look at (6.11) again. In the first summation, we have $a < i \leq x$; therefore, the boxes (i,b) are below the box (a,b), that is, they are in $F - S$ (since (a,b) is an inner corner of S), and so we can apply (6.14) to replace the terms $P[xy|ib]$ by the terms $P[xy|iy]P[xy|xb]$. This leads to

$$\sum_{a<i\leq x} P[xy|ib] = \sum_{a<i\leq x} P[xy|iy]P[xy|xb] = P[xy|xb] \sum_{a<i\leq x} P[xy|iy] \tag{6.15}$$

$$= P[xy|xb]P[xy|ay](h_{a,y} - 1) \tag{6.16}$$

where in the last step we simply used (6.12).

If we transform the second summation in (6.11), we get that

$$\sum_{b<j\leq y} P[xy|aj] = P[xy|ay]P[xy|xb](h_{x,b} - 1). \tag{6.17}$$

Finally, we replace the two sums in (6.11) by the expressions we just computed in (6.15) and (6.17) to get

$$P[xy|ab] = \frac{P[xy|ay]P[xy|xb](h_{a,y} - 1 + h_{x,b} - 1)}{h_{a,b} - 1} = P[xy|ay]P[xy|xb],$$

as it is clear that $h_{a,y} + h_{x,b} - 1 = h_{a,b}$. See Figure 6.7 for an illustration of this.

Now that we proved our statement for (a,b), we can remove (a,b) from S to get S', and choose an inner corner of the new shape $F - S'$. We can then repeat the whole procedure until (6.14) is proved for all boxes of F. ∎

In order to continue our proof of the Frobenius formula, we need some new notions. Let F be the Ferrers shape of a partition $\lambda = (a_1, a_2, \cdots, a_k)$. Recall that these parts are written in non-increasing order. Choose p and q so that $p > k$ and $q > a_1$. Then the *(p,q)-complementary shape* $F_{p,q}^c$ is the set of boxes that are in the rectangle spanned by $(1,1)$ and (p,q), but not in F.

Example 6.17

Let F be the Ferrers shape of $(5,4,3)$, and let $(p,q) = (4,6)$. Then F is shown in Figure 6.9, and $\overline{F_{p,q}}$ is represented by the shaded boxes. ⬜

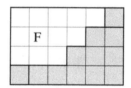

FIGURE 6.9

A shape and its complementary shape.

Note that the *outer corners* of F, that is, the boxes that we can add to F to get another Ferrers shape, are the same as the *inner corners* of $F_{p,q}^c$, that is, elements that can be removed from $F_{p,q}^c$. For the rest of this proof, if an outer corner of $F_{p,q}^c$ is denoted by $\overline{K}$, and the box immediately on its left is an inner corner of F, then that box will be denoted by K.

We define a *special complementary hook walk* in the complementary shape $F_{p,q}^c$ as a hook walk that starts at (p, q) (hence "special"), and uses steps north and west (hence "complementary"). These walks can stop at any of the inner corners of $F_{p,q}^c$.

Finally, we define the distance between two boxes in the obvious way, that is

$$d((x, y), (v, z)) = |x - v| + |y - z|.$$

In what follows, we will be interested in computing the probability that a special complementary hook walk stops in a given inner corner of $F_{p,q}^c$. The alert reader probably suspects that we are interested in this because our random procedure that will produce each SYT with the correct probability will be defined recursively, by ensuring that it puts the maximal element in a certain corner. The alert reader is right, of course.

The following Lemma shows that the choice of p and q is not that crucial.

LEMMA 6.18

Let $\lambda = (a_1, a_2, \cdots, a_k)$ be a partition of n corresponding to the Ferrers shape F, and let $p > k$ and $q > a_1$. Then the probability that a special complementary hook walk in $F_{p,q}^c$ will end at inner corner $\overline{K}$ of $F_{p,q}^c$ is

$$P[\overline{K}] = \frac{\prod_R d(\overline{K}, R)}{\prod_{\overline{R}} d(\overline{K}, \overline{R})}, \tag{6.18}$$

where $\overline{R}$ ranges over all inner corners of $F_{p,q}^c$, and R ranges over all outer corners of F.

Note that this means in particular that the probability $P[\overline{K}]$ is *independent* of the choice of p and q. Indeed, the inner corners of $F_{p,q}^c$ are determined by the outer corners of F, and so neither the numerator nor the denominator of the right-hand side of (6.18) depends on p and q.

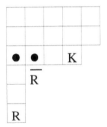

FIGURE 6.10
A generic term that does not cancel.

PROOF First, we will prove our statement for a regular (not a complementary) Ferrers shape. So instead of special complementary hook walks, we will be looking at hook walks starting at $(1, 1)$.

Let $K = (x, y)$ be the corner in which we want our hook walk to finish. Then we need to compute the probability $P[xy|11]$. Applying (6.14), then (6.9) and (6.10), we get

$$P[xy|11] = P[xy|1y]P[xy|x1] = \prod_{1 \leq i < x} \frac{h_{i+1,y}}{h_{i,y} - 1} \cdot \prod_{1 \leq j < y} \frac{h_{x,j+1}}{h_{x,j} - 1}.$$

Taking a closer look at the fractions on the right-hand side, we see that many of them are equal to 1. For instance, look at the generic term $\frac{h_{x,j+1}}{h_{x,j}-1}$. The horizontal parts of these hooks end in (x, y). The vertical parts are different if and only if the jth column of our shape ends in a corner R. It is in this case, and only in this case, that $\frac{h_{x,j+1}}{h_{x,j}-1} \neq 1$. See Figure 6.10 for an illustration. In this case, the denominator is equal to $d(K, R)$, whereas the numerator is equal to $d(K, \overline{R})$, where $\overline{R}$ is the box just below the end of column $j + 1$.

If we change our shape to a complementary shape, our hook walk to a special complementary walk, and K to $\overline{K}$, we get the statement of the lemma. ∎

Example 6.19
Considering the Ferrers shape in Figure 6.10, let $K = (3, 4)$, and let us look at the term $\frac{h_{3,2}}{h_{3,1}-1}$. The value of this term is $3/6$. On the other hand, we have $d(K, R) = 6$ and $d(K, \overline{R}) = 3$, verifying our argument. □

The previous lemma provided a formula for $P[\overline{K}]$, but it was a rather complicated one. The following lemma will bring that formula closer to what we need.

LEMMA 6.20
Let F be a Ferrers shape on m boxes, and let F' be a shape obtained from F

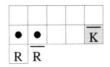

FIGURE 6.11
How corners cause non-canceling terms.

by adding an outer corner $\overline{K}$. Then

$$P[\overline{K}] = \frac{f^{F'}}{(m+1)f^F}.$$

PROOF   We can express both $f^{F'}$ and f^F by the hooklength formula. After that, $m!$ cancels, and so do all hooklengths that belong to hooks *not* being in the same row or column as $\overline{K}$.

What happens with the remaining terms? Let $\overline{K} = (x, y)$. Looking at the row of $\overline{K}$, we see that the hook $H_{x,j}$ of F is as long as the hook $H'_{x,j+1}$ of F', *unless* the vertical parts of these hooks are different, that is, unless there is a corner R at the end of column j of F. In that case, instead of a cancellation, we get a $d(\overline{K}, R)/d(\overline{K}, \overline{R})$ factor. An analogous argument applies for the column of $\overline{K}$. However, by Lemma 6.18, the product of these terms $d(\overline{K}, R)/d(\overline{K}, \overline{R})$ is precisely $P[\overline{K}]$, proving our claim.   ∎

Example 6.21
Let F be the Ferrers shape of the partition $(5, 4, 1)$, and let $\overline{K} = (2, 5)$. The reader is invited to verify using Figure 6.11 that the hooks of F and F' have the same length if the peak of the hook is not in the second row and not in the fifth column. The reader is also invited to verify cancellations involving $h_{x,j}$ of F and $h'_{x,j+1}$ of F', provided that column j of F does not end in a corner.

If $j = 1$, then column j of F ends in corner R. Then we have $h_{2,1} = 5$ in F and $h'_{2,2} = 4$ in F'. On the other hand, we have $d(\overline{K}, R) = 5$, and $d(\overline{K}, \overline{R}) = 4$, verifying our argument.   ▢

Now we are ready to describe the random procedure that will produce each SYT of shape F with probability $f^F/n!$, for all shapes F on n boxes.

Choose p and q so that they are larger than n. We construct a series of SYT $T_0, T_1, \cdots, T_n$ so that T_i has i boxes as follows. Obviously, T_0 is the empty tableau. For brevity, let $Z_i = T^c_{i_{p,q}}$, the complementary shape of T_i in the rectangle spanned by (p, q). For $i \geq 1$, we get T_i from T_{i-1} by inserting the entry i to one of the outer corners of T_{i-1}. The crucial question is *which outer corner*? This corner is chosen by taking the *endpoint* of a special complementary walk in Z_{i-1}.

THEOREM 6.22
Let T be a Standard Young Tableau on n boxes having shape F. Then the random procedure defined in the previous paragraph produces T with probability $f^F/n!$.

PROOF   We prove our statement by induction on n, the initial case being trivial. Let T_* be the SYT that we obtain from T by omitting n, and let F_* be its shape. Then, by our induction hypothesis, the probability that our algorithm constructs T' is $P[T_*] = f^{F_*}/(n-1)!$. When T is constructed, first T_* must be constructed (and the probability of this is $f^{F_*}/(n-1)!$), and then a randomly chosen special complementary walk in the complementary shape of T_* has to end in the corner $\overline{K} = F - F_*$. The probability of the latter is, by Lemma 6.20, $P[\overline{K}] = f^F/(n \cdot f^{F_*})$. Therefore, we have

$$P[T] = P[T_*]P[\overline{K}] = \frac{f^{F_*}}{(n-1)!} \cdot \frac{f^F}{n \cdot f^{F_*}} = \frac{f^F}{n!},$$

completing the proof.   ∎

As we have explained after Example 6.12, Theorem 6.22 immediately implies Theorem 6.11.

6.2   Expectation

We need a little bit more machinery in order to use stronger probabilistic tools. First of all, we formalize the common sense notion of probability we used in the previous section.

A *random variable* is a function $X : \Omega \to \mathbf{R}$ that associates numbers to the elements of our sample space. Most of the random variables we are going to work with will have a finite range, that is, the set of values they take will be finite. The sum and product of two random variables, and a constant multiple of a variable, is defined in the way that is usual for ordinary functions. As the reader has probably noticed, this book is mostly devoted to permutations; therefore, our sample space Ω will most often be S_n, and the random variable X will most often be some permutation statistic. Two random variables X and Y are called *independent* if for all i and j, the equality

$$P[X = i, Y = j] = P[X = i] \cdot P[Y = j]$$

holds.

One of the most important statistics of a random variable is its *expectation*, defined below.

DEFINITION 6.23 *Let $X : \Omega \to \mathbf{R}$ be a random variable that has a finite range S. Then the number*

$$E(X) = \sum_{i \in S} i \cdot P(X = i)$$

is called the expectation *of X on Ω.*

Other names for $E(X)$ include "expected value," "expected number," "mean," "average," or "weighted average." The latter is meant with the individual probabilities as weights. If there is no danger of confusion as to what the variable X is, its expectation $E(X)$ is sometimes denoted by μ.

For certain random variables, the expectation is easy to compute directly. This is often the case for variables defined by symmetric permutation statistics.

Example 6.24

Recall that $d(p)$ is the number of descents of the permutation p. Let n be a fixed positive integer, and let $Z : S_n \to \mathbf{R}$ be the random variable defined by $Z(p) = d(p) + 1$. That is, Z counts the ascending runs of p. Then $E(Z) = (n + 1)/2$. ⬚

PROOF Recall that $A(n, i)$ denotes the number of n-permutations with $i - 1$ descents. Directly from the definition of $E(Z)$, we have

$$E(Z) = \sum_{i=1}^{n} i \cdot \frac{A(n, i)}{n!}.$$

Noting that $A(n, i) = A(n, n + 1 - i)$, we see that

$$iA(n, i) + (n - i + 1)A(n, n + 1 - i) = \frac{n + 1}{2} \cdot (A(n, i) + A(n, n - i + 1)),$$

and the proof follows after summation over i. ∎

6.2.1 Application: Finding the Maximum Element of a Sequence

The fact that X was symmetric played a crucial role in the above argument. Nevertheless, there is another interesting general phenomenon of which the above example was a special case. Recall from Chapter 1 that the Eulerian numbers $A(n,k)$ are not only symmetric, but also unimodal; therefore the sequence $A(n,k)$ has either its one or two maxima in the middle.

Recall that the Eulerian polynomials $A_n(z)$ are defined by the equation $A_n(z) = \sum_{k=1}^n A(n,k)z^k$. Therefore, $A'_n(z) = \sum_{k=1}^n kA(n,k)z^{k-1}$, and

$$A'_n(1) = \sum_{k=1}^n kA(n,k) = n!E(Z).$$

In other words, $E(Z) = \frac{A'_n(1)}{n!} = \frac{A'_n(1)}{A_n(1)}$. Noting that we did not use anything specific about the Eulerian polynomials, we conclude that the expectation of a permutation statistic on S_n can be obtained by substituting 1 into the derivative of the relevant generating function, and dividing the result by $n!$.

Let $A(z) = \sum_{i=1}^n a_i z^i$ be the ordinary generating function of some sequence $a_1, a_2, \cdots, a_m$ that enumerates n-permutations according to some statistic. The expression $A'(1)/n!$ or $\frac{A'(1)}{A(1)}$ may ring a bell for the alert reader. Indeed, recalling Darroch's theorem (Theorem 3.25), we remember that if $A(z)$ has real zeros only, then its sequence of coefficients has at most two maximal elements, and these are at distance less than one from $\frac{A'(1)}{A(1)}$. So we have proved the following corollary of Darroch's theorem.

COROLLARY 6.25
Let s be a permutation statistic on S_n, and let $a_i = |p \in S_n : s(p) = i|$. Let Z be the random variable corresponding to s. Assume that the ordinary generating function $A(z) = \sum_{i=1}^n a_i z^i$ has real zeros only. Then the sequence $a_1, a_2, \cdots, a_n$ has either one or two maximal elements, and they are at distance less than one from $E(Z)$.

In other words, the expectation of Z does not only help us to understand the average behavior of Z, but also (if the real zeros condition holds) provides near-perfect information about the location of the maxima of Z.

6.2.2 Linearity of Expectation

One of the reasons for which the expectation of a variable is a very useful statistic is the following theorem.

THEOREM 6.26
Let X and Y be two random variables defined over the same finite sample space Ω. Then $E(X + Y) = E(X) + E(Y)$.

Note that we do not assume *anything* about X and Y other than the fact that they are defined over the same sample space. It does not matter how (if at all) they were related to each other; whether they are dependent or independent variables. In other words, the *mean* of the sum of two variables is *insensitive* to dependency relations, in case you were wondering about the chapter title.

PROOF   Let $x_1, x_2, \cdots, x_n$ be the values that X takes with a positive probability, and let $y_1, y_2, \cdots, y_m$ be the values that Y takes with a positive probability. Then it follows from the definition of expectations that

$$E(X+Y) = \sum_{i=1}^{n}\sum_{j=1}^{m}(x_i + y_j)P(X = x_i, Y = y_j)$$

$$= \sum_{i=1}^{n}\sum_{j=1}^{m}x_i P(X = x_i, Y = y_j)$$

$$+ \sum_{i=1}^{n}\sum_{j=1}^{m}y_j P(X = x_i, Y = y_j)$$

$$= \sum_{i=1}^{n}x_i P(X = x_i) + \sum_{j=1}^{m}y_j P(Y = y_j)$$

$$= E(X) + E(Y).$$

∎

We point out that if a is a positive constant, then it is easy to prove that $E(aX) = aE(X)$. This fact, together with Theorem 6.26 is often referred to by saying that E is a linear operator.

The application of Theorem 6.26 often involves the method of *indicator random variables* as in the following example.

Example 6.27
Let $n \geq 2$. The expected number of 2-cycles in a random n-permutation is $1/2$. □

SOLUTION   Let $X : S_n \to \mathbf{R}$ be the random variable giving the number of 2-cycles of a permutation. Let i and j be two distinct elements of $[n]$, and let $X_{i,j} : S_n \to \mathbf{R}$ be the random variable defined by

$$X_{i,j}(p) = \begin{cases} 1 \text{ if } i \text{ and } j \text{ form a 2-cycle in } p, \\ 0 \text{ otherwise.} \end{cases}$$

It is then easy to compute the expectation of $X_{i,j}$. Indeed,

$$E(X_{i,j}) = 0 \cdot P(X_{i,j} = 0) + 1 \cdot P(X_{i,j} = 1) = \frac{(n-2)!}{n!} = \frac{1}{n(n-1)}. \quad (6.19)$$

The observation that makes the variable $X_{i,j}$ useful for us is that summing $X_{i,j}(p)$ over all 2-element subsets $(i,j) \subseteq [n]$ we get $X(p)$. Therefore, by Theorem 6.26, we obtain

$$E(X) = E\left(\sum_{(i,j)\subseteq[n]} X_{i,j}\right) = \sum_{(i,j)\subseteq[n]} E(X_{i,j}) = \binom{n}{2}E(X_{1,2}) = \frac{1}{2},$$

as the choice of i and j is clearly insignificant. ∎

The variables $X_{i,j}$, or in general, variables taking values 0 and 1 depending on the occurrence of an event, are called indicator random variables, and are often very useful.

The following theorem is not a very subtle, but very general, tool in proving that it is unlikely that a variable is larger than a multiple of its expectation. (We will state the theorem in a special case that is relevant in our combinatorial applications, but a more general treatment is possible.)

THEOREM 6.28

[Markov's Inequality] Let X be any nonnegative random variable that has a finite range, and let $\alpha > 0$. Then the inequality

$$P[X > \alpha\mu] < \frac{1}{\alpha}$$

holds.

In Section 6.4 we are going to prove a very similar inequality (Theorem 6.48), which is often called Chebyshev's inequality. The ambitious reader is invited to wait for the proof of that inequality, then to try to prove Markov's inequality in a similar manner. Then the reader can check the proof that we will provide as the solution of Exercise 13.

Markov's inequality is particularly useful when the expectation of a variable is very low compared to its maximum value.

Example 6.29

Let $a > 1$. Then the number of n-permutations with more than $a \cdot \ln(n+1)$ cycles is less than $n!/a$. ☐

SOLUTION   This is immediate if we recall that by Lemma 3.26, the average n-permutation has $H(n) = \sum_{i=1}^{n} \frac{1}{i} < \ln(n+1)$ cycles. ∎

Note that this example is particularly striking when n is a very large number. Let n be large enough so that n is roughly equal to $10000\ln(n+1)$. Let

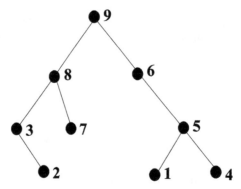

FIGURE 6.12
The tree $T(p)$ for $p = 328796154$.

$a = 100$. Then the probability that a randomly selected n-permutation has more than $100 \ln(n + 1)$, or in other words, more than roughly $n/100$ cycles is less than $1/100$. Speaking in more general terms, it is very unlikely that a randomly selected n-permutation will have at least cn cycles, regardless of how small the positive constant c is.

6.3   Application: Rank in Decreasing Binary Trees

Let $p = p_1 p_2 \cdots p_n$ be a permutation. Let us recall that the *decreasing binary tree* of p, which we denote by $T(p)$, is defined as follows. The root of $T(p)$ is a vertex labeled n, the largest entry of p. If a is the largest entry of p on the left of n, and b is the largest entry of p on the right of n, then the root will have two children, the left one will be labeled a, and the right one labeled b. If n is the first (resp. last) entry of p, then the root will have only one child, and that is a left (resp. right) child, and it will necessarily be labeled $n - 1$ as $n - 1$ must be the largest of all remaining elements. Define the rest of $T(p)$ recursively, by taking $T(p')$ and $T(p'')$, where p' and p'' are the substrings of p on the two sides of n, and affixing them to a and b.

Note that $T(p)$ is indeed a binary tree, that is, each vertex has 0, 1, or 2 children. Also note that each child is a left child or a right child of its parent, even if that child is an only child. Given $T(p)$, we can easily recover p by reading T according to the tree traversal method called *in-order*. In other words, first we read the left subtree of $T(p)$, then the root, and then the right subtree of $T(p)$. We read the subtrees according to this very same rule. See Figure 6.3 for an illustration.

The following definitions can be applied to all varieties of rooted trees, even though we will only explore it for decreasing binary trees. First, in a rooted tree, we say that a *descending path* is a path that starts at a vertex v, and then each vertex in the path is a descendant of the preceding vertex. In other words, descending paths never go *backwards*.

DEFINITION 6.30   *In a rooted tree, we say that vertex v is of rank k if the shortest descending path from v to any leaf of the tree consists of k edges.*

In other words, if v is of rank k, then the edge-distance between v and the closest leaf is k. So leaves are of rank 0, neighbors of leaves are of rank 1, and so on.

We are interested in the following question. Let us say that n goes to infinity. From the set of all vertices of all decreasing binary trees on n vertices, we select a vertex uniformly at random. So each vertex has $1/(n!n)$ chance to be selected. What is the probability that the selected vertex is of rank k?

6.3.1   Two simple initial cases

Let us first handle the case of $k = 0$. That is, we are interested in the probability that a randomly selected vertex is a leaf. In order to alleviate notation, let us assume that all permutations p in this discussion are of length n, and so $T(p)$ has n vertices.

Let $z(p)$ denote the number of leaves in the tree $T(p)$, and let $E(z)$ denote the expectation of $z(p)$ taken over all permutations p of length n.

PROPOSITION 6.31
For all integers $n \geq 2$, the equality $E(z) = \frac{n+1}{3}$ holds.

PROOF   Let $p = p_1 p_2 \cdots p_n$. Let $2 \leq i \leq n-1$. Then it is straightforward to prove, for instance by induction on n, that the vertex corresponding to p_i is a leaf if and only if it is smaller than both of its neighbors, and that event has probability $1/3$. On the other hand, if $i \in \{1, n\}$, then p_i is a leaf if and only if it is smaller than its only neighbor, an event of probability $1/2$. Therefore, if we denote by $z_i(p)$ the indicator variable of the event that p_i is a leaf, then by linearity of expectation we get

$$E(z) = \sum_{i=1}^{n} E(z_i) = (n-2) \cdot \frac{1}{3} + 2 \cdot \frac{1}{2} = \frac{n+1}{3}.$$

∎

It is perhaps a little bit surprising that the formula for entries of rank 1 is just as simple as the formula proved in Proposition 6.31. Let $Y(p)$ denote the number of vertices of p that are of rank 1.

THEOREM 6.32
Let $n \geq 4$. Then the equality $E(Y) = \frac{3(n+1)}{10}$ holds.

PROOF   Let $a_{n,1}$ be the total number of vertices in all decreasing binary trees on n vertices that are of rank 1. Note that if $n > 1$, then each leaf must have a unique parent, and that parent must always be a vertex of rank 1. However, some vertices of rank 1 are parents of *two* leaves. We will now determine the number d_n of such vertices, which will then yield a formula,

$$a_{n,1} = \frac{(n+1)!}{3} - d_n \tag{6.20}$$

for $a_{n,1}$, where $n \geq 4$.

Let p_i be a vertex that is of rank 1 and has two leaves as children. Let us assume for now that $3 \leq i \leq n-2$ holds. Then p_i is larger than both of its neighbors, and both of those neighbors p_{i-1} and p_{i+1} are leaves, so they are smaller than both of their neighbors, meaning that $p_{i-1} < p_{i-2}$, and $p_{i+1} < p_{i+2}$. On the other hand p_i must be smaller than both of its second neighbors; otherwise, its children could not be p_{i-1} and p_{i+1}. This means that out of the 120 possible permutations of the mentioned five entries, only four are possible, since p_i must be the middle one in size, its neighbors must be the two smallest entries, and its second neighbors must be the two largest entries. So if $Z_i(p)$ is the indicator variable of the event that p_i has two leaves as children (in which case p_i is always of rank 1), then for $i \in [3, n-2]$, we get $E(Z_i) = \frac{4}{120} = \frac{1}{30}$. If $i = 1$ or $i = n$, then p_i cannot have two children. Finally, if $i = 2$ or $i = n-1$, then an analogous argument shows that $E(Z_i) = \frac{2}{24} = \frac{1}{12}$. Therefore, since $Z = \sum_{i=2}^{n-1} Z_i$ denotes the number of vertices that have two leaves as children (and are therefore of rank 1), then by linearity of expectation we have

$$E(Z) = \sum_{i=2}^{n-1} E(Z_i) = 2 \cdot \frac{1}{12} + (n-4) \cdot \frac{1}{30} = \frac{n+1}{30}.$$

Therefore, $d_n = (n+1)!/30$, so formula (6.20) implies that

$$a_{n,1} = \frac{(n+1)!}{3} - \frac{(n+1)!}{30} = \frac{3}{10} \cdot (n+1)!,$$

which proves our claim.   ∎

At this point, the reader should think about which entries of p correspond to vertices of $T(p)$ that are of rank 1.

6.3.2   Higher values of k

If $k \geq 2$, then finding the total number of vertices of rank k is significantly more complicated. The main reason for this is that, in this case, then the unique parent of a vertex of rank $k-1$ does not have to be a vertex of rank k; it can be a vertex of rank ℓ, where $0 < \ell \le k$. For instance, in the tree $T(p)$ shown in Figure 6.3, vertex 3 is of rank 1, and its parent, vertex 8, is also of rank 1.

6.3.2.1   A simple, but useful Lemma

Let $a_{n,k}$ be the total number of vertices of rank k in all decreasing binary trees of size n. It is then clear that $a_{n,k+1} \le a_{n,k}$ since each vertex of rank $k+1$ must have at least one child of rank k. While finding the exact value of $a_{n,k}$ is beyond the scope of this introductory discussion, the following lemma will turn out to be useful for us, even if its bound is far from being optimal.

LEMMA 6.33
For each positive integer k, there exists a positive constant γ_k so that if n is large enough, then

$$\frac{a_{n,k}}{n \cdot n!} \ge \gamma_k.$$

In other words, for any *fixed* k, the probability that a randomly selected vertex of a randomly selected decreasing binary tree of size n is of rank k is larger than γ_k. In this early discussion, we will not try to find the best value of γ_k.

Before we prove Lemma 6.33, we need a simple notion. A *perfect binary tree* is a binary tree in which every non-leaf vertex has two children, and every leaf is at the same distance from the root. So a perfect binary tree in which the root is of rank ℓ has $1 + 2 + \cdots + 2^\ell = 2^{\ell+1} - 1$ vertices.

We will now compute the expected number of vertices p_i that are of rank k for which the subtree rooted at p_i is a perfect binary tree. The expected number of such vertices is obviously a lower bound for the expected number of vertices of rank k.

Let Q_k be the probability that for a randomly selected permutation p of length $2^{k+1} - 1$, the tree $T(p)$ is a perfect binary tree (disregarding the labels). It is then clear that $Q_0 = 1$, and

$$Q_{k+1} = \frac{1}{2^{k+1} - 1} Q_k^2. \tag{6.21}$$

So $Q_1 = 1/3$, and $Q_2 = 1/63$. In particular, Q_k is always a positive real number.

PROPOSITION 6.34
Let $n \ge 2^{k+1} - 1$, let $p = p_1 p_2 \cdots p_n$ be a permutation, and let us assume that $2^k + 1 \le i \le n - 2^k$. (In other words, i is not among the smallest 2^k indices or the largest 2^k indices in p.) Let P_k be the probability that the vertex p_i of

$T(p)$ is of rank k, and the subtree of $T(p)$ rooted p_i is a perfect binary tree. Then the equation

$$P_k = Q_k \cdot \frac{2}{(2^{k+1}+1)2^{k+1}}$$

holds for $k \geq 1$. In particular, P_k is a positive real number that does not depend on n or i.

PROOF   The subtree rooted at the vertex p_i of $T(p)$ will be a perfect binary tree with its root being a vertex of rank k if the following two independent events occur.

1. The string $p_{[i,k]}$ of $2^{k+1} - 1$ consecutive entries of p whose middle entry is p_i correspond to a decreasing tree that is a perfect binary tree, and

2. All entries in $p_{[i,k]}$ are less than *both* entries bracketing $p_{[i,k]}$, that is, both p_{i-2^k} and p_{i+2^k}.

The first of these events occurs at probability Q_k, and the second one occurs at probability $\frac{2}{(2^{k+1}+1)\cdot 2^{k+1}}$, proving our claim. ∎

So $P_0 = 1/3$, and $P_1 = 1/30$, as we computed in the proofs of Proposition 6.31 and Theorem 6.32. Furthermore,

$$P_2 = Q_2 \cdot \frac{2}{8 \cdot 9} = \frac{1}{63} \cdot \frac{1}{36} = \frac{1}{2268}.$$

Now we are in a position to prove Lemma 6.33.

PROOF   (of Lemma 6.33) Let $V_i(p)$ be the indicator random variable of the event that the subtree of p that is rooted at p_i is a perfect binary tree whose root is of rank k. Then it follows from the definition of P_k that

$$E(V_i(p)) = P_k.$$

If $V(p)$ denotes the number of vertices of p that are of rank k and whose subtrees are perfect binary trees, then the linear property of expectation yields

$$E(V(p)) = (n - 2^{k+1})P_k,$$

since we do not allow i to be among the smallest 2^k indices or the among the largest 2^k indices. Therefore, (after division by n), the total number $a_{n,k}$ of vertices at of rank k in all decreasing binary trees of size n satisfies

$$\frac{a_{n,k}}{n \cdot n!} \geq \left(1 - \frac{2^k}{n}\right)P_k \geq \frac{P_k}{2}$$

for $n \geq 2^{k+1}$. This completes the proof, since we can set $\gamma_k = P_k/2$. ∎

6.3.3 A System of Differential Equations

In order to determine the exact value of $a_{n,k}$ for $k \geq 2$, we turn to exponential generating functions. We recall that the exponential generating function for the number of permutations of length n, and equivalently, decreasing binary trees on n vertices, is $\sum_{n \geq 0} n! \frac{z^n}{n!} = 1/(1-z)$.

For $k \geq 0$, let $A_k(z) = \sum_{n \geq 0} a_{n,k} \frac{z^n}{n!}$ denote the exponential generating function of the numbers of all vertices of rank k in all decreasing binary trees of size n. Let $B_k(z)$ denote the exponential generating function for such trees in which the root is of rank k. In both $A_k(z)$, and $B_k(z)$, we set the constant term to 0 since empty trees have no roots. Note that $A_k(z) = \sum_{n \geq 1} \frac{a_{n,k}}{n!} z^n$, so in particular, the coefficient of z^n in A_k is the *expected number* of vertices of rank k in a randomly selected decreasing binary tree of size n.

LEMMA 6.35
The following differential equations hold. First, $B_0(z) = z$. Furthermore,

$$B_k'(z) = 2B_{k-1}(z) \cdot \left(\frac{1}{1-z} - B_1(z) - B_2(z) - \cdots - B_{k-2}(z) \right) - B_{k-1}(z)^2$$

if $k > 0$.

PROOF Let T be a decreasing binary tree counted by $B_k(z)$. Let us remove the root of T. On the one hand, this yields a structure counted by $B_k'(z)$. On the other hand, this yields an ordered pair of decreasing binary trees such that one of them has a root of rank $k-1$, and the other one has a root of rank ℓ, with $\ell \geq k-1$. By the Product Formula for exponential generating functions, such pairs are counted by the first product on the right-hand side. At the end of the right-hand side, we must subtract $B_{k-1}(z)^2$ as ordered pairs in which *both* trees have a root of rank $k-1$ are double-counted by the preceding term. ∎

Example 6.36
Let $k = 1$. Then Lemma 6.35 yields

$$B_1'(z) = 2B_0(z) \cdot \left(\frac{1}{1-z} \right) - B_0(z)^2$$

$$= \frac{2z}{1-z} - z^2.$$

Therefore, using the equality $B_1(0) = 0$, we deduce that

$$B_1(z) = 2 \ln \left(\frac{1}{1-z} \right) - 2z - \frac{z^3}{3}.$$

⬚

LEMMA 6.37

For $k \geq 0$, the linear differential equation

$$A'_k(z) = \frac{2}{1-z} \cdot A_k(z) + B'_k(z)$$

holds.

PROOF   Let (T, v) be an ordered pair so that T is a decreasing binary tree on n vertices, and v is a vertex of T that is of rank k. Now remove the root of T. If the root was v itself, then we get a structure counted by $B'_k(z)$, just as we did in the proof of Lemma 6.35. Otherwise, we get an ordered pair (R, S) of structures, one of which is a decreasing binary tree, and the other one of which is an ordered pair of a binary search tree and a vertex of that tree that is at rank k. This explains the first summand of the right-hand side by the Product Formula. ∎

Example 6.38

Setting $k = 1$, we see that $A_1(z)$ is the unique solution of the linear differential equation

$$A'_1(z) = \frac{2}{1-z} \cdot A_1(z) + \frac{2z}{1-z} - z^2 - 2$$

with initial condition $A_1(0) = 0$. This yields

$$A_1(z) = \frac{-\frac{1}{5}z^5 + \frac{1}{2}z^4 - z^3 + z^2}{(1-z)^2}.$$

☐

For larger values of k, we can *theoretically* proceed in the same way. We emphasize *theoretically*, because significant practical difficulties will arise as the solutions to our differential equations become more complicated. The following example shows some of these difficulties.

Example 6.39

Let $k = 2$. Then Lemma 6.35 yields

$$B'_2(z) = 4\frac{\ln(1/(1-z))}{1-z} + 4z\ln(1/(1-z)) - \frac{2}{3}\frac{z^3}{1-z} - \frac{2}{3}z^4 \qquad (6.22)$$

$$- 4\frac{z}{1-z} - 4\ln(1/(1-z))^2 + \frac{4z^3}{3}\ln(1/(1-z)) - \frac{z^6}{9}. \qquad (6.23)$$

Now we can solve the differential equation provided by Lemma 6.37 with $k = 2$, to get

$$
\begin{aligned}
A_2(z) = {} & \frac{1721}{8100(1-z)^2} - \frac{z^7}{81} + \frac{z^6}{324} - \frac{5z^5}{54} + \frac{2z^4}{9}\ln(1/(1-z)) + \frac{23z^4}{324} \\
& - \frac{4z^3}{45}\ln(1/(1-z)) + \frac{349z^3}{2025} + \frac{14z^2}{15}\ln(1/(1-z)) + \frac{979z^2}{2700} \\
& - \frac{8z}{5}\ln(1/(1-z)) + \frac{4219z}{4050} - \frac{4z}{3}\ln(1/(1-z))^2 \\
& + \frac{4}{3}\ln(1/(1-z))^2 - \frac{1721}{8100} - \frac{22}{15}\ln(1/(1-z)).
\end{aligned}
$$

☐

While $A_2(z)$ has many summands in the above form, it is not difficult to prove that the coefficients of $\frac{1721}{8100} \cdot \frac{1}{(1-z)^2}$ grow faster than the coefficients of all other summands. Therefore, as n goes to infinity, the equality

$$
c_2 = \lim_{n\to\infty} \frac{a_{n,2}}{n!n} = \frac{1721}{8100} \approx 0.2124691358
$$

holds.

This result raises many questions. In general, what kind of function will $A_3(z)$ be? Will it always be a sum of many summands, one of which will have significantly larger coefficients than the others? Will that summand always be a *rational* function? We have to introduce a new notion to address these questions.

6.3.3.1 A class of functions

Let $\mathbf{PL}(z)$ be the class of functions $f : \mathbf{R} \to \mathbf{R}$ which are of the form

$$
f(z) = \sum_{i=1}^{m} a_i(1-z)^{b_i}\ln\left(\frac{1}{1-z}\right)^{c_i}, \tag{6.24}
$$

where the coefficients a_i are rational numbers, while the exponents b_i and c_i are non-negative integers. Roughly speaking, $\mathbf{PL}(z)$ is the class of functions that are "polynomials in $1 - z$ and $\ln\left(\frac{1}{1-z}\right)$".

Note that the class $\mathbf{PL}(z)$ would not change if in its definition (6.24) we replaced $(1 - z)$ by z. We use form (6.24) because that form is more directly applicable in the situations that will arise in this section.

A few facts about $\mathbf{PL}(z)$ that are straightforward to prove using integration by parts will be useful in the upcoming discussion.

PROPOSITION 6.40
The class $\mathbf{PL}(z)$ is closed under integration with respect to z.

PROOF   We need to show that $\int (1-z)^b \ln\left(\frac{1}{1-z}\right)^c \, dz \in \mathbf{PL}(z)$. We prove this by induction on c, the initial case of $c = 0$ being obvious. Integration by parts yields

$$\int (1-z)^b \ln\left(\frac{1}{1-z}\right)^c \, dz = -\ln\left(\frac{1}{1-z}\right)^c \cdot \frac{(1-z)^{b+1}}{b+1} \tag{6.25}$$

$$+ \int \frac{(1-z)^b}{b+1} \cdot c \ln\left(\frac{1}{1-z}\right)^{c-1} \, dz. \tag{6.26}$$

By the induction hypothesis, the integral on the right-hand side is in $\mathbf{PL}(z)$, proving our claim.   ∎

PROPOSITION 6.41
Let $b \geq 0$ and $c \geq 0$ be integers. Then

$$\int (1-z)^b \ln\left(\frac{1}{1-z}\right)^c \, dz = (1-z)^{b+1} \cdot g(z) + p(z),$$

where p is a polynomial function, and $g(z) \in \mathbf{PL}(z)$. The integral on the left-hand side is taken with constant term 0.

PROOF   Induction on c, the initial case being that of $c = 0$. If $b = 0$, then the statement is true, since $\int 1 \, dz = x = (1-z) \cdot (-1) + 1$. If $b > 0$, then the statement is true, since $\int (1-z)^b \, dz = (1-z)^{b+1} \cdot \frac{-1}{n+1}$.

   The induction step directly follows from (6.25) and from the induction hypothesis.   ∎

LEMMA 6.42
For all $k \geq 0$, we have
$$B_k(z) \in \mathbf{PL}(z).$$

PROOF   We prove the statement by induction on k. It is obvious that $B_0(z) = z$, and we saw in Example 6.36 that $B_1(z) = 2\ln\left(\frac{1}{1-z}\right) - 2z - \frac{z^3}{3}$. So the statement is true for $k = 0$ and $k = 1$. Now let us assume that the statement of the lemma holds for all positive integers less than k. It then follows from Lemma 6.35 that the summands of $B_k'(z)$ are all in $\mathbf{PL}(z)$, except possibly some summands of the form $a_i \cdot \frac{1}{1-z} \cdot \ln\left(\frac{1}{1-z}\right)^{c_i}$, where a_i is a rational number and c_i is a non-negative integer. The integral of each such summand is in $\mathbf{PL}(z)$ by Fact 10.2, and integrals of the other summands (those that are in $\mathbf{PL}(z)$) are in $\mathbf{PL}(z)$ by Proposition 6.40. Therefore, as $\mathbf{PL}(z)$ is closed under addition, our claim is proved.   ∎

While the power series $A_k(z)$ are in general not in $\mathbf{PL}(z)$, the following weaker statement does hold for them.

THEOREM 6.43
For all nonnegative integers k, we have

$$A_k(z) = \frac{p_k(z)}{(1-z)^2} + f(z), \tag{6.27}$$

where $f(z) \in \mathbf{PL}(z)$, and $p_k(z)$ is a polynomial function with rational coefficients that is not divisible by $(1-z)$.

PROOF Lemma 6.37 provides a linear differential equation for $A_k(z)$. Solving that equation, we get

$$A_k(z) = \frac{\int B_k'(z)(1-z)^2\,dz}{(1-z)^2} + \frac{C}{(1-z)^2}, \tag{6.28}$$

where the integral on the right-hand side is meant with 0 as constant term.

We saw in the proof of Lemma 6.42 that the summands of $B_k'(z)$ are all in $\mathbf{PL}(z)$, except possibly some summands of the form $a_i \cdot \frac{1}{1-z} \cdot \ln\left(\frac{1}{1-z}\right)^{c_i}$. Therefore, the summands of $(1-z)^2 B_k'(z)$ are all in $\mathbf{PL}(z)$. Even more strongly, each summand of $(1-z)^2 B_k'(z)$ is of the form $a_i(1-z)^{b_i} \ln\left(\frac{1}{1-z}\right)^{c_i}$, with $b_i \geq 1$. Therefore, Proposition 6.41 implies that the integral of each summand is of the form $(1-z)^{b_i+1}g_i(z) + p_{\langle i \rangle}(z)$, where $g_i(z) \in \mathbf{PL}(z)$, and $p_{\langle i \rangle}(z)$ is a polynomial function with rational coefficients. As $b_i + 1 \geq 2$, this implies that $\int B_k'(z)(1-z)^2\,dz = (1-z)^2 g(z) + q_k(z)$, where $g(z) \in \mathbf{PL}(z)$ and $q_k(z)$ is a polynomial function with rational coefficients, and our claim is proved. $\blacksquare$

Determining the value of c_3 is conceptually the same as determining c_2. However, the computation becomes much more cumbersome. Lemma 6.35 provides a formula for $B_2'(z)$ as a sum. According to Maple, that sum has 52 summands of the form $a_i z^{b_i} (\ln(1/(1-z)))^{c_i}$. Using that expression for $B_2'(z)$, we can compute $A_3(z)$ using Lemma 6.37. Maple obtains a solution that has 59 summands. However, only 17 of these 59 summands contribute to $p_3(z)$, and therefore, only these 17 summands influence c_3. The value we obtain for c_3 is

$$c_3 = \frac{250488312501647783}{2294809143026400000} \approx 0.1091543117.$$

Details of this computation can be found in [71].

While larger values of k could theoretically be handled in the same way, in the practice these larger values of k lead to differential equations that simply contain functions with too many summands for Maple to handle. However, alternate forms of those equations can be obtained with some trickery. That allowed Boris Pittel and the present author to compute the values of c_4 and c_5 in [76]. They obtained that c_4 equals

$$\frac{12205846414165366219629011323264630441299990228351242558015678732 3}{3353377025022449199852900725670960067418280803797231788288000000000},$$

a fraction whose denominator has 67 digits, and whose approximate value is 0.0364. The approximate value of c_5 is 0.0074. Furthermore, in its simplest form, the rational number c_5 has a denominator that consists of 274 digits.

Note that $\sum_{i=0}^{5} c_i \approx 0.9975$, meaning that for large n, about 99.75 percent of all vertices of all decreasing binary trees of size n are at distance five or less from the closest leaf.

6.4 Variance and Standard Deviation

While the expectation of a random variable X contains information about the average behavior of X, it does not describe how much the different values of X can differ. A constant variable can have the same expectation as one that does not take the same value twice. If we want to obtain information about the behavior of X from this more subtle viewpoint, we must use more subtle operators, such as *variance*.

DEFINITION 6.44   *Let X be a random variable. Then*

$$Var(X) = E((X - E(X))^2)$$

is called the variance *of X.*

Example 6.45
Let $X(p)$ be number of 2-cycles of a randomly selected n-permutation p, and let $n \geq 4$. Then

$$Var(X) = \frac{1}{2}.$$

□

SOLUTION   The following observation that is immediate from the linearity of expectation is often useful in variance computations.

$$Var(X) = E(X^2) - 2E(X)E(X) + E(X)^2 = E(X^2) - E(X)^2. \qquad (6.29)$$

In our case, Example 6.27 shows that $E(X)^2 = 1/4$. So all we have to do is to compute $E(X^2)$.

In order to accomplish that, define the indicator random variables $X_{i,j}$ as in Example 6.27. Note that $X_{i,j}^2 = X_{i,j}$, and therefore, $E(X_{i,j}^2) = E(X_{i,j})$. Then we have

$$E(X^2) = E\left(\left(\sum_{(i,j)\subseteq[n]} X_{i,j}\right)^2\right) = \sum_{(i,j)\subseteq[n]} E(X_{i,j}) + \sum_{\{i,j\}\neq\{i',j'\}} E(X_{i,j}X_{i',j'})$$

$$= E(X) + \sum_{\substack{\{i,j\},\{i',j'\}\subseteq[n] \\ \{i,j\}\cap\{i',j'\}=\emptyset}} E(X_{i,j}X_{i',j'}),$$

where the second term is explained by the fact that no entry of p can be part of more than one cycle.

Finally, we have to compute $E(X_{i,j}X_{i',j'})$ in the case when (i,j) and (i',j') are disjoint sets. In this case, we have $P(X_{i,j}X_{i',j'} = 1) = (n-4)!/n!$; therefore, $E(X_{i,j}X_{i',j'}) = \frac{(n-4)!}{n!}$. On the other hand, the number of (ordered) disjoint pairs of 2-element subsets of $[n]$ is $\binom{n}{2}\binom{n-2}{2}$. Substituting these into the last equation, we get

$$E(X^2) = \frac{1}{2} + \frac{(n-4)!}{n!}\binom{n}{2}\binom{n-2}{2} = \frac{3}{4}.$$

Therefore,

$$\mathrm{Var}(X) = E(X^2) - E(X)^2 = \frac{3}{4} - \frac{1}{4} = \frac{1}{2}.$$

∎

Note that if $2 \leq n < 4$, then $\mathrm{Var}(X) = 1/4$, since in that case, p cannot have two disjoint 2-cycles, and the term $\frac{(n-4)!}{n!}\binom{n}{2}\binom{n-2}{2}$ equals 0, not 1/4. Finally, for $n = 1$, we obviously have $E(X) = \mathrm{Var}(X) = 0$.

Example 6.46
Let $Y(p)$ be the number of descents of the n-permutation $p = p_1 p_2 \cdots p_n$, and let $n \geq 2$. Then

$$\mathrm{Var}(Y) = \frac{n+1}{12}.$$

◻

SOLUTION   Define the indicator variable $Y_i : S_n \to \mathbf{R}$ by

$$Y_i(p) = \begin{cases} 1 \text{ if } p_i > p_{i+1}, \\ 0 \text{ if not.} \end{cases}$$

It is then clear that $\sum_{i=1}^{n-1} Y_i = Y$, that $E(Y_i) = 1/2$, and that $E(Y_i^2) = 1/2$. It is immediate from Example 6.24 that $E(Y) = (n-1)/2$. Therefore, we have

$$\text{Var}(Y) = E(Y^2) - E(Y)^2 = E(Y^2) - \frac{(n-1)^2}{4},$$

and

$$E(Y^2) = E\left(\left(\sum_{i=1}^{n-1} Y_i\right)^2\right) = \sum_{i=1}^{n-1} E(Y_i^2) + \sum_{i<j} 2E(Y_iY_j)$$

$$= \frac{n-1}{2} + \sum_{i<j-1} 2E(Y_iY_j) + \sum_{i=1}^{n-2} 2E(Y_iY_{i+1}).$$

In the $\binom{n-2}{2}$ pairs where $i < j-1$, the probability that i and j are both descents is $1/4$. In the pairs where $j = i+1$, this probability is $1/6$ as only one of the six possible patterns on the entries p_i, p_{i+1}, p_{i+2} has the required property. Substituting this into the last equation, we get

$$E(Y^2) = \frac{n-1}{2} + \frac{(n-2)(n-3)}{4} + \frac{n-2}{3}.$$

Therefore

$$\text{Var}(Y) = \frac{n-1}{2} + \frac{(n-2)(n-3)}{4} + \frac{n-2}{3} - \frac{(n-1)^2}{4} = \frac{n+1}{12}. \qquad (6.30)$$

∎

The variance of X is always nonnegative, as it is the expectation of a non-negative variable. This makes the following definition meaningful.

DEFINITION 6.47   *The standard deviation of the random variable Y is the value $\sqrt{\text{Var}(Y)}$.*

If there is no danger of confusion as to which variable is referred to, the standard deviation is often denoted by σ.

The following classic theorem shows that it is unlikely for a variable to differ from its expectation by a multiple of its standard deviation.

THEOREM 6.48
[Chebyshev's Inequality] Let $\lambda > 0$, and let X be any random variable that has a finite range. Then the inequality

$$P[|X - \mu| \geq \lambda\sigma] \leq \frac{1}{\lambda^2}$$

holds.

PROOF   Note that

$$E((X - \mu)^2) = \sum_i i \cdot P[(X - \mu)^2 = i] \geq \sum_{i \geq \lambda^2 \sigma^2} i \cdot P[(X - \mu)^2 = i]$$

as the last term is obtained by omitting some non-negative summands from the previous one. Since each summand in the last term satisfies $i \geq \lambda^2 \sigma^2$, this implies

$$\sigma^2 \geq \lambda^2 \sigma^2 \cdot \sum_{i \geq \lambda^2 \sigma^2} P[(X - \mu)^2 = i] = \lambda^2 \sigma^2 \cdot P[|X - \mu| \geq \lambda \sigma].$$

Dividing by $\lambda^2 \sigma^2$, we get the inequality that was to be proved.   ∎

Chebyshev's inequality is useful when the standard deviation of a variable is small compared to its expectation, or compared to its maximal value.

Example 6.49
For every positive ϵ, and every positive a, there exists an integer N so that if $n > N$, then the number of n-permutations that have less than $(0.5 - a)n$ descents or more than $(0.5 + a)n$ descents is less than $\epsilon \cdot n!$.   □

SOLUTION   Let p be a randomly selected n-permutation, and let $X(p) = d(p)$. Then $E(X) = (n-1)/2$, and $\text{Var}(X) = \sqrt{(n+1)/12}$. So the variance is of a smaller order of magnitude than the expectation, and therefore, constant multiples of the variance are still very small compared to the expectation. Chebyshev's inequality implies that for *all* real numbers λ, we have

$$P\left[|X - \mu| > \lambda \cdot \sqrt{(n+1)/12}\right] < \frac{1}{\lambda^2}. \tag{6.31}$$

Now let us select λ so that $\frac{1}{\lambda^2} < \epsilon$. Then select N so large that if $n > N$, then $an > \lambda\sqrt{(n+1)/12}$. Then $P[|X - \mu| > an]$ is less than the left-hand side of (6.31), while ϵ is larger than the right-hand side of (6.31).   ∎

The phenomenon described by Example 6.49 is often referred to by saying that the number of descents of a permutation is *concentrated around its mean*. In general, Chebyshev's theorem implies that a variable is concentrated around its mean if its standard deviation is of a smaller order of magnitude than its mean.

6.4.1   Application: Asymptotically Normal Distributions

While this book only discusses *discrete probability*, the reader must have seen
the definition and basic notions of *continuous probability* in a Calculus class.
We will therefore not repeat the basic definitions here. Recall that the variable
Z has *normal distribution* if it has density function

$$f(x) = \frac{1}{\sigma\sqrt{2\pi}} e^{-(x-\mu)^2/(2\sigma^2)}.$$

Here μ is the mean and σ^2 is the variance of Z. In this case, the graphs of
$f(x)$, and its antiderivative $F(x)$ (the distribution function of Z) look *normal*
as can be seen in Figure 6.13.

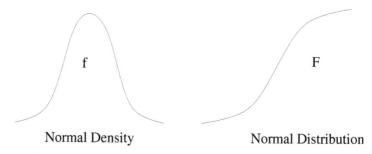

Normal Density   Normal Distribution

FIGURE 6.13
Normal density and normal distribution.

Why do we mention this in a book about discrete combinatorial objects?
The answer is because a notion of *convergence* makes normality relevant for
sequences of discrete distributions as follows.
We say that the sequence of variables Z_n *converges to the variable Z in
distribution* if

$$\lim_{n\to\infty} F_n(a) = F(a)$$

for every real number a where F is continuous. Here $F_n(a)$ (resp. $F(a)$) is the
distribution function of Z_n (resp. Z). That is, $F_n(a) = P[X_n \leq a]$. If $\mu = 0$
and $\sigma = 1$, and Z has normal distribution, then we say that Z has *standard
normal distribution*.
Let X_n be a random variable, and let $a_n(k)$ be a triangular array of non-
negative real numbers, $n = 1, 2, \cdots$, and $0 \leq k \leq m(n)$ so that

$$P[X_n = k] = p_n(k) = \frac{a_n(k)}{\sum_{i=0}^{m(n)} a_n(i)}.$$

Set $g_n(z) = \sum_{k=0}^{m(n)} p_n(k)z^k$. In other words, $g_n(z)$ is a generating function satisfying $g_n(1) = 1$. Such generating functions are sometimes called *probability generating functions*.

Let us introduce the following notation that is standard in most probability texts.

- Let $\bar{Z} = Z - E(Z)$,

- Let $\tilde{Z} = \bar{Z}/\sqrt{\text{Var}(Z)}$, and

- Let $Z_n \to N(0,1)$ mean that Z_n converges in distribution to a variable with standard normal distribution.

The following theorem, which is proved in most probability textbooks, connects the real roots-only property to normal distributions.

THEOREM 6.50
Let Z_n and $g_n(z)$ be as above. If $g_n(z)$ has real roots only, and

$$\sigma_n = \sqrt{\text{Var}(Z_n)} \to \infty,$$

then $\tilde{Z}_n \to N(0,1)$.

Example 6.51
Let Z_n be the number of cycles of a randomly selected permutation of length n. Then Z_n converges to a normal distribution, that is, $\tilde{Z}_n \to N(0,1)$. ⬚

SOLUTION Recall from Chapter 3 that

$$g_n(z) = \frac{1}{n!}z(z+1)\cdots(z+n-1),$$

so the real roots criterion holds. Furthermore, you are asked to compute σ_n in Exercise 35. Once σ_n is computed, it becomes obvious that $\sigma_n \to \infty$, completing the proof. ∎

The fact that the distribution of the number of cycles of a random n-permutation converges to a normal distribution is sometimes referred to by the sentence, "the number of cycles of a random permutation is asymptotically normal".

The reader is invited to use Theorem 6.50 to prove that the number of *descents* of a random permutation is asymptotically normal. This implies that the same holds for all Eulerian permutation statistics. A somewhat more advanced application of Theorem 6.50 shows that the number of blocks of a random partition of the set $[n]$ is asymptotically normal.

Finally, we point out that the real zeros property is not necessary for asymptotic normality to hold. An example for this is d-descents of permutations, which are defined in Exercise 14. Inversions of permutations are a special case of d-descents. When the real zeros property does not hold, but σ_n is large in some sense, a logical next step to prove asymptotic normality is often the *Janson dependency criterion*. That criterion, and some of its applications to permutation enumeration, can be found in [62] and [67]. The main idea is that if the probability generating function of Z_n has real roots only, then Z_n can be decomposed into the sum of some independent random variables. If such a decomposition is not possible, but Z_n can be decomposed into the sum of some random variables amongst which there are not too many dependency relations compared to $\text{Var}(Z_n)$, then the Janson dependency criterion may be applicable.

6.5   Application: Longest Increasing Subsequences

Let p be a randomly selected n-permutation, and let $X_n(p)$, or, when there is no danger of confusion, $X(p)$ denote the *length of the longest increasing subsequence* of p. What can we say about $E(X)$?

We start with a classic result due to Erdős and Szekeres that will provide us with a lower bound for $E(X)$.

PROPOSITION 6.52
Let $n = km + 1$. Then any n-permutation p contains either an increasing subsequence of length $k + 1$, or a decreasing subsequence of length $m + 1$.

PROOF   Extending the idea of Theorem 4.21, we define the *2-way rank* of an entry of a permutation. The 2-way rank of the entry p_i of p is (r, s) if the longest increasing subsequence of p ending at p_i is of length r, and the longest decreasing subsequence of p ending at p_i is of length s. For example, the 2-way rank of the sixth entry of 7215436 is $(2, 4)$.

It is clear that no two entries of p can have the same 2-way rank as the rightmost of two entries would have higher first coordinate or higher second coordinate than the 2-way rank of the leftmost one. Therefore, all $n = km+1$ entries of p must have different 2-way ranks. Since there are only km distinct 2-way ranks with $r \in [k]$ and $s \in [m]$, the proposition is proved by the Pigeonhole Principle.   ∎

COROLLARY 6.53
For all positive integers n, the inequality

$$E(X) \geq \frac{1}{2} \cdot \sqrt{n}$$

holds, where the expectation is taken over all n-permutations.

PROOF For any n-permutation p, either p or its reverse p^r has an increasing subsequence of length at least $\sqrt{n}$ by Proposition 6.52. Therefore, $X(p) + X(p^r) \geq \sqrt{n}$. ∎

A little algebraic manipulation can lead to a better lower bound. We know that for any n-permutation p,

$$\max_{i \in n} r(p_i) \cdot \max_{j \in n} s(p_j) \geq n;$$

otherwise, we could not have n different 2-way ranks in p. For any positive real numbers a and b, we have $a + b \geq 2\sqrt{ab}$, and therefore

$$\max_{i \in n} r(p_i) + \max_{j \in n} s(p_j) \geq 2\sqrt{n},$$

proving (after taking reverses), the following stronger corollary.

COROLLARY 6.54
For all positive integers n, the inequality

$$E(X) \geq \sqrt{n}$$

holds, where the expectation is taken over all n-permutations.

The problem of determining $E(X)$ more precisely has been the subject of vigorous research in the last eighty years. The above results should at least provide an intuitive justification as to why the following question, also known as Ulam's problem, was asked in this form.

QUESTION 6.55 *Let n go to infinity. Does the limit*

$$\lim_{n \to \infty} \frac{E(X_n)}{\sqrt{n}} \tag{6.32}$$

exist, and if yes, what is it?

Corollary 6.54 shows that if (6.32) exists, then it is at least 1. The existence of (6.32) was first proved by Hammersley [210]. Numerical evidence then seemed to suggest that this limit is probably close to 2. This result has originally been proved in two parts, by Logan and Schepp [245] showing the lower bound $\lim_{n\to\infty} \frac{E(X_n)}{\sqrt{n}} \geq 2$, and by Vershik and Kerov [319] showing the upper bound $\lim_{n\to\infty} \frac{E(X_n)}{\sqrt{n}} \leq 2$.

Knowing the expectation of X_n only begins the story. The solution of Ulam's problem opened the door to even deeper research of the distribution of X_n. The interested reader should consult [11] for a survey of results, and [28] for some more recent, spectacular improvements in this area.

Exercises

1. Let i, j, k, and ℓ be four elements of $[n]$, and let p be a randomly selected n-permutation, where $n \geq 4$.

   (a) What is the probability that the four elements are all in one cycle of p?

   (b) What is the probability that these four elements are in four different cycles?

2. (−) What is the probability that the three largest entries of an n-permutation are in the same cycle?

3. Continuing Exercise 1, what is the probability that the four elements are

   (a) in two different cycles?

   (b) in three different cycles?

4. Find the expectation and variance of the number of excedances of randomly selected n-permutations.

5. Let $n \geq k$. Find the expected number of k-cycles in a randomly selected n-permutation.

6. Find the variance of the number of k-cycles of randomly selected n-permutations.

7. Let $Z(p)$ be the number of inversions of the n-permutation p, and let $n \geq 2$. Compute the variance of Z.

8. Let $U(p)$ be the number of *weak excedances* of the n-permutation p. Compute $E(U)$ and $\mathrm{Var}(U)$.

9. Is there a Ferrers shape on 20 boxes that has three hooks of length 7?

10. Let X be the number of leaves of a randomly selected rooted plane tree on $n + 1$ vertices. Find $E(X)$.

11. We want to build a tree with vertex set $[n]$ satisfying as many constraints as possible from a finite set of constraints. The constraints are all of the type $\{(a, b)(c, d)\}$, meaning that the unique path in the tree connecting a to b should not intersect the unique path connecting c to d.

    Prove that no matter how many constraints we have, we can always find a tree having vertex set $[n]$ that satisfies at least one third of them.

12. What is the average number of alternating runs of all n-permutations?

13. Prove Markov's inequality.

14. Let X be defined on the set A of all 1234-containing n-permutations, and let $X(p)$ denote the number of 1234-copies in p. Let Y be defined on the set B of all 1324-containing n-permutations, and let $Y(p)$ denote the number of 1324-copies of p. What is larger, $E(X)$ or $E(Y)$?

15. Let $Y(p)$ denote the length of the cycle containing the entry 1 in a randomly selected involution of length n. Find $E(Y)$.

16. Find a noninductive proof for the fact that the average number of cycles of a randomly selected n-permutation is $H(n) = \sum_{i=1}^{n} \frac{1}{i}$.

17. Let a be a positive real number, and let p be a randomly selected n-permutation. Prove that
$$P\left[i(p) - \frac{1}{2}\binom{n}{2} > a\binom{n}{2}\right] \to 0$$
as n goes to infinity.

18. Let X_n be defined on the set of *derangements* of length n, and let $X_n(p)$ be the number of cycles of p. Prove that for $n \geq 4$, we have
$$(n-1)D(n-2)E(X_{n-2}+1) + (n-1)D(n-1)E(X_{n-1}) = D(n)E(X_n).$$

19. Let Y_n be defined on the set of derangements of length n, and let $Y_n(p)$ be the size of the cycle containing the maximal entry of p. Find a recurrence relation for $E(Y_n)$.

20. We know from Section 1.2 that for any fixed n, the sequence $G(n, k)$ is unimodal. Where is the peak of that sequence, that is, for which k is $G(n, k)$ maximal?

21. Let p and q be two randomly selected n-permutations, and let $G_{p,q}$ be the graph defined in Problem Plus 5. Let $Z(p, q)$ be the number of Hamiltonian cycles in $G_{p,q}$. Prove that
$$E(Z) = \frac{n+1}{n}.$$

22. Let X be a random variable whose range is finite, and assume that X takes only nonnegative values. Prove that

$$P[X = 0] \leq \frac{\text{Var}(X)}{E(X)^2}.$$

23. Find an alternative proof for the result of Example 6.27 using the result of Example 6.3.

24. We extend the notion of *descents* to Standard Young Tableaux as follows. We say that i is a descent of the SYT of P if $i + 1$ appears in a row of P that is strictly below the row containing i. So for instance, the SYT consisting of one single row has no descents, while the SYT on n boxes consisting of one single column has $n - 1$ descents.

    Let n be a fixed positive integer. For $i \in [n-1]$, let A_i be the event that i is a descent in a randomly selected SYT on n boxes. Find a short, direct proof of the fact that $P[A_i]$ is independent of i. Do not use the result of the next exercise.

25. Let λ be a Ferrers shape on n boxes. Let $i \in [n-1]$, and let $A_{i,\lambda}$ be the event that i is a descent in a randomly selected SYT of shape λ. Prove that $P[A_{i,\lambda}]$ is independent of the choice of i.

26. A deck of cards is bijectively labeled by the elements of $[n]$, and is originally arranged in increasing order of the labels. We split the deck into a smaller decks of consecutive cards, with empty decks allowed, then we randomly merge the small decks together. Note that before merging, the smaller decks contained their cards in increasing order. This sequence of operations is called a *riffle shuffle*. Prove that the probability that we obtain a given permutation $p \in S_n$ is

$$P_{a,n}[p] = \frac{\binom{a+n-ri(p)}{n}}{a^n},$$

    where $ri(p)$ denotes the number of *rising sequences* of p. A rising sequence is an increasing subsequence of *consecutive* integers. For instance, 1324756 has three rising sequences, namely, 12, 3456, and 7.

27. Deduce Theorem 1.8 from the result of the previous exercise.

28. (+) A set F of n-permutations is called *min-wise independent* if for all $X \subseteq [n]$, and all $x \in X$ we have

$$P[\min\{p(X)\} = p_x] = \frac{1}{|X|},$$

    where $p = p_1 p_2 \cdots p_n$ is a randomly selected permutation in F. That is, each element of X is equally likely to be the index of the smallest entry among the entries indexed by X.

Small families of min-wise permutations are important tools in efficiently detecting identical or near-identical websites. Prove that there exists a min-wise independent family of size less than 4^n.

29. Let F be a min-wise independent family of n-permutations. Find a lower bound for $|F|$.

30. Let F be a min-wise independent family of n-permutations, and let $i \in [n]$ be a fixed integer. Let $X \subseteq [n]$, and let $x \in X$. Denote P_x the probability that for a randomly chosen permutation $p = p_1 p_2 \cdots p_n \in F$, the entry p_x is the ith smallest among the entries in the set $p(X)$. Prove that

$$P_x = \frac{1}{|X|}.$$

Note that the definition of min-wise independence automatically assures this in the special case when $i = 1$.

31. Let $X(p)$ denote the number of very tight copies of the pattern 12 in a randomly selected n-permutation. (In other words, $X(p)$ is the number of indices i so that in $p = p_1 p_2 \cdots p_n$, we have $p_{i+1} = p_i + 1$.) Find $E(X)$.

32. Let $X(p)$ be the smallest index i that is a descent of the n-permutation p, and set $X(p) = n$ if $p = 12 \cdots n$. Find $E(X)$ and $\mathrm{Var}(X)$ if n is fixed.

33. (−) Stirling permutations were defined in Exercise 42 of Chapter 1. Let $p = p_1 p_2 \cdots p_{2n}$ be a Stirling permutation of size $2n$. Let us say that i is a *plateau* of p if $p_i = p_{i+1}$. Let us say that i is an ascent of p if $i = 0$ or if $p_i < p_{i+1}$. Finally, let us say that i is a descent of p if $i = 2n$ or $p_i > p_{i+1}$. (We can imagine a 0 at the front and end of p, in order to motivate counting $i = 0$ as an ascent and $i = 2n$ as a descent.) Let $P\ell(p)$ (resp. $A(p)$, $D(p)$) denote the number of plateaux (resp. ascents, descents) of p. Prove that

$$E(P\ell) = E(A) = E(D) = \frac{2n + 1}{3}.$$

34. Let $X_n(p)$ denote the number of *odd* cycles of the randomly selected n-permutation p, and let $Y_n(p)$ denote the number of randomly selected *even* cycles of p. Compute

$$\lim_{n \to \infty} E(X_n - Y_n).$$

35. Let $X(p)$ be the number of cycles of the randomly selected n-permutation p. Find $\mathrm{Var}(X)$.

36. (+) Let $X(p)$ be the number of peaks of the permutation p. Let $E(X)$ be the expectation of X taken over all *132-avoiding* n-permutations. Compute $E(X)$ and compare the result with the corresponding result for *all* n-permutations.

37. Consider all c_{n-1} rooted plane unlabeled trees on n vertices, and choose one of their total nc_{n-1} vertices uniformly at random.

    (a) What is the probability a_n that the selected vertex is a leaf?

    (b) Interpret the result of part (a) in terms of permutations.

38. Continuing the previous exercise, what is the probability that the selected vertex is a neighbor of a leaf?

39. Consider all rooted plane trees whose vertices are bijectively labeled with the elements of $[n]$ so that each vertex has a label that is smaller than the label of its parent. Let T_n be the number of such trees. Find an explicit formula for T_n.

40. Let ℓ_n be the total number of leaves in all T_n trees of size n defined in Exercise 39. Find an explicit formula for ℓ_n.

41. (+) *Decreasing nonplane* 1-2 trees were defined in Exercise 15 of Chapter 1, where we proved that they are counted by the Euler numbers. Consider all vertices of all E_n such trees on vertex set $[n]$, and choose one vertex uniformly at random. Let a_n be the probability that the selected vertex is a leaf. Find $\lim_{n \to \infty} a_n$.

42. (+) Continuing the preceding exercise, let b_n the probability that the chosen vertex is a neighbor of a leaf. Find $\lim_{n \to \infty} b_n$.

43. Let $X(p)$ be the number of all inversions in the permutation p. Compute $E(X)$ over the set of all 312-avoiding permutations of length n.

44. Let $X(p)$ be the number of all inversions in the permutation p. Compute $E(X)$ over the set of all 321-avoiding permutations of length n.

Problems Plus

1. Let $i \in [n-2]$, and let λ be a given Ferrers shape on n boxes. Let X_i be the indicator variable of the event that both i and $i+1$ are descents in a randomly selected SYT of shape λ. Is it true that $E(X_i)$ is independent of i? (Descents of SYT are defined in Exercise 24.)

2. Let X_i be defined as in the previous Problem Plus. Find a formula for $E(X_i)$.

3. Let $1 \leq j < i - 1 \leq n - 2$, and let $X_{i,j}$ be the indicator variable of the event that both i and j are descents in a randomly selected SYT of shape λ. Is it true that $E(X_{i,j})$ is independent of i?

4. Let $X_{i,j}$ be defined as in the previous Problem Plus. Find a formula for $E(X_{i,j})$.

5. Let $p = p_1 p_2 \cdots p_n$ be an n-permutation, and let G_p be the directed graph on vertex set $[n]$ in which there is an edge from i to j if $p_i = j$.

   Now let p and q be two randomly selected n-permutations, and consider the graph $G_{p,q}$ on vertex set $[n]$ whose edges are the edges of G_p and the edges of G_q. Let n go to infinity. What is the probability that $G_{p,q}$ contains a directed Hamiltonian cycle? (A Hamiltonian cycle of a graph is a cycle that contains all vertices of the graph.)

6. Keep the notation of the previous Problem Plus, and let p, q, and r be three randomly selected n-permutations. Let $G_{p,q,r}$ be the graph that consists of the edges of G_p, G_q, and G_r on vertex set $[n]$. Let n go to infinity. What is the probability that $G_{p,q,r}$ contains a directed Hamiltonian cycle?

7. What is the expected number $E_{d,\lambda}$ of descents in a randomly selected SYT of a given shape λ on n boxes?

8. We have defined descents in Standard Young Tableaux in Exercise 24. Continuing that line of thinking, we can extend the notion of *major indices* to SYT by setting $maj(T) = \sum_{i \in D(T)} i$. Here $D(T)$ denotes the set of all descents of the Standard Young Tableaux T.

   What is the expectation $E_{maj,\lambda}$ of the major index in a randomly selected SYT of a given shape λ on n boxes?

9. Let p be a permutation obtained by the riffle shuffle algorithm given in Exercise 26. Find a formula for the probability that p is of a given type.

10. Let $a = 2$, and let us repeat the riffle shuffle algorithm of Exercise 26 m times. This means that starting from the second application, the smaller decks will not necessarily contain their cards in increasing order. Let p be a fixed n-permutation. Find a formula for the probability that this procedure results in p.

11. Stirling permutations were defined in Exercise 42 of Chapter 1. Let $D_n(p)$ denote the number of descents of the Stirling permutation p of size $2n$. Find an explicit formula for $Var(D_n)$.

12. Let $A_n(z) = \sum_p z^{D(p)}$, where p ranges all Stirling permutations of size $2n$.

   (a) Prove that $A_n(z)$ has real roots only.

   (b) If we select a Stirling permutation p of size $2n$ at random, what is the most likely value for $D_n(p)$?

13. Let us select a random element (p, q) of the direct product $S_n \times S_n$. What is the probability that p and q have the same number of cycles?

14. Let $p = p_1 p_2 \cdots p_n$ be a permutation. We say that the pair (i, j) is a d-descent in p if $i < j \leq i + d$, and $p_i > p_j$. In particular, 1-descents correspond to descents in the traditional sense, and $(n - 1)$-descents correspond to inversions. For fixed n and d, let $X_{n,d}(p)$ denote the number of d-descents of the n-permutation p. Find $\mathrm{Var}(p)$.

15. Let q and q' be two permutation patterns of length k, and let $X_{n,q}(p)$ (resp. $X_{n,q'}(p)$) denote the number of copies of q (resp. q') contained in p.

   (a) Prove that the equality $\mathrm{Var}(X_{n,q}) = \mathrm{Var}(X_{n,q'})$ does not necessarily hold.

   (b) Prove that for all patterns q of length k, there exists a constant c_k so that $\mathrm{Var}(X_{n,q}) = c_k n^{2k-1} + O(n^{2k-2})$.

16. Let q be a 132-avoiding permutation pattern of a fixed length k, let $X_q(p)$ be the number of copies of q in the n-permutation p, and let $E(X_q)$ be the expectation of X_q taken over *all 132-avoiding* permutations of length n.

   For which q will $E(X_q)$ be minimal? For which q will $E(X_q)$ be maximal?

17. Keep the notation of the previous problem. Prove that

$$E(X_{312}) = E(X_{231}) = E(X_{213}).$$

   Try to find a bijective proof.

18. Let p be a 132-avoiding permutation, and let $T(p)$ the decreasing binary tree of p. Note that as p is 132-avoiding, we can remove the labels of $T(p)$ without losing any information about p. Let us remove all edges from connecting left children to their parents from $T(p)$. We call the remaining components, which are all paths descending to the right, the *spines* of p. The *spine structure* of p is the size of its spines, in descending order. See Figure 6.14 for an illustration.

   Prove that if p and q are two 132-avoiding permutations, and their spine structures are identical, then $E(X_p) = E(X_q)$, where the expectations are taken over *all 132-avoiding* permutations of length n.

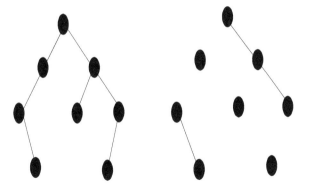

FIGURE 6.14
The decreasing binary tree $T(p)$ for $p = 65783412$ and its spines, leading to the spine structure $(3, 2, 1, 1)$.

19. Can the statement of the preceding Problem Plus be reversed? That is, if $E(X_p) = E(X_q)$ for two 132-avoiding permutations p and q, does it follow that p and q have identical spine structures?

20. A *k-alternating subsequence* in a permutation $p = p_1 p_2 \cdots p_n$ is a subsequence $p_{i_1} p_{i_2} \cdots p_{i_k}$ so that

$$p_{i_1} > p_{i_2} < p_{i_3} > p_{i_4} < \cdots ,$$

and, for all j,

$$|p_{i_j} - p_{i_{j+1}}| \geq k.$$

Let $as_k(p)$ be the length of the longest k-alternating subsequence of p. Find $E(as_k)$, where the expectation is taken over all permutations of length n.

Solutions to Problems Plus

1. Yes, this is true, and can be proved similarly to Exercise 25. See [214] for an analysis of all cases.

2. Let $\lambda = (a_1, a_2, \cdots, a_d)$, and let the conjugate of λ be denoted by $\lambda' = (a_1', a_2', \cdots, a_m')$. It is proved in [214] that

$$E(X_i) = d_{\lambda'} = \sum_{h \geq j \geq k} \frac{a_i'(a_j' - 1)(a_k' - 2)}{n(n-1)(n-2)}. \tag{6.33}$$

3. Yes, this is true, and can be proved again similarly to Exercise 25. See [214] for details.

4. Let us keep the notation of the solution of Problem Plus 2. Furthermore, let us introduce the new notation

$$e_\lambda = \sum_{h \geq j \geq k \geq l} \frac{a_i(a_j - 1)(a_k - 2)(a_l - 3)}{n(n-1)(n-2)(n-3)}.$$

It is then proved in [214] that

$$E(X_{i,j}) = c_{\lambda'} - d_\lambda - d_{\lambda'} + e_\lambda + e_{\lambda'},$$

where $c_{\lambda'}$ is defined in (6.34) below.

5. This probability converges to zero as n goes to infinity as is proved in [125].

6. This probability converges to 1 as n goes to infinity as is shown in [180].

7. Let the given shape λ have rows of length $a_1, a_2, \cdots, a_k$ and columns of length $a_1', a_2', \cdots, a_k'$. It then follows from a more general result of Peter Hästö [214] that

$$c_{\lambda'} = E_{d,\lambda} = \frac{n-1}{2} \left[1 + \sum_{j=1}^{k} \frac{a_i(a_i - 1)}{n(n-1)} - \sum_{j=1}^{k} \frac{a_i'(a_i' - 1)}{n(n-1)} \right] \quad (6.34)$$

$$= (n-1) \sum_{j \leq t} \frac{a_t'(a_j' - 1)}{n(n-1)}. \quad (6.35)$$

Note that by the result of Exercise 25 and the linearity of expectation, we get that for a given $i \in [n-1]$,

$$P[i \in D(T)] = \sum_{j \leq t} \frac{a_t'(a_j' - 1)}{n(n-1)} \quad (6.36)$$

where T is a randomly selected SYT of shape λ.

8. It is straightforward from (6.36) and the linearity of expectation that

$$E_{maj,\lambda} = \binom{n}{2} \frac{1}{2} \left[1 + \sum_{j=1}^{k} \frac{a_i(a_i - 1)}{n(n-1)} - \sum_{j=1}^{k} \frac{a_i'(a_i' - 1)}{n(n-1)} \right]$$

$$= \binom{n}{2} \sum_{j \leq t} \frac{a_t'(a_j' - 1)}{n(n-1)},$$

where the a_i are defined as in the solution of the previous Problem Plus. Note that here, just as in (6.34), the two sums within the brackets cancel if our Ferrers shape is self-conjugate.

9. It is proved in [147] that the probability that p is of type $(n_1, n_2, \cdots, n_n)$ is

$$\frac{\prod_{j=1}^{n} \binom{f_{ja}+n_j-1}{n_j}}{a^n},$$

where f_{ja} is the number of aperiodic circular words of length j over an alphabet of a letters. In other words, f_{ja} is the number of ways to design aperiodic necklaces using j beads having colors $1, 2, \cdots, a$. The numbers f_{ja} are fairly well-studied, and it is known that

$$f_{ja} = \frac{1}{j} \sum_{d|j} \mu(d) a^{j/d},$$

where d ranges all positive divisors of j, and μ is the number-theoretical Möbius function, that is, $\mu(d) = 0$ if d is divisible by a perfect square larger than 1, otherwise $\mu(d) = (-1)^k$, where k is the number of distinct prime divisors of d. See [302] for this fact, and the definition of the Möbius function in a more general setting.

10. It is proved in [31] that this probability is

$$P_{n,m} = \frac{\binom{2^m+n-ri(p)}{n}}{2^{mn}}.$$

See Exercise 26 for the definition of $ri(p)$.

11. It is proved in [63] that $\mathrm{Var}(D_n) = \frac{2n^2-2}{18n-9}$. The main tool of the proof is the following recurrence relation. Let $s_n = E(D_n^2)$. Then $s_{n+1} = \frac{2n-1}{2n+1} \cdot s_n + \frac{4n+4}{3}$.

12. (a) This result is proved in [63] by an induction argument. We have $A_1(z) = z$, and $A_2(z) = 2z^2 + z = z(2z+1)$. It is not difficult to prove that the recurrence relation

$$A_n(z) = (z - z^2)A'_{n-1}(z) + (2n - 1)zA_{n-1}(z) \qquad (6.37)$$

or, equivalently,

$$A_n(z) = z(1-z)^{2n}\frac{d}{dz}\left((1-z)^{1-2n}A_{n-1}(z)\right) \qquad (6.38)$$

holds. Let us assume that the roots of $A_{n-1}(z)$ are real, distinct and non-positive. Clearly, $A_n(z))$ vanishes at $z = 0$. Furthermore, by Rolle's theorem, (6.38) shows that $A_n(z)$ has a root between any pair of consecutive roots of $A_{n-1}(z)$, providing $n-1$ real roots for $A_n(z)$. So the last root must also be real, since complex roots of polynomials with real coefficients must come in conjugate pairs.

We must still show that the last root of $A_n(z)$ must be on the right of the rightmost root of A_{n-1}. Consider (6.37) at the rightmost root z_0 of A_{n-1}. As z_0 is negative, so is $z_0 - z_0^2$, and so $A_n(z_0)$ and $A'_{n-1}(z_0)$ have opposite signs. The claim now follows, since in $-\infty$, the polynomials $A_n(z)$ and $A'_{n-1}(z)$ must converge to the same (infinite) limit, as their degrees are of the same parity. As $A'_{n-1}(z)$ has no more roots on the right of z_0, the polynomial $A_n(z)$ must have one.

(b) As $A_n(z) = \sum_i a_i z^i$ has real roots only, it follows from Theorem 3.25 that if its largest coefficient is a_i, then $|i - E(D_n)| < 1$. We computed in Exercise 33 that $E(D_n) = (2n+1)/3$. So if $(2n+1)/3$ is an integer, then the most likely value of $D_n(p)$ is $m = (2n+1)/3$, otherwise it is one of the two integers bracketing m.

13. This question was answered by Herb Wilf [331] who proved that the requested probability is asymptotically equal to $\frac{1}{2\sqrt{\pi \ln n}}$. The proof uses the fact that if $g(z) = \sum_{k=0}^{n-1} a_k z^k$, then

$$\sum_{k=0}^{n-1} |a_k|^2 = \frac{1}{2\pi} \int_0^{2\pi} |g(e^{it})| \, dt.$$

Wilf then applies the above formula to the generating polynomial $\sum_{k=1}^n c(n,k)z^k = z(z+1)\cdots(z+n-1)$ of the signless Stirling numbers of the first kind, and computes the integral on the right-hand side using an advanced technique called the method of Laplace for integrals.

14. It is proved in [62] that if $n \geq 2d$, then

$$\text{Var}(X_n) = \frac{6dn + 10d^3 - 3d^2 - d}{72}, \tag{6.39}$$

while if $n/2 < d \leq n - 1$, then $\text{Var}(X_n)$ equals

$$\frac{2n^3 - 6n^2 + 4n - 12d^3 - 21d^2 - 9d - 12n^2 d + 24nd^2 + 30nd + 18}{72}.$$

$$\tag{6.40}$$

15. (a) It is easy to verify that $\text{Var}(X_{4,123}) \neq \text{Var}(X_{4,132})$.

(b) This result is proved in [67].

16. The minimum is taken for the increasing permutation and the maximum is taken by the decreasing permutation of length k. This family of questions was initiated by Joshua Cooper, and the proofs of these results (due to the present author) can be found in [66].

17. The first equality is trivial (take inverses), but the second one is not. See [69] for both a straightforward computational proof and a more elucidating bijective proof. We mention that Cheyne Homberger [216] has proved the following. The average number of 231-copies in all *132*-avoiding n-permutations is equal to the average number of 231-copies in all *123*-avoiding n-permutations.

18. This result is due to Kate Rudolph [277].

19. Yes. This fact was conjectured by Kate Rudolph [277], and proved soon afterwards by Lynn Chua and Krishanu Roy Sankar [114].

20. Tommy W. Cai [107] has proved that $E(as_k) = \frac{4(n-k)+5}{6}$, proving a conjecture of Drew Armstrong.

7

Permutations and the Rest. Algebraic Combinatorics of Permutations.

7.1 Robinson–Schensted–Knuth Correspondence

This chapter is devoted to the connections between permutations and various other objects in combinatorics. This is certainly a huge topic, and we can therefore only skim the surface of a few selected areas. Our goal is to give the reader an overview of some main lines of research to aid the decision of what literature to consult next.

In the first section, we present the famous Robinson–Schensted–Knuth correspondence that connects the combinatorics of permutations and the combinatorics of Standard Young Tableaux. There are several excellent books [185, 279], devoted entirely or mostly to the fascinating subject of Young Tableaux, and we will not try to parallel them. Instead, we will be concentrating on those parts of the area that are most directly connected to the enumeration of permutations.

The Robinson–Schensted–Knuth correspondence provides a direct bijective proof for Theorem 6.11, showing once again that the number of pairs of SYT of the same shape, consisting of n boxes each, is $n!$. This is achieved by a bijection risk from the set of all n-permutations onto that of such pairs. (There is no risk involved, but when pronounced, that word is shorter than, say, RSK.) In addition, the bijection has a very rich collection of interesting properties, such as turning natural parameters of permutations into natural parameters of SYT.

To start, let $\pi = \pi_1 \pi_2 \cdots \pi_n$ be an n-permutation. We are going to construct a pair of Standard Young Tableaux $\mathrm{risk}(\pi) = (P, Q)$. The two tableaux will have the same shape. The two tableaux will be constructed together, in n steps, but according to different rules. In the P-tableau, some entries will move after they are placed, while this will not happen in the Q-tableau. We will call our tableaux P and Q throughout the procedure, but if we want to emphasize that they are not completely built yet, we call them P_i and Q_i to show that only i steps of their construction have been carried out.

We are going to describe the bijection risk step-by-step, and we will illustrate each step by the example of $\pi = 52314$.

DOI: 10.1201/9780429274107-7

P

| 5 |

2314

Q

| 1 |

FIGURE 7.1

The situation after Step 1 of creating risk(52314).

P

| 2 |
| 5 |

314

Q

| 1 |
| 2 |

FIGURE 7.2

The situation after Step 2 of creating risk(52314).

Step 1   Take the first entry of π, and put it in the top left corner of the tableau P that we are in the process of creating. Then put the entry 1 in the top left corner of the tableau Q that we are creating, encoding the fact that the top left corner was the first position of P that was filled. See Figure 7.1 for an illustration.

Step 2   Now take π_2, the second entry of π. If $\pi_2 > \pi_1$, then simply put π_2 to the second position of the first line of P. We then write 2 in the second box of the first line of Q, to encode that this was the box of P that got filled second.

   If $\pi_2 < \pi_1$ (as in our running example), then we cannot do this as the first row of P cannot contain entries in decreasing order. Therefore, in this case, π_2 will take the place of π_1, and π_1 will descend one line, to take the first position of the second row of P. To encode this, we write 2 to the first box of the second line of Q.

   See Figure 7.2 for an illustration.

Step i We then continue the process the same way. Let us assume that the first i entries of π have already been placed, and that we have created a pair (P_i, Q_i) of partial Standard Young Tableaux on i boxes each, and of the same shape. Now we have to place π_{i+1}. Generalizing the rules that we have seen in the special case of $i = 2$, we look for the leftmost entry y in the first row that is larger than π_{i+1}. If there is no such entry, we simply put π_{i+1} at the end of the first line. If there is such an entry y, then π_{i+1} will take the position of the entry y, while y will descend one line, and will be inserted in the second row of P_i according to the same rules. In other words, π_{i+1} displaces the smallest entry y of the first row that is larger than π_{i+1}. Note that this preserves the Standard Young Tableau structure, since the first row will remain increasing, and the column into which π_{i+1} is inserted will remain increasing. Indeed, π_{i+1} displaced an entry y that was larger than π_{i+1}, and π_{i+1} is larger than its left neighbor in the first row.

FIGURE 7.3

The situation after Steps 3, 4, and 5 of creating risk(52314).

When y is looking for its place in the second row, the same rules apply to y as applied for π_{i+1} in the first row. That is, if there is no element larger than y in the second row, then y will be placed at the end of the second row; otherwise, y will displace the smallest entry z in the second row that is larger than y, and this entry z will descend to the third row, to look for a position, subject to the same rules. For the same reasons as we explained in the previous paragraph, these steps will preserve the Standard Young Tableau structure of P_{i+1}. When this procedure ends, the resulting tableau P_{i+1} will have a box in a position where P_i did not. We will write $i + 1$ into that position in the Q-tableau, to encode the fact that that position was the $(i + 1)$st position of P to get filled. This preserves the Standard Young Tableau structure of the Q-tableaux, since $i + 1$ gets inserted at the end of a row and at the end of a column.

Repeating this placement procedure n times, we get a pair $(P, Q) = (P_n, Q_n)$ of Standard Young Tableaux of the same shape, consisting of n boxes each. The fact that P_n and Q_n are indeed Standard Young Tableaux is easy to prove by induction since we have seen the insertion of each new entry preserves the Standard Young Tableau structure of both P_i and Q_i. The shapes of P and Q are identical, as in each step we added a new box to the same position in each of them. We then set risk$(\pi) = (P, Q)$.

See Figure 7.3 to see how the complete image risk(52314) is obtained.

THEOREM 7.1

The map risk *defined above is a bijection from* S_n *to the set of pairs of Standard Young Tableaux* (P, Q) *having identical shape and consisting of* n *boxes each.*

PROOF It suffices to show that risk has an inverse, that is, for any pair (P, Q) of Standard Young Tableaux having identical shapes and consisting of

n boxes each, there exists exactly one $\pi \in S_n$ so that $\text{risk}(\pi) = (P, Q)$.

We prove our statement by induction on n, the initial case of $n = 1$ being trivial.

In order to prove the inductive step, we show how to recover the last entry π_n of π from (P, Q). The position of the entry n in the Q-tableau reveals which position a of the P-tableau was filled last. As n is at the end of a row in Q, this position a is at the end of a row in P. If this is the first row, then there was no displacement involved in the last step of the creation of P and Q, and π_n is simply the content $c(a)$ of position a.

If a is at the end of row i, then $c(a)$ got to a after being displaced from its position b in row $i - 1$. Fortunately, we can easily recover b. Indeed, if u was the entry that displaced $c(a)$ from position b, then in the $(i-1)$st row, $c(a)$ was the smallest entry larger than u. Therefore, u is the largest entry in the $(i-1)$st row that is smaller than $c(a)$, and the position of u is b. Now we can argue as in the previous paragraph. That is, if $i = 2$, that is, b is in the first row of P, then u could not have come from a higher row, so u must have come directly from π, forcing $\pi_n = u$. Otherwise, u got to b after being displaced from the $(i-2)$nd row. In the latter case, we repeat the above argument to find the position in the $(i-2)$nd row from which u was displaced, and the entry that displaced it, and so on.

This procedure ends when we reach the first row and find out which entry started Step n of the tableau-creating procedure. That entry is, obviously, π_n.

Once we have determined π_n, we remove the box containing π_n from P and the box containing n from Q. By the definition of the Q-tableaux, this leaves us with two Standard Young Tableaux P' and Q' on $n - 1$ boxes that are of identical shape. (The entries of P' are not necessarily the elements of $[n - 1]$, but that does not matter, as they are precisely the entries of the partial permutation $\pi' = \pi_1 \pi_2 \cdots \pi_{n-1}$.) By our induction hypothesis, we can recover π' from the pair (P', Q'), completing our induction proof. ∎

Example 7.2
Let P and Q be as shown in Figure 7.4. Then the position of 8 in Q tells us that the last box to be added to P was the box containing 7. This is the box a of the above argument. Its content $c(a) = 7$ got there after being displaced from the end of row 2. In that row, it had to be at the end, so it was displaced by the entry 4. The entry 4 in turn had to be displaced from row 1, by the largest entry there smaller than 4. That entry is 3, so we have $\pi_8 = 3$. □

If $\text{risk}(\pi) = (P, Q)$, we will often write $P(\pi)$ and $Q(\pi)$ for the two tableaux of $\text{risk}(\pi)$.

As we have mentioned, many parameters of π are encoded in $\text{risk}(\pi)$. The length of the longest increasing subsequence in π is, for example, very easy to read off $\text{risk}(\pi)$.

P

1	3	6
2	4	
5	7	
8		

Q

1	4	5
2	6	
3	8	
7		

P'

1	4	6	3
2	7		
5			
8			

Q'

1	4	5
2	6	
3		
7		

FIGURE 7.4
Recovering (P', Q') from (P, Q).

LEMMA 7.3
If the length of the longest increasing subsequence of π is k, then $P(\pi)$ has k columns.

PROOF   Recall from the proof of Theorem 4.21 that π_i is called an entry of rank j if π_i is at the end of a subsequence of π that is of length j, but there is no increasing subsequence of length $j + 1$ in π that ends in π_i.

We prove a stronger statement by showing that the first position each entry of rank j of π takes during the construction of $P(\pi)$ by the risk algorithm is the jth position of the first row. This will show that the length of the first row, therefore, the number of columns, of $P(\pi)$ is indeed the maximal rank in π.

Our proof of this stronger statement is by induction on j. The case of $j = 1$, corresponds to left-to-right minima. These entries must enter the first row at the first box as they are smaller than any entry previously placed. So the initial step is complete.

Now let us assume that we know the statement is true for entries of rank j, and let us prove it for entries of rank $j + 1$. Let x be such an entry. Let x' be the entry of rank j on the left of x that is closest to x. By our induction hypothesis, x' entered the first row at box j. So when x starts looking for its position in the first row, box j is either taken by x' or (at this point we cannot yet exclude it) another, smaller entry that displaced x'. In either case, box j contains an entry that is smaller than x, so x cannot enter the first row anywhere before position $j + 1$.

On the other hand, x cannot enter the first row anywhere after position $j+1$ either. Indeed, assume this happens; this implies that when x starts looking for a position, there is an entry $y < x$ in position $j + 1$ of the first row. By our induction hypothesis, that entry cannot have rank less than $j + 1$. That is a contradiction, however, for we could affix x to the end of any increasing subsequence ending in y, yielding that the rank of x is at least $j + 2$. ∎

The result of Lemma 7.3 provides an obvious alternative proof of the fact that $Av_n(123 \cdots k) \leq (k-1)^{2n}$. There is, however, a much more refined application of this result to permutation enumeration, one that we promised in Chapter 5.

THEOREM 7.4
For any fixed positive integer k, the sequence $Av_n(12 \cdots k)$ is P-recursive.

Note that this is the only result known that proves Conjecture 5.4 for an infinite number of patterns.

PROOF   We have just seen in Lemma 7.3 that permutations not having increasing subsequences of length k can be associated with pairs of Standard Young Tableaux having at most $k-1$ columns. In other words, $S_n(12 \cdots k) = \sum_F f_F^2$, where F runs through all Ferrers shapes of size n that have at most $k - 1$ columns, and f_F denotes the number of Standard Young Tableaux of shape F. Let $F = (m_1, m_2, \cdots, m_{k-1})$ be such a Ferrers shape. Then $m_1 \geq m_2 \geq \cdots \geq m_{k-1} \geq 0$, and $\sum_{i=1}^{k-1} m_i = n$, with m_i denoting the size of the ith column. The hooklength formula now implies (see Exercise 23 for details) that

$$f_F = \left[\prod_{1 \leq i < j \leq k-1} (m_i - m_j + j - i) \right] \cdot \frac{(m_1 + m_2 + \cdots + m_{k-1})!}{(m_1 + k - 2)! \cdots (m_{k-1})!}. \qquad (7.1)$$

We can easily see that the right-hand side is P-recursive in each of its variables; therefore, by repeated applications of Lemma 5.10, it is also P-recursive in the sum of these variables. This implies that

$$\sum_{m_1 + \cdots + m_{k-1} = n} f_F^2 = Av_n(12 \cdots k) \qquad (7.2)$$

is P-recursive in n. ∎

This proof used the connection, established in Lemma 7.3, between the length of the *first* row of $P(\pi)$ and the length of the *longest* increasing subsequence of π. There is a far-reaching generalization of this observation, proved

1	3	4	8
2	5	7	9
6			

FIGURE 7.5
The tableau $P(261735984)$.

by Curtis Greene and Daniel Kleitman, and, independently, by Sergey Fomin, who was 19 years old at that time.

THEOREM 7.5
[199] Let π be a permutation, let $P(\pi)$ have k rows, and let a_i denote the length of the ith row of $P(\pi)$. Then for all $i \in [k]$, the maximum size of the union of i increasing subsequences in π is equal to $a_1 + a_2 + \cdots + a_i$.

Example 7.6
Let $\pi = 261735984$. Then $P(\pi)$ is shown in Figure 7.5.

We see from $P(\pi)$ that $(a_1, a_2, a_3) = (4, 4, 1)$. On the other hand, the longest increasing sequence of π is of length four (2678), the largest union of two increasing sequences is of size eight (2678,1359), and the largest union of three increasing sequences is of size nine. □

Note that, in particular, Theorem 7.5 implies that if π and π' are two n-permutations so that $P(\pi) = P(\pi')$, then the maximum size of the union of i increasing subsequences in π is equal to the same parameter in p'. In other words, this parameter depends only on the P-tableau of a permutation.

PROOF   (of Theorem 7.5) This is a special case of the famous Greene–Fomin–Kleitman theorem for partially ordered sets. See [199] for a proof. A more accessible reference is [279].   ∎

It seems that the increasing subsequences of π are quite well encoded by $P(\pi)$. It is natural to ask whether there are similarly strong results about the *decreasing* subsequences of π. Fortunately, the answer is in the affirmative, because of the following theorem of Schensted [280] describing the P-tableau of the *reverse* π^r of π in terms of the *transpose* (the reflected image through the NW-SE diagonal) $P(\pi)^T$ of $P(\pi)$.

THEOREM 7.7
For any n-permutation π, the equality $P(\pi)^T = P(\pi^r)$ holds.

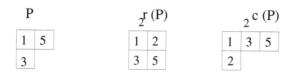

FIGURE 7.6

The P-tableaux of 31524 and 42513.

$$P \qquad\qquad r_2(P) \qquad\qquad c_2(P)$$

1	5
3	

1	2
3	5

1	3	5
2		

FIGURE 7.7

Row and column insertion of 2 into P.

Example 7.8

Let $\pi = 31524$, then $\pi^r = 42513$, and the corresponding P-tableaux are indeed conjugates of each other as shown in Figure 7.6. ▯

We need to introduce some machinery developed in [280] before we can prove Theorem 7.7. Let P be a *partial* Standard Young Tableau, that is, a tableau that we have at some point as our P-tableau during the construction of risk(π). That is, P has rows and columns that are strictly increasing, but the integers written in the boxes of P can form any subset of $[n]$, not just an initial segment.

Now let us say the next entry of π to be inserted to P by the risk algorithm is x. The partial tableau that we obtain from P once x is inserted will be denoted by $r_x(P)$, where r refers to *row insertion*. Before you ask what other insertion we could possibly talk about, we define *column* insertion just as we did (row) insertion, except that instead of rows, we use columns. That is, the entry x to be inserted arrives into the first *column*, displaces the smallest entry a that is larger than x; this entry a then enters the second *column*, and proceeds analogously. The partial tableau obtained from P by column inserting x will be denoted by $c_x(P)$.

Example 7.9

Let $\pi = 3512746$, and let P be the partial tableau obtained after three steps of the risk algorithm. Then $x = 2$, and the tableaux P, $r_2(P)$ and $c_2(P)$ are shown in Figure 7.7. ▯

The crucial property of row and column insertion is that they *commute* in the following strong sense.

PROPOSITION 7.10

Let P be a partial tableaux, and let u and v be two distinct positive integers that are not contained in P. Then we have

$$c_v(r_u(P)) = r_u(c_v(P)).$$

The proof is not overly difficult, but is a little bit cumbersome, as there are several cases to consider. The cases are based on what u and v are, and where they will be inserted. We do not want to break our line of thought here, and will therefore give the proof in Exercise 19.

Now we are ready to prove Theorem 7.7.

PROOF (of Theorem 7.7) Let $\pi = u_1 u_2 \cdots u_n$, then $\pi^r = u_n u_{n-1} \cdots u_1$. To alleviate notation, let $r_{u_i} = r_i$, and let $c_{u_j} = c_j$, and let us omit extraneous parentheses. We claim that

$$c_n c_{n-1} \cdots c_1(\emptyset) = r_1 r_2 \cdots r_n(\emptyset). \tag{7.3}$$

Proving (7.3) is sufficient as the left-hand side is clearly $P(\pi)^T$, and the right-hand side is $P(\pi^r)$. We prove (7.3) by induction on n, the initial case of $n = 1$ being obvious. As our initial tableau is empty, it does not make any difference whether we row or column insert an entry into it. Therefore,

$$r_1 r_2 \cdots r_n(\emptyset) = r_1 \cdots r_{n-1} c_n(\emptyset) = c_n r_1 \cdots r_{n-1}(\emptyset),$$

where the last equality holds because of Proposition 7.10. Applying the induction hypothesis by replacing $r_1 \cdots r_{n-1}(\emptyset)$ by $c_{n-1} \cdots c_1(\emptyset)$, we get (7.3).

∎

The following theorem is perhaps even more interesting, and has a plethora of applications. It tells us what happens if we *interchange* the two tableaux that make up risk(π).

THEOREM 7.11

Let $\pi \in S_n$, and risk(π) $= (P, Q)$. Then we have risk(π^{-1}) $= (Q, P)$.

This classic theorem is due to Marcel-Paul Schützenberger [282]. His original proof used induction. An elegant geometric proof that in fact proved a more general statement was given by X. Viennot in [320]. That proof can be found in English in [279].

Example 7.12

Let $\pi = 312$, then $\pi^{-1} = 231$, and as it is easy to verify, risk(π) and risk(π^{-1}) are shown in Figure 7.8. ▯

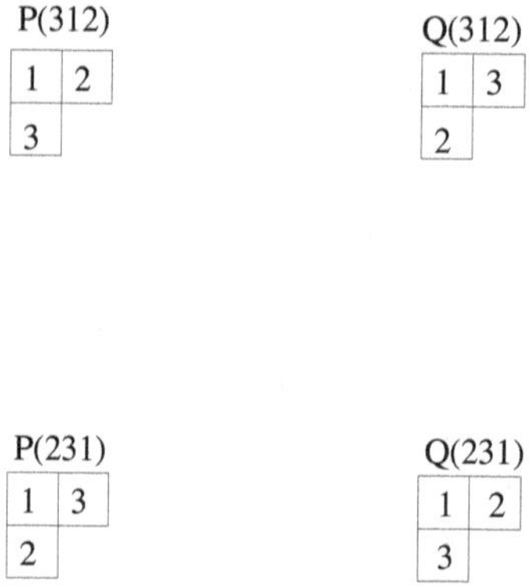

P(312)

1	2
3	

Q(312)

1	3
2	

P(231)

1	3
2	

Q(231)

1	2
3	

FIGURE 7.8
The images of 312 and its inverse, 231.

In particular, π is an involution if and only if $P(\pi) = Q(\pi)$. That is, the Robinson–Schensted–Knuth algorithm provides a bijection from the set of all n-involutions onto the set of all Standard Young Tableaux on n boxes. It is therefore easy to enumerate all these tableaux.

COROLLARY 7.13
The number of all Standard Young Tableaux on n boxes is equal to

$$\sum_{|F|=n} f_F = \sum_{i=0}^{\lfloor n/2 \rfloor} \binom{n}{2i} \cdot (2i-1)!!.$$

PROOF It is easy to see that the right-hand side is equal to the number of all involutions on $[n]$. Indeed, first choose the $2i$ entries that will be parts of 2-cycles in $\binom{n}{2i}$ ways, then take a fixed point-free involution on them. This latter can be done in $(2i-1)!!$ ways as we have seen in Corollary 3.55. ∎

COROLLARY 7.14
The equality

$$\sum_{n \geq 0} f_F \cdot \frac{z^n}{n!} = \exp\left(z + \frac{z^2}{2}\right)$$

holds.

The bijection g between involutions of length n and SYT on n boxes turns the number of fixed points of the involutions to a simple parameter of the corresponding SYT. See Problem Plus 2 for that result. The existence of g

also helps counting involutions avoiding monotone patterns, as illustrated by Problem Plus 11.

There is an interesting, close connection between descents of permutations, and descents of the corresponding SYT. Recall that we say that i is a *descent* of a Standard Young Tableau Z if i appears in a row in Z that is strictly above the row in which $i + 1$ appears in Z.

THEOREM 7.15
Let $\pi \in S_n$, and let $i \in [n-1]$. Then i is a descent of π if and only if i is a descent of $Q(\pi)$.

PROOF First let us assume that $i \in D(p)$, that is, $\pi_i > \pi_{i+1}$. We need to show that the insertion of π_{i+1} results in the addition of a new box to the P-tableaux that is below the box resulting from the insertion of π_i.

As $\pi_i > \pi_{i+1}$, we know that π_{i+1} gets inserted to the first row of $P(\pi)$ weakly on the left of π_i. If the insertion of π_i ended in the first row, then we are done, as π_{i+1} will then have to displace an entry from the first row.

In any case, the entry a_1 displaced from the first row by π_{i+1} is smaller than the entry b_1 displaced from the first row by π_i. Therefore, even if the insertion of π_i ends in the second row, that of π_{i+1} has to go on to at least one more row. We can then repeat this argument for a_1 and b_1 instead of π_{i+1} and π_i, and the second row instead of the first row, and then iterate it for further rows. We then see that the insertion of π_{i+1} will always end below that of π_i.

Now let us assume that $i \notin D(p)$, that is, $\pi_i < \pi_{i+1}$. Then reversing inequalities in the above argument shows that the insertion of π_{i+1} will end weakly above that of π_i. This implies that in $Q(\pi)$, the entry $i + 1$ will be weakly above the entry i, proving the second half of our claim. ∎

7.2 Posets of Permutations

7.2.1 Posets on S_n

There are various ways to define a partial order on the set of all n-permutations for a fixed n, or on the set of all finite permutations for that matter. The first two permutation posets that we mention, the *Bruhat order* and the *weak Bruhat order*, are ubiqitous in algebraic combinatorics as they can be generalized from S_n to a larger set of groups called *Coxeter groups*. The interested reader should consult [40] for these generalizations, as well as for further information about these posets on permutations.

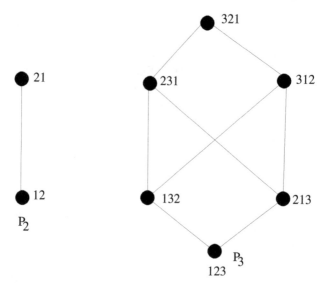

FIGURE 7.9

The Bruhat orders on S_2 and S_3.

7.2.1.1 Bruhat Order

DEFINITION 7.16   *Let P_n be the partially ordered set of all n-permutations in which $p < q$ if p can be obtained from q by a series of operations, each of which interchanges the two entries of an inversion. Then P_n is called the Bruhat order on S_n.*

For the sake of brevity, an operation that interchanges the two entries of an inversion will be called a *reduction*. The Bruhat order is sometimes called the *strong* Bruhat order for reasons that will become obvious shortly.

Example 7.17

Figure 7.9 shows the posets P_2 and P_3.   ⬚

As the reader probably knows, in a poset, we say that y covers x if $x < y$, but there is no z so that $x < z < y$, or visually, when y is "immediately above" x. The reader should spend some time justifying some of the covering relations of P_3, in order to become familiarized with this partial order. For instance, why does 231 cover 132?

Recall that a poset is called *graded* if all of its maximal (non-extendible) chains have the same length, where the length of a chain is the number of its elements minus one.

PROPOSITION 7.18
The Bruhat order P_n is a graded poset.

PROOF We claim that if y covers x in P_n, then y has exactly one more inversion than x. This will obviously imply that all maximal chains of P_n have length $\binom{n}{2}$.

Let $x < y$, and let us assume that y covers x. By Definition 7.16, $x < y$ means that x can be obtained from y by a series of reductions. However, as x is *covered* by y, all series of reductions that turn y into x must consist of one single reduction. Let that reduction be the transposition $(y_i y_j)$, where $i < j$, while $y_i > y_j$, and let us assume that this reduction results in decreasing the number of inversions of y by more than one. That means that there is an index k so that $i < k < j$ and $y_i > y_k > y_j$. In that case, however, the permutation $z = y(y_i y_k)$ would satisfy $x < z < y$, contradicting the assumption that y covers x. ∎

In a finite graded poset, the *rank* of an element is the length of any maximal chain ending at that element. It follows from the above proof that $rank_{P_n}(p) = i(p)$, where $i(p)$ is the number of inversions of $p \in S_n$.

We are going to present a classic result that provides a characterization of the Bruhat order. To that end, we need an additional definition. Let p be an n-permutation. For each $(a, b) \in [n] \times [n]$, we define

$$p(a, b) = |\{i \in [a] \text{ so that } p_i \geq b\}|.$$

Example 7.19
Let $p = 31452$, and let $(a, b) = (3, 2)$. Then $p(a, b) = p(3, 2)$ is the number of entries among the first three entries of p that are at least 2, that is to say, $p(3, 2) = 2$. ☐

Visually, we can think of $p(a, b)$ as the number of dots in the $a \times (n - b + 1)$ rectangle at the top left corner of the diagram of p, where the diagram of p is illustrated in Figure 7.10.

THEOREM 7.20
Let p and q be two n-permutations. Then $p \leq_{P_n} q$ if and only if $p(a, b) \leq q(a, b)$ for all $(a, b) \in [n] \times [n]$.

See Theorem 2.1.5 in [40] for a proof. It is easy to prove that the condition is necessary, and the reader is asked to do so in Exercise 13.
This characterization of the Bruhat order leads to the following theorem.

FIGURE 7.10

The diagrams of $p = 24153$ and $p^{-1} = 31524$.

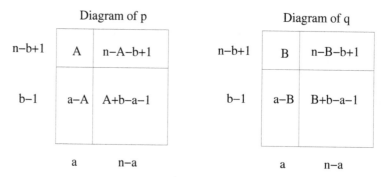

FIGURE 7.11

The number of dots in various rectangles.

THEOREM 7.21

Let p and q be two n-permutations. Then $p \leq_{P_n} q$ if and only if $p^{-1} \leq_{P_n} q^{-1}$.

PROOF   As $(p^{-1})^{-1} = p$, it suffices to prove one implication. Using Theorem 7.20, it suffices to prove that if $p(a,b) \leq q(a,b)$ for all $(a,b) \in [n]^2$, then $p^{-1}(a,b) \leq q^{-1}(a,b)$ for all $(a,b) \in [n]^2$. Let us represent permutations with their diagrams as shown in Figure 7.10. It is obvious from the definition of inverse that the diagram of p^{-1} is obtained from that of p by reflection through the $x = y$ diagonal.

As we mentioned right after Example 7.19, the number $p(a,b)$ is just the number of dots in a rectangle of shape $(n-b+1) \times a$ in the top left corner of the diagram of p. So our condition means that no matter how large a rectangle we take in the top left corner of the diagram of p, the number of dots in that rectangle is never more than the number of dots in the corresponding rectangle in the diagram of q.

All we need to prove is that the same inequality will hold for the rectangles that are in the bottom right corners of the two diagrams. Indeed, taking inverses turns these rectangles into top left corner rectangles. As each row and each column of our diagrams contains exactly one dot, it is easy to compute the number of dots in these rectangles, as shown in Figure 7.11. Clearly, if $A \leq B$, then $b - a - 1 + A \leq b - a - 1 + B$, and our theorem is proved. ∎

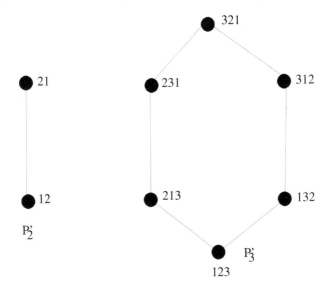

FIGURE 7.12
The weak Bruhat orders on S_2 and S_3.

7.2.1.2 Weak Bruhat Order

There is another partial ordering of permutations that is based on transpositions.

DEFINITION 7.22 *Let P'_n be the partially ordered set of all n-permutations in which $p < q$ if p can be obtained from q by a series of operations, each of which interchanges two consecutive entries that form an inversion. Then P'_n is called the* weak Bruhat order *on S_n.*

Example 7.23
Figure 7.12 shows the posets P'_2 and P'_3. ☐

It is obvious that y covers x in P'_n if and only if x can be obtained from y by an *adjacent transposition*, that is, a reduction that interchanges two consecutive entries. This implies that the weak Bruhat order is graded, and $rank(x) = i(x)$.

As P'_n is graded, all its maximal chains have the same length, namely $\binom{n}{2}$. This means that in any maximal chain, the entries i and j are interchanged *exactly once*.

Now that we know that the maximal chains of P'_n are so similar to each other, we could ask how many such maximal chains exist. This question turns

7	6	5	8	2
4	3	1
9

FIGURE 7.13
A balanced tableau.

out to be remarkably interesting. It was first answered by Richard Stanley
[294], who proved his own conjecture showing that the number of maximal
chains of P'_n is equal to the number of Standard Young Tableaux of shape
$(n-1, n-2, \cdots, 2, 1)$. The proof in [294] uses symmetric functions, so the
quest for a combinatorial proof has continued. At this point, we mention
that a chain connecting permutation p to the identity permutation is called
a *reduced decomposition* of p, so the task at hand is finding the number of
reduced decompositions of $n(n-1)\cdots 21$.

A year later Curtis Greene and Paul Edelman defined a new class of tableaux,
called *balanced tableaux*, and then found a remarkably simple bijection between
these tableaux of the same staircase shape and the maximal chains of P'_n. We
will now give an overview of their results. See [151] for a short summary of
all their results, and see [152] for the proofs of these results.

Recall the definition of a hook from Chapter 6. For a box a in a Ferrers
shape, let ℓ_a denote the *leg length* of a, that is, the number of boxes that are
in the same column as a, and are *weakly below* a.

DEFINITION 7.24   *Let F be a Ferrers shape that consists of m boxes.
A balanced tableau of shape F is a tableau whose boxes are bijectively filled
with the entries of $[m]$ so that for each box a, the content of a is the ℓ_ath
largest entry in the hook H_a.*

Note that there is some variation in the literature as to whether the content
of a should be the ℓ_ath *largest* or ℓ_ath *smallest* entry in its own hook, but for
staircase shapes, this causes no confusion. Indeed, the hooklength of a in such
a shape is always $2\ell_a - 1$, so the ℓ_ath *largest* entry of that hook is also its ℓ_ath
smallest.

Example 7.25
Figure 7.13 shows a balanced tableau on nine boxes.   ⬚

THEOREM 7.26
*The number of maximal chains of the weak Bruhat order P'_n is equal to the
number of balanced tableaux of shape $(n-1, n-2, \cdots, 2, 1)$.*

3	4	2
5	6	
1		

FIGURE 7.14
The tableau $f(C)$.

PROOF We are going to construct a bijection f from the set MC_n of all maximal chains of P_n' to the set $BalSt(n)$ of balanced tableaux of the staircase shape $(n-1, n-2, \cdots 2, 1)$.

Let $C \in MC_n$ be the maximal chain whose kth edge corresponds to the adjacent transposition interchanging the entries i and j, with $i > j$. Then we define $f(C)$ to be the tableau whose box in position $(n+1-i, j)$ contains the entry k. In other words, column j of $f(C)$ describes the transpositions that moved j and a larger entry, while row $n+1-i$ of $f(C)$ describes the transpositions that moved i and a smaller entry. Here "describes" means "tells when it happened."

Example 7.27
If $n = 4$, and c is the chain 1234, 2134, 2143, 2413, 4213, 4231, 4321, then $f(C)$ is the balanced tableau shown in Figure 7.14. For instance, in the first step, c interchanges the entries $i = 2$ and $j = 1$, so the number 1 is placed into position $(4 + 1 - 2, 1) = (3, 1)$ of $f(C)$. $\quad$ □

We first show that f indeed maps into $BalSt(n)$. Let us look at the box $(n+1-i, j)$. We need to prove that its content k is indeed the $(i-j)$th largest entry in its hook $H_{n+1-i,j}$.

The entries below the box $(n+1-i, j)$ tell us when the interchanges (x, j) took place, where $x < i$. The entries on the right of the box $(n+1-i, j)$ tell us when the interchanges (i, y) took place, with $y > j$. Originally, i and j were at distance $i - j$ from each other, so there had to be exactly $i - j - 1$ interchanges of the above types before i and j could be interchanged. So the content k of the box $(n+1-i, j)$ is the $(i-j)$th largest of its hook, as it should be. Therefore $f(C) \in BalSt(n)$.

It is clear that f is an injection. Indeed, if $C \neq C'$, and the kth edge of C and C' are different, then the position of k in $f(C)$ will be different from that of k in $f(C')$.

We still need to prove that f is a surjection, that is, that for all $B \in BalSt(n)$, there exists a $C \in MC_n$ so that $f(C) = B$. Take $B \in BalSt(n)$, and try to find its preimage under f. Then B specifies an order in which we should carry out the $\binom{n}{2}$ transpositions on the decreasing permutation. What we have to show is that the balanced property of B assures that in every step, we will be asked to carry out an *adjacent* transposition.

In order to prove this statement, we need the following, somewhat surprising characterization of balanced tableaux, given by Greene and Edelman. It shows that in fact, staircase shaped balanced tableaux are even more "balanced" than we might think. Let $c(i,j) = c_B(i,j)$ denote the content of box (i,j) of a given tableau B. We will omit B when there is no danger of confusion.

LEMMA 7.28

[152] Let B be a tableau of staircase shape $(n-1, n-2, \cdots, 1)$ whose boxes are bijectively filled with the elements of $[\binom{n}{2}]$. Then B is balanced if and only if, for all (i,j) satisfying $i + j \leq n$, and for all $s > i$, exactly one of $c(s,j)$ and $c(i, n-s+1)$ is larger than $c(i,j)$.

To remember the two boxes whose content brackets $c(i,j)$, note that in the first one, the second coordinate is fixed, the first is changed to s, in the second one, the first coordinate is fixed, the second is changed to $n - s + 1$.

PROOF   The "if" part is not surprising. Indeed, $H_{i,j} - (i,j)$ can be partitioned into disjoint unions of pairs $\{c(s,j), c(i, n-s+1)\}$, and if $c(i,j)$ is larger than exactly one element from each pair, then it is larger than half of the entries in $H_{i,j} - (i,j)$, and so B is balanced.

It is surprising, however, that this seemingly stronger condition is also necessary, that is, if B is balanced, then this condition has to hold. Let us assume that this is not true, and let us further assume, without loss of generality, that there exists $s > i$ so that $c(i,j) > c(s,j)$, and also, $c(i,j) > c(i, n-s+1)$. Then, by the balanced property of B, there exists another index t so that $c(i,j) < c(t,j)$, and also, $c(i,j) < c(i, n-t+1)$. Let us assume, again without loss of generality, that $t > s$.

Let us now assume that $H_{i,j}$ is a *minimal counterexample* to our statement. (It is easy to check that in that case, we must have $h_{i,j} > 5$.) The reader is invited to follow our argument in Figure 7.15. For brevity, in this figure we set $x' = n - x + 1$, and also, we marked entries known to be larger than $c(i,j)$ by black dots, and entries known to be smaller than $c(i,j)$ by gray dots.

As we assumed that $H_{i,j}$ was a minimal counterexample, our statement must hold for $H_{s,j}$, and $H_{i,t'}$. Therefore, as our statement holds for $H_{s,j}$, we must have

$$c(s, t') < c(s, j) < c(i, j).$$

On the other hand, as our statement holds for $H(i, t')$, we must have

$$c(s, t') > c(i, t') > c(i, j).$$

As the last two chains of inequalities clearly contradict each other, our lemma is proved.   ∎

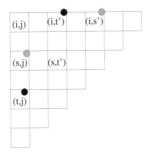

FIGURE 7.15

The entry $c(s, t')$ leads to a contradiction.

Now we can return to the proof of the surjectivity of f. Let $B \in BalSt(n)$, and let us start building up a maximal chain C by decoding B. That is, in step m, let us interchange the entries i and j, where $m = c(n - i + 1, j)$ in B. We must show that each step will actually define an adjacent transposition. Let us assume that this is not the case, and let k be a minimal counterexample. That is, let us assume that step k asks us to interchange x and y, so that $x < y$, but x and y are not in consecutive positions.

First we show that if $x < z < y$, then z cannot still be between x and y. Indeed, look at $H_{n+1-y,x}$. This hook has k written in its peak, and its length is $2(y - x) - 1$. As B is balanced, there are $y - x - 1$ entries in $H_{n+1-y,x}$ that are less than k. This means that before step k, there have been $y - x - 1$ interchanges of the types (x, z), with $x < z$, and (z, y), with $z < y$. So no $z \in (x, y)$ could be still located between x and y. Now let $v < x < y$, and let us assume that v has been interchanged with x at some point before step k. This means that $c(n + 1 - x, v) < k$. However, by Lemma 7.28, $c(n+1-x, v)$ has to be larger than exactly one of $c(n+1-y, v)$ and $c(n + 1 - x, y) = k$. Because of the previous sentence, the only way for this to happen is $c(n + 1 - y, v) < c(n + 1 - x, v) < k$. So in particular, $c(n+1-y, v) < k$, meaning that y and v were also interchanged before step k, so v is not located between x and y.

Finally, an analogous argument shows that if $x < y < u$, and u and y were interchanged before step k, then so were u and x. Therefore, no element can be located between x and y, so x and y are adjacent as claimed. Consequently, the chain C can always be built up so that $f(c) = B$, and our proof is complete.

∎

We have mentioned that Stanley [294] has proved that the number of all maximal chains in the weak Bruhat order of S_n is the number of all *Standard Young Tableaux* of shape $(n - 1, n - 2, \cdots, 2, 1)$. Now we have seen the proof of Edelman and Greene showing that this number is also equal to the number of *balanced* tableaux of shape $(n - 1, n - 2, \cdots, 2, 1)$. This certainly means that there are as many balanced tableaux of this shape as SYT of this shape. It turns out that a much more general statement is true.

THEOREM 7.29

[152] Let F be any Ferrers shape, and let b^F be the number of all balanced tableaux of shape F. Then

$$b^F = f^F.$$

In other words, there are as many balanced tableaux of *any given shape* as there are SYT of that same shape. See [152] for a proof of this theorem. The proof follows from some sophisticated bijections. There are some sporadic cases in which the bijection is simple, but in the general case it is not.

Theorem 7.29 shows that the number of balanced tableaux of a given shape is also given by the hooklength formula. It would be interesting to find a probabilistic proof of this fact.

7.2.2 Posets on Pattern-Avoiding Permutations

Let P_n^A be the partially ordered set of 132-avoiding n-permutations ordered by *strict containment* of the descent sets. That is, in P_n^A we have $p < q$ if $D(p) \subset D(q)$.

It is then clear that P_n^A is ranked, and we know from Problem Plus 1 of Chapter 4 that there are $\frac{1}{n}\binom{n}{k}\binom{n}{k+1}$ elements of rank k in P_n^A. In particular, this means that P_n^A has as many elements of rank i as of rank $n-1-i$, in other words, P_n^A is *rank-symmetric*.

There is a much deeper notion of symmetry in posets.

DEFINITION 7.30 *We say that a poset P is self-dual if it has an anti-automorphism, that is, if there exists a bijection $f : P \to P$ so that $p \leq_P q$ if and only if $f(q) \leq f(p)$.*

A bijection f having the property required in the above definition is often called *order-reversing* because of what it does.

Example 7.31

For any positive integer n, the Boolean algebra B_n of all subsets of $[n]$, ordered by containment, is self-dual. ⬜

SOLUTION   Let f be the bijection that maps each subset of $[n]$ to its complement. Then f is clearly order-reversing, proving our claim. ∎

THEOREM 7.32
The poset P_n^A is self-dual.

PROOF (of Theorem 7.32) Let $p \in P_n^A$. Recall Exercise 27 of Chapter 4. Define $f(p)$ to be the 132-avoiding permutation whose unlabeled binary tree is obtained from the unlabeled binary tree $T(p)$ of p by reflecting $T(p)$ through a vertical axis. By part (b) of the mentioned exercise, this reflection will turn left edges into right edges, and so ascents into descents, and vice versa. In particular, the vertex that was in the ith position from the left will now be in the ith position from the right. Therefore, by part (c) of the mentioned exercise, if $p_i > p_{i+1}$, then we will have $f(p_{n-i}) < f(p_{n+1-i})$. So $i \in D(p)$ if and only if $n - i \notin D(f(p))$ for $i \in [n-1]$.

In other words, the descent set of $f(p)$ is precisely the *reverse complement* of the descent set of p, implying that f is order-reversing. ∎

Let Q_n^A be the poset of 321-avoiding n-permutations ordered by strict containment of the set of excedances.

PROPOSITION 7.33
The posets P_n^A and Q_n^A are isomorphic.

PROOF For each $S \subseteq [n-1]$, let $E_n^{321}(S)$ be the set of 321-avoiding n-permutations with excedance set $S \subseteq [n-1]$.

Let also $D_n^{132}(\alpha(S))$ be the set of 132-avoiding n-permutations with descent set equal to $\alpha(S)$, the reverse-complement of S.

We construct a bijection $s \colon E_n^{321}(S) \to D_n^{132}(\alpha(S))$ illustrated by Example 7.34. If $p \in E_n^{321}(S)$, then, as seen earlier in the definition of θ, the entries p_j with $j \notin S$ form an increasing subsequence. This, and the definition of excedance imply that p_j is a *right-to-left minimum* (that is, smaller than all entries on its right) if and only if $j \notin \mathrm{Exc}(p) = S$.

Now let $p' = p_n p_{n-1} \cdots p_1$ be the reverse of p. Note that p' is a 123-avoiding permutation that has a left-to-right minimum at position $i \leq n$ exactly if $n + 1 - i \notin S$.

Recall from the proof of Lemma 4.4 that there is exactly one 132-avoiding permutation p'' which has the same set of left-to-right minima and has them at these same positions. Namely, p'' is obtained by keeping the left-to-right minima of p' fixed, and successively placing in the remaining positions, from left to right, the smallest available element that does not alter the left-to-right minima. We set $s(p) = p''$. From the proof of Theorem 7.32 we see that $i \in D(p'')$ if and only if $n - i \notin S$ for $i \in [n-1]$, in other words, when $i \in \alpha(S)$, and so p'' belongs indeed to $D_n^{132}(\alpha(S))$.

It is easy to see that s is invertible. Clearly, p' can be recovered from p'' as the only 123-avoiding permutation whose left-to-right minima have the same values and positions as those of p''. (All entries that are not left-to-right minima are to be written in decreasing order.) Then p can be recovered as the reverse of p'.

The bijections $s\colon E_n^{321}(S) \to D_n^{132}(\alpha(S))$ for all choices of $S \subseteq [n-1]$ produce an order-reversing bijection from Q_n^A to P_n^A. But P_n^A is self-dual, so the proof is complete. ∎

Example 7.34

Take $p = 3\,4\,1\,6\,2\,9\,5\,10\,7\,8 \in E_{10}^{321}(S)$ for $S = \{1, 2, 4, 6, 8\}$. Then its reversal $p' = 8\,7\,10\,5\,9\,2\,6\,1\,4\,3$ has left-to-right minima 8, 7, 5, 2, 1 in positions 1, 2, 4, 6, 8. We obtain $s(p) = p'' = 8\,7\,9\,5\,6\,2\,3\,1\,4\,10$, a permutation in $D_{10}^{132}(\{1, 3, 5, 8\})$. ☐

7.2.3 Infinite Poset of Permutations

Let P be the poset of *all finite permutations* ordered by pattern containment. That is, in this poset, $p \leq q$ if and only if p is contained in q as a pattern. This means that permutation classes defined in Chapter 4 are precisely the *ideals* of P.

We have seen in Chapters 4 and 5 that if we want to find permutations that avoid all of the patterns contained in a given set S, then our task is getting progressively harder if new elements are being added to S. It is therefore reasonable to ask whether this task will eventually become impossible. That is, let N be an arbitrary positive integer. Is it possible to find N permutations so that none of them contains any other as a pattern? Or, even more strongly, is it possible to find an *infinite* antichain in the poset P?

This question was attacked, and the affirmative answer discovered and re-discovered several times, during the last third of the twentieth century. Here we present a construction that may be chronologically the first. The construction (without proof of the antichain property) was published by Tarjan [310] in 1972.

THEOREM 7.35
The poset P contains an infinite antichain.

The above result could be reformulated by saying that P is not a *well-quasi-ordering*.

PROOF   Let $n \geq 2$, and let

$$p_n = 2\,(4n-1)\,4\,1\,6\,3\,8\,5 \cdots (4n-2)\,(4n-5)\,(4n)\,(4n-3).$$

In other words, p_n has $4n$ entries, and consists of two parts, the increasing subsequence $246 \cdots (4n)$ in the odd positions, and the odd entries in the even positions in increasing order, except for $4n - 1$, which is moved up into the second position. We claim that the p_n form an infinite antichain.

Assume the contrary, that is, that $p_k \leq p_n$ for some $k < n$. Which entries of p_n could play the roles of the entries of p_k? The role of $4k - 1$ has to be played by $4n-1$; otherwise, we could only find at most two smaller entries on its right. Therefore, the entry 2 of p_n must play the role of the entry 2 of p_k. This, however, totally ties our hand in making the remaining $4k-2$ selections. Indeed, the entry 1 of p_n must be chosen to play the role of 1 as that is the only entry smaller than 2; therefore, the entry 4 of p_n has to be chosen to play the role of 4 as that is the only entry not yet selected that precedes 1. These forced selections continue, and we have to choose the leftmost $4k - 2$ entries of p_n to play the roles of the first $4k - 2$ entries of p_n. Then we must choose $4n$ to be the next-to-last entry as that is the only entry larger than $4n - 1$, forcing us to choose $4n - 3$ to play the role of the last entry of p_k. However, this last entry is too large. Indeed, it is larger than the entry $4k - 2$ that we choose when we selected the $(4k - 3)$th entry of our purported copy of p_k. This is a contradiction, as in p_k the last entry is smaller than the entry in position $4k - 3$.   ∎

7.3   Simplicial Complexes of Permutations

We have defined simplicial complexes in Problem Plus 1 of Chapter 1, but we repeat that definition for easy reference.

DEFINITION 7.36   *A simplicial complex Δ is a family of subsets of an underlying set S so that if $F \in \Delta$ and $G \subset F$, then $G \in \Delta$.*

In other words, a simplicial complex is an ideal of the Boolean algebra with underlying set S. The sets that belong to the collection Δ are called the *faces* of Δ. If $S \in \Delta$ has i elements, then we call S an $(i - 1)$-dimensional face. The dimension of Δ is, by definition, the dimension of its maximal faces.

Example 7.37
Let P be a finite partially ordered set. Then the collection of all *chains* in P forms a simplicial complex, called the *chain complex* of P.   ⬚

Indeed, every induced subposet of a chain is a chain.

If we can prove that certain objects, say, n-permutations with k descents, are in bijection with k-element faces of a given simplicial complex, then that can have additional algebraic significance. In fact, additional algebraic interpretations can be found for the numbers enumerating our objects. The algebraically inclined reader is invited to consult [95] for details.

7.3.1 Simplicial Complex of Restricted Permutations

As we have mentioned, and in some cases, shown, there are many objects enumerated by the Catalan numbers. Among all of these, we now choose 231-avoiding permutations and show how they form a simplicial complex. The reason for this choice is that 231-avoiding permutations fit well in a more general class of permutations, called t-stack-sortable permutations, to which we will return in the next chapter. The following theorem explains what we mean when we say that the set of these permutations form a simplicial complex.

THEOREM 7.38

There exists a simplicial complex (with an underlying set of $\binom{n}{2}$ elements) whose $(k-1)$-dimensional faces, that is, k-element faces, are in bijection with 231-avoiding n-permutations having k ascents.

PROOF   We have seen in Exercise 17 of Chapter 4 that 231-avoiding n-permutations are in bijection with northeastern lattice paths from $(0,0)$ to (n, n) that never go above the main diagonal. We have also seen in that same exercise that this bijection turns the ascents of the permutations into north-to-east turns of lattice paths.

Now let Δ be the simplicial complex of all sets S of points for which there exists a northeastern lattice path r so that the set of all north-to-east turns of r is equal to S. (Of course, we still have to show that Δ is indeed a simplicial complex.) It is clear that given S, we can recover r. So the faces of this simplicial complex Δ are indeed in bijection with 231-avoiding n-permutations. The previous paragraph shows why $(k-1)$-dimensional faces correspond to permutations with k ascents.

Finally, we prove that Δ is indeed a simplicial complex. Let $S \in \Delta$. Then there exists a subdiagonal northeastern lattice path r so that the set of north-to-east turns of r is equal to S. That is possible if and only if the points of S form a chain in $\mathbf{N}^2$ (in the natural cordinate-wise ordering), are between certain limits, and have all different horizontal and vertical coordinates. However, if that is the case, then that must be true for all subsets $T \subset S$, implying that there is a northeastern lattice path t whose set of north-to-east turns is T. That means $T \in \Delta$, and the proof is complete. ∎

Example 7.39

Let $n = 4$, and let $S = \{(2,1), (3,2)\}$. Then the northeastern lattice paths whose set of north-to-east turns is contained in S are shown in Figure 7.16.

□

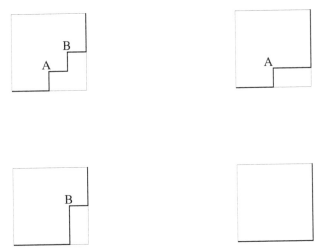

FIGURE 7.16
Each subset of S defines a lattice path.

7.3.2 Simplicial Complex of All n-Permutations

There are several ways to define a simplicial complex whose k-element faces are in bijection with n-permutations having k descents. One of these was presented in Theorem 1.27, and others can be found in [189] or [153]. Here we present such a simplicial complex based on the bijective representation of permutations by labeled lattice paths. We have discussed this representation in the proof of Theorem 1.27, but for easy reference, we include it here.

Let $p = p_1p_2 \cdots p_n$ be an n-permutation having k descents, and let us say that $D(p) = \{d_1, d_2, \cdots, d_k\}$. Then it is not difficult to prove that we can bijectively represent p by a northeastern lattice path $F(p)$ consisting of k vertical and $n - k$ horizontal edges, where the edges are labeled according to the following rules.

Let the edges of $F(p)$ be denoted $a_1, a_2, \cdots, a_n$, and let e_i be the label of a_i. Then the following has to hold.

(i) The edge a_1 is horizontal and $e_1 = 1$,

(ii) If the edges a_i and a_{i+1} are both vertical, or both horizontal, then $e_i \geq e_{i+1}$,

(iii) If a_i and a_{i+1} are perpendicular to each other, then $e_i + e_{i+1} \leq i + 1$.

The set of labeled lattice paths of length n satisfying these conditions is denoted $\mathcal{P}(n)$.

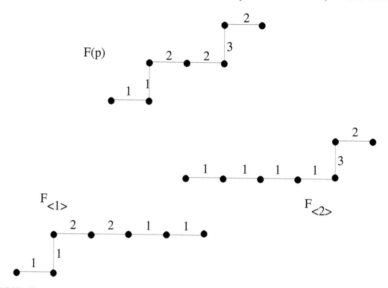

FIGURE 7.17
Decomposing $F(p)$ into $(F_{\langle 1 \rangle}, F_{\langle 2 \rangle})$.

We are going to decompose $F(p)$ into a k-tuple of northeastern lattice paths, $(F_{\langle 1 \rangle}, F_{\langle 2 \rangle}, \cdots, F_{\langle k \rangle})$, each of which will consist of one vertical step and $n-1$ horizontal steps. The unique vertical step of $F_{\langle i \rangle}$ will be in the same position as the ith vertical step of $F(p)$. In other words, the unique vertical edge of $F_{\langle i \rangle}$ will be its $(d_i + 1)$st edge, corresponding to the ith descent of p. We still have to specify the labels of the edges of $F_{\langle i \rangle}$. Let $e_{i,j}$ be the label of the jth edge of $F_{\langle i \rangle}$. We then set

$$
e_{i,j} = \begin{cases} 1 \text{ if } j \leq d_i, \\ e_j \text{ if } d_i + 1 \leq j \leq d_{i+1}, \\ 1 \text{ if } d_{i+1} + 1 \leq j. \end{cases}
$$

It is straightforward to check that this is a valid definition, that is, rules (i)-(iii) are satisfied.

Example 7.40
Figure 7.17 shows how we decompose $F(p)$ if $p = 612435$.   ☐

Now let Δ be the family of k-tuples $(F_{\langle 1 \rangle}, F_{\langle 2 \rangle}, \cdots, F_{\langle k \rangle})$ that can be obtained from n-permutations the way described above, for some k. Note that instead of saying "k-tuples," we might as well say "sets" as the order of the

$F_{\langle i \rangle}$ within each such set is completely determined by the position of the single vertical edge in each $F_{\langle i \rangle}$.

THEOREM 7.41
The collection of sets Δ defined in the previous paragraph is a simplicial complex.

PROOF The proof is very similar to that of Theorem 7.38. Let $F' = (F_{\langle 1 \rangle}, F_{\langle 2 \rangle}, \cdots, F_{\langle k \rangle})$ be a k-tuple of lattice paths from $\mathcal{P}(n)$ having one single vertical edge each. Define the numbers b_i so that the vertical edge of $F_{\langle i \rangle}$ is the b_ith edge of $F_{\langle i \rangle}$. Then $F' \in \Delta$ if and only if $b_1 < b_2 < \cdots < b_k$. If F' has this property, then obviously so do all its subwords (subsets), so $F' \in \Delta$ implies $F'' \in \Delta$, for all $F'' \subset F'$. ∎

Exercises

1. (−) Prove that $I_n(123 \cdots k) \leq (k-1)^n$.

2. Noncrossing partitions of $[n]$ are defined in Exercise 24. Find a bijection from the set of 321-avoiding permutations with k excedances to the set of noncrossing partitions of $[n]$ having $k + 1$ blocks.

3. Prove, without the use of the hooklength formula, that the number of Standard Young Tableaux of shape $2 \times n$ is C_n.

4. What is the average number of descents of a $2 \times n$ Standard Young Tableau?

5. Let π be a permutation of length 80 that avoids the pattern $12 \cdots 9$, and let us assume that $P(\pi)$ has ten rows. How long is the fifth row of $P(\pi)$?

6. Find a formula for the number of northeastern lattice paths from $(0,0)$ to (n, n) that never go above the main diagonal and are symmetric about the diagonal $x + y = n$.

7. Prove that $I_n(123) = \binom{n}{\lfloor n/2 \rfloor}$.

8. Prove that $I_n(123 \cdots k)$ is P-recursive in n.

9. Let $D_{k,r}(n)$ denote the number of n-permutations p in which the longest increasing subsequences have k elements and for which r is the largest natural number so that there exist r disjoint increasing subsequences of

maximum size in p. Prove that then $D_{k,r}(n)$ is a P-recursive sequence of n.

10. Let π be a permutation so that the first row of $P(\pi)$ is of length a, and the second row of $P(\pi)$ is of length b. Is it true that π has two disjoint increasing subsequences s_1 and s_2 so that s_1 has length a and s_2 has length b?

11. (–) Prove that

$$f^F = \sum_{F'} f^{F'},$$

where F' ranges over the set of Ferrers shapes that can be obtained from F by omitting a box (which is necessarily an inner corner).

12. Let π be any permutation other than the increasing or decreasing one. Find a simple way to create another permutation σ so that $P(\pi) = P(\sigma)$.

13. Prove part of Theorem 7.20 by showing that if $p(a,b) > q(a,b)$ for some $(a,b) \in [n] \times [n]$, then $p \leq q$ cannot hold.

14. For what positive integers n is the number of involutions of length n even?

15. Is there an infinite antichain in the poset P of finite permutations ordered by pattern containment that consists of 123-avoiding permutations?

16. (–) Keeping in mind that the reverse or the complement of an involution is not necessarily an involution, prove that nevertheless, for all positive integers n, we have

$$I_n(123\cdots k) = I_n(k\cdots 321).$$

17. We can view a permutation p as a poset P_p as follows. The elements of P_p are the entries of p, and $x <_P y$ if (x,y) is a non-inversion of p. Prove that P_p and $P_{p^{-1}}$ are isomorphic.

18. Let k be any positive integer. For some n, find k different n-permutations p_i so that all the P_{p_i} are isomorphic to one another.

19. Prove Proposition 7.10.

20. Let us consider all $n!$ vectors that are obtained by permuting the coordinates of the n-dimensional vector

$$\begin{pmatrix} 1 \\ 2 \\ \cdots \\ n \end{pmatrix}.$$

Denote by π_n the *convex hull* of all of these vectors in $\mathbf{R}^n$, that is, the smallest convex set that contains all of these vectors. It is clear that π_n is a polyhedron, and therefore, it is often called the *permutahedron*. How many edges does π_n have?

21. Prove that

$$\sum_{k=\lfloor (n+1)/2 \rfloor}^{n} \left(\frac{2k-n+1}{n+1} \binom{n+1}{k+1} \right)^2 = \frac{\binom{2n}{n}}{n+1}.$$

22. (+) Prove that the number of dissections of a convex $(n+2)$-gon by d nonintersecting diagonals is equal to the number of Standard Young Tableaux of shape $(d+1, d+1, 1, 1, 1, \cdots, 1)$, where the number of rows of length 1 is $n-1-d$.

23. Complete the proof of Theorem 7.4 by showing that the hooklength formula indeed implies (7.1).

24. Recall the definition of a *crossing* in a set partition from Problem Plus 9 of Chapter 5. Let us call a partition *noncrossing* if it has no crossings.

   (a) Define a bijection from the set of noncrossing partitions of $[n]$ having k blocks to the set of 132-avoiding n-permutations having $k-1$ descents.

   (b) Define the partially ordered set $NC(n)$ of noncrossing partitions of $[n]$ by refinement. That is, $\pi \leq_{NC(n)} \pi'$ if all blocks of π' are unions of blocks of π. Compare this poset to the poset P_n^A defined in Section 7.2.2.

25. Recall that a partially ordered set P is called a *lattice* if for any two elements $x, y \in P$, the set

$$\{z \in P | x \leq z \text{ and } y \leq z\}$$

has a (unique) minimum element $x \vee y$, and the set

$$\{u \in P | x \geq u \text{ and } y \geq u\}$$

has a (unique) maximum element $x \wedge y$.

Is the weak Bruhat order P_n' a lattice?

26. Prove that there exists a bijection $f : P_n' \to P_n'$ so that $x \wedge f(x) = 123 \cdots n$ and $x \vee f(x) = n(n-1) \cdots 21$. See the previous exercise for the relevant definitions.

27. A finite lattice L is called *complemented* if for any element $x \in L$, there exists a *unique* element $y \in L$ so that $x \wedge y = 0$ and $x \vee y = 1$, where 0 denotes the minimum element of L, and 1 denotes the maximum element of L.

Is P'_n a complemented lattice?

28. Let I_n be the induced subposet of the (strong) Bruhat order P_n whose elements are the *involutions* of length n. Is I_n a lattice?

29. Let I_n be defined as in the previous exercise. Is I_n self-dual?

30. The *dimension* of the poset P is the smallest positive integer d so that P is the intersection of d total orderings. What is the dimension of the poset P_p defined in Exercise 17?

31. (+) Find an asymptotic formula for the number of 2-dimensional posets on n labeled elements.

32. We have defined the permutahedron in Exercise 20. Clearly, π_n is in fact an $(n-1)$-dimensional polyhedron, as all its vertices lie within the hyperplane $\sum_{i=1}^{n} x_i = \binom{n+1}{2}$. So in particular, π_4 is three-dimensional. Do all faces of π_4 have the same number of edges?

33. Let p be a randomly selected involution of length n. Let $i \in [n-1]$. What is the probability that i is a descent of p?

34. Decide if the following statements are true or false. In all three statements, p and q are two n-permutations, and their respective sets of inversions are $I(p)$ and $I(q)$.

   (a) If $I(p) \leq I(q)$, then $p \leq q$ in the Bruhat order.
   (b) If $p \leq q$ in the Bruhat order, then $I(p) \leq I(q)$.
   (c) If $p \leq q$ in the weak Bruhat order, then $I(p) \leq I(q)$.

35. Let a_n be the number of involutions of length n. Prove that

$$n! < a_n^2 < n! p(n),$$

where $p(n)$ is the number of partitions of the integer n.

36. Let a_n denote the number of involutions of length n. Prove that

$$\binom{2n}{n} n! = \sum_{r=0}^{2n} \binom{2n}{r} (-1)^r a_r a_{2n-r}. \tag{7.4}$$

37. Let us call two n-permutations $p = p_1 p_2 \cdots p_n$ and $q = q_1 q_2 \cdots q_n$ *colliding* if there is at least one index i so that $|p_i - q_i| = 1$. Let S be a set of n-permutations so that each pair of elements of S is colliding. Prove that S has at most $\binom{n}{\lfloor n/2 \rfloor}$ elements.

38. Let $\rho(n)$ be the size of the largest set of n-permutations in which every pair of elements is colliding. Prove that $\rho(n) \geq \rho(n-1) + \rho(n-2)$.

39. Prove that for all positive integers n, the equality $\mathrm{Av}_n(231, 312) = I_n(231, 312)$ holds.

40. Prove that there are no pairs (p, q) of patterns of length three or longer, other than those in the preceding exercise so that $\mathrm{Av}_n(p, q) = I_n(p, q)$ holds for all p and q.

Problems Plus

1. Let $D_k(n)$ be the number of n-permutations in which the longest increasing subsequences have k elements and they *all* have at least one element in common. Prove that $D_k(n)$ is a P-recursive function of n.

2. We have seen in Theorem 7.11 that the Robinson–Schensted–Knuth correspondence naturally defines a bijection between inversions of length n and SYT of size n. Let f be this bijection, and assume that we are told the *shape* of $f(p)$ (so not the content of each box). How can we figure out the number of fixed points of p from this information?

3. What is the number of 321-avoiding fixed point-free involutions of length $2n$?

4. What is the number of 123-avoiding fixed point-free involutions of length $2n$?

5. Prove that the number of 123-avoiding involutions of length $2n+1$ having exactly one fixed point is

$$I_n^{(1)}(123) = \binom{2n+1}{n}.$$

6. Let $I_n^{(k)}(q)$ denote the number of q-avoiding involutions of length n having exactly k fixed points. Prove that

$$I_n^{(k)}(321) = I_n^{(k)}(132) = I_n^{(k)}(213) = \begin{cases} \frac{k+1}{n+1}\binom{n+1}{(n-k)/2} & \text{for } n+k \text{ even,} \\ 0 & \text{for } n+k \text{ odd.} \end{cases}$$

7. Let P be a poset having n elements, and let a_1 denote the length of its longest chain. Let a_2 be defined so that the largest number of elements

that the union of two chains can contain in P is $a_1 + a_2$. Similarly, for $i \geq 2$, let a_i be defined so that the largest number of elements that the union of i chains can contain in P is $a_1 + a_2 + \cdots + a_i$. Continue defining the a_i as long as they are positive.

(a) Prove that $a_1 \geq a_2 \geq \cdots \geq a_k$, where a_k is the last a_i that has been defined (in other words, $\sum_{i=1}^{k} a_i = n$). Therefore, there exists a Ferrers shape $F = (a_1, a_2, \cdots, a_k)$. Let b_i be the length of *column i* of F. Prove that for any positive integer j, the sum $b_1 + b_2 + \cdots + b_j$ is equal to the largest number of elements that can be covered by j *antichains* in P.

(b) Why is the result of part (a) a generalization of Theorem 7.5?

8. Let us try to generalize the result of the previous Problem Plus as follows. Let G be a graph on n vertices, and let a_i be defined so that $a_1 + a_2 + \cdots + a_i$ is equal to the largest number of vertices that can be contained in the union of i cliques (complete subgraphs) in G. Again, define the a_i as long as they are positive.

(a) Is it true that $a_1 \geq a_2 \geq \cdots \geq a_k$?

(b) Assuming that the answer to the question of part (a) is yes, define b_i as in part (b) of the previous Problem Plus. Is it true that the sum $b_1 + b_2 + \cdots + b_j$ is equal to the largest number of elements that can be contained in j *anticliques* (independent sets of points, in other words, empty subgraphs) in G?

9. For what class of graphs will statements (a) and (b) of the previous Problem Plus follow directly from the result of Theorem 7.5 ?

10. Find a formula for $I_n(1234)$.

11. (a) Find a formula for $I_n(12345)$.
    (b) Find a formula for $I_n(123456)$.

12. Prove that for all positive integers and n, and for all *even* positive integers k, we have

$$\binom{2n}{n} Av_n(12 \cdots k{+}1) = \sum_{r=0}^{n} \binom{2n}{r} (-1)^r I_r(12 \cdots k{+}1) I_{2n-r}(12 \cdots k{+}1).$$

13. Let $x_k(m)$ denote the number of *fixed point-free* involutions of length m with no *decreasing subsequences* of more than k elements. Prove that for all positive integers n, and for all *odd* positive integers k the equality

$$\sum_{r=0}^{2n} \binom{2n}{r} x_k(r) x_k(2n - r) = \sum_{r=0}^{2n} (-1)^r \binom{2n}{r} I_r(12 \cdots k + 1) I_{2n-r}(12 \cdots k + 1)$$

(7.5)

holds.

14. Prove that for all positive integers n, the equality

$$\sum_{r=0}^{2n}(-1)^r\binom{2n}{r}I_r(1234)I_{2n-r}(1234) = I_{2n}(12345) \qquad (7.6)$$

holds.

15. Let $F(2n)$ be the induced subposet of the Bruhat order P_{2n} whose elements are *fixed point-free* involutions of length $2n$.

    The *rank-generating function* of a finite graded poset P is the polynomial $W_P(x) = \sum_{i=0}^{k} r_i x^i$, where r_i is the number of elements of P that are of rank i. Prove that

    $$W_{F(2n)} = [\mathbf{1}][\mathbf{3}]\cdots[\mathbf{2n-1}].$$

16. Find an asymptotic formula for the number of isomorphism classes of two-dimensional posets on n elements. Here $[\mathbf{m}] = 1 + q + \cdots + q^{m-1}$.

17. Let $k \geq 2$ be an integer, and let us say that two n-permutations $p = p_1 p_2 \cdots p_n$ and $q = q_1 q_2 \cdots q_n$ are k-*different* if there exists an index i so that $p_i \neq q_i$, while $1 \leq p_i, q_i \leq k$. Let $K(k)$ be the largest integer for which there exists a $K(k)$-element set of n-permutations that are pairwise k-different. Prove that $K(k) \geq (k+1)!/2$.

18. Recall that in Exercise 52 of Chapter 4, we defined $u_n(q)$ to be the number of permutations that avoid q and have a *unique longest increasing subsequence*, or ULIS. Prove that $\lim_{n\to\infty} (u_n(321))^{1/n} = 4$.

Solutions to Problems Plus

1. We claim that our condition that all maximum-length increasing subsequences have a common element is equivalent to the seemingly weaker condition that any two maximum-length increasing subsequences intersect. If we can prove this, then our problem will be reduced to the special case $r = 1$ of Exercise 9.

   To prove our claim, let us assume that p is a permutation in which any two increasing subsequences of maximum length k intersect. We construct a directed graph G_p associated to p. The vertices of G_p are the entries of p and there is an edge from the entry i to the entry j if and only if $i < j$ and i is on the left of j. So an increasing subsequence of length k in p corresponds to a directed path of length k in G_p. Now let us remove all edges not in any maximum-length-path from G_p, and add a "source" s and a "sink" t to get the graph G'_p. That is, s and t are vertices so that s has indegree zero, and there is an edge from s to all left-to-right minima of p, while t has outdegree zero and there is an edge to t from all right-to-left maxima of p. So each increasing subsequence of size k corresponds to an $s \to t$ path of maximum size in a natural way. Now suppose these directed paths of maximum length do not have a vertex in common. Then we can delete any vertex v and still have an $s \to t$ path in G'_p. In other words, G'_p is $2 - (s, t)$-connected, which implies, by the famous theorem of Menger (see for example [246]) that there are at least two vertex-disjoint $s \to t$ paths in G'_p. This is equivalent to saying that there are two increasing subsequences of size k in p that are disjoint, which is a contradiction and the proof is complete.

2. The number of fixed points of p is equal to the number of *columns of odd length* in $f(p)$. See [32] for a proof of this fact, and for a reference for the original paper of Schützenberger where the result was proved.

3. We have seen in the previous Problem Plus that the number of fixed points of an involution equals the number of odd columns of its P-tableau. Therefore, a fixed point-free involution has no odd columns. If, in addition, such an involution avoids 321, then its P-tableau cannot have columns longer than 2. This implies the shape of this P-tableau must be $2 \times n$. As an involution is completely determined by its P-tableau, we conclude from Exercise 3 that the number of such SYT is C_n.

4. Let us first try to proceed as in the previous Problem Plus. In any case, the rows of the P-tableau of such an involution must be of length at most two, so the P-tableau has two columns. Because of the fixed point-free criterion, odd columns are not allowed. So (except for the trivial, one-column case), the P-tableau will have two columns, of length $2(n - k)$, and $2k$, where $k \leq n/2$. It is then routine to compute by the hooklength formula that the number of these tableaux, and therefore, the number of fixed point-free 123-avoiding involutions of length $2n$ is

$$\sum_{k=0}^{n/2} \binom{2n}{2k} \frac{2n - 4k + 1}{2n - 2k + 1}.$$

This is not a particularly simple formula.

However, it has recently been proved in [146] that in fact, the number of these involutions is $\binom{2n-1}{n}$. The proof uses bijections to certain lattice paths called Dyck paths.

5. It follows from the solution of Exercise 7 that $I_{2n+1}(q) = \binom{2n+1}{n}$. As a 123-avoiding involution of odd length has to have exactly one fixed point, the result follows.

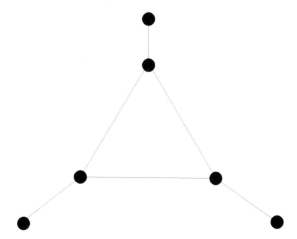

FIGURE 7.18
A counterexample.

6. It is clear that $I_n^{(k)}(132) = I_n^{(k)}(213)$ for any n and k as 231 is the reverse complement of 132, and the reverse complement of an involution is an involution. The enumeration formula for 132 is proved in [205], and the bijection between the sets enumerated by $I_n^{(k)}(132)$ and $I_n^{(k)}(321)$ is given in [146].

7. (a) This is the famous Greene–Kleitman–Fomin theorem, which has several proofs. The two earliest ones are [200] and [176]. A different proof, using the concept of *orthogonal* families of chains and antichains, was given in [179].

   (b) For any permutation p, we can take its permutation poset P_p. The chains of P_p will be in bijection with the increasing subsequences of p, and the antichains of P_p will be in bijection with the decreasing subsequences of p. Note that in this special case, the proof of our claim follows from Theorem 7.7.

8. (a) No, this is not true. See Figure 7.18 for a counterexample. It is easy to see that for the graph shown in that figure, we have $a_1 = 3$, $a_2 = 1$, and $a_3 = 2$.

   (b) This is not true either. See Figure 7.19 for a counterexample. One checks easily that $a_1 = a_2 = 2$, and $a_3 = 1$. This would have to imply $b_1 = 3$, but there is no anticlique of size three in this graph.

9. The class of graphs we are looking for is that of *comparability graphs of posets*. If P is a poset, then its comparability graph $G(P)$ is the graph whose vertex set is the set of elements of P, and two vertices are

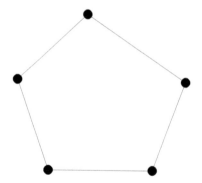

FIGURE 7.19

A counterexample.

connected by an edge if the corresponding elements of P are comparable. Then the cliques of $G(P)$ correspond to the chains of P, and the anticliques of $G(P)$ correspond to the antichains of P.

10. The numbers $I_n(1234)$ are equal to the Motzkin numbers, that is,

$$I_n(1234) = \sum_{i=0}^{n/2} C_i \binom{n}{2i}.$$

This result was first proved by Amitaj Regev in [272], who used symmetric functions in his argument. A simpler proof is given in [297], Exercise 7.16.b, but that proof still uses symmetric functions. In recent years, there are a plethora of results on the subject, that together yield that $I_n(1234) = M_n$, and do not use symmetric functions. In fact, it is known that

$$M_n = I_n(2143) = I_n(1243) = I_n(1234).$$

The first of the above three equalities was bijectively proved in [207], the second one was bijectively proved in [206], and the third one follows from a result proved in [220]. That result is $I_n(12 \oplus q) = I_n(21 \oplus q)$, for any pattern q. Therefore,

$$I_n(1243) = I_n(2134) = I_n(1234),$$

where the first equality is true because if p is an involution, then so is the *reverse complement* of p.

We point out that [220] contains the even stronger result that $I_n(123 \oplus q) = I_n(321 \oplus q)$ for any pattern q. It would be interesting to find a more direct, combinatorial proof.

11. (a) The number $I_n(12345)$ is in fact the number of Standard Young Tableaux on n boxes having at most four columns. It is proved in

[197] that

$$I_n(12345) = \begin{cases} C_k^2 \text{ if } n = 2k \text{ is even,} \\ C_k C_{k+1} \text{ if } n \text{ is odd.} \end{cases}$$

(b) It is proved in [197] that

$$I_n(123456) = 6 \sum_{i=0}^{\lfloor n/2 \rfloor} \binom{n}{2i} C_i \frac{(2i+2)!}{(i+2)!(i+3)!}.$$

This is the largest k for which an exact formula is known for $I_n(12\cdots k)$.

12. This result is due to Herbert Wilf, who gave a generating function proof in [329]. Such an elegant result certainly asks for a direct bijective proof, but none is known to this day.

13. This result is due to the present author [64]. The crucial idea of the proof is the following. Let B_n be the set of ordered pairs (p, q), where p is an involution on a subset s_p of $[2n]$, and q is an involution on the set $[2n] - s_p$, the complement of s_p in $[2n]$.

Let $B(n, k, r)$ be the subset of B_n consisting of pairs (p, q) so that neither p nor q has a decreasing subsequence longer than k. Note that here p is an involution of length r and q is an involution of length $2n - r$. It is then clear that

$$|B(n, k, r)| = \binom{2n}{r} I_r(12 \cdots k + 1) I_{2n-r}(12 \cdots k + 1),$$

the summand of the right-hand side of (7.5) indexed by r. Let $B(n, k) = \cup_r B(n, k, r)$.

Let f be the involution defined on B_n that takes the largest fixed i point present in $p \cup q$ and if i is part of p, then moves it to q, and if i is part of q, then moves it to p. Let $f_{n,k}$ be the restriction of f to the set $B(n, k)$. It can then be shown that $f_{n,k}$ maps into $B(n, k)$. So the only pairs $(p, q) \in B(n, k)$ whose contribution to the right-hand side of (7.5) is not canceled by the contribution of $f_{n,k}(p, q)$ are the pairs for which $f(p, q)$ is not defined. It follows from the definition of $f_{n,k}$ that these are the pairs in which both p and q are fixed point-free involutions.

14. Apply the result of the previous Problem Plus with $k = 3$. It follows from the result of Problem Plus 2 that if p is a fixed-point free involution, then the Standard Young Tableau corresponding to p has no odd columns. Therefore, $x_{2m+1}(r) = x_{2m}(r)$. So $x_3(r) = x_2(r)$, and so the result of the previous Problem Plus reduces to

$$\sum_{r=0}^{2n} \binom{2n}{r} x_2(r) x_2(2n - r) = \sum_{r=0}^{2n} (-1)^r \binom{2n}{r} I_r(1234) I_{2n-r}(1234).$$

Note that $x_2(r)$ is just the number of Standard Young Tableaux in which each column is of length two (since the columns are of even length not more than two). The number of such SYT is the Catalan number C_r (where r has to be even) as shown in Exercise 3. This shows that the two sides of (7.6) are equal to $\sum_i \binom{n}{i}\binom{2n}{2i} C_i C_{n-i}$. It is shown in Problem Plus 11, part (a), that this is precisely the formula for $I_n(12345)$.

15. This result is proved in [139] where the authors define an interesting notion of weight on fixed point-free involutions.

16. It is proved in [332] that the number of these classes is $(1 + o(1))n!/2$. The result was also proved by El-Zahar and Sauer [163].

17. This result is proved in [232] in a more general context. A construction simply consist of all *even* permutations of length $k + 1$. Indeed, if two such permutations were not k-different, then they would differ only in a 2-cycle of the form $(i\ k + 1)$, but then they could not both be even. This lower bound is known to be optimal for $k = 2$ and $k = 3$, but is known to be not optimal for $k = 4$.

18. This result was proved in [81]. Let $n = 2m + 1$. Then it is shown in that paper that $u_n(321) \geq C_m^2$. The proof uses the Robinson-Schensted correspondence, and a bijection of Claesson and Kitaev [121].

8

Get Them All. Algorithms and Permutations.

8.1 Generating Permutations

8.1.1 Generating All n-Permutations

If we want to write a computer program to test a conjecture concerning permutations, we need to have an efficient method to generate all n-permutations for our machine. If the conjecture only concerns permutations with some restrictions, we can save a lot of time and effort by having a fast way to generate only those permutations with the required property.

One can certainly list all permutations lexicographically. That is, let us define a partial order on the set of all n-permutations as follows. Let $p = p_1 p_2 \cdots p_n < q_1 q_2 \cdots q_n$ if for the smallest index i for which $p_i \neq q_i$, we have $p_i < q_i$. The total order defined on S_n is called the *lexicographic* order. So for instance, $34152 < 35412$ since the smallest index for which p_i and q_i are different is $i = 2$, and $p_2 < q_2$.

It is obvious that the smallest element of S_n in the lexicographic order is the identity permutation $12 \cdots n$. Therefore, in order to construct an algorithm to list all n-permutations in the lexicographical order, it suffices to have a method to find the permutation *immediately following* a given permutation p in the lexicographic order. Such a method is provided by the following proposition. Recall that in a poset we say that q covers p if $p < q$, and there is no r so that $p < r < q$.

PROPOSITION 8.1

Let $p = p_1 p_2 \cdots p_n$ be a permutation. Let i be the largest ascent of p. Then the permutation q covering p in the lexicographical order is given by $q = p_1 p_2 \cdots p_{i-1} q_i q_{i+1} \cdots q_n$, where q_i is the smallest element in $\{p_{i+1}, p_{i+2}, \cdots p_n\}$ that is larger than p_i, and the string $q_{i+1} \cdots q_n$ contains the remaining entries of $p_i p_{i+1} \cdots p_n$ in increasing order.

Example 8.2
If $p = 2415763$, then $i = 4$, and $q = 2416357$. ☐

DOI: 10.1201/9780429274107-8

PROOF   Let q be as above, then $p < q$ as $p_i < q_i$. Let us assume that there exists an n-permutation $r = r_1 r_2 \cdots r_n$ so that $p < r < q$. Then $r_1 r_2 \cdots r_{i-1} = p_1 p_2 \cdots p_{i-1}$ must hold, otherwise one of $p < r$ and $r < q$ could not be true. As we must have $p_i \leq r_i \leq q_i$, and q_i is the smallest remaining integer larger than p_i, we must have either $p_i = r_i$, or $r_i = q_i$. Both are impossible, however. Indeed, in the first case, we could not have $p < r$, as the rest of p is decreasing, and in the second case, we could not have $r < q$, as the rest of q is increasing. ∎

Proposition 8.1 provides an obvious algorithm that generates all permutations of length n, one-by-one, in the lexicographic order. The algorithm will stop when the index i, that is, the largest ascent of p, cannot be found; that is, when $p = n(n-1)\cdots 1$, the maximum element of our total order.

8.1.2   Generating Restricted Permutations

A huge amount of research has been done on objects counted by the Catalan numbers. These objects, of which Richard Stanley lists over 150 different kinds in [297], have been enumerated according to various statistics, leading to interesting open problems. It is therefore desirable to be able to generate these objects efficiently. We will show how to do this with 231-avoiding permutations.

First, we define a total order H_n on the set of all 231-avoiding n-permutations as follows. Let $p = LnR$ and $q = L'nR'$ be two n-permutations, where L denotes the string on the left of n in p, and R denotes the string on the right of n in p, and L' and R' are defined analogously for q. Note that L and R could be empty. Now let $|L|$ denote the length of L. We say that $p \leq_{H_n} q$ if

(a) $|L| > |L'|$, or

(b) $|L| = |L'|$, and $L < L'$ in $H_{|L|}$, or

(c) $|L| = |L'|$, and $L = L'$, and $R \leq R'$ in $H_{|R|}$.

Example 8.3
Let $n = 8$. Then $p = 21348576 \leq 32184567 = q$ as L is of length four, while L' is of length three, so rule (a) applies. Similarly, if $r = 32148567$, then $p \leq r$ as $2134 \leq 3214$, by the repeated application of rule (b) ($213 \leq 321$). Finally, if $s = 21348765$, then $p \leq s$ by rule (c). ⬚

Example 8.4
For $n = 4$, the list of all elements of H_n in increasing order is 1234, 2134, 1324, 3124, 3214, 1243, 2143, 1423, 1432, 4123, 4213, 4132, 4312, 4321. ⬚

It is straightforward to see that H_n is indeed a total order as any two elements of H_n are comparable. Futhermore, H_n has minimum element $12\cdots n$ and maximum element $n\cdots 21$. Note that the order H_n is quite *left-heavy*, that is, a little change made at the beginning of a permutation can influence the rank of the permutation much more than a little change at the end of the permutation.

This observation suggests that we try to locate the unique element q covering $p = LnR$ in H_n by the following algorithm.

Start with $p = LnR$. We will describe how to transform certain parts of p. Parts that are not mentioned in a given step are left unchanged by that step. We point out that 0-element and 1-element strings are considered monotone decreasing.

(i) If R is not monotone decreasing, then replace p by R, redefine L and R for this new, shorter p, and go to (i) again.

(ii) If R is monotone decreasing, but L is not, then replace p by L, redefine L and R for this new, shorter p, and go to (i) again.

(iii) If L and R are both monotone decreasing, then move the first (leftmost) entry of L to the position immediately after the maximal entry of p. This creates new strings L and R; order the entries of L and the entries of R increasingly. The result of all these changes is the permutation q.

The above algorithm will output a permutation for all inputs $p \neq n(n-1)\cdots 1$.

Example 8.5

- Let $p = 21348576$. Then $L = 2134$, and $R = 576$. So R is not monotone decreasing. Therefore step (i) applies.

- Now we look at $p = 576$, and see that in it, both L and R are monotone decreasing. Therefore, step (iii) applies, and we obtain the string 756.

- Concatenating this with the unchanged part of the original permutation, we get that $q = 21348756$.

□

Example 8.6

- Let $p = 215439876$, then $L = 21543$, and $R = 876$. So R is monotone decreasing, but L is not. So step (ii) applies.

- Now $p = 21543$, so $L = 21$ and $R = 43$. Both L and R are monotone decreasing, so step (iii) applies.

- In step (iii), we move 2 beyond 5, and rearrange L and R in increasing order. This gives the string 15234, which, concatenated with the unchanged part 9876, gives $q = 152349876$.

$\square$

THEOREM 8.7

For any 231-avoiding n-permutation p other than the decreasing permutation, the algorithm described above outputs the unique n-permutation q that covers p in H_n.

PROOF   Note that being 231-avoiding is equivalent to the property that all entries of L are smaller than all entries of R, and that this holds recursively for L and R, and their recursively defined subwords. Steps (i) and (ii) of the algorithm will preserve this property as they operate within L and within R. Step (iii) moves the *largest* entry of L to the *leftmost* position of R, meaning that R now starts with its smallest entry. This again assures that the "entries of L are smaller than entries of R" property is preserved. Therefore q is indeed 231-avoiding.

Now we show that q indeed covers p. We are going to prove this by induction on n. The initial cases of $n = 1$ and $n = 2$ are obvious. Let us assume that we know that the statement is true for all positive integers less than n.

If $p = LnR$ is such that at least one of L and R is not monotone decreasing, then it is straightforward to see by the induction hypothesis that q covers p. If L and R are both monotone decreasing, then the only way to find permutations that are larger than p in H_n is by moving n closer to the front. If we want to find a permutation that *covers* p, then we must move n up by one position. As our permutations are 231-avoiding, this means that the entry that gets beyond n in this move must be the *largest* entry of L. As L is monotone decreasing, its largest entry is its leftmost one, just as prescribed in Step (iii). Finally, to get the smallest possible permutation with the new position of n, we must order L and R increasingly. This completes the proof. $\blacksquare$

Note that by the obvious bijections (reverse, complement), versions of the above algorithm will generate all 132-avoiding, 213-avoiding, or 312-avoiding permutations. It is less obvious how to generate all 321-avoiding or 123-avoiding permutations efficiently.

8.2 Stack-Sorting Permutations

The task of *sorting*, that is, arranging n distinct elements in increasing order efficiently, is a central problem of computer science. There are various sorting algorithms, like *merge sort*, or *heap sort*, that can handle this task in $O(n \log n)$ steps. It is also known that in the worst case, we indeed need this many steps. The interested reader is invited to consult a book on the Theory of Algorithms, such as [230], to learn these results.

In this chapter we discuss some combinatorial sorting algorithms. In these algorithms, our goal is still to arrange n distinct objects (for the sake of brevity, the elements of $[n]$) in increasing order, but our hands will be tied by certain rules.

Let $p = p_1 p_2 \cdots p_n$ be a permutation, and let us assume that we want to rearrange the entries of n so that we get the identity permutation $123 \cdots n$. This does not seem to be a mountainous task. However, this task will become more difficult if we are told that all we can use in our sorting efforts is a vertical *stack* that can hold entries in increasing order only (largest one at the bottom). We are even told how we have to use the stack (though this will rather help us than hold us back, as we will explain after Example 8.9).

We note that there are alternative combinatorial sorting tools that lead to interesting problems; for instance, there are linear sorting devices that can receive or release entries on both ends. We will discuss some of these in Section 8.4. In the present section, we will devote all our attention to a certain greedy approach, which was first studied in detail by Julian West [325], though Tarjan did some work on the subject before that. Chronologically, this is not the earliest approach, but this is the one that has received the most attention. Since this approach seems to be the prevailing one, we will often refer to it simply as "stack-sorting," instead of the longer, and more precise term "greedy stack-sorting." We will mention some other approaches in later sections, and in the exercises.

In order to stack-sort $p = p_1 p_2 \cdots p_n$ in the greedy way, we consider the entries of the input permutation p one-by-one. First take p_1, and put it in the stack. Second, we take p_2. If $p_2 < p_1$, then it is allowed for p_2 to go in the stack on top of p_1, so we put p_2 there. If $p_2 > p_1$, however, then first we take p_1 out of the stack, and put it to the first position of the output permutation, and *then* we put p_2 into the stack. We continue in this way: at step i, we compare p_i with the element $r = p_{a_{i-1}}$ currently on the top of the stack. If $p_i < r$, then p_i goes on the top of the stack; if not, then r goes to the leftmost empty position of the output permutation, and p_i gets compared to the new element that is currently on the top of the stack. The algorithm ends when all n entries passed through the stack and are in the output permutation $s(p)$.

We are now ready to announce the most important definition of this section.

output	stack	input
		3142
	3	142
	1 3	42
1	3	42
13	4	2
13	2 4	
132	4	
1324		

FIGURE 8.1
Stack-sorting $f = 3142$.

DEFINITION 8.8 *If the output permutation $s(p)$ defined by the above algorithm is the identity permutation $123\cdots n$, then we say that p is stack-sortable.*

Example 8.9
Figure 8.1 shows the stages of stack-sorting the permutation 3142. We conclude that $s(p) = 1324$, therefore 3142 is not stack-sortable. □

If you think that our sorting algorithm is too arbitrary, in that it requires the entries in the stack to be in increasing order, or that it tells us when an entry should enter or leave the stack, consider the following. If at any point of time, the entries in the stack were not in increasing order, then we could surely not obtain the identity permutation at the end since no entry x currently below an entry y in the stack could pass y in the output. So the monotone increasing requirement for the stack is not a real restriction. Now that we know this, let us look at the other requirement. That requirement says that if $p_i < r$, where r is currently on the top of the stack and p_i is the next entry of the input, then p_i goes on the top of the stack, and if not, then r

goes to the output. We claim that this is not a real restriction either. Indeed, if p_i is the larger one, then it cannot go on top of r as that would destroy the increasing property of the stack. If p_i is the smaller one, then r should not go to the output before p_i does as that would create an inversion in the output. This shows that if a permutation is sortable by a stack, our algorithm will output the identity permutation, and the mentioned criteria are not limiting our possibilities.

What permutations are stack-sortable? This is one of the very few easy questions of this area.

PROPOSITION 8.10
A permutation is stack-sortable if and only if it is 231-avoiding.

PROOF   Let p be stack-sortable, and let us assume that entries b, c, and a of p form a 231-pattern in this order. Then b would enter the stack at some point, but would have to leave it before the larger entry c could enter it. As a could only enter the stack after c, it is clear that a would arrive to the output after b did, implying that $s(p)$ would contain the inversion ba, which is a contradiction.

Conversely, let us assume that p is not stack-sortable. Then the image $s(p)$ contains an inversion yx. That means $y > x$, so y must have entered the stack before x did, and so y must have left the stack before x even arrived there. What forced y to leave the stack before x arrived? By our algorithm, this could only be an entry z that is larger than y and is located between y and x in p. However, in that case yzx was a 231-pattern in p.   ∎

Note that we have in fact proved the following statement that will be useful in later applications.

COROLLARY 8.11
The entries $u < v$ of p will appear in $s(p)$ in decreasing order if and only if there is a 231-pattern in p whose leftmost entry is v and whose rightmost entry is u.

8.2.1   2-Stack-Sortable Permutations

We have seen that only $C_n < 4^n$ permutations of length n are stack-sortable out of $n!$ total permutations, meaning that greedy stack sorting in itself is not very efficient. However, we can repeat the entire sorting algorithm by sending our output $s(p)$ through the stack again.

DEFINITION   8.12   *The permutation p is called 2-stack-sortable if $s(s(p)) = s^2(p) = 123 \cdots n$. Similarly, p is called t-stack-sortable if $s^t(p) = 123 \cdots n$.*

It is not hard to see that no n-permutation will require more than $n - 1$ applications of s to be sorted. In other words, all n-permutations are $(n-1)$-sortable. See Exercise 5 for a proof of this fact.

Encouraged by the simple result of Proposition 8.10, one might try to characterize 2-stack-sortable permutations by classic pattern avoidance. These efforts, however, are bound to fail. Indeed, sets of permutations that consist of all permutations avoiding a given set of patterns are always *permutation classes*. That is, if p is in such a set, then so are all the substrings of p. However, this is not true for 2-stack-sortable permutations.

Example 8.13

The permutation $q = 35241$ is 2-stack-sortable. Indeed, $s^2(p) = s(32145) = 12345$. On the other hand, its substring $p = 3241$ is not 2-stack-sortable, as $s^2(p) = s(2314) = 2134$. $\square$

For 2-stack-sortable permutations, some characterization is nevertheless possible. This is the content of the next lemma.

LEMMA 8.14

[325] A permutation is 2-stack-sortable if and only if it does not contain a 2341-pattern, and it does not contain a 3241-pattern, except possibly as part of a 35241-pattern.

PROOF   Let us assume first that p is 2-stack-sortable. Then by Proposition 8.10, its image $s(p)$ must be 231-avoiding. However, if $bcda$ were a 2341-pattern in p, then bca would be a 231-pattern in $s(p)$. Similarly, if $cbda$ were a 3241-pattern in p that is not part of a 35241-pattern, then $s(p)$ would contain a 231-pattern. This pattern could be bca (if there is no 231-pattern starting with c and ending with b) or cfa (if f is the largest entry between c and b in p so that cfb is a 231-pattern).

Let us assume now that p is not 2-stack-sortable. Then $s(p)$ is not stack-sortable; therefore, it must contain a 231-pattern. Let bca be such a pattern. Then it follows from Corollary 8.11 that there is a 231-pattern bda and a 231-pattern cea in p. If b is on the left of c in p, then we see that $bcea$ is a 2341-pattern in p. If b is on the right of c in p, then again by Corollary 8.11, there cannot be any entry between c and b in p that is larger than c. However, that implies that $cbea$ is a 3241-pattern in p that is not part of a 35241-pattern, completing our proof. ∎

The problem of finding an exact formula for the number $W_2(n)$ of 2-stack-sortable n-permutations is a difficult and fascinating one.

THEOREM 8.15

For all positive integers n, the identity

$$W_2(n) = \frac{2(3n)!}{(n+1)!(2n+1)!}$$

holds.

The above formula was first conjectured in a Ph.D. thesis by Julian West [325], who pointed out that these numbers also enumerate nonseparable rooted planar maps on $n+1$ edges. The first proof was provided five years later by Doron Zeilberger [338], who used a computer to find the solution to a degree-9 functional equation. Two other proofs [149], [204] have been found later. Both construct fairly complicated bijections between the set of 2-stack-sortable n-permutations and the aforementioned planar maps. The latter have been enumerated by Tutte [313] in 1963. Finally, a bijection between 2-stack-sortable permutations, and a certain class of labeled trees, the so-called $\beta(1,0)$-trees, was discovered [127], but a simple proof is yet to be found. Even a simple proof for the much weaker claim that $W_2(n) \leq \binom{3n}{n}$ could be very useful as it could provide useful insights towards finding upper bounds for the numbers of t-stack-sortable permutations, for $t > 2$.

We close our introduction to 2-stack-sortable permutations with a fascinating fact that has not been satisfactorily explained yet. It is a classic result of Kreweras that does not look related at first sight.

THEOREM 8.16

The number of lattice paths starting at $(0,0)$, ending at $(i,0)$, and using $3n+2i$ steps, each of which is equal to either $(1,1)$, or $(0,-1)$, or $(-1,0)$, that never leave the first quadrant is

$$\frac{4^n(2i+1)}{(n+i+1)(2n+2i+1)}\binom{2i}{i}\binom{3n+2i}{n}.$$

This result has several proofs. The most direct one is by Olivier Bernardi [36]. See [88] for a history of the problem and for other proofs. If we set $i = 0$, then we get that the number of such lattice paths that end at $(0,0)$ in $3n$ steps is precisely $2^{2n-1} \cdot W_2(n)$. A direct proof of this fact could lead to a simple proof for the formula for the numbers $W_2(n)$.

8.2.2 t-Stack-Sortable Permutations

Let $W_t(n)$ be the number of t-stack-sortable n-permutations, that is, permutations p for which $s^t(p) = 12\cdots n$. It is very difficult to enumerate these permutations if $t > 2$. See Exercise 9 for an upper bound that holds for all t, but is far from what seems to be the truth.

Encouraged by the exact formulae $W_1(t) = \frac{\binom{2n}{n}}{n+1}$ and $W_2(t) = \frac{2\binom{3n}{n}}{(n+1)(2n+1)}$, we might think that in the general case, we could hope for a formula like

$$W_t(n) = \frac{\binom{(t+1)n}{n}}{p(n)}, \tag{8.1}$$

where $p(n)$ is a polynomial with rational coefficients. Numerical evidence, however, does not support this conjecture. Indeed, if (8.1) held for $t = 3$ and some polynomial $p(n)$, then the number $W_3(n)$ could have no prime factor larger than $4n$. This contradicts the available data since $W_3(6) = 606$, which has prime factor 101. The first few values of $W_3(n)$ can be found in the *Encyclopedia of Integer Sequences* as sequence A134664. The fact that $W_3(6) = 606$ follows from Exercise 8.

Next, one could ask if at least the inequality $W_t(n) \leq \binom{(t+1)n}{n}$ holds. Recently, Colin Defant, Andrew Elvey Price, and Anthony J. Guttmann [138] proved that the answer to that question is negative. They showed that the exponential growth rate of the sequence $W_3(n)$ is at least 9.4854, which is larger than the exponential growth rate of the sequence $\binom{4n}{n}$, which is $256/27 \approx 9.4815$.

The following simple observation narrows the field of permutations that have a chance to be t-stack-sortable.

PROPOSITION 8.17
If the permutation p is t-stack-sortable, then p avoids the pattern $23 \cdots (t + 2)1$.

PROOF If p contains such a pattern, then the entries forming a copy of such a pattern will form a $23 \cdots t1(t+1)$-pattern in $s(p)$, a $23 \cdots (t-1)1t(t+1)$-pattern in $s^2(p)$, and so on. Finally, they will form a $23145 \cdots (t + 1)$-pattern in $s^{t-1}(p)$, implying that $s^{t-1}(p)$ contains a 231-pattern, and is not stack-sortable. So its image, $s^t(p)$, is not the identity permutation. ∎

8.2.2.1 Equivalent Algorithm

Consider the following modification of the t-stack sorting operation. Instead of passing a permutation through a stack t times, we pass it through t stacks placed next to each other in series as follows. The first stack operates as the usual stack except that when an entry x leaves it, it does not go to the output right away. It goes to the next stack if $x < j$, where j is the entry on the top of the next stack, or if the next stack is empty. If $j < x$, then x cannot move until j does.

The general step of this algorithm is as follows. Let $S_1, S_2, \cdots, S_t$ be the t stacks, with a_i being the entry on top of stack S_i. If the next entry x of the input is smaller than a_1, we put x on top of S_1. Otherwise, we find the smallest i so that a_i can move to the next stack (that is, that $a_i < a_{i+1}$ or S_{i+1} is empty), and move a_i on top of S_{i+1}. If we do not find such i, or if $S_1, S_2, \cdots, S_{t-1}$ and the input have all been emptied out, then we put the entry on the top of S_t into the output.

We will call this algorithm the *right-greedy* algorithm, as it always makes the rightmost move possible as the entries move from the input on the right to the output on the left.

We can describe the movement of the entries of the input permutation p through the stack by *stack words*. If $t = 1$, then there are just two kinds of moves, an entry either moves in the stack or out of the stack. Let us denote these steps with letters A and B, respectively. Then the movement of all entries of p is described by a stack word consisting of n copies of A and n copies of B in which for all i, the ith A precedes the ith B. The number of such words is well-known to be $C_n = \binom{2n}{n}/(n+1)$. On the other hand, if p is stack-sortable, then its output is the identity, so given the stack word of p, we can uniquely recover p. It is easy to prove by strong induction that each word that satisfies the conditions described in this paragraph is indeed the stack word of a stack-sortable permutation, so this is a stack word proof of the fact that $W_1(n) = C_n$.

In general, if there are t stacks, then there are $t+1$ different kinds of moves. Therefore, the movement of p through the $t + 1$ stacks can be described by a word of length $(t + 1)n$ that consists of n copies of each of $t + 1$ different letters.

In what follows, we will consider the case of $t = 3$, when there are four letters, to prove a significantly sharper upper bound for the exponential order of the sequence $W_3(n)$. Let A denote the move of an entry from the input to the first stack, let B denote the move of an entry from the first stack to the second stack, let C denote the move of an entry from the second stack to the third stack, and let D denote the move of an entry from the third stack to the output.

Note that we will also call stack words 3-stack words or 2-stack words when we want to emphasize the number of stacks that are used to sort a given word. Also note, for future reference, that for all t, the descents of p are in bijective correspondence with the AA factors of p. (An XY factor of a word is just a letter X immediately followed by a letter Y.)

We will identify 3-stack-sortable permutations with their stack words. We can do that since if p is 3-stack-sortable, then its image under the 3-stack sorting algorithm is the identity permutation, so given the stack word of p, we can uniquely recover p.

PROPOSITION 8.18

Let w be a 3-stack word of a permutation. Then all of the following hold.

1. *There is no BB factor in w.*

2. There is no CC factor in w.

3. There is no BAB factor in w.

4. There is no CBA^jC factor in w, where $j \geq 0$.

PROOF   Each of these statements holds because otherwise the entries in the second or third stack would not be increasing from the top of the stack to the bottom of the stack.   ∎

PROPOSITION 8.19
Let w be a 3-stack word of a permutation. Then all of the following hold.

1. There is no DA factor in w.

2. There is no DB factor in w.

3. There is no CA factor in w.

PROOF   Each of these statements holds because of the greediness of our algorithm. For instance, a D cannot be followed by an A, since the move corresponding to D did not change the content of the first stack, so if the A move was possible after the D move, it was possible before the D move, and therefore, *it would have been made* before the D move. Analogous considerations imply the other two statements.   ∎

Note that the conditions given in Propositions 8.18 and 8.19 are necessary, that is, they must hold in 3-stack words of all permutations, but they are *not* sufficient. In other words, if a word satisfies all these conditions, it is not necessarily the 3-stack word of a permutation.

Let w be a 3-stack word of a *3-stack-sortable* permutation p, and let $v = v(w)$ be the subword of w that consists of the letters B, C and D in w. In other words, $v = v(w)$ is the word obtained from w by removing all copies of the letter A. This can create BB factors in v, even though there were no BB factors in w.

Note that v describes how the stack sorted image $s(p)$ of p traverses the second and third stacks. Note that as p is 3-stack-sortable, $s(p)$ is 2-stack-sortable. So v is the *2-stack word* of the 2-stack-sortable permutation $s(p)$ over the alphabet $\{B, C, D\}$. Therefore, there are $W_2(n) = \frac{2}{(n+1)(2n+1)}\binom{3n}{n}$ possible choices for v.

Furthermore, every descent of $s(p)$ bijectively corresponds to a BB-factor of v. The number of 2-stack-sortable permutations of length n with $k - 1$ descents is known (see Problem Plus 8.1) to be

$$W_2(n, k-1) = \frac{(n+k-1)!(2n-k)!}{k!(n+1-k)!(2k-1)!(2n-2k+1)!}. \tag{8.2}$$

LEMMA 8.20

The number $W_3(n)$ of 3-stack-sortable permutations of length n satisfies the inequality

$$W_3(n) \le \sum_{k=1}^{(n+1)/2} \frac{(n+k-1)!(2n-k)!}{k!(n+1-k)!(2k-1)!(2n-2k+1)!} \cdot \binom{2n-2k}{n-1}.$$

PROOF Let us count all such permutations with respect to the number of descents of their stack sorted image $s(p)$. If $s(p)$ has $k-1$ descents, then its 2-stack word v has $k-1$ factors BB. In order to recover the 3-stack word w of p, we must insert n copies of A into v so that we get a valid 3-stack word. As BB factors and BAB factors are not allowed in w, we must insert two copies of A into the middle of every BB factor, and we also have to put one A in front of the first B. After that, we still have $n-2(k-1)-1 = n-2k+1$ copies of A left. We can insert these in n possible slots only, namely on the left of the first B, and immediately following any B except the last one. (This is because Proposition 8.19 tells us that there are no CA or DA factors in w.) Therefore, by a classic balls-and-boxes argument, the number of ways to place all copies of A is at most

$$\binom{n-2k+1+n-1}{n-1} = \binom{2n-2k}{n-1}.$$

As there are $W_2(n, k-1)$ choices for v, the proof is complete by summing over all possible values of k. ∎

It is easy to prove that $\lim_{n\to\infty} \sqrt[n]{W_3(n)}$ exists. Now we are ready to prove an upper bound for that limit.

THEOREM 8.21

The inequality

$$\lim_{n\to\infty} \sqrt[n]{W_3(n)} \le 12.53296$$

holds.

PROOF As Lemma 8.20 provides an upper bound for $W_3(n)$ as a sum of less than n summands, it suffices to prove that the largest of those summands is of exponential order 12.539547. In order to do that, we use Stirling's formula that states that $m! \sim (m/e)^m \sqrt{2\pi m}$, so $\lim_{m\to\infty} \sqrt[m]{m!} = m/e$. Let $w_3(n, k)$ denote the number of 3-stack-sortable permutations p of length n so that $s(p)$ has k descents. Setting $k = nx$, with $x \in (0, 1]$, and applying Stirling's formula to each factor of the bound in Lemma 8.20, this leads to the equality

$$g(z) := \lim_{n \to \infty} \sqrt[n]{w_3(n, zn)}$$

$$= \frac{(1+z)^{1+z} \cdot (2-z)^{2-z} \cdot (2-2z)^{2-2z}}{z^z \cdot (1-z)^{1-z} \cdot (2z)^{2z} \cdot (2-2z)^{2-2z} \cdot (1-2z)^{1-2z}}$$

$$= (1+z) \cdot (2-z)^{2-z} \cdot z^{-3z} \cdot (1-z)^{-1+z} \cdot (1-2z)^{2z-1} \cdot \left(\frac{z+1}{4}\right)^x.$$

The function g takes its maximum when $g'(z) = 0$, which occurs when

$$z = \frac{1}{12} \cdot (27 + 12 \cdot \sqrt{417})^{1/3} - \frac{13}{4 \cdot (27 + 12 \cdot \sqrt{417})^{1/3}} + \frac{1}{4} \approx 0.2883918927.$$

For that value of z, we get $g(z) = 12.53296$, completing the proof. ∎

Colin Defant [132] proved Theorem 8.21 using a different method. His method was more complicated, but he was able to use it prove that the inequality

$$\lim_{n \to \infty} \sqrt[n]{W_4(n)} \le 21.97225$$

holds as well.

8.2.2.2　Symmetry

Let $W_t(n, k)$ be the number of t-stack-sortable permutations with k descents. We propose to fix n and t, and investigate the sequence $W_t(n, k)_{0 \le k \le n-1}$. For instance, if $n = 5$, and $t = 3$, then we find the sequence $1, 25, 62, 25, 1$, while for $n = 4$ and $t = 1$, we get the sequence $1, 6, 6, 1$. A look at further numerical evidence suggests several interesting properties of these sequences. The simplest one is *symmetry*, that is, it seems that $W_t(n, k) = W_t(n, n-1-k)$, or in other words, there seem to be as many t-stack-sortable n-permutations with k descents as with k ascents. In fact, for $t = n - 1$, the statement is obvious, considering Exercise 5, while for $t = 1$ and $t = 2$ we can verify that the statement is true using the known explicit formulae. See Problem Plus 1 of Chapter 4 for the relatively simple formula for $t = 1$, and see Problem Plus 1 of this chapter for the more complex formula for $t = 2$. These are not sporadic special cases.

THEOREM 8.22
For all fixed n and t, we have $W_t(n, k) = W_t(n, n-1-k)$.

PROOF　There seems to be no trivial reason for this symmetry. Indeed, the usual symmetries of permutation classes (reverse, complement) that turn ascents into descents do not preserve the t-stack-sortable property, even when

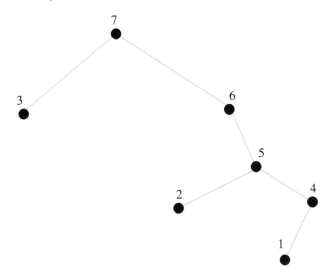

FIGURE 8.2
The decreasing binary tree of $p = 3762514$.

$t = 1$. In order to prove our theorem, we need to find a more subtle symmetry that turns ascents into descents, *and* preserves the t-stack-sortable property for *any* t.

We will use the notion of *decreasing binary trees*, that we have discussed in Section 6.3. For easy reference, here is the definition again.

DEFINITION 8.23   *Let* $p = LnR$ *be an* n-*permutation, where* L *and* R *denote the (possibly empty) substrings of* p *on the left and right of* n. *The decreasing binary tree* $T(p)$ *of* p *is the binary plane tree with root* n, *left subtree* $T(L)$, *and right subtree* $T(R)$.

Example 8.24
The decreasing binary tree of the permutation 3762514 is shown in Figure 8.2. Here we have $s(p) = 3214567$.   ☐

It is easy to prove (Exercise 31 of Chapter 1) that decreasing binary trees on n nodes are in bijection with n-permutations.

There are two reasons for which decreasing binary trees are fitting to the task at hand. First, just as their remote cousins, the binary plane trees, they encode the number of descents. Indeed, the reader is invited to prove the simple fact that the number of descents of p is equal to the number of right edges (that is, edges that go from northwest to southeast) of $T(p)$. Second, the stack sorted image $s(p)$ can easily be read of $T(p)$, thanks to the following proposition.

PROPOSITION 8.25
Let $p = LnR$ be an n-permutation. Then we have

$$s(p) = s(L)s(R)n. \tag{8.3}$$

PROOF   By the definition of the stack sorting algorithm, the entry n can only enter the stack when the stack is empty, that is, when all entries on the left of n passed through the stack. This shows that $s(p)$ will start with the string $s(L)$. Once L passed through the stack, n will enter. However, as n is larger than any other entry, n will not be forced out from the stack by any other entry. So n will stay at the bottom of the stack until all other entries have passed through it. This shows why $s(p)$ will continue with the string $s(R)$, and then end in the entry n.   ∎

This important "left-right-n" property could be used to *define* the stack sorting operation. Here is what it means in the context of decreasing binary trees.

COROLLARY 8.26
Given $T(p)$, we can read off $s(p)$ from $T(p)$ by reading the nodes of $T(p)$ in postorder. That is, first we read the left subtree of the root, then the right subtree of the root, and then the root itself. Each subtree is read by this same rule.

PROOF   Immediate by induction on n, using Proposition 8.25. See Figure 8.2 for an example.   ∎

Now we are ready to define the bijection that will prove Theorem 8.22. We will define this bijection f in terms of decreasing binary trees as this approach shows the "real reason" for the nice symmetries we are going to prove. As there is a one-to-one correspondence between these trees and n-permutations, f can easily be interpreted in terms of permutations as well.

DEFINITION 8.27   *Let $T(n, k)$ be the set of decreasing binary trees on n vertices that have k right edges, and let $T(p) \in T(n, k)$. For each vertex v of $T(p)$ do as follows.*

(a) *If v has zero or two children, leave the subtrees of v unchanged.*

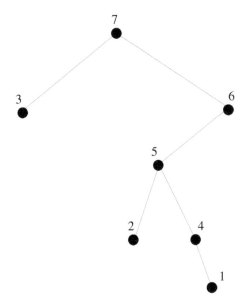

FIGURE 8.3
The tree $f(T(p))$ for $p = 3762514$.

(b) If v has a left subtree only, then turn that subtree into a right subtree.

(c) If v has a right subtree only, then turn that subtree into a left subtree.

Let $f(T(p))$ be the tree we obtain from $T(p)$ in this way.

Example 8.28
Let $p = 3762514$, then $T(p)$ is the tree shown in Figure 8.2.

To construct $f(T(p))$, note that the nodes corresponding to 3, 7, 2, 5, and 1 have either zero or two children, so they are left unchanged. The node 6 has a right subtree only, so that subtree is turned into a left subtree, while the node 4 has a left subtree only; therefore, that subtree is turned into a right subtree. We end up with the tree $f(T(p))$ shown in Figure 8.3.

We then read off that the corresponding permutation is $f(p) = 3725416$.
□

It is then clear that $f(T(p)) \in T(n, n-1-k)$, so f has the desired effect on the number of descents of the permutation p.

You could say, "Fine. So you have proved in an ever so sophisticated way that there are as many n permutations with k descents as there are with k ascents. I can do it by just reversing all permutations." At this point, you would be right in saying so. However, our bijection f *preserves the t-stack-sortable property for any t*, implying that the restriction of f to the set of

t-stack-sortable n-permutations is just what we need to prove our theorem. This is the content of the next lemma.

For a permutation p, we will write $f(p)$ for the permutation whose decreasing binary tree is $f(T(p))$. In other words, $f(T(p)) = T(f(p))$.

LEMMA 8.29
For any fixed n and t, the n-permutation p is t-stack sortable if and only if $f(p)$ is t-stack-sortable.

PROOF   We claim that

$$s(p) = s(f(p)), \tag{8.4}$$

that is, $s(p)$ is $(t-1)$-stack-sortable if and only if $s(f(p))$ is, which clearly implies the statement of the Lemma.

Intuitively speaking, what (8.4) says is that pushing some lonely left edges to the right or vice versa does not change the postorder reading of $T(p)$. Let us make this argument more precise.

We prove (8.4) by induction on n. For $n = 1$ and $n = 2$, the formula obviously holds. Now let us assume that we know the statement for all non-negative integers less than n.

If the root of $T(p)$ has two children, then $p = LnR$, and the postorder reading of $T(p)$ is just the concatenation of the postorder reading of $T(L)$, the postorder reading of $T(R)$, and n. By our induction hypothesis, the postorder reading of $T(L)$ is the same as that of $T(f(L))$, and the postorder reading of $T(R)$ is the same as that of $T(f(R))$. Therefore, as the root of $T(f(p))$ has two children, and they are roots of the trees $T(f(L))$ and $T(f(R))$, the postorder reading of $T(p)$ and that of $T(f(p))$ are identical as they are concatenations of identical strings.

If the root of $T(p)$ has a left child only, then $p = Ln$, and the postorder reading of $T(p)$ is that of $T(L)$, with n added to the end. In this case, the root of $T(f(p))$ has only one child, and that is a right child. This child is the root of a subtree isomorphic to $T(L)$. This is no problem, however. As the root has only one child, the postorder reading of $T(f(p))$ is just the postorder reading of this one subtree, that is, $T(L)$, with n added to end. This proves our claim. If the root of $T(p)$ has a right child only, the argument is the same with "left" and "right" interchanged.   ∎

Therefore, the restriction of f to the set of t-stack-sortable n-permutations maps to the set of t-stack-sortable n-permutations. We have seen that f maps permutations with k descents to permutations with k ascents, so our theorem is proved.   ∎

8.2.3 Unimodality

A useful feature of the proof of Theorem 8.22 was that we could prove a relatively strong statement about t-stack-sortable permutations, without even knowing what they are, in the sense of an explicit characterization. In this subsection, we are continuing this line of work.

THEOREM 8.30
For all fixed n and t, the sequence $W_t(n, k)_{0 \leq k \leq n-1}$ is unimodal.

Before we prove the theorem, we point out that while the special cases $t = 1$ and $t = 2$ can again be verified by the aforementioned explicit formulae, the special case $t = n - 1$ is not nearly as obvious as it was when we proved symmetry.

PROOF   As we saw in Theorem 8.22, the sequence at hand is symmetric, and therefore, it suffices to prove that $W_t(n, k) \leq W_t(n, k + 1)$ for $k \leq \lfloor (n - 3)/2 \rfloor$.

We resort to decreasing binary trees again, and the outline of our proof will also be somewhat similar to that of Theorem 8.22. We will first define an injection $z : T(n, k) \to T(n, k + 1)$. Then we will show that z preserves the t-stack-sortable property, completing our proof.

Let T be any decreasing binary tree on n nodes. We define a total order of the nodes of T as follows. Let us say that a node v of T is on *level j* of T if the distance of v from the root of T is j. Then our total order consists of listing the nodes on the highest level of T going from left to right, then the nodes on the second highest level left to right, and so on, ending with the root of T.

Let T_i be the subgraph of T induced by the smallest i vertices in this total order. Then T_i is either a tree or a forest with at least two components. If T_i is a tree, then $f(T_i)$ is a tree as described in Definition 8.27. If T_i is a forest, then we define $f(T_i)$ as the plane forest whose hth component is the image of the hth component of T_i.

Now let $k \leq \lfloor (n - 3)/2 \rfloor$, and let $T \in T(n, k)$. We define $z(T)$ as follows. Take the sequence $T_1, T_2, \cdots, T_n$. Denote by $\ell(T)$ the number of left edges of the forest T, and by $r(T)$ the number of right edges of the forest T. Find the smallest index i so that $\ell(T_i) - r(T_i) = 1$. We will now explain why such an index always exists. If T_2 is a left edge, then $\ell(T_2) - r(T_2) = 1 - 0 = 1$, and we are done. Otherwise, at the beginning we have $\ell(T_2) - r(T_2) < 1$, while at the end we have

$$\ell(T_n) - r(T_n) > ((n - 1) - \lfloor (n - 3)/2 \rfloor) - \lfloor (n - 3)/2 \rfloor > 1,$$

because of the restriction on k. So at the beginning, $\ell(T_j) - r(T_j)$ is too small, while at the end, it is too large. On the other hand, it is obvious that as i changes from 2 to n, at no step could $\ell(T_j) - r(T_j)$ "skip" a value as it could

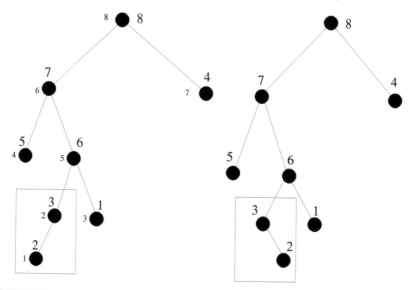

FIGURE 8.4

A decreasing binary tree and its image under z.

only change by 1. Therefore, for continuity reasons, it has to be equal to 1 at some point, and we set i to be the smallest index for which this happens.

Now apply f to the forest T_i, and leave the rest of T unchanged. Let $z(T)$ be the obtained tree. Before we prove that z is an injection, let us consider an example.

Example 8.31

Let T be the decreasing binary tree shown on the left of Figure 8.4. The small numbers denote the ranks of the vertices in the total order defined above, and the large numbers are the entries of the permutation associated with p.

Then the smallest index i for which $\ell(T_i) - r(T_i) = 1$ is $i = 2$, so we have to apply f to the tree T_2, the tree that is encapsulated in Figure 8.4. The image $z(T)$ is then shown on the right-hand side of Figure 8.4. ⬚

Let us return to the proof of the injectivity of z.

LEMMA 8.32

For all $k \leq \lfloor (n-3)/2 \rfloor$, the function z described above is an injection from $T(n,k)$ into $T(n,k+1)$.

PROOF   It is clear that z maps into $T(n, k+1)$ as z consists of the application of f to a subgraph T_i in which left edges outnumber right edges by one. We know that f turns left edges into right edges and vice versa, so z indeed increases the number of right edges by 1.

To see that z is an injection, note that $z(T_i)$ is the smallest subforest of $z(T)$ that consists of the first few nodes of $z(T)$ in the total order of all nodes and in which right edges outnumber left edges by one.

Now let $U \in T(n, k+1)$. If U does not have a subforest that consists of the first few nodes of U in the total order of all nodes and in which the right edges outnumber the left edges by one, then by the previous paragraph, U has no preimage under z. Otherwise, the unique preimage $z^{-1}(U)$ can be found by finding the smallest such subforest, applying f to it, and leaving the rest of U unchanged. ∎

So the injection z provides yet another proof of the fact that the Eulerian numbers form a unimodal sequence. In order to show the relevance of z to t-stack-sortable permutations, we still have to show that z preserves the t-stack-sortable property. This will be done again in a way that is quite similar to the proof of Theorem 8.22. Denote by $s(T)$ the postorder reading of the decreasing binary tree T.

LEMMA 8.33
For any decreasing binary trees T for which $z(T)$ is defined, we have

$$s(T) = s(z(T)).$$

PROOF   The only place T and $z(T)$ differ is in their first i vertices in the total order. However, even those parts agree in their postorder readings as they are images of each other under the postorder-preserving bijection f. ∎

The proof of Theorem 8.30 is now immediate. Restrict z to the set of decreasing binary trees corresponding to t-stack-sortable n-permutations with k descents, where $k \leq \lfloor (n-3)/2 \rfloor$. By the two preceding lemmas, this restriction maps injectively into the set of trees corresponding to t-stack-sortable n-permutations with $k+1$ descents, proving our claim. ∎

Petter Brändén [94] has provided an alternative proof for Theorem 8.30 as part of a more general framework.

8.2.3.1 Log-Concavity

Having seen that the sequence $W_t(n, k)_{0 \leq k \leq n-1}$ is unimodal for any fixed n and k, it is natural to ask whether these sequences are log-concave as well. Surprisingly, a general answer for this question is not known. It is rather rare in enumerative combinatorics to have a sequence that arises naturally from some enumeration problem, is known to be symmetric and unimodal, and is not known (but widely thought) to be log-concave. So our sequences $W_t(n, k)_{0 \leq k \leq n-1}$ are examples of a rare phenomenon.

There are some sporadic results, though. For $t = 1$ and $t = 2$, the result is immediate from the explicit formulae for our numbers. For $t = n - 1$, the statement is equivalent to the log-concavity of the Eulerian numbers. See Exercise 13 for the special case of $t = n - 2$.

8.2.3.2   Real Zeros

Numerical evidence seems to suggest the even stronger statement that for any fixed n and t, the polynomial

$$W_{n,t}(x) = \sum_{k=0}^{n-1} W_t(n, k)x^{k+1} \tag{8.5}$$

has real zeroes only. This observation has not been proved to hold in all cases yet. The only simple special case is when $t = n - 1$, because in that case the statement simply claims that the Eulerian polynomials have real zeroes only.

The special cases of $t = 1$ and $t = 2$ are not obvious, even though we have exact formulae for the numbers $W_t(n, k)$. In both of these special cases, Petter Bränden proved that $W_{n,t}(x)$ has real roots only [93]. In that same article, Bränden developed the analytic tools to prove that $W_{n,n-2}(x)$ has real roots only, marking the fourth special case for which the real-roots-only conjecture is known to be true.

8.3   Pop-stack sorting

Pop-stack sorting is the following version of stack-sorting. The entries of the permutation to be sorted, p, enter a specialized stack, called a *pop-stack* one-by-one, just as in stack-sorting, as long as the entries in the pop-stack form an increasing sequence from top to bottom. However, when this is no longer possible, that is, when the next entry of p is larger than the entry on the top of the pop-stack, then the *entire pop-stack* must go into the output. We call the permutation p *pop-stack-sortable* if the image pop(p) of p by this deterministic algorithm is the identity permutation.

Example 8.34
The permutation $p = 3215476$ is pop-stack-sortable as the pop-stack will empty out after the third, fifth, and seventh entry enters it.   ◻

Example 8.35
The permutation $p = 4213756$ is not pop-stack-sortable, since pop(p) starts with the subsequence 124.   ▯

It is not difficult to characterize pop-stack-sortable permutations in terms of pattern avoidance.

THEOREM 8.36
A permutation p is pop-stack-sortable if and only if p avoids both 231 and 312.

PROOF   Let $p = r_1 r_2 \cdots r_k$ be the unique decomposition of p into non-extendible descending runs (decreasing sequences of consecutive entries). Then pop(p) $= a_1 a_2 a_k$, where a_i is the reverse of r_i. Therefore, pop(p) $= 12 \cdots n$ if and only if *all entries* in a_i are smaller than *all* entries of a_j, whenever $i < j$. That happens when p is a *layered* permutation. (See Definition 4.94.) The solution of Exercise 13 of Chapter 5 shows that a permutation is layered if and only if it avoids both 231 and 312.   ∎

If p is not the identity permutation, then pop(p) has less inversions than p. This implies that applying the pop-stack sorting operation to p a sufficient number of times will result in the identity permutation, no matter what p is. But *how many times* do we have to send p through the pop-stack to sort it? In other words, what is the smallest integer t so that $\text{pop}^t(p) = 12 \cdots n$, for all n-permutations p? As before, $\text{pop}^t(p)$ is defined recursively, by the rule $\text{pop}^t(p) = \text{pop}(\text{pop}^{t-1}(p))$, and we say that p is t-pop-stack-sortable if $\text{pop}^t(p)$ is the identity permutation.

Recall that for the regular stack sorting operation s, the answer to the analogous question is $n - 1$. So it might seem plausible to think that for the pop-stack sorting operation, the answer will be a higher number, because pop sorts only 2^{n-1} permutations of length n in one step, while s sorts almost 4^n permutations of length n in one step.

The following theorem shows that this intuition is false.

THEOREM 8.37
Every permutation of length n is $(n-1)$-pop-stack-sortable.

In other words, in the worst case scenario, pop-stack sorting is just as efficient as stack sorting! Somewhat surprisingly, the theorem is a direct consequence of a result of the combinatorial geometer Peter Ungar [314] from 1982. In what follows, we present a version of that proof that was given by Michael Albert and Vincent Vatter [10].

The main idea of the proof is the following. We will show that after $n - 1$ applications of pop, the k smallest entries of p will precede the $n - k$ largest entries of p, for all k. This clearly implies that $\text{pop}^{n-1}(p)$ is the identity permutation. In order to do this, we will sometimes disregard all properties of the entries of p other than being larger than k, by replacing them by 0s and 1s. We will also define a nondeterministic sorting operation on binary words called *tumble* so that $\text{pop}(p)$ will be just one of the possible outputs of the tumble operation on p. Finally, we will define a *deterministic* operation on binary words called *flip* which will lead to the worst possible output of tumble, and show that even flip is strong enough to sort all binary words of length n in $n-1$ iterations.

In order to proceed in that direction, we introduce the following notation. For a permutation $p = p_1 p_2 \cdots p_n$, set

$$p \mid_k (i) = \begin{cases} 0 \text{ if } p \leq k, \\[2mm] 1 \text{ if } p > k. \end{cases}$$

Example 8.38
Let $p = 356142$, and let $k = 3$, then $p \mid_3 = 011010$.   ☐

In other words, $p \mid_k$ is a binary word that encodes which positions of p contain entries larger than k. Let us call $p \mid_k$ the k-*truncation* of p.

Let w be a binary word, that is, a word over the alphabet $\{0,1\}$. A *factor* in a word is just a substring of consecutive entries. We define *tumble*, a nondeterministic operator on the set of all finite binary words as follows. The *tumble* of w is the set $T(w)$ of all words that can be obtained from w by the following steps.

1. Choose a set of non-overlapping $1^i 0^j$ factors with i and j being positive so that the chosen factors together cover all 10-factors of w.

2. Reverse all those chosen $1^i 0^j$ factors in place.

Example 8.39
Let $w = w_1 w_2 \cdots w_8 = 11100101$. Reversing the strings $w_2 w_3 w_4$ and $w_6 w_7$, we get the word $10110011 \in T(w)$. Reversing the strings $w_1 w_2 w_3 w_4 w_5$ we get the word $00111011 \in T(w)$. The reader is invited to verify that $T(w)$ consists of six words.   ☐

The following result shows that the k-truncation of $\text{pop}(p)$ is always in the tumble of the k-truncation of p.

PROPOSITION 8.40
For all permutations p, and all positive integers k, the relation

$$(\text{pop}(p)) \mid_k \in T(p \mid_k)$$

holds.

PROOF   First we consider $T(p \mid_k)$. Let $p_{\langle 1 \rangle}, p_{\langle 2 \rangle}, \cdots, p_{\langle \ell \rangle}$ be the decomposition of p into maximal (nonextendible) decreasing sequences. Then clearly,

$$p \mid_k = (p_{\langle 1 \rangle}, p_{\langle 2 \rangle}, \cdots, p_{\langle \ell \rangle}) \mid_k = (p_{\langle 1 \rangle}) \mid_k (p_{\langle 2 \rangle}) \mid_k \cdots (p_{\langle \ell \rangle}) \mid_k .$$

As the $p_{\langle m \rangle}$ are *maximal* decreasing subsequences, the last entry of $p_{\langle m \rangle}$ is smaller than the first entry of $p_{\langle m+1 \rangle}$. Therefore, all the 10-factors of $p \mid_k$ occurs *within* one of the factors $p_{\langle m \rangle} \mid k$. Also note that each of the k-truncations $p_{\langle m \rangle} \mid_k$ are of the form $1^j 0^j$, 1^i, or 0^j, depending on whether the first and last entries of the decreasing run $p_{\langle m \rangle}$ are more than k or not.

When constructing $T(p \mid_k)$, one possible choice of factors to reverse is the set of all factors of $p_{\langle m \rangle} \mid k$ that are of the form $1^i 0^j$, since their union contains all 10-factors of $T(p \mid_k)$.

Let us now compute pop(p) $\mid_k$. We have

$$
\begin{aligned}
\text{pop(p)} \mid_k &= \text{pop}(p_{\langle 1 \rangle} (p_{\langle 2 \rangle} \cdots p_{\langle \ell \rangle}) \mid_k \\
&= (p_{\langle 1 \rangle}^{rev}) \mid_k (p_{\langle 2 \rangle}^{rev}) \mid_k \cdots (p_{\langle \ell \rangle}^{rev}) \mid_k \\
&= (p_{\langle 1 \rangle}^{rev}) \mid_k (p_{\langle 2 \rangle}^{rev}) \mid_k \cdots (p_{\langle \ell \rangle}^{rev}) \mid_k \in T(p \mid_k).
\end{aligned}
$$

The last word is an element of $T(p \mid_k)$ since reversing factors $p_{\langle m \rangle}$ that are of the form 0^j or 1^j is the same as not reversing them, and we showed in the preceding paragraph that if we reverse all $p_{\langle m \rangle}$ factors that are of the form $1^i 0^j$, then we get a word in $T(p \mid_k)$. This completes the proof. ∎

COROLLARY 8.41
For all permutations p, and all positive integers k and t, the relation

$$\text{pop}^t(p) \mid_k \in T^t(p \mid_k)$$

holds.

In other words, the k-truncation of the output of t runs through the pop stack can also be obtained by k-truncating the same permutation, then applying the tumble operator to it t times. For easier parsing of the following proof, note that pop acts on *permutations* while T acts on *k-truncations of permutations*, that is, *words* over the alphabet $\{0, 1\}$.

PROOF   We use induction on t, the initial case of $t = 1$ being done in Proposition 8.40. For the induction step, note that

$$\text{pop}^t(p) \mid_k = \left(\text{pop}(\text{pop}^{t-1}(p))\right) \mid_k \in T\left((\text{pop}^{t-1}(p)) \mid_k\right) \subseteq T(T^{t-1}(p \mid_k)),$$

where we used the induction hypothesis in the last step. ∎

For a binary word w, let $F(w)$ denote the word obtained from w by reversing (”-flipping”) all 10-factors of w. For instance, if $w = 011010$, then $F(w) = 010101$. We will call $F(w)$ the *flip* of w. It is obvious from this definition that $F(w) \in T(w)$ for all w.

Now we introduce a partial order on the set of all finite binary words. If v and w are two such words, then we say that $v \leq w$ if for all i, the ith zero of v, if it exists, occurs in the same position or earlier than the ith zero of w. For instance, $00110 \leq 01010$ in this partial order, while the words 10001 and 01010 are incomparable. The following proposition shows that the flip operation preserves this partial order.

PROPOSITION 8.42
If $v \leq w$, then $F(v) \leq F(w)$.

PROOF   Let $v \leq w$. If $F(v) \leq F(w)$ does not hold, then there is an i so that the ith 0 of $F(w)$ occurs before the ith zero of $F(v)$. Crucially, flipping a word moves each letter of that word by at most one position to the left. So, it has to be the case that the ith 0 of w did move to the left, but the ith zero of v did not, and that originally they were both in the same position, say in the rth position. So, the ith zero 0 w is immediately preceded by 1, but the ith zero of v is not. However, that means that in the $(r-1)$st position of v, we have the $(i-1)$st zero of v; whereas in the $(r-1)$st position of w, we have a 1, meaning that the $(i-1)$st zero of w occurs before position $r-1$. This contradicts the assumption that $v \leq w$.   ∎

Sorting a binary word means turning it into the word $0^a 1^b$. Note that any tumble of a binary word w takes each of the zeros in w at least as far left as F does. Therefore, the number of flips it takes to sort w is an upper bound for the number of tumble operations it takes to sort w. Furthermore, for any word w that consists of a copies of 0 and b copies of 1, the inequality $w \leq 1^b 0^a$ holds. Therefore, by repeated applications of Proposition 8.42, it follows that if m flips sort $1^b 0^a$, then m flips, and; therefore, m tumbles, sort any word w that consists of a copies of 0 and b copies of 1. This motivates the following lemma.

LEMMA 8.43
It takes precisely $a + b - 1$ flips to sort $1^b 0^a$.

PROOF   Note that throughout subsequent applications of the flip operator, zeros never leapfrog each other. So it suffices to show that it takes $a + b - 1$ flips for the rightmost zero to get into its final position, which is the ath position. The rightmost 0 will not move during the first $a - 1$ flips, after which it will move one notch to the left during each flip. As that last zero must move b notches to the left, the proof is complete. ∎

As we pointed out in the paragraph right before Lemma 8.43, this means that all binary words of length n can be sorted in $n-1$ tumbles. That includes the k-truncation of any permutation p. In other words, if $w \in T^{n-1}(p \mid_k)$, then

$$w \le F^{n-1}(p \mid_k) = 0^k 1^{n-k}.$$

Applying Corollary 8.41 with $t = n - 1$, we get that

$$\text{pop}^{n-1}(p) \mid_k = 0^k 1^{n-k},$$

as we needed. This shows that in $\text{pop}^{n-1}(p)$, the entries that are at most k precede the entries that are larger than k, and that happens for every k. Therefore, $\text{pop}^{n-1}(p)$ must be the identity permutation, and we have proved Theorem 8.37.

8.4 Variations of Stack-Sorting

As we mentioned in Section 8.2, the stack sorting operation s we have been discussing is not the earliest one in this line of research. The first comprehensive study of stack-like sorting devices is due to Donald Ervin Knuth, [230], Section 2.2.1. His approach to stacks was slightly different than ours. He considered the following problem.

Let us *start* with the permutation $123 \cdots k$, instead of ending in it, and let us try to obtain as many permutations from it as possible, using the stack. This time we do *not* require that the entries in the stack be in increasing order, as that would prevent us from obtaining any permutation different from the identity. Therefore, at any point of time, we have two choices. Either we put the entry on the top of the stack to the output, or we put the next entry of the input into the stack.

If an n-permutation p can be obtained from $12 \cdots n$ in this way, then we call it *obtainable*.

Example 8.44
Let $p = 231$. Then p is obtainable as we can proceed as shown in Figure 8.5.

It is then proved in [230] by a simple direct argument that p is obtainable if and only if it is 312-avoiding. We suggest that the reader try to find such a proof. On the other hand, as we have studied the stack sorting operation

Output	Stack	Input
		123
	1	23
	2 1	3
2	1	3
2	3 1	
23	1	
231		

FIGURE 8.5
Obtaining $p = 231$.

s so extensively, we will reduce this problem to the characterization of stack-sortable permutations.

PROPOSITION 8.45
The n-permutation p is obtainable if and only if p^{-1} is stack sortable.

PROOF   Let p^{-1} be stack-sortable. This means that a certain sequence of movements of the entries of p^{-1} through the stack (namely, the sequence defined by s) turns p^{-1} into $id = 12 \cdots n$. Carrying out the same sequence of movements on $p^{-1} \cdot p$ will turn $p^{-1} \cdot p$ into $id \cdot p = p$. In other words, we could say that we relabeled the entries of p^{-1} so that it became id, before sending it through the stack. So p is obtainable.

Now let us assume that p is obtainable. That means that a certain sequence of movements through the stack turns id into p. Then that same sequence of movements turns p^{-1} into id. Note that this implies that at any given point of time, the entries of the stack were in increasing order with the smallest one on the top. Indeed, if $i < j$, and j had been above i in the stack, then j would be before i in the output, so the output could not be id. So the sequence of movements that turns p^{-1} into id is precisely the stack sorting operation s, so $s(p^{-1}) = id$, and our proof is complete. ∎

As a permutation is 312-avoiding if and only if its inverse is 231-avoiding, we see again that p is obtainable if and only if it is 312-avoiding.

In both approaches to stacks we have seen, the crucial property of the stack was that *the entries entered the stack at its top, and left it at its top.* This was the property that enabled us to change the order of some entries by passing them through the stack. There are other sorting devices with slightly different restrictions. One of them is a *queue*, which differs from the stack because the entries enter at the top of the queue, and leave at the bottom of the queue. So the entries will leave the queue in the order they were received (as is the case, hopefully, with a dentist's waiting room). Therefore, the queue is not an exciting sorting device as the only permutation it can sort is the identity permutation itself.

Knuth has considered several generalizations of stacks and queues in his ground-breaking work [230]. In all of these, his approach was to look for the set of permutations that could be obtained by some kind of sorting device. The most powerful sorting device considered in [230] is the *double-ended queue*, or *deque*. This is a sorting device that has the capabilities of both a stack and a queue, and even a little bit more. That is, entries can enter at both ends of the deque, and they can leave at both ends of the deque. Again, there is no restriction on the order of the entries that are in the deque. Let us call an n-permutation p *general deque obtainable* if it can be obtained from $12 \cdots n$ using a deque. The adjective *general* is inserted to avoid confusion as some authors use the word deque for a more restricted sorting device. That is, p is deque-obtainable if it can be obtained from $12 \cdots n$ by a movement sequence, in each step of which we either place the next entry of the input to either end of the deque, or place the entry currently at the top or bottom of the deque into the output.

Example 8.46
The permutation 3412 is deque obtainable. See Figure 8.6 for a movement sequence leading to this permutation. ▯

In order to have an equivalent of Proposition 8.45, we need to define when we call a permutation *general deque-sortable*, or *gd-sortable*, and what the *gd-sorting* algorithm is. As the general deque is more flexible a tool than the stack, we can allow more flexibility in the deque sorting algorithm than in the stack sorting algorithm. That is, our sorting algorithm will not be deterministic, and the elements in the stack will not have to be in increasing order. They will have to form a *unimodal* sequence, though. The goal of the deque sorting algorithm is again to turn a permutation p into the permutation $id = 12 \cdots n$. We define a generic step of the gd-sorting algorithm as follows.

Nondeterministic General Deque Sorting.
START At any given step, if the deque is empty, we just place the next entry

	Output	Deque	Input
Start			1234
After Steps 1–2		1 2	
After Steps 3–4		3 1 2 4	
After Steps 5–6	34		
End	3412		

FIGURE 8.6
Obtaining $p = 3412$ by a deque.

of the input in the deque. If there is only one entry a in the deque, then we can place the next entry of the input on either side a in the deque.

Otherwise, we compare the next element of the input, p_i, to the entries at the top and bottom of the deque, x, and y.

(a) If p_i is smaller than exactly one of x or y, then p_i is placed to the end of the deque where the element larger than p_i is.

(b) If $p_i < x$ and $p_i < y$, then we can place p_i to either end of the deque, and go back to **START**.

(c) If $p_i > x$ and $p_i > y$, then we move the smaller of x and y to the output, and go back to **START**.

If there are no more entries in the input, then we empty the deque, always moving the smaller of the two entries at the ends of the deque to the output **END**.

Note that this algorithm assures that the entries in the deque indeed always form a unimodal sequence.

DEFINITION 8.47   *An n-permutation p is called* general deque sortable *if it can be turned into id using the above nondeterministic algorithm.*

The only nondeterministic step of this algorithm is (b). Definition 8.47 means that if there is *any* series of choices at step (b) that will result in the identity permutation at the end, then p is called gd-sortable. See Exercise 26 for a deterministic version of this same algorithm.

PROPOSITION 8.48
A permutation p is general deque obtainable if and only if p^{-1} is general deque sortable.

PROOF   Analogous to the proof of Proposition 8.45.   ∎

It seems that general deque sorting is a significantly more complex procedure than 2-stack sorting. We will see that this intuition holds true in an important aspect. In another aspect, however, it fails, as shown by the following result that was proved in a slightly different form by Pratt [264].

LEMMA 8.49
The set of general deque sortable permutations forms a class.

In other words, if q is general deque sortable, and $p \leq q$ in the pattern containment order, then p is also general deque sortable. Recall that this was *not* true for 2-stack sortable permutations.

PROOF   (of Lemma 8.49) As taking inverses preserves the pattern containment relation, it is sufficient to prove that the set of general deque obtainable permutations forms a class. If q is gd-obtainable with a sequence S of movements, and $p \leq q$, then p is gd-obtainable by the subsequence of S that belongs to a given copy of p contained in q. Indeed, the presence of the remaining entries of q is not necessary for the entries of that p-copy (it does not help them in any way) in order to move through the deque.   ∎

In other words, gd-sortable permutations form an ideal I_{gd} in the poset P of all finite permutations ordered by pattern containment, and permutations that are not gd-sortable form a dual ideal $\bar{I}_{gd}$ in P. The minimal elements of $\bar{I}_{gd}$ (or any ideal, for that matter), form an antichain A. If A were finite, then we could characterize $\bar{I}_{gd}$ as the set of permutations avoiding all of a finite number of patterns, namely the patterns in A. However, this turns out to be not the case.

THEOREM 8.50
The antichain of all minimal elements of $\bar{I}_{gd}$ is infinite.

PROOF   See Problem Plus 6.   ∎

No simpler characterization is known. So even if the gd-sortable permutations form a closed class, and the 2-stack-sortable permutations do not, the latter are easier to characterize, and even, enumerate. It is probably not surprising after this theorem that the enumeration of gd-sortable permutations of length n is not resolved yet. However, if we take away from the very strong abilities of the deque, we can succeed in the analogous task.

We will say that a deque is *input-restricted* if entries can only enter at the top of the deque, but can leave at both ends. Similarly, we say that a deque is *output-restricted* if entries can only leave at the bottom of the deque, but can enter at both ends.

The reader probably senses some symmetry between the two sorting devices, and as we will see, the reader is right. So for a while we will *restrict* our attention to input restricted deques. In line with our previous definitions, we will say that an n-permutation p is *ir-obtainable* if we can obtain it by sending the identity permutation $12\cdots n$ through an input-restricted deque. For the definition of *ir-sortability*, we first need to define the sorting algorithm. This algorithm is a somewhat simplified version of the gd-sorting algorithm, which was explained immediately preceding Definition 8.47. The simplification is that this is a *deterministic algorithm*.

Deterministic IR Deque Sorting.
START At any given step, if the deque is empty, we just place the next entry of the input in the deque.

Otherwise, we compare the next element of the input, p_i, to the entries at the top and bottom of the deque, x, and y.

(a) If $p_i < x$, then put p_i on top of x in the deque, and go back to **START**.

(b) If $p_i > x$, and the deque is not monotone decreasing starting at the top, then move the *smaller* of x and y in the output, and go back to **START**. (If there is exactly one entry in the deque, that is considered a monotone decreasing deque.)

(c) If $p_i > x$, and the deque is monotone decreasing starting at the top, then put p_i on top of x in the deque, and go back to **START**.

If there are no more entries in the input, then we empty the deque, moving always the smaller of the two entries at the ends of the deque to the output. The output permutation is denoted $ir(p)$. **END**.

Note that this algorithm assures that the elements of the deque always form a unimodal sequence. Now we can call an n-permutation p *ir-sortable* if $ir(p) = id$. By now it should be obvious that p is ir-obtainable if and only if p^{-1} is ir-sortable.

Knuth [230] has computed the number of ir-sortable n-permutations by a generating function argument. His work was completed before the pattern avoidance *explosion*, that is, the bonanza of enumerative results on permutations avoiding certain patterns. We will now show how to reduce the problem of enumerating ir-sortable permutations into a problem of enumerating pattern avoiding permutations.

Let us first try to characterize the set of ir-sortable permutations. It is easy to check that all permutations of length three or less are ir-sortable. The problems start at length four, when we cannot ir-sort 4231, and we cannot ir-sort 3241. This is not an accident.

LEMMA 8.51
If p contains 4231 or 3241, then p is not ir-sortable.

PROOF   Let us assume p contains 4231, and let $a < b < c < d$ form a copy of 4231 in p. In other words, these four entries follow in the order $dbca$ in p. So d will enter the deque first. If p is to be ir-sortable, then b has to enter the deque before d can leave it; otherwise, d precedes b in the output. Now b is above d, and therefore b has to leave the deque before c arrives, let alone before a arrives. This means that b will precede a in the output.

Similarly, let us assume that p contains 3241, and let $x < y < v < z$ form a copy of 3241 in p. So the order of these four entries in p is $vyzx$. Then v enters the deque first. Then, if p is to be ir-sortable, y enters before v can leave. Now y is above v, so it has to leave the deque before z arrives, meaning that y will precede x in the output. ∎

Having seen how difficult the problem of characterizing permutations that are sortable by some device can be, the reader might be bracing for some complicated counterpart of Lemma 8.51. If this is the case, then the reader will be happily surprised.

THEOREM 8.52
The permutation p is ir-sortable if and only if *it avoids both 4231 and 3241.*

PROOF   As Lemma 8.51 proves the "only if" part, we only have to prove the "if" part.

We prove the contrapositive. Let p be such that $ir(p) \neq id$. Let $a < b$ be two entries that form an inversion in $ir(p)$, that is, b precedes a in $ir(p)$. It is straightforward to see that in this case, b has to precede a in p, too. (Recall that the deque is unimodal.) What forced b to leave the deque before a arrived? The only step of the sorting algorithm when an entry has to leave the deque is (b). Therefore, b had to leave the deque at a point when b was the smaller of the two entries b and c at the two ends of the deque, the deque was not monotone decreasing, and an entry x larger than both b and c was the next entry of the input. At that point, one of the following two situations had to occur.

1. The entry b was at the top of the deque, and c was at the bottom. That means c precedes b in p, and so $cbxa$ is a 3241-copy in p.

2. The entry c was at the top of the deque, and b was at the bottom. As we know that the deque was not monotone decreasing, there had to be an entry y in the deque that was larger than c. Then y precedes c in p, and the subsequence $ycxa$ is either a 3241-pattern, or a 4231-pattern, depending on which one of x and y is larger.

∎

COROLLARY 8.53
Let r_n be the number of ir-sortable n-permutations, with $r_0 = 0$. Then the equality

$$\sum_{n \geq 0} r_n z^n = \frac{1 - z - \sqrt{1 - 6z + z^2}}{2}$$

holds.

PROOF　　　This follows from the solution of Exercise 9 of Chapter 4. So the ir-sortable permutations are enumerated by the large Schröder numbers. ∎

Note that this implies that $r_{n+1} = \sum_{i=0}^{n} \binom{2n-i}{i} C_{n-i}$. What can be said about *output-restricted* deques? These are deques that allow entries to leave at the top only, but allow entries to enter at either end. If we want to sort a permutation p by such a deque, it is obvious that we have to keep the deque increasing from top to bottom, which makes it easy to construct rest of the sorting algorithm. The reader should try to solve Exercises 24 and 25 for the details.

Exercises

1. Take a permutation of length n. Assign a direction (Left or Right) to each entry, such as in $2_L 3_R 4_L 1_R 5_L$. Call an entry x *winning* if its neighbor in the direction given to x exists, is y, and $y < x$. So for instance, in $2_L 3_R 4_L 1_R 5_L$, the entries 4 and 5 are winning.

   Now consider the following algorithm. Start with the permutation $p_1 = 1_L 2_L \cdots n_L$. Find the largest winning entry x, and interchange it with y. Finally, change the direction of all entries z if $z > x$. Call the obtained new permutation p_2. Then apply this same procedure to p_2, and call the resulting permutation p_3, and so on. Prove that this procedure will generate all $n!$ permutations of length n in the first $n!$ steps.

2. Which n-permutation will be generated *last* by the above algorithm?

3. Find an algorithm to generate all n-permutations with a fixed number k of inversions.

4. (−) Find an algorithm to generate all n-permutations with a fixed major index m.

5. (−) Prove that all n-permutations are $(n-1)$-stack-sortable.

6. Is there a permutation p so that $s(p) = 2413657$?

7. Characterize all n-permutations that are not $(n-2)$-stack-sortable. Then enumerate these permutations.

8. Characterize all n-permutations that are not $(n-3)$-stack-sortable. Then enumerate these permutations.

9. Prove that $W_t(n) \le (t+1)^{2n}$.

10. In the proof of Theorem 7.32 we have seen a bijection that (after taking reverses) maps 231-avoiding, so stack-sortable, n-permutations with k descents into such permutations with $n-1-k$ descents. Was that bijection the same as the bijection f of Theorem 8.22 in the special case $t = 1$?

11. Consider the following modification of the t-stack sorting operation. Instead of passing a permutation through a stack t times, we pass it through t stacks placed next to each other in series as follows. The first stack operates as the usual stack except that when an entry x leaves it, it does not go to the output right away. It goes to the next stack if $x > j$, where j is the entry on the top of the next stack. If $j < x$, then x cannot move until j does.

The general step of this algorithm is as follows. Let $S_1, S_2, \cdots, S_t$ be the t stacks, with a_i being the entry on top of stack S_i. If the next entry x of the input is smaller than a_1, we put x on top of S_1. Otherwise, we find the smallest i so that a_i can move to the next stack (that is, that $a_i < a_{i+1}$ or S_{i+1} is empty), and move a_i on top of S_{i+1}. If we do not find such i, or if $S_1, S_2, \cdots, S_{t-1}$ and the input have all been emptied out, then we put the entry on the top of S_t into the output.

(a) Let $t = 2$. What is the image of 231 under this operation?

(b) Characterize all permutations p whose image $s_t(p)$ under the operation defined above is the identity permutation.

The above sorting algorithm is sometimes called *left-greedy* [22] as we always make the leftmost move possible as the entries move from the input on the right to the output in the left.

12. Prove that $W_t(n)$ cannot be larger than the number of Standard Young Tableaux of shape $(t + 1) \times n$ in which row i does not contain two consecutive integers, for $2 \leq i \leq t$.

13. Prove that for any fixed n, the sequence $W_{n-2}(n, k)_{0 \leq k \leq n-1}$ is log-concave.

14. Prove that for any fixed t, and any even positive integer n, the number $W_t(n)$ is even.

15. Define the bijection f of Definition 8.27 without using decreasing binary trees.

16. A *fall* in a permutation $p = p_1 p_2 \cdots p_n$ is an index i that satisfies the inequalities $p_{i-1} > p_i > p_{i+1}$. A *rise* is an index i that satisfies the inequalities $p_{i-1} < p_i < p_{i+1}$. Prove that the number of t-stack-sortable n-permutations with k rises is equal to the number of t-stack-sortable n-permutations with k falls.

17. Let P_{no-t} be the poset of all finite permutations that are not t-stack-sortable, ordered by pattern containment. Does P_{no-t} contain an infinite antichain?

18. Call a permutation p *sorted* if there is a permutation q so that $s(q) = p$. What is the maximum number of descents that a sorted n-permutation can have?

19. Let p and q be two n-permutations so that $s(p) = pr$ and $s(q) = qr$ for some $r \in S_n$, where pr means the product of p and r in S_n. Prove that the decreasing binary trees $T(p)$ and $T(q)$ differ only in their labels, not in their underlying trees.

20. Is there a permutation pattern r so that if p contains r, then p cannot be sorted?

21. For $p \in S(n)$, let b_p be the unique n-permutation so that $pb_p = s(p)$, where the left-hand side refers to the product in S_n. Let

$$B = \{b_p | p \in S_n\}.$$

How many elements does B have?

22. How many permutations $b \in S_n$ are there so that there exists *exactly one* $p \in S_n$ so that $b = b_p$?

23. For which n-permutations p is $s^{-1}(p)$ the largest (consists of the largest number of permutations)?

24. Define the deterministic sorting algorithm *outres* that uses an output-restricted deque.

25. (a) Characterize permutations that are or-sortable.

   (b) Deduce that the number of or-sortable n-permutations is equal to the number of ir-sortable n-permutations.

26. Define a deterministic version of the general deque sorting algorithm that turns the n-permutation p into $12\cdots n$ if and only if p is gd-sortable.

27. Let us call an n-permutation p *separable* if

   (a) $p = 1$, or

   (b) $p = LR$, where L and R are both separable permutations (after relabeling), and the entries of L are either the smallest $|L|$ elements of $[n]$, or the largest $|L|$ elements of $[n]$.

   Find a characterization of separable permutations in terms of pattern avoidance.

28. We know that if $\operatorname{pop}(p) = 12\cdots n$, then $s(p) = 12\cdots n$. However, is it true that if $\operatorname{pop}^t(p) = 12\cdots n$ for some t, then $s^t(p) = 12\cdots n$ also holds?

29. Which permutations are sortable by t parallel queues? That is, we have t queues, the entries can only enter them at the top, and leave at the bottom, but for each entry of the input, we are free to decide into which queue that entry will enter. Once an entry leaves its queue, it goes immediately into the output.

30. Consider the following two modifications of the sorting algorithm of Exercise 11.

   (a) Sorting by t stacks *in series*. This is an undeterministic algorithm defined as follows. In any given step, move one entry from the top of a stack S_i to the top of the next stack S_{i+1} so that the new stack remains increasing from top to bottom. (Recall that S_0 is the input and S_{t+1} is the output.) In other words, omit the requirement that in each step we have to do this with the stack of the smallest index i for which this is possible.

   (b) *Left-greedy* sorting through t stacks. That is, in each step of this deterministic algorithm, we find the *largest* index i so that we can move the entry from the top of S_i to S_{i+1} without violating any constraints. That is, the $S_1, S_2, \cdots, S_t$ all have to remain increasing from top to bottom, and the output has to be a string $123\cdots m$

for some m. If no such move is possible, then we say that the algorithm fails.

Prove that if $t = 2$, then the algorithms defined in (a) and (b) are equivalent, that is, they sort the same set of permutations.

31. Find a formula for the number of stack-sortable permutations of length n with exactly k peaks. Recall that a peak in a permutation is an entry that is larger than both of its neighbors.

32. Let $p = p_1 p_2 \ldots p_n$ be a permutation, and set $p_{n+1} = n + 1$. If i is a descent of p, then let $r_i(p)$ be the permutation obtained from p by moving a_i to the right as little as possible so that a_i is between entries a_j and a_{j+1} satisfying $a_j < a_i < a_{j+1}$. For instance, $r_3(314265) = 312465$.

Let $d_1 < d_2 < \ldots < d_k$ be the descents of p. Prove that

$$r_{d_k}\left(r_{d_{k-1}} \cdots \left(r_{d_1}(p)\right)\right) = s(p). \tag{8.6}$$

33. Find the ordinary generating function for the numbers a_n of 2-stack-sortable n-permutations that avoid the pattern 132.

34. Let $W_t(z) = \sum_{n \geq 0} W_t(n) z^n$. Prove that $W_t(z)$ is not a rational function.

35. A nonnegative integer m is called a *fertility number* (with respect to the stack sorting operation) if there exists a permutation p so that there are exactly m permutations q satisfying $s(q) = p$. Prove that if A and B are fertility numbers, then so is AB.

Problems Plus

1. Prove that for all n and k, the equality

$$W_2(n, k) = \frac{(n + k)!(2n - k - 1)!}{(k + 1)!(n - k)!(2k + 1)!(2n - 2k - 1)!} \tag{8.7}$$

holds.

2. A $\beta(1, 0)$-tree is a rooted plane tree whose vertices are labeled with *positive* integers according to the following rules.

   (a) The label of each leaf is 1.

   (b) The label of each internal node is at most the sum of the labels of its children.

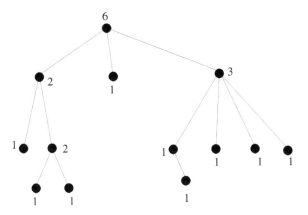

FIGURE 8.7
A $\beta(1,0)$-tree.

   (c) The label of the root is the sum of the labels of its children.

See Figure 8.7 for an example.

Prove that the number of $\beta(1,0)$-trees on $n+1$ vertices is equal to $W_2(n)$.

3. Prove that there exists a simplicial complex Δ so that the set of $(k-1)$-dimensional faces of Δ is in bijection with the set of 2-stack-sortable n-permutations having k ascents.

4. Sorted permutations were defined in Exercise 18. Let p be a sorted permutation, and denote $s^{-1}(p)$ the set of its preimages under the stack sorting operation. Prove that there is always a permutation $q \in s^{-1}(p)$ that has strictly more inversions than all other elements of $s^{-1}(p)$.

5. Find an algorithm that decides whether a given permutation is sorted.

6. Find a sufficient and necessary condition for a permutation to be gd-obtainable.

7. Recall the definition of sorting by t stacks in series from Exercise 30. Let $t = 2$. Characterize the set of permutations that are sortable by this procedure. We will call these permutations *two-series-sortable*.

8. Find a formula for the number of permutations of length n that are two-series-sortable.

9. Modify the sorting procedure of Exercise 11 by making it undeterministic as follows. In any given step, we can either

   (a) Put the next element of the input into any stack S_i, for any $i \in [t]$, or

(b) Put the entire content of stack S_j into the output for some $j \in [t]$ (without changing the order of the entries).

Note that the stacks S_i are in fact pop-stacks in this model. This collection of pop-stacks is referred to as t *pop-stacks in parallel*. Let us call a permutation p sortable by t pop-stacks in parallel if there is a sequence of steps of the above two kinds that turns p into the identity permutation.

Is it true that the set of all permutations that are sortable by t pop-stacks in parallel forms a closed class?

10. Let $f(n)$ be the number of all n-permutations that are sortable by two pop-stacks in parallel. Prove that $f(n) = n!$ if $n \leq 3$, and

$$f(n) = 6f(n-1) - 10f(n-2) + 6f(n-3)$$

if $n > 3$.

11. In part (b) of Exercise 30, we defined the left-greedy algorithm. The *right-greedy* algorithm is just the algorithm defined in Section 8.2.2.1. Is it true that if a permutation is sortable by the right-greedy algorithm on t stacks, then it is also sortable by the left-greedy algorithm on t stacks?

12. (a) Let $B_t(n, k)$ be the number of t-stack-sortable n-permutations that have exactly k peaks. Prove that

$$W_{n,t}(z) = \sum_{k=0}^{n/2} \frac{B_t(n, k)}{2^{n-1-2k}} z^k (1 + z)^{n+1-2k}, \qquad (8.8)$$

where $W_{n,t}(z)$ was defined in (8.5) as the generating polynomial of t-stack-sortable n-permutations with respect to their number of descents.

(b) Deduce Theorem 8.30.

13. Prove that for $t = 1$, $t = 2$, and $t = n - 2$, the polynomial $W_{n,t}(z)$ has real roots only.

14. Let us call a permutation *uniquely sorted* if it is sorted and has exactly one preimage under the stack sorting operation. Find a sufficient and necessary condition for p to be uniquely sorted.

15. We defined fertility numbers in Exercise 35 of this chapter. Are all nonnegative integers fertility numbers with respect to the stack sorting operation?

16. (a) Prove that if $n \geq 4$ is even, then n points in the plane that are not all collinear determine at least n different directions, and this bound is sharp.

(b) Prove that if $n \geq 4$ is even, then $n+1$ points in the plane that are not all collinear determine at least n different directions, and this bound is sharp.

17. Let p be any permutation. What can we say about the relation between $|s^{-1}(p)|$ and $|s^{-1}(s(p))|$?

18. If $s^t(p)$ is the identity permutation, but $s^{t-1}(p)$ is not, then we say that the *depth* of p in the stack sorting tree is t. Let $\mathcal{D}(n)$ be the average depth of all permutations of length n. So $\mathcal{D}(1) = 0$, $\mathcal{D}(2) = 1/2$, and $\mathcal{D}(3) = 1$. What can we say about the fraction $\mathcal{D}(n)/n$ as n goes to infinity?

19. Let us say that permutation p is *t-sorted* if there exists a permutation π so that $s^t(\pi) = p$. At most how many descents can a t-sorted permutation of length n have?

20. Let us assume that n and t are of the same parity. Characterize all t-sorted permutations of length n that have a maximum number of descents. Then enumerate such permutations.

Solutions to Problems Plus

1. We have mentioned in the text that 2-stack sortable n-permutations are in bijection with nonseparable rooted planar maps on $n+1$ edges. That bijection, which can be found in [149], turns the number of descents of the permutation into the number of vertices of the map, and turns the number of ascents of the permutation into the number of faces of the map. However, rooted nonseparable planar maps have been enumerated according to their number of vertices and faces in [101], and have been found to be counted by the expression on the right-hand side of (8.7). An alternative solution will be mentioned in the solution of the next Problem Plus.

2. See [128] for a short proof. Also note that $\beta(1,0)$-trees on $n+1$ vertices having k leaves are in bijection with 2-stack-sortable n-permutations having k descents. As the former have been enumerated in [127] and have been shown to be counted by (8.7), this provides an alternative proof for the previous Problem Plus.

3. This result was published in [56]; see that paper for the details that are omitted here. Denote by $D^{\beta(1,0)}_{n+1,k}$ the set of all $\beta(1,0)$-trees on $n+1$ nodes having k internal nodes. Our plan is as follows. To each $\beta(1,0)$-tree $T \in D^{\beta(1,0)}_{n+1,k}$ we will associate a k-tuple $(T_1, T_2, \cdots, T_k) \in [D^{\beta(1,0)}_{n+1,1}]^k$ of $\beta(1,0)$-trees –, in an injective way. By the solution of the previous Problem Plus, this is equivalent to the statement to be proved.

First, we specify the order in which we will treat the k internal nodes of T. To that end, we extend the notion of postorder from binary plane trees to plane trees the obvious way. That is, for each node, read its subtrees from left to right, then the node itself, and do this recursively for all nodes. This rule linearly orders all nodes of T, and in particular, turns our set $\{V_1, \cdots, V_k\}$ of internal nodes into the k-tuple $(V_1, \cdots, V_k)$ of internal nodes.

Let $i \in [k]$ and let V_i be the ith internal node of our $\beta(1,0)$-tree T. Let V_i have d_i descendents, excluding itself. Denote by l_i the number of nodes of T that precede V_i in the postorder reading of T. Similarly, denote by r_i the number of nodes of T that follow V_i in the postorder reading of T.

Define T_i as the unique $\beta(1,0)$-tree with one internal node Z_i so that Z_i has d_i descendents, and the root of T_i has ℓ_i leaf-children on the left of Z_i and r_i leaf-children on the right of Z_i. The only node whose label has to be defined is the only internal node Z_i, and we set $label_{T_i}(Z_i) = label_T(V_i)$.

We show that we can indeed always set $label_{T_i}(Z_i) = label_T(V_i)$, that is, $label_T(V_i)$ is never too big for the label of Z_i. Indeed, Z_i has d_i children, all leaves, so any positive integer at most as large as d_i is a valid choice for the label of Z_i. On the other hand, V_i has d_i descendents in T, so $label_T(V_i) \leq d_i$, and therefore $label_T(V_i)$ is a valid choice for $label_{T_i}(Z_i)$. Our decomposition map h is then defined by $h(T) = (T_1, T_2, \cdots, T_k)$. See Figure 8.8 for an example of this map.

One can then show by induction on k that the map

$$h : D^{\beta(1,0)}_{n+1,k} \to [D^{\beta(1,0)}_{n+1,1}]^k$$

defined by $h(T) = (T_1, T_2, \cdots, T_k)$ is an injection.

Note that h is not a surjection. Indeed, for $(T_1, T_2, \cdots, T_k)$ to have a preimage, we must have $\ell_1 < \ell_2 < \cdots < \ell_k$. This also shows that k-tuples of $\beta(1,0)$-trees with one internal node, and k-element sets of $\beta(1,0)$-trees with one internal node are equivalent for our purposes.

Finally, we show that the defining property of simplicial complexes holds for our construction. That is, we show that if there exists a $\beta(1,0)$-tree T so that $h(T) = \{T_1, T_2, \cdots, T_k\}$, then for any subset

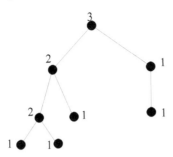

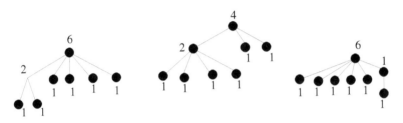

FIGURE 8.8
Decomposing a $\beta(1,0)$-tree.

$I = \{i_1, i_2, \cdots, i_j\} \subset [k]$, there exists a $\beta(1,0)$-tree T_I so that $h(T_I) = \{T_{i_1}, T_{i_2}, \cdots, T_{i_j}\}$.

It is clear that all integers $\ell_{i_1}, \ell_{i_2}, \cdots, \ell_{i_j}$ are different as even the integers $\ell_1, \ell_2, \cdots, \ell_k$ are all different. Relabel the elements of I so that they are in increasing order. This gives us a j-tuple $(T_{i_1}, T_{i_2}, \cdots, T_{i_j})$ of $\beta(1,0)$-trees with one internal node.

Let us assume first that I has $k-1$ elements, with i being the missing element. Then we construct T_I from T as follows. Let V_i be the ith internal node of T. Remove V_i from T, and connect all its children to the father F_i of V_i, preserving their left-to-right order. Say the rightmost child of V_i was R_i. Now add V_i back to the tree so that it is a child of F_i and it immediately follows R_i in the left-to-right list of the children of F_i. Note that now V_i became a sibling of one of its former children, and it became a leaf. Therefore, we have to change the label of V_i to 1, but we do not have to change any other labels. Indeed, there is no other node who lost descendents in this operation. (The only exception is when V_i was a child of the root. In that case, we may have to change the label of the root, but that will not cause any problems.) Call the $\beta(1,0)$-tree obtained this way T_I. See Figure 8.9 for an example of this procedure.

It is straightforward to verify that $h(T_I) = (T_{i_1}, T_{i_2}, \cdots, T_{i_{k-1}})$. If I

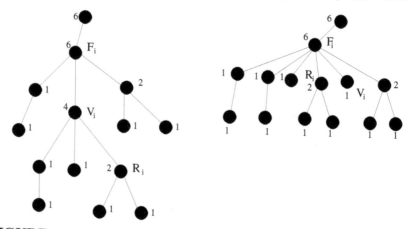

FIGURE 8.9
Removing an internal node.

has less than $k - 1$ elements, then we construct T_I by iterating this procedure.

4. This result was proved by M. Bousquet-Mélou, and can be found in [88]. She defines two operations on decreasing binary trees that increase the number of inversions in the corresponding permutation, but do not change the postorder reading of the tree. One operation concerns vertices that have a left subtree only, and turns that subtree into a right subtree. The second operation concerns vertices v that have a left child a and a nonempty right subtree T so that the *leftmost* vertex y of T is larger than a. Then the operation takes the subtree rooted at a and moves it so that it is the right subtree of y. See Figure 8.10 for an example.

   The author then proves by induction on the size of the trees that these operations will always lead to a unique tree T_{can}. She calls that permutation corresponding to that tree a *canonical* permutation. Note that neither of the two operations defined above can be carried out on T_{can}, which shows some properties of this tree.

5. A recursive algorithm to decide this question was developed by M. Bousquet-Mélou [87]. She uses a decomposition first mentioned in [338]. Let p be an n-permutation that has $k - 1$ descents, or, in other words, k ascending runs. Let these ascending runs be $r_1, r_2, \cdots, r_k$. We have seen in the text that taking the stack sorted image of a permutation q is nothing else but taking the decreasing binary tree $T(q)$ of q and then reading it in postorder. Now let us assume that $s(q) = p$. Then any time we pass in p from one ascending run to the next, in the postorder reading of T_q, we must jump from the left subtree of a vertex to

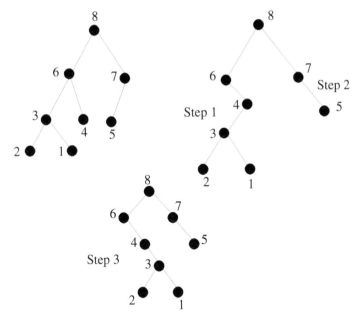

FIGURE 8.10
Finding the canonical preimage of 21346578.

the right subtree of a vertex, as that is the only way we can obtain a descent. This observation justifies the following algorithm.

To obtain a permutation q so that $s(q) = p$, take the ascending run decomposition $r_1, r_2, \cdots, r_k$ of p, and for each i, take the decreasing binary tree that is corresponding to the *reverse* of r_i. Denote these trees by T_i, for $i \in [k]$. Attach a left-half edge to the top of these trees. Now try to build up q right to left as follows. In Step 1, find the smallest element in r_k that is larger than the largest element in r_{k-1}, and attach T_{k-1} to that vertex as a left subtree. If there is no such element (which would mean that p did not end in n), then p is not sorted. Otherwise, continue this way. In step i, find the smallest element in the leftmost path of the tree already created that has no left child and is larger than the largest element of T_{k-i}. If there is no such element, then the previous paragraph implies that p is not sorted. Otherwise, we continue this procedure till an n-vertex tree is created. In that case, p is sorted as $p = s(q)$, where the permutation corresponding to the created tree is q. Note that the permutation q will be the canonical preimage of p. See Figure 8.11 for an example.

6. The following was proved in [264] in a slightly different form. The set of minimal non-gd-sortable permutations is $a_1 = 52341$, $a_2 = 5274163$, $a_3 = 7\,2\,9\,4\,1\,6\,3\,8\,5$, $a_4 = 9\,2\,11\,4\,1\,6\,3\,8\,5\,10\,7$, and so on, and

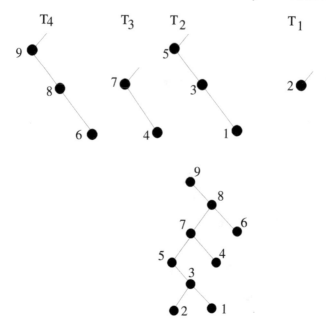

FIGURE 8.11
Finding a preimage for $p = 213547689$.

all permutations that can be obtained from the b_i by interchanging the
two maximal entries, and/or interchanging the first two entries. Note in
particular that in the a_i, the even entries are fixed points and the odd
entries are cyclically translated (except in a_1).

7. It is proved in [22] that these are the permutations that avoid all patterns
   of the form $q_m = 2(2m - 1)416385 \cdots 2m(2m - 3)$, for each $m \geq 2$. In
   other words, in position $2i - 1$, the pattern q_m contains the entry $2i$,
   and in position $2i$ the pattern q_m contains the entry $2i - 3$ if $i > 1$, and
   the entry $2m - 1$ is in the second position of q_m.

8. It is proved in [22] that the number $A(n)$ of these permutations is equal
   to $\mathrm{Av}_n(1342)$. We have proved an exact formula for $\mathrm{Av}_n(1342)$ in Theorem 4.56.

9. Yes. In fact, more is true. It is proved in [21] that for any t, there exists
   a finite set F_t of permutation patterns so that p is sortable by t pop-
   stacks in parallel if and only if p avoids all patterns in F_t. So the class
   of these permutations is not just an ideal, but an ideal generated by a
   finite number of elements. The proof is constructive, so the elements of
   F_t can actually be found. For $t = 2$, there are seven of them.

10. This result is due to Michael Atkinson [21].

11. Yes, that is true as was proved by Rebecca Smith [287].

12. (a) That result is due to Petter Brändén [94], who proved that the descent generating polynomial of certain classes of permutations has a non-negative expansion in the basis $B = \{z^k(1+z)^{n+1-2k}\}_k$ of polynomials, where $k = 0, 1, \cdots, \lfloor n - 1/2 \rfloor$. It turns out that t-stack sortable permutations are one example of the classes for which Brändén's results hold.

    (b) This follows from the existence of the expansion explained above, since all the elements of the basis B are symmetric and unimodal polynomials that have the same center of symmetry.

13. The special cases of $t = 2$ and $t = n - 2$ were proved by Petter Brändén in [93] who built his own powerful machinery to prove the real zeros property for a much larger class of polynomials. The special case of $t = 1$ is implicit in the works of Francesco Brenti, but an explicit proof can be found in [91].

14. A permutation of length n is uniquely sorted if and only if it is sorted and has exactly $(n-1)/2$ descents. This is a result of Colin Defant [134].

15. No. Colin Defant [135] showed that the smallest fertility number that is congruent to 3 modulo 4 is 27.

16. These questions were the motivation for the combinatorial geometer Peter Ungar that somewhat surpisingly lead to an argument proving Theorem 8.37. See [314] for proofs.

17. It is proved in [137] that $|s^{-1}(p)| \leq |s^{-1}(s(p))|$.

18. It is proved in [137] that

$$0.62433 \sim \lambda \leq \lim\inf \mathcal{D}(n)/n \leq \lim\sup \mathcal{D}(n)/n$$

$$\leq 0.6(7 - 8\log 2) \approx 0.87289.$$

    Here λ is the Golomb-Dickman constant , which is $1/n$ times the expected length of the longest cycle of a permutation of length n selected uniformly at random out of all permutations of length n.

19. Colin Defant [136] proved that this number is at most $\lceil (n - t)/2 \rceil$.

20. Colin Defant [136] proved that such a permutation $p = p_1 p_2 \cdots p_n$ is t-sorted and has $\lceil (n-t)/2 \rceil$ descents if and only if its left-to-right maxima are $p_1, p_3, p_5, \cdots p_{n-t+1} p_{n-t+2} \cdots p_n$. Then it follows that $p_j = j$ if $j \geq n - t + 1$. From this, it is easy to prove by induction on $n - t$ that the requested number is $(n - t - 1)!!$.

9

How Did We Get Here? Permutations as Genome Rearrangements.

9.1   Introduction

Let p and q be two permutations, written in one-line notation. In this section, we will attempt to transform p into q using only certain kinds of allowed steps. If k is the smallest integer so that p can be transformed into q using k allowed steps, then we say that the distance $d(p, q)$ of p and q is k.

The motivation for this research comes from molecular biology, where genomes of various species can be modeled by permutations, and the distance between these permutations can be interpreted as the evolutionary distance between species. A classic paper of the field, by Shridhar Hannenhalli and Pavel Pevzner [211], is entitled *Transforming Cabbage into Turnip*, which in turn explains the cover page of this book, since the genes constituting the genomes are modeled by the entries of the permutations corresponding to the genomes.

The notions of distance we study in this book will all be *left-invariant*, that is, they will have the property that for any n-permutations p, q, and r,

$$d(p, q) = d(r \circ p, r \circ q). \tag{9.1}$$

Here $r \circ p$ means the composition of the bijective functions $r : [n] \to [n]$ and $p : [n] \to [n]$, with r applied first. The left-invariant property described in (9.1) is explained by the fact that the distance between two genomes depends only on those two genomes, and not on the labeling of the genes in each genome. So relabeling the genes in one genome, and then relabeling the genes of the other genome in the same way will not change the distance between the two genomes.

Setting $r = q^{-1}$, formula (9.1) means that $d(p, q) = d(q^{-1} \circ p, id)$. So if we know how to determine the distance of a generic permutation from the identity, then we also know how to find the distance of any two permutations from each other.

This reduces our distance-measuring problem to a *sorting* problem. The question we attempt to answer is now the following. Let us assume that the notion of distance is fixed. Given an arbitrary permutation p of length n, how many moves (of a specified kind) are necessary to transform p into the

DOI: 10.1201/9780429274107-9

increasing permutation $123 \cdots n$? What is the permutation that is the *most difficult* to sort, that is, whose distance from the increasing permutation is the largest? How *efficient* is a given sorting method for the average permutation? Can we find an *algorithm* that finds the shortest distance from p to the increasing permutation?

9.2   Block Transpositions

Perhaps the most well-studied biologically motivated sorting operation is that of *block transpositions*. In a block transposition, two consecutive, adjacent blocks (substrings) of entries are interchanged so that the order of entries within each block is unchanged. Formally speaking, the permutation

$$p = p_1 p_2 \cdots p_i p_{i+1} \cdots p_j p_{j+1} \cdots p_k p_{k+1} \cdots p_n \tag{9.2}$$

is turned into the permutation

$$q = p_1 p_2 \cdots p_i p_{j+1} \cdots p_k p_{i+1} \cdots p_j p_{k+1} \cdots p_n. \tag{9.3}$$

Example 9.1

Below is a block transposition with $i = 2$, $j = 4$ and $k = 7$.

$$245613987 \longrightarrow 24 \mid 56 \mid 39 \mid 87 \longrightarrow 24 \mid 139 \mid 56 \mid 87.$$

◻

We point out that this use of the word *transposition* is different from its use in group theory.

Note that one can view a block transposition as an operation that interchanges two adjacent blocks, or as an operation that takes one block and places it somewhere else in the permutation.

Let t be the smallest positive integer so that there exists a sequence of t block transpositions that turn p into q. We then say that $t = \mathrm{btd}(p, q)$ is the *block transposition distance* of the n-permutations p and q. The reader is invited to verify that the block transposition distance is a *metric*, that is, $\mathrm{btd}(p, p) = 0$ for all p, and $\mathrm{btd}(p, q) = \mathrm{btd}(q, p)$ for all p and q (since the inverse of a block transposition is also a block transposition), and for all p, q and r, the triangle inequality

$$\mathrm{btd}(p, q) + \mathrm{btd}(q, r) \geq \mathrm{btd}(p, r)$$

holds.

We are now going to formally prove a simple but important property of the block transposition distance that we mentioned in the previous section.

PROPOSITION 9.2
The block transposition distance of two permutations is left-invariant. *That is, if p, q, and r are n-permutations, then*

$$t = \text{btd}(p, q) = \text{btd}(r \circ p, r \circ q).$$

PROOF   Induction on k, the case of $t = 0$ being trivial, and the case of $t = 1$ being easy since if $\text{btd}(p, q) = 1$, then there exist indices i, j, and k so that q is obtained from p as shown in (9.2) and (9.3). However, in that case, the same block transposition will turn

$$r \circ p = p_{r_1} p_{r_2} \cdots p_{r_i} p_{r_{i+1}} \cdots p_{r_j} p_{r_{j+1}} \cdots p_{r_k} p_{r_{k+1}} \cdots p_{r_n}$$

into

$$r \circ q = p_{r_1} p_{r_2} \cdots p_{r_i} p_{r_{j+1}} \cdots p_{r_k} p_{r_{i+1}} \cdots p_{r_j} p_{r_{k+1}} \cdots p_{r_n}.$$

Now let us assume that the statement is true for t and prove it for $t + 1$. Let $\text{btd}(p, q) = t + 1$, and let q' be a permutation so that $\text{btd}(p, q') = t$ and $\text{btd}(q', q) = 1$. Then, by our induction hypothesis, $\text{btd}(r \circ p, r \circ q') = t$ and $\text{btd}(r \circ q', r \circ q) = 1$, so there is a sequence of $t + 1$ block transpositions that take $r \circ p$ to $r \circ q$. This proves that $\text{btd}(r \circ p, r \circ q) \leq t + 1 = \text{btd}(p, q)$.

In order to prove that $\text{btd}(r \circ p, r \circ q) \geq \text{btd}(p, q)$, repeat this argument with the roles reversed, (that is, p replaced by $r \circ p$ and q replaced by $r \circ q$), and noting that $p = r^{-1} \circ (r \circ p)$ and $q = r^{-1} \circ (r \circ q)$.   ∎

With a slight abuse of language, let $\text{btd}(p) := \text{btd}(p, \text{id})$, the *block transposition distance* of p, be the smallest integer k so that there exists a sequence of k block transpositions that takes p into the identity permutation.

As stated in the introduction of this chapter, our goal is to find good estimates for $\text{btd}(p)$, as well as $\text{td}(n) = \max_{p \in S_n} \text{btd}(p)$. In other words, we would like to find the smallest number m so that *all* n-permutations can be sorted by at most m block transpositions. It would also be interesting to find out which permutations are the *most difficult* to sort. Note that $\text{td}(n)$ is called the *block transposition diameter* of S_n, since that number is the longest distance between two points in the graph whose vertices are the elements of S_n, and in which two vertices p and p' are adjacent if there is a block transposition that turns p into p'.

The following simple observation provides a lower bound on the block transposition distance $\text{btd}(p)$ of a permutation from the identity.

PROPOSITION 9.3

No block transposition can decrease the number of descents of a permutation by more than two.

In other words, if p has $d(p)$ descents, then $\mathrm{btd}(p) \geq d(p)/2$.

PROOF   Keeping the notation of (9.2) and (9.3), we see that the only positions in which a descent could be turned into an ascent by a block transposition are i, j, and k. In order for the number of descents to decrease by three, such a change would have to occur in each of those three positions. That could only happen if all of the inequalities

1. $p_i > p_{i+1}$,

2. $p_j > p_{j+1}$,

3. $p_k > p_{k+1}$,

4. $p_i < p_{j+1}$,

5. $p_k < p_{i+1}$,

6. $p_j < p_{k+1}$

held. That is impossible, since that would imply that

$$p_i > p_{i+1} > p_k > p_{k+1} > p_j > p_{j+1} > p_i.$$

∎

Based on this observation, and our intuition, it would be plausible to think that the *decreasing* permutation $n \cdots 21$ is perhaps the most difficult to sort, since it has the highest possible number of descents. This motivates the precise computation of $\mathrm{btd}(n \cdots 21)$.

LEMMA 9.4

[167] For all integers $n \geq 3$, the equality $\mathrm{btd}(n \cdots 21) = \lceil (n+1)/2 \rceil$ holds.

In particular, the block transposition diameter $\mathrm{td}(n)$ of the symmetric group S_n satisfies the inequality $\mathrm{td}(n) \geq \lceil (n+1)/2 \rceil$.

PROOF   First, we show that $\lceil (n+1)/2 \rceil$ block transpositions are necessary if we want to transform the decreasing n-permutation into the increasing one. The decreasing permutation has $n-1$ descents. The first block transposition away from the decreasing permutation will decrease the number of descents by exactly one, and the last block transposition arriving at the identity will also decrease the number of descents by exactly one. Hence, the other block transpositions are responsible for decreasing the number of descents by $n-3$, and Proposition 9.3 shows that that takes at least $(n-3)/2$ additional moves.

Now we show that $\lceil (n+1)/2 \rceil$ block transpositions are sufficient to transform the decreasing n-permutation into the increasing one. First, note that it suffices to prove this claim for odd n. Indeed, if n is even, we can sort the decreasing permutation $n(n-1)\cdots 32$ using $\lceil (n+1)/2 \rceil - 1$ block transpositions, then use the remaining block transposition to move the entry 1 to the front.

So we can assume without loss of generality that $n = 2k+1$. Then let the first block transposition move the two-element block $(k+1)k$ to the front of the decreasing permutation, let the second block transposition move the two-element block $(k+2)(k-1)$ into the middle of the block that last moved to the front, and so on. After k block transpositions of this kind, we obtain the permutation

$$(k+1)\ (k+2)\cdots 2k\ 1\ 2\cdots k\ (2k+1),$$

which we then turn into the identity permutation by interchanging the block of the first k entries and the block of the next k entries. ∎

Example 9.5
We can sort $p = 7654321$ in four block transpositions as follows.

1. 765 | 43 | 21,

2. 4376 | 52 | 1,

3. 45237 | 61,

4. 456 | 123 | 7,

5. 1234567.

   ◻

The best known *upper bound* for td(n) is given by the following theorem of Henrik Eriksson, Kimmo Eriksson, Johan Karlander, Lars Svenson, and Johan Wästlund.

THEOREM 9.6
[167] Let $n \geq 9$, and let p be an n-permutation. Then

$$btd(p) \leq \lfloor (2n-2)/3 \rfloor.$$

Therefore, td(n) $\leq \lfloor (2n-2)/3 \rfloor$.

The proof of this theorem will be given in Problem Plus 9.6. We point out that if $n < 9$, then the slightly weaker statement btd(p) $\leq \lceil 2n/3 \rceil$ holds.

Theorem 9.6 and Lemma 9.4 imply that

$$\lceil (n+1)/2 \rceil \leq \text{td}(n) \leq \lfloor (2n-2)/3 \rfloor,$$

where the lower bound is assured by the decreasing permutation. That, of course, does not mean that the decreasing permutation is indeed the most difficult to sort, but for several years, no n-permutation p was known to satisfy $\text{btd}(p) > \text{btd}(n \cdots 21)$. That changed when in 2004 Isaac Elias and Tzvika Hartman [157] presented the following result.

PROPOSITION 9.7

Let i be a nonnegative integer. Then the permutation

$$EH(i) = 4\ 3\ 2\ 1\ 5\ 13\ 12 \cdots 6\ 14\ 17\ 16\ 15 \cdots 14+4i\ 17+4i\ 16+4i\ 15+4i$$

has block transposition distance $10 + 2i$.

In other words, $EH(i)$ consists of the following parts, in the following order left to right.

1. The decreasing sequence 4321,

2. the fixed point 5,

3. the decreasing sequence of consecutive integers from 13 to 6,

4. the repeated segments $14+j\ 17+j\ 16+j\ 15+j$ for all integers j satisfying $0 \leq j \leq i$.

As $EH(i)$ is of length $n = 17 + 4i$, we see that $\text{btd}(EH(i)) = 10 + 2i > \lceil (n+1)/2 \rceil = 9 + 2i$. So the decreasing permutation is *not* the most difficult one to sort by block transpositions; $EH(i)$ is more difficult. Note that this construction exists if $n \geq 17$ and $n = 4k + 1$ for some integer $k \geq 4$.

9.3　Block Interchanges

Let $p = p_1 p_2 \cdots p_n$ be a permutation. A *block interchange* is an operation that interchanges two blocks of consecutive entries without changing the order of entries within each block. Unlike in block transpositions, the two blocks *do not need to be adjacent*. Interchanging the blocks $p_i p_{i+1} \cdots p_{i+a}$ and $p_j p_{j+1} \cdots p_{j+b}$ with $i + a < j$ results in the permutation

$$p_1 \cdots p_{i-1} \mid p_j p_{j+1} \cdots p_{j+b} \mid p_{i+a+1} \cdots p_{j-1} \mid p_i p_{i+1} \cdots p_{i+a} \mid p_{j+b+1} \cdots p_n.$$

For instance, if $p = 3417562$, then interchanging the block of the first two entries with the block of the last three entries results in the permutation 5621734.

Let bid(p, q) denote the *block interchange distance* of p and q, that is, the smallest positive integer t so that there is a sequence of t block interchanges that takes p into q.

The reader is invited to verify that the block interchange distance is a metric and that it is left-invariant.

With a slight abuse of language, let bid(p) denote the *block interchange distance* of p from the identity permutation, that is, the number of block interchanges that are necessary to sort p.

It turns out that sorting by block interchanges is much better understood than sorting by block transpositions. The following definition is crucial to understanding the strong results related to sorting by block interchanges, and it turns out to be useful in studying block transposition distances as well.

DEFINITION 9.8   Let $p = p_1 p_2 \cdots p_n$ be an *n-permutation. The cycle graph $G(p)$ of p is a directed graph on vertex set $\{0, 1, \cdots, n, n + 1\}$ and $2n + 2$ edges that are colored either black or gray as follows. Set $p_0 = 0$, and $p_{n+1} = n + 1$.*

1. *For $1 \leq i \leq n + 1$, there is a black edge from p_i to p_{i-1}, and*

2. *For $0 \leq i \leq n$, there is a gray edge from i to $i + 1$.*

The cycle graph was defined by Vineet Bafna and Pavel Pevzner in [27]. It has many variations, which are equivalent for most purposes, but some of which are easier to use for a given problem than others.

It is straightforward to show that $G(p)$ has a unique decomposition into *edge-disjoint directed cycles in which the colors of the edges alternate*. Indeed, note that each vertex $1, 2, \cdots, n$ has one edge of each color leaving that vertex, and one edge of each color entering that vertex. There is a unique way for a cycle to leave the vertex 0, namely by going to 1 using a gray edge. After that, there is a unique way to leave 1 by a black edge, namely by going to the entry immediately preceding the entry 1 in p, and so on. When the first alternating cycle is completed, we disregard its edges, and find the subsequent alternating cycles of $G(p)$ similarly.

For the rest of this chapter, when we say "cycles of $G(p)$", we always mean the alternating cycles in the alternating cycle decomposition of $G(p)$. Note that a vertex *can* occur twice in the same cycle of $G(p)$, if it is first visited by an edge of a certain color, then by an edge of the other color.

Let $c(G(p))$ be the number of directed cycles in this decomposition of $G(p)$. Note that $c(G(p))$ is **not** equal to the number of cycles of p in the traditional sense that was discussed in Chapter 3. In order to avoid confusion, we will use the notation $c(\Gamma(p))$ for that notion. That is, $c(\Gamma(p))$ is the number of cycles of p in the traditional sense. In other words, $c(\Gamma(p))$ is the number of cycles in the directed graph $\Gamma(p)$ on vertex set $[n]$ that has n edges, one edge from i to p_i for each i. Note that $G(p)$ may have as many as $n+1$ alternating cycles, but $\Gamma(p)$ can have at most n cycles. Also note that $G(p)$ has $2n+2$ edges, while $\Gamma(p)$ has n edges. The edges of $G(p)$ are of two kinds, and the alternating cycles of $G(p)$ are alternating between edges of the two kinds.

At this point, the reader is advised to try to solve the simple Exercises 1 and 2 that concern $c(G(p))$ in the case when p is the increasing or decreasing permutation.

See Figures 9.1 and 9.2 for three examples. Black edges are represented by solid, thick lines, and gray edges are represented by thin, dotted lines.

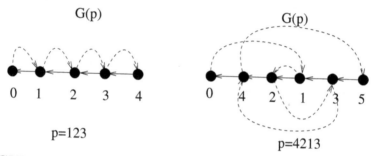

FIGURE 9.1

The graphs $G(p)$ for $p = 123$ and $p = 4213$. One sees that $c(G(123)) = 4$ and $c(G(4213)) = 1$.

The following result of D. A. Christie [116] explains the importance of cycle graphs.

THEOREM 9.9

The number of block interchanges needed to sort the n-permutation p is

$$\mathrm{bid}(p) = \frac{n+1-c(G(p))}{2}.$$

Note that Theorem 9.9 implies that $n+1$ and $c(G(p))$ are always of the same parity.

Since $c(G(p)) \geq 1$ for all non-empty permutations p, Theorem 9.9 implies that the diameter of S_n with respect to the block interchange distance is at most $\lfloor n/2 \rfloor$.

In order to prove Theorem 9.9, we need two lemmas, the first of which is the following.

G(p)

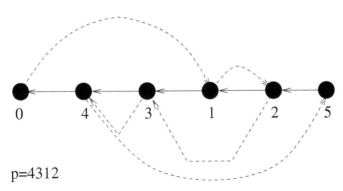

p=4312

FIGURE 9.2
The graph $G(p)$ for $p = 4312$. One sees that $c(G(4312)) = 3$.

LEMMA 9.10
Let p be any n-permutation that is not the identity permutation. Then there exists a block interchange that increases $c(G(p))$ by two.

PROOF   As $p = p_1 p_2 \cdots p_n$ is not the identity permutation, p contains at least one inversion, that is, there are two indices i and j in $[n]$ so that $p_i > p_j$ while $i < j$. If there are several such pairs, then choose the pair for which p_j is minimal, and if there are still several eligible pairs, then choose the one for which p_i is maximal. For easier notation, set $x = p_j$ and $y = p_i$. It then follows from our definitions that $x - 1$ is on the left of y, and $y + 1$ is on the right of x. (It is possible that $x = 1$, but then imagining the entry $x - 1 = 0$ in front of p will not cause any problems, because, as we will see below, $x - 1$ will not actually be part of the blocks to be interchanged.)
So p has the form

$$p = \cdots x - 1 \mid \cdots y \mid \cdots \mid x \cdots \mid y + 1 \cdots .$$

Now apply the block interchange indicated by the vertical bars, that is, interchange the block starting immediately after $x-1$ and ending in y and the block starting in x and ending immediately before $y + 1$, to get the permutation

$$p' = \cdots x - 1 \mid x \cdots \mid \cdots \mid \cdots y \mid y + 1 \cdots .$$

Let us call this the *canonical* block interchange of p. We claim that $c(G(p')) = c(G(p)) + 2$.
In order to prove this claim, note the block interchange defined above does not change any gray edges of $G(p)$, and changes either three or four black

edges of $G(p)$, depending on whether y and x were in adjacent positions in p or not. We are going to show that the claim holds in both cases. We will repeatedly use the simple fact that if in $G(p)$ there is an alternating path from u to v, then u and v are part of the same alternating cycle of $G(p)$. You are asked to prove this fact in Exercise 7.

1. Let us first assume that x immediately follows y in p. That means that the alternating cycle decomposition of $G(p)$ contains the alternating path (starting with a black edge) $a\ x - 1\ x\ y\ y + 1\ b$, where a is the entry immediately on the right of $x - 1$, and b is the entry immediately on the left of $y + 1$ in p. In particular, those six entries were in the same cycle C. However, in p', the two-vertex cycles $x - 1\ x$ and $y\ y + 1$ are formed, and the rest of C forms a separate cycle, with the newly created black edge ab closing the cycle. So the cycle C of $G(p)$ is replaced by three cycles in $G(p')$, while the other cycles are unchanged. See Figure 9.3 for an illustration.

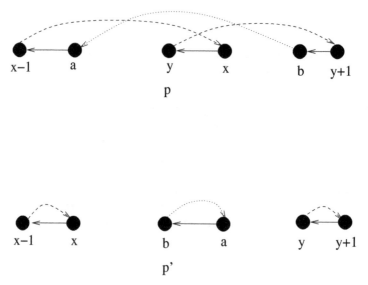

FIGURE 9.3
The canonical block interchange of p when y and x are adjacent. The arrow from b to a represents an alternating path, not necessarily an edge.

2. Let us now assume that x and y are not neighbors in p, and introduce the notation a, b, c, and d for certain entries of p, so that $x - 1\ a$, $y\ b$, $c\ x$, and $d\ y + 1$ are adjacent pairs of entries in p. Using that notation, $G(p)$ will always contain the alternating path $P_a = a\ x - 1\ x\ c$ and the

alternating path $P_b = b\ y\ y+1\ d$, but P_a and P_b may or may not be in the same cycle.

(a) If P_a and P_b are part of the same alternating cycle C_{ab} of $G(p)$, that is, when there is an alternating path A_{da} from d to a ending in a gray edge, then in $G(p')$, the cycle C_{ab} will be replaced by three cycles, namely the two-vertex cycles $x-1\ x$ and $y\ y+1$, and the cycle that contains b, d, the path A_{da}, and c. Other cycles will not change. See Figure 9.4 for an illustration.

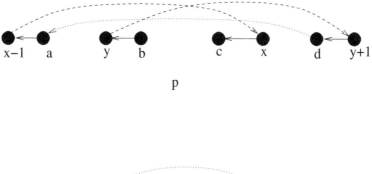

p

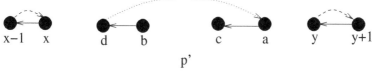

p'

FIGURE 9.4
The canonical block interchange of p when y and x are not adjacent, but are part of the same cycle of $G(p)$.

(b) If P_a and P_b are part of two distinct cycles C_a and C_b of $G(p)$, then in $G(p')$, each of these cycles will break up into two cycles. Indeed, C_a breaks up into the two-vertex cycle $x-1\ x$ and the rest of C_a, completed by the new black edge ac, while C_b breaks up into the two-vertex cycle $y\ y+1$ and the rest of C_b, completed by the new black edge bd. So in this case, two cycles of $G(p)$ turn into four cycles of $G(p')$. See Figure 9.5 for an illustration.

∎

A remarkable property of block interchanges is that while they can be used to increase $c(G(p))$ by two, they will *never* increase $c(G(p))$ by more than two.

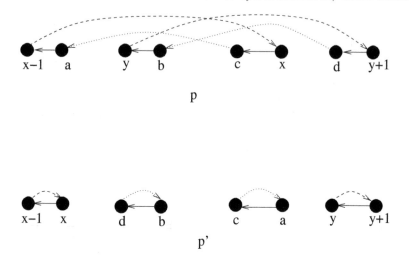

FIGURE 9.5
The canonical block interchange of p when y and x are not adjacent, and are not part of the same cycle of $G(p)$.

LEMMA 9.11
No block interchange increases $c(G(p))$ by more than two.

PROOF   A block interchange B changes at most four black edges of $G(p)$ (the black edges connecting the interchanged blocks to non-moving blocks). Before B is executed, those black edges are part of at least one cycle, and after B is executed, they are part of at most four cycles. So the only way that B could possibly increase the number of cycles of $G(p)$ by more than two would be if B broke one cycle into four smaller cycles. That would mean that the inverse transformation B^{-1} of B, which is also a block interchange, would change one black edge from each of four different cycles, and would turn those four cycles into one cycle.

However, that is impossible. Indeed, consider the block interchange, shown in Figure 9.6 that changes four black edges, namely the black edges aa', $b'b$, cc' and $d'd$, and assume that in $B^{-1}(p) = p'$, the four new black edges, that is, ca', $b'd$, ac', and $d'b$ are in the same cycle C.

Let us start walking around C starting with ca'. What is the first newly created edge we encounter? It must be ac'. Indeed, if it were $b'd$, that would mean that the edges aa' and $b'b$ were in the same cycle of p, and if it were $d'b$, that would mean that the edges aa' and $d'd$ were in the same cycle of p. So the first new edge we encounter in C after ca' is ac'. The next new edge we encounter cannot be $b'd$, since that would imply that the edges cc' and $b'b$ were in the same cycle of p. It cannot be $d'b$ either, since that would imply that the edges cc' and $d'd$ were in the same cycle of p. This contradiction

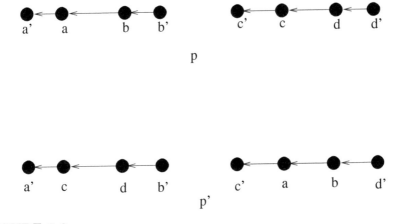

FIGURE 9.6
The four new black edges cannot all be part of the same cycle.

proves that the four new edges of p' cannot all be part of the same cycle.

Therefore, as block interchanges are reversible, block interchanges that turn some cycles into four different cycles must start with two cycles, not one. ∎

Now the proof of Theorem 9.9 is immediate.

PROOF (of Theorem 9.9) The cycle graph of the identity permutation has $n+1$ cycles, and as we mentioned (see Exercise 2), the identity is the only permutation with that property. By Lemma 9.11, $\text{bid}(p) \geq (n+1-c(G(p)))/2$, and by Lemma 9.10, $\text{bid}(p) \leq (n+1-c(G(p)))/2$, since that many block interchanges can transform p into a permutation with $n+1$ cycles in its cycle graph, that is, into the identity permutation. ∎

9.3.1 Average Number of Block Interchanges Needed to Sort p

Now that we know how to sort any given permutation by block interchanges, we can address the more global question of how efficient sorting by block interchanges is. Theorem 9.9 tells us that every n-permutation can be sorted by at most $\lfloor n/2 \rfloor$ block interchanges, but what can we say about the *average* permutation?

It follows from Theorem 9.9 that we could answer this question if we could count all n-permutations whose cycle graph has a given number of alternating cycles. This motivates the following definition.

DEFINITION 9.12 *The number of n-permutations p satisfying $c(G(p)) = k$ is called a* Hultman number, *and is denoted by* $\mathcal{S}_H(n,k)$.

The name *Hultman numbers* was given by Anthony Labarre [148] and honors Axel Hultman.

Example 9.13
For $n = 3$, we have $\mathcal{S}_H(3,4) = 1$ and $\mathcal{S}_H(3,2) = 5$. Indeed, all 3-permutations except the identity have two alternating cycles in their cycle graph (because they have block interchange distance 1). $\quad \Box$

It turns out that for a permutation p, there is a rather close, but nontrivial, connection between $c(G(p))$, which is the number of alternating cycles of the cycle graph of p, and $c(\Gamma(p))$, which is the number of cycles of p in the traditional sense. The following result of Doignon and Labarre [148] makes this explicit.

THEOREM 9.14
The Hultman number $\mathcal{S}_H(n,k)$ is equal to the number of ways to obtain the cycle $(12 \cdots n(n+1)) \in S_{n+1}$ as a product qr of permutations, where $q \in S_{n+1}$ is any cycle of length $n+1$, and the permutation $r \in S_{n+1}$ has exactly k cycles, that is, $c(\Gamma(r)) = k$.

The proof of this theorem is given in Problem Plus 2.

The following immediate consequence of Theorem 9.14 is more suitable for our purposes.

COROLLARY 9.15
The Hultman number $\mathcal{S}_H(n,k)$ is equal to the number of $(n+1)$-cycles q so that the product $(12 \cdots n(n+1))q$ is a permutation with exactly k cycles, that is, $c(\Gamma((12 \cdots n(n+1))q)) = k$.

PROOF   If $(12 \cdots n(n+1))q = w$, where w has k cycles, and q is an $(n+1)$-cycle, then multiplying both sides of the last equation by q^{-1} from the right, we get the equation

$$(12 \cdots n(n+1)) = wq^{-1}.$$

The claim of the Corollary is now immediate from Theorem 9.14, since q^{-1} is a cycle of length $n+1$.   ∎

Example 9.16
We have seen in Example 9.13 that $\mathcal{S}_H(3,2) = 5$. So by Corollary 9.15, there must be five permutations q of length four that consist of a single cycle so that $(1234)q$ has exactly two cycles.

In order to see that this is the case, note that there are six 4-cycles on [4]. One of them, the inverse of (1234), that is, the permutation (4321), is clearly ineligible for the role of q since $(1234)q = (1)(2)(3)(4)$ has four cycles. The other five 4-cycles are all eligible for the role of q, since for each of them, $(1234)q$ will be the product of two odd permutations, hence an even permutation. Hence $(1234)q$ cannot be one 4-cycle, or a 2-cycle and two fixed points, and it also cannot be the identity permutation since q is not the inverse of (1234). So $(1234)q$ must have two cycles in it for each of the five 4-cycles q that are not equal to (4321).   ☐

So finding the average of the numbers $c(G(p))$ over all n-permutations p is equivalent to finding the average of the numbers $c(\Gamma((12\cdots n(n+1))q)$, where q is an $(n+1)$-cycle. This is significant, because this translates our problem into the language of counting traditional cycles in permutations, which is a very well-studied area.

Let us consider the product $s = (12\cdots n)z$, where z is a cycle of length n. We will monitor the changes in $c(\Gamma(s))$ as the two factors of s grow in length. Let a and b be two entries satisfying $z(a) = b$, and let us insert the entry $n+1$ into z to get the permutation z' so that $n+1$ is inserted "between" the entries a and b in the following sense.

$$z'(i) = \begin{cases} z(i) \text{ if } i \notin \{a, n+1\}, \\ n+1 \text{ if } i = a, \text{ and} \\ b \text{ if } i = n+1. \end{cases}$$

See Figure 9.7 for an illustration.

The following proposition describes how the insertion of the new entries changes the number of cycles of the Γ-graph, that is, the number of cycles of s in the traditional sense.

PROPOSITION 9.17
Let a, b, and z' be defined as above, and let $s' = (12\cdots(n+1))z'$. Then the following equalities hold.

1. *If $2 \le a$, and $a-1$ and $z(1)$ are not in the same cycle of s, then $c(\Gamma(s')) = c(\Gamma(s)) - 1$ if $2 \le a$.*

2. *If $2 \le a$, and $a-1$ and $z(1)$ are in the same cycle of s, then $c(\Gamma(s')) = c(\Gamma(s)) + 1$.*

3. *If $a = 1$, then $c(\Gamma(s')) = c(\Gamma(s)) + 1$.*

PROOF   Let us assume first that $a \ge 2$, and that $a-1$ is in a cycle C_1 of s, and $z(1)$ is in a different cycle C_2 of s. Let $C_1 = ((a-1)b\cdots)$ and let $C_2 = (z(1)\cdots n)$. After the insertion of $n+1$ into z, the obtained permutation

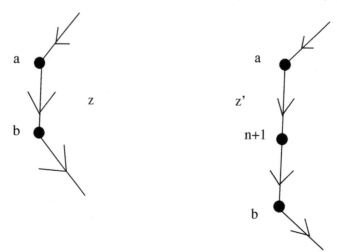

FIGURE 9.7
How z' is obtained from z.

$s' = (12 \cdots (n+1))z'$ sends $a-1$ to $n+1$, then $n+1$ to $z(1)$, then leaves the rest of C_2 unchanged till its last entry. Then it sends n back to $z'(n+1) = b$, from where it continues with the rest of C_1 with no change. So in s', the cycles C_1 and C_2 are united, the entry $n+1$ joins their union, and there is no change to the other cycles of s. See Figure 9.8 for an illustration.

Let us now assume that $a \geq 2$, and that $a-1$ and $z(1)$ are both in the same cycle C of s. Then $C = ((a-1)b \cdots nz(1) \cdots)$. After the insertion of $n+1$ into z, the obtained permutation $s' = (12.. \cdots (n+1)z'$ sends $a-1$ to $n+1$, then $n+1$ to $z(1)$, cutting off the part of C that was between $a-1$ and n. So C is split into two cycles, the cycle $C' = ((a-1)(n+1)z(1) \cdots)$ and the cycle $C'' = (b \cdots n)$. Note that $s'(n) = b$ since $z'(n+1) = b$. See Figure 9.9 for an illustration.

Finally, if $a = 1$, then $s'(n+1) = (n+1)$, so the entry $n+1$ forms a 1-cycle of s', and the rest of the cycles of s do not change.   ∎

Proposition 9.17 shows that inserting $n+1$ into a position of z will sometimes decrease and sometimes increase the number of cycles of the product $s' = (12 \cdots (n+1))z$. The question is, of course, how many times will an increase and how many times will a decrease occur. In light of Proposition 9.17, this is the same question as asking how often the entries $z(1)$ and $a-1$ are in the same cycle of s. Furthermore, since z is just an arbitrary n-cycle, and $a-1$ is an arbitrary element of a fixed n-cycle, this is equivalent to the following question.

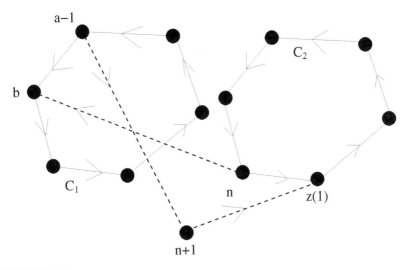

FIGURE 9.8
If $a - 1$ and $z(1)$ are in different cycles of s, those cycles will turn into one.

QUESTION   **9.18**   *Let TC_n denote the set of ordered pairs (x, y) of n-permutations that consist of one n-cycle each. Let i and j be two fixed elements of the set $[n] = \{1, 2, \cdots, n\}$. Select an element (x, y) of TC_n at random. What is the probability that the product xy contains i and j in the same cycle?*

Example 9.19
Let $n = 3$, then TC_n has four elements, namely $((123), (123))$, $((123), (132))$, $((132), (123))$, and $((132), (132))$. For the pair listed first, we have $xy = (132)$, for the pairs listed second and third, we have $xy = (1)(2)(3)$, and for the pair listed fourth, we have $xy = (123)$. So for any two distinct elements $i, j \in [3]$, the probability that i and j are in the same cycle in the product xy of the two permutations in a randomly selected element of TC_2 is $\frac{2}{4} = \frac{1}{2}$.   ∎

Based on this and other numerical evidence, the present author conjectured in 2008 that for odd integers $n \geq 3$, the answer to Question 9.18 was $1/2$. For even integers, numerical evidence suggested that the answer to Question 9.18 was somewhat less than $1/2$, so the present author asked what the precise answer was.

Question 9.18 was answered by Richard Stanley [303] in the same year. He gave a complicated proof using various high-powered techniques of abstract algebra and analysis, such as symmetric functions, character theory, and non-elementary integration. Stanley's result is the following.

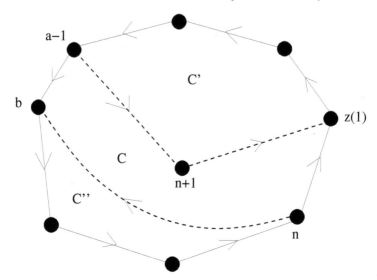

FIGURE 9.9
If $a - 1$ and $z(1)$ are in the same cycle of s, that cycle will split into two cycles.

THEOREM 9.20
[303] Let i and j be two fixed, distinct elements of the set $[n]$, where $n > 1$. Let (x, y) be a randomly selected element of TC_n. Let $p(n)$ be the probability that i and j are in the same cycle of xy. Then

$$p(n) = \begin{cases} \frac{1}{2} & \text{if } n \text{ is odd, and} \\ \frac{1}{2} - \frac{2}{(n-1)(n+2)} & \text{if } n \text{ is even.} \end{cases}$$

So $p(n) = 1/2$ indeed holds for odd integers $n \geq 3$. Note that among all values that a probability can take, $1/2$ is probably the one that is the most intriguing in that if a probability of an event is proved to be one half, then it is completely natural to ask for a combinatorial, in fact, bijective, proof of that fact. Such a proof, which is still not simple, has recently been given in [35].

It is now not difficult to describe how the average of the Hultman numbers grow. Let a_n be the average number of cycles (in the traditional sense) in all permutations of the set $\{xy | (x, y) \in TC_n\}$.

LEMMA 9.21
Let $n \geq 1$. Then $a_1 = 1$, and the numbers a_n grow as follows.

1. *If $n = 2m + 2$, then $a_n = a_{n-1} + \frac{1}{n-1}$,*

2. *If $n = 2m + 1$, then $a_n = a_{n-1} + \frac{1}{n-1} - \frac{1}{m(m+1)}$.*

PROOF

1. We apply Proposition 9.17, with n replaced by $n - 1$, which is an odd number. So z' is a cycle of length n obtained from a cycle z of length $n - 1$ through the insertion of the maximal element n into one of $n - 1$ possible positions. If $a \neq 1$, then $a - 1$ and $z(1)$ are equally likely to be in the same cycle or not in the same cycle of s. Therefore, an increase of one or a decrease of one in $c(G(s))$ is equally likely. If, on the other hand, $a = 1$, which occurs in $1/(n-1)$ of all cases, then $c(G(s))$ increases by one. So

$$a_n = \frac{n - 2}{n - 1}a_{n-1} + \frac{1}{n - 1}(a_{n-1} + 1) = a_{n-1} + \frac{1}{n - 1}.$$

2. We again apply Proposition 9.17, with n replaced by $n - 1$, which is now an even number, namely $n - 1 = 2m$. If $a \neq 1$, which happens in $(n - 2)/(n - 1)$ of all cases, then the probability of $a - 1$ and $z(1)$ falling into the same cycle of s is $\frac{1}{2} - \frac{2}{(n-2)(n+1)} = \frac{1}{2} - \frac{1}{(2m-1)(m+1)}$ by Theorem 9.20. By Proposition 9.17, in these cases $c(G(s))$ grows by one. If $a = 1$, which occurs in $1/(n-1)$ of all cases, then $c(G(s))$ always grows by one. So

$$a_n = \frac{n - 2}{n - 1} \cdot \left(\frac{1}{2} - \frac{1}{(2m - 1)(m + 1)} \right)(a_{n-1} + 1)$$
$$+ \frac{n - 2}{n - 1} \cdot \left(\frac{1}{2} + \frac{1}{(2m - 1)(m + 1)} \right)(a_{n-1} - 1)$$
$$+ \frac{1}{n - 1}(a_{n-1} + 1),$$

which is equivalent to the statement of the lemma as can be seen after routine rearrangements.

∎

It is now easy to prove an explicit formula for a_n.

THEOREM 9.22
For all positive integers n, the equality

$$a_n = \frac{1}{\lfloor (n + 1)/2 \rfloor} + \sum_{i=1}^{n-1} \frac{1}{i}$$

holds.

PROOF   This is a direct consequence of Lemma 9.21 and the fact that

$$\sum_{i=1}^{t} \frac{1}{i(i+1)} = \sum_{i=1}^{t} \left(\frac{1}{i} - \frac{1}{i+1} \right)$$

$$= 1 - \frac{1}{t+1}.$$

Indeed, if we repeatedly apply Lemma 9.21 to compute a_i from a_{i-1} starting at a_2, then at each step, we will have a new summand equal to $1/(i-1)$, and in every other step we will also have a summand equal to $\frac{-1}{j(j+1)}$. So, assuming that $n = 2m$ (the case of odd n is completely analogous), we compute

$$a_n = a_{n-1} + \frac{1}{n-1}$$

$$= a_{n-2} + \frac{1}{n-1} + \frac{1}{n-2} - \frac{1}{(m-1)m}$$

$$= a_2 + \sum_{i=1}^{n-1} \frac{1}{i} - \sum_{j=1}^{m} \frac{1}{(m-1)m}$$

$$= \frac{1}{\lfloor (n+1)/2 \rfloor} + \sum_{i=1}^{n-1} \frac{1}{i}.$$

∎

Recall the fact, which we proved in several different ways in this book, that the average number of cycles of $\Gamma(p)$, taken over all n-permutations, is the harmonic number $H(n) = \sum_{i=1}^{n} \frac{1}{i}$. Theorem 9.22 shows that a_n is just a little bit more than that.

Theorem 9.22 now enables us to state and prove an explicit formula for the average distance between two randomly selected n-permutations, or equivalently, the average block interchange distance of an n-permutation from the identity permutation.

THEOREM 9.23
[65] The average number of block interchanges needed to sort an n-permutation is

$$b_n = \frac{n - \frac{1}{\lfloor (n+2)/2 \rfloor} - \sum_{i=2}^{n} \frac{1}{i}}{2}.$$

PROOF   By Theorem 9.14 and Theorem 9.22, the average value of $c(G(p))$ of a randomly selected n-permutation p is $a_{n+1} = \frac{1}{\lfloor (n+2)/2 \rfloor} + \sum_{i=1}^{n} \frac{1}{i}$. However, Theorem 9.9 states that the number of block interchanges needed to sort p is $(n+1-c(G(p)))/2$. Taking the average value of this expression over all n-permutations proves our claim. ∎

9.3.1.1 Variations of the Cycle Graph

The concept of cycle graphs is a very important one, and as such, it has been introduced several times, in slightly different (but essentially equivalent) versions. One such version identifies vertices $p_0 = 0$ and $p_{n+1} = n + 1$ so that these two vertices become one, and leaves all the other rules unchanged, except that $i + 1$ and $i - 1$ are now meant modulo n. So the new vertex (let us call it 0), now has four edges adjacent to it, namely one edge of each color leaving it, and one edge of each color arriving at it, just like any other vertex.

This "circular" version $G'(p)$ of the cycle graph is sometimes more convenient to use than the linear version $G(p)$ that we studied in this section. For now, Figure 9.10 shows the graph $G'(4312)$ as an example. As the vertices 0 and $n + 1$ have been contracted into one vertex called 0, the vertices of this graph can conveniently be arranged in a circle.

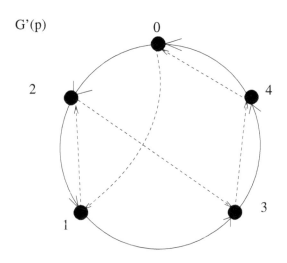

G'(p)

p=4312

FIGURE 9.10
The graph $G'(p)$ for $p = 4312$.

In Exercise 6, you are asked to prove that $G'(p)$ has the same number of alternating cycles as $G(p)$, no matter what p is.

Exercises

1. (–) Prove that if n is odd, then $c(G(n \cdots 21)) = 2$, and if n is even, then $c(G(n \cdots 21)) = 1$.

2. (–) Prove that the identity permutation is the only n-permutation p so that $c(G(p)) = n + 1$.

3. Prove that if $p = p_1 p_2 \cdots p_n$ is not the identity permutation, then there exists a block interchange that increases $c(G(p))$ by two that moves a block containing the smallest entry p_i for which $p_i \neq i$.

4. Let p be an n-permutation. Prove the inequality

$$\mathrm{btd}(p) \geq \frac{n + 1 - c(G(p))}{2}.$$

5. Let us call a cycle of $G(p)$ an *odd* cycle if the number of black (equivalently, gray) edges of that cycle is odd. In other words, a cycle is odd if it consists of $2m$ edges, where m is an odd integer. Let $c_{odd}(G(p))$ denote the number of odd cycles of $G(p)$.

   Prove that no block transposition can increase $c_{odd}(G(p))$ by more than two.

   Note that this fact is not a trivial consequence of Lemma 9.11. It could possibly happen (but it does not) that a block transposition turns two even blocks into four odd blocks.

6. (–) Prove that for any permutation p, the graphs $G(p)$ and $G'(p)$ have the same number of alternating cycles.

7. Prove that if in $G(p)$ there is an alternating path from u to v, then u and v are part of the same alternating cycle of $G(p)$.

8. (–) Prove that no block transposition on p can change the parity of $c(G(p))$.

9. (–) The *breakpoint graph* $B(p)$ of the n-permutation p is defined as follows. Its vertex set has $2n + 2$ elements, with the vertices being $L_0, R_0, L_1, R_1 \cdots, L_n, R_n$. We arrange these vertices so that they form a $(2n+2)$-gon, and $L_0, R_0, \cdots$ are the labels of the vertices, going counterclockwise. The edges of $B(p)$ are given as follows.

   (a) For each index $i \in [0, n]$, there is a black edge from R_{p_i} to $L_{p_{i+1}}$, where $n + 1$ is interpreted as 0, and

   (b) for each index $i \in [0, n]$, there is a gray edge from L_{i+1} to R_i, where $n + 1$ is interpreted as 0.

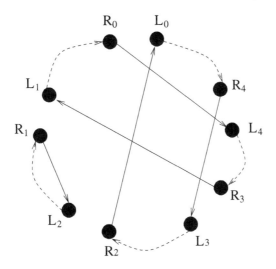

FIGURE 9.11
The breakpoint graph $B(4312)$.

See Figure 9.11 for an illustration.

Prove that for any permutation p, the graphs $G(p)$ and $B(p)$ have the same number of alternating cycles.

10. (−)

Let p be a permutation, and let C be a cycle of $G(p)$ with $2k + 1$ black edges. Then C is called *good* if it is possible to split C into $2k + 1$ cycles that have one black edge each using k block transpositions so that the other cycles of $G(p)$ remain invariant. If C is not good, then C is called *bad*.

Keep the definition of the graph $B(p)$ from the preceding exercise. Let d_{r-1} be the decreasing permutation of length $r - 1$. Prove that

(a) if r is even, then $B(d_{r-1})$ is a bad odd cycle,

(b) if $r = 4k + 1$, then $B(d_{r-1})$ is the union of two bad odd cycles, and

(c) if $r = 4k + 3$, then $B(d_{r-1})$ is the union of two even cycles.

11. Sort the permutation 864297531 with five block transpositions.

12. Let us say that the permutation $p = p_1 p_2 \cdots p_n$ has a bond in $i \in [0, n]$ if $p_{i+1} = p_i + 1$, where $p_0 = 0$ and $p_{n+1} = n + 1$. State and prove a slightly stronger version of Theorem 9.6 in terms of the number of bonds of p.

13. Sort the permutation 963852741 with five block transpositions.

14. Let p be an n-permutation, and let us sort p by block interchanges in which each block consists of *one entry*. Find a formula for the number of such operations needed to sort p.

15. (–) Let p be an n-permutation, and let is(p) be the length of the longest increasing subsequence of p. Prove that btd$(n) \leq n -$ is(p), and show an example for an infinite family of permutations for which this upper bound is tight.

16. The *edit distance* of a permutation p from the identity is the number of block transpositions *in which one block is of size 1* needed to sort p. Equivalently, this is the number of steps needed to sort p if a step consists of choosing one entry of p and placing it somewhere else. Express the edit distance of p in terms of a simple parameter of p.

Problems Plus

1. Prove Theorem 9.6.

2. Prove Theorem 9.14.

3. Refine Theorem 9.14.

4. Find an explicit formula for $\mathcal{S}_H(n, 1)$.

5. Let $n \geq 3$, and let i, j, and k be three distinct elements of $[n]$. Let TC_n be defined as in Question 9.18. What is the probability that if we choose a random element $(x, y) \in TC_n$, then the entries i, j, and k are in the same cycle of the product xy?

6. Let $X_n(p)$ denote the block transposition distance of the n-permutation p from the identity. Prove that there exists a constant c so that

$$E(X_n) \geq \frac{n}{2} - \frac{\ln n}{4} - c.$$

7. Let $X(p) = c(G(p))$. Find Var(X), where the variance is taken on the set of all n-permutations.

8. Let a_n be the number of derangements p of length $n + 1$ that can be obtained as $(12 \cdots (n + 1))q$, where q is a cycle of length $n + 1$. Find a formula for the numbers a_n.

9. Let a be a partition of the integer $n + 1$. Find a formula for the number of permutations of length $n + 1$ that can be obtained as $(12 \cdots (n + 1))q$, where q is a cycle of length $n + 1$.

10. In *cut-and-paste sorting* of permutations, one can move a block of entries of a permutation p and insert it somewhere else in p, *possibly reversed*. Show that for any positive integers n, there exists an n-permutation p that takes at least $\lfloor n/2 \rfloor$ cut-and-paste moves to sort.

11. Prove that for all n-permutations p, the inequality $\mathrm{btd}(p) \leq n - c_{odd}(\Gamma(p))$ holds, where $c_{odd}(\Gamma(p))$ is the number of odd cycles of $\Gamma(p)$.

12. A *prefix transposition* is a block transposition that displaces the leftmost entry of a permutation. Prove that for infinitely many values of n, there exist n-permutations that take $\lfloor (3n+1)/4 \rfloor$ prefix transpositions to sort.

13. A *prefix exchange* is a block interchange that swaps that first entry of a permutation and any other entry of the permutation. The *prefix exchange distance* of the permutation p, denoted by $\mathrm{pexc}(p)$ is the minimum number of prefix exchanges needed to sort p. Find an explicit formula for $\mathrm{pexc}(p)$.

14. Let $\sigma^{(k)}$ denote the product of k maximal cycles (of length n) chosen uniformly at random. Prove that

$$P(\sigma^{(k)} \text{ is a cycle}) = \frac{1 + (-1)^{n-1}}{n+1}. \tag{9.4}$$

15. Keeping the notation of the previous Problem Plus, find a formula for $P(\sigma^{(2)} \text{ is a cycle})$.

16. Find a formula for $P(\sigma^{(2)} \text{ is a derangement})$.

17. Let n be a fixed positive integer, and let ℓ_k denote the number of permutations p of length n whose longest increasing subsequence is of length k and whose tableaux $risk(p)$ (defined in Chapter 7) are of the shape of a hook. Recall that this means that $P(p)$ and $Q(p)$ are tableaux that are unions of their first row and first column.

Prove that the sequence $\ell_1, \ell_2, \cdots, \ell_n$ is log-concave. Explain the connection of this problem to sorting.

18. Let n be a fixed positive integer, and let i_k denote the number of involutions p of length n whose longest increasing subsequence is of length k and for which $risk(p)$ (defined in Chapter 7) are of the shape of a hook.

Prove that the sequence $i_1, il_2, \cdots, i_n$ is log-concave.

Solutions to Problems Plus

1. It is proved in [167] by an exhaustive case-by-case analysis that if p is not the identity permutation, then it is possible to find two block transpositions B and B' so that one of $BB'p$, $B'pB$, and $pB'B$ has at least three bonds. The claim is then straightforward by induction.

2. Let $p = p_1 p_2 \cdots p_n$ be an n-permutation, and let $G'(p)$ be the circular version of its cycle graph. Define the permutation

$$p' : \{0, 1, \cdots, n\} \to \{0, 1, \cdots, n\}$$

by setting $p'(i) = j$ if there is a black edge in $G'(p)$ from $i + 1$ to j (where $n + 1$ is identified with 0). Furthermore, set

$$p^* = (0 \ p_n \ p_{n-1} \cdots 2 \ 1).$$

It is then straightforward to verify that

$$p' = p^* \circ (0 \ 1 \ \cdots n) = (0 \ 1 \ \cdots n) \cdot p * .$$

Furthermore, $c(\Gamma(p')) = G'(p)$. Indeed, each cycle of $\Gamma(p')$ (that is, each cycle of p' in the traditional sense) corresponds to an alternating cycle of $G'(p)$, since if $p'(i) = j$, then there is a gray edge from i to $i+1$, then a black edge from $i + 1$ to j. It is not difficult to show that the map $g : p \to p'$ is bijective, completing the proof. This argument is due to Jean-Paul Doignon and Anthony Labarre, and can be found in [148].

3. We have seen in the solution of the previous Problem Plus that there is a bijection g from S_n to the set of permutations of length $n+1$ that can be written as a product of $(0 \ 1 \ n)$ and another cycle of length $n + 1$ so that each cycle of length k in $G'(p)$ is mapped into an alternating cycle with k black edges in $\Gamma(p')$.

   Therefore, if $a_1, a_2, \cdots, a_{n+1}$ is a sequence of non-negative integers satisfying $\sum_{i=0}^{n+1} i a_i = n+1$, then the number of n-permutations p for which $G'(p)$ has exactly a_i alternating cycles with i black edges for each i is equal to the number of permutations q of length $n + 1$ that can be written as a product of $(0 \ 1 \cdots \ n)$ and another cycle of length $n + 1$ such that $\Gamma(q)$ has exactly a_i cycles of length i.

4. If n is odd, then $\mathcal{S}_H(n, 1) = 0$. If n is even, then $\mathcal{S}_H(n, 1) = \frac{2}{n+2} n!$. See [148] for a proof.

5. Let $t(n)$ denote the requested probability. It is proved in [303] that

$$t(n) = \begin{cases} \frac{1}{3} + \frac{1}{(n-2)(n+3)} & \text{if } n \text{ is odd, and} \\ \frac{1}{3} - \frac{3}{(n-1)(n+2)} & \text{if } n \text{ is even.} \end{cases}$$

6. Finding the average value of $c_{odd}(G(p))$ is helpful here. By the result of Problem Plus 3, knowing that average is equivalent to knowing the average of the numbers $c(\Gamma(q))$, where q is a permutation of length $n+1$ that can be obtained as a product $(12 \cdots (n+1))r$, with r being a cycle of length $n+1$.

This average has been computed by Richard Stanley [301], who showed that if $n = 2m$, then the average of $c(\Gamma(q))$ taken over the set of n-permutations that can be obtained by multiplying $(12 \cdots n)$ by an n-cycle is

$$\sum_{k=1}^{m} \frac{1}{2k} - \frac{2}{n!} \left(\sum_{k=1}^{m-1} (n - 2k)!(2k - 1)! \right) - \frac{1}{n}.$$

Stanley also states that a similar formula exists when n is odd. In other words, the average number of *even* cycles of $G(p)$ is just a little bit less than half of the average number of *all* cycles of $G(p)$. The latter was computed in Theorem 9.22, where it was found to be close to $\ln n$. Our claim can now be proved by a routine computation.

7. It is proved in [238] that

$$\mathrm{Var}(X) = \left(H(n) + \frac{1}{\lfloor (n+2)/2 \rfloor} \right) \cdot \left(1 - \frac{1}{(n+2)!} \right).$$

Here $H(n)$ denotes the nth harmonic number.

8. We can assume without loss of generality that $q = (1q_2 \cdots q_{n+1})$. Then p will be a derangement if and only if $q_i \neq q_{i+1} + 1$ module $n + 1$ for any i. From this observation, a little bit of work yields the recurrence relation

$$a_n = (n - 2) \cdot a_{n-1} + (n - 1) \cdot (2 \cdot a_{n-2} + a_{n-3})$$

for $n \geq 3$, while $a_0 = 1$, $a_1 = 0$, and $a_2 = 1$. Let $A(z) = \sum_{n \geq 0} a_n \frac{z^n}{n!}$. Then the above recurrence relation leads to the functional equation $A(z) = e^{-z}(1 - \ln(1 - z))$, which in turn yields the formula $a_n = (-1)^n + \sum_{k=0}^{n-1} (-1)^k \binom{n}{k}(n - k - 1)!$.

9. There are two ways of computing these numbers, both of which are complicated, though sometimes possible. See [148] for a relatively direct argument and see [293] for a high-powered argument.

10. Let us say that i is a *parity adjacency* for the permutation $p = p_1p_2 \cdots p_n$ if p_i and p_{i+1} are of different parity. It can be shown [129] that no cut-and-paste move increases the number of parity adjacencies of p by more than two. The result now follows if we consider $p_0 = 0$ and $p_{n+1} = n+1$, and set p to be an n-permutation with only one parity adjacency.

11. This result is due to Anthony Labarre [239], who proved it as a consequence of a more complex framework (and not simply showing that in each step, $c_{odd}\left(\Gamma(p)\right)$ can be increased by one by a block transposition).

12. This result is due to Anthony Labarre [240], who showed that all permutations p for which each cycle in the alternating cycle decomposition of $G(p)$ contains an even number of black edges have this property.

13. Let $f(p)$ be the number of fixed points of the permutation $p = p_1 p_2 \cdots p_n$. It is proved in [2] that

$$\text{pexc}(p) = n - c\left(\Gamma(p)\right) - 2f(p) - \begin{cases} 0 \text{ if } p_1 = 1, \\ \\ 2 \text{ if } p_1 \neq 1. \end{cases}$$

14. If n is even, then the claim is obvious, since the product of two n-cycles is an even permutation, so it cannot be an n-cycle. If n is odd, then the result is equivalent to the fact that the number of ways to write an n-cycle as a product of two n-cycles is $2(n-1)!/(n+1)$. This result has been discovered many times. See [108] and its references for an elementary proof.

15. It is proved in [76] that

$$P(\sigma^{(k)} \text{ is a cycle}) = \frac{1}{N} \sum_{r=0}^{N-1} (-1)^{(k+1)r} \binom{N-1}{r}^{-k+1}.$$

16. It is proved in [76] that

$$P(\sigma^{(2)} \text{ is a derangement}) = N \sum_{\tau=0}^{N-1} \frac{(-1)^\tau}{(N-\tau)\tau!} + \frac{(-1)^N}{(N-1)!}.$$

17. The connection to sorting is that we can reverse these permutations (this will not change the fact that the shape of their tableaux is a hook), and then the longest increasing subsequences will become longest decreasing subsequences. The length of those can be used to determine the edit distance of a permutation from the identity, as you are asked to show in Exercise 16.

See [77] for a proof of log-concavity.

18. This result is proved in [77].

10

Do Not Look Just Yet. Solutions to Odd-Numbered Exercises.

10.1 Solutions for Chapter 1

1. Note that

$$\alpha(S) = \binom{n}{s_1}\binom{n-s_1}{s_2-s_1}\binom{n-s_2}{s_3-s_2}\cdots\binom{n-s_k}{n-s_k}$$

$$= \frac{n!}{s_1!(n-s_1)!} \cdot \frac{(n-s_1)!}{(s_2-s_1)!(n-s_2)!} \cdots \frac{(n-s_k)!}{(n-s_k)!} =$$

$$\frac{n!}{s_1! \cdot (s_2-s_1)! \cdot \cdots \cdot (n-s_k)!} = \binom{n}{s_1, s_2-s_1, \cdots, n-s_n}.$$

3. If $p = p_1 p_2 \cdots p_n$ has k descents, then its complement p^c clearly has $n-1-k$ descents. Here p^c is the n-permutation defined by $(p^c)_i = n+1-p_i$.

5. Look at the sequence $\{b_i\}_i$ where $b_i = a_i/a_{i-1}$. Then $\{a_i\}_i$ is log-concave if and only if $\{b_i\}_i$ is weakly decreasing, while $\{a_i\}_i$ is unimodal if and only if once $\{b_i\}_i$ gets to a number that is not larger than 1, it never grows back above 1. As this second condition on $\{b_i\}_i$ is obviously weaker than the first, the statement is proved.

7. We prove the statement by induction on n. Our claim is true for $n = 0$ as $z^0\binom{z}{0} = 1$, and $n = 1$ as $z = 0 \cdot \binom{z+1}{1} + 1 \cdot \binom{z}{1}$. Now let us assume that we know the statement for n. Multiply both sides of (1.2) by z to get

$$z^{n+1} = \sum_{k=0}^{n} A(n,k)\binom{z+n-k}{n} z. \qquad (10.1)$$

Now note that

$$\binom{z+n-k}{n} z = k\binom{z+n+1-k}{n+1} + (n-k+1)\binom{z+n-k}{n+1}. \qquad (10.2)$$

DOI: 10.1201/9780429274107-10

Therefore, the right-hand side of (10.1) equation becomes

$$\sum_{k=0}^{n} A(n,k)k\binom{z+n+1-k}{n+1} + A(n,k)(n-k+1)\binom{z+n-k}{n+1} =$$

$$\sum_{k=0}^{n} k\binom{z+n+1-k}{n+1}A(n,k) + (n-k-1)\binom{z+n-k+1}{n+1}A(n,k-1)$$

$$= A(n+1,k)\binom{z+n+1-k}{n+1},$$

where the last step uses the result of Theorem 1.7, with $k-1$ playing the role of k. Comparing this to the left-hand side of (10.1) proves the statement.

9. The crucial observation is that the sum on the right-hand side of the equation to be proved has $n-k+2$ terms. This suggests that we compare it to what we get if we apply Corollary 1.19 to compute $A(n, n-k+1) = A(n,k)$. (The summation 1.19 would have only $n-k+1$ terms, but an additional term in which $r=0$ can be added to it without changing its value.) That Corollary gives

$$A(n, n-k+1) = \sum_{r=1}^{n-k+1} S(n,r)r!\binom{n-r}{n+1-k-r}(-1)^{n+1-k-r}$$

$$= \sum_{r=1}^{n-k+1} S(n,r)r!\binom{n-r}{k-1}(-1)^{n+1-k-r},$$

which, after the substitution $h = n-r$, is just what was to be proved.

11. This is a classic result due to Frobenius [183]. Compare the coefficients of z^i on the two sides. On the left-hand side, it is $A(n,i)$. On the right-hand side, it is

$$\sum_{k=0}^{n-1}\binom{n-k}{i-1}S(n,k)k!(-1)^{n-k-i+1}.$$

Setting $h = n-k$, this is precisely the result of Exercise 9.

13. For each permutation $p \in S_n$, let $f(p)$ be the reverse complement of p, that is, the permutation whose ith entry is $n+1-p_{n+1-i}$. This sets up a bijection from the set of excedances of p onto the set of weak excedances of $(p^c)^r$ *that are less than* n. Indeed, if i is an excedance of p, then $i > 1$, so $n+1-i < n$. On the other hand, n is always a weak excedance of any permutation, so in particular, of $(p^c)^r$. This proves

that the number of n-permutations with $k-1$ excedances is the same as that of n-permutations with k weak excedances. Our statement is then proved by Theorem 1.36.

15. Let $T(z) = \sum_{n\geq 1} T_n \frac{z^n}{n!}$. The reader is invited to verify that $T'(z) = 1+T(z)+T^2(z)/2$ by removing the root of a decreasing non-plane 1-2 tree and analyzing the three possible cases. Solving that differential equation with the initial condition $T(0) = 0$, we get $T(z) = \tan z + \sec z - 1$.

17. Let us insert the entry n into a permutation of length $n-1$ with $k-1$ distinct r-falls so that the number of r-descents does not change. To do this, we can do one of three things: we insert n between two entries p_i and p_{i+1} so that $p_i \geq p_{i+1} + r$, or we insert n in the last position, or we insert n so that it immediately precedes one of the $r-1$ entries that are larger than $n-r$. Altogether, this will give us

$$(k-1+1+r-1)A(n-1,k,r) = (k+r-1)A(n-1,k,r)$$

permutations enumerated by $A(n,k,r)$ in which n is not part of any r-falls.

Now let us insert entry n into a permutation of length $n-1$ with $k-2$ distinct r-falls so that the number of r-descents increases by one. We know from the previous paragraph that there are $k+r-2$ possible insertions of n into each permutation enumerated by $A(n-1,k-1,r)$ that do not increase the number of r-falls. So there are $n-k-r+2$ that do, providing us with $(n-k-r+2)A(n-1,k-1,r)$ permutations enumerated by $A(n,k,r)$ in which n is part of an r-fall. Summing over the two cases, we complete the proof.

19. No, they are not the same unless $r=1=\ell$. Indeed, just by looking at the definitions of these numbers, we have $\sum_{k=1}^{n} A(n,k,r) = n!$ for any fixed r, but we also have $\sum_{k=1}^{n} A_t(n,k,\ell) = n!^\ell$ for any fixed ℓ.

21. Let us first assume that z is a positive integer. Then the left-hand side is just the number of ways to color the elements of $[n]$ so that the color of each element is chosen independently from the set $[z]$. The right-hand side is the same, counted by the number of colors actually used. Indeed, if m colors are used, then they define a partition of $[n]$ into m blocks in $S(n,m)$ ways. Then the colors of the blocks can be chosen in $z(z-1)\cdots(z-m+1) = (z)_m$ ways. This proves the claim if z is a positive integer.

Otherwise, note that both sides of 21 are polynomials in z. They agree for infinitely many values of z (all positive integers), so they must be identical.

23. No, that is false. Indeed, let $P(z) = 1+z+3z^2$ and let $Q(z) = 1+z+4z^2$. Then $P(z)Q(z) = 1 + 2z + 8z^2 + 7z^3 + 12z^4$ fails to be unimodal.

25. (a) Let us assume that p has two alternating runs and starts with an ascent. Such permutations increase on the left of n and decrease on the right of n. Therefore, we can choose the set of entries that precede n in $2^{n-1}-2$ ways, and each such choice will correspond to one permutation. As p^c has the same number of alternating runs as p, this proves $G(n,2) = 2(2^{n-1}-2) = 2^n - 4$, if $n \geq 2$.

(b) One way to find a formula for $G(n,3)$ is by using Lemma 1.38 with $k=3$, and the above result for $G(n,2)$, to get that for $n \geq 3$, we have

$$G(n,3) = 3G(n-1,3) + (2^n - 8) + 2(n-3),$$

$$G(n,3) - 3G(n-1,3) = 2^n + 2n - 14,$$

and then by solving this recurrence using the ordinary generating function of the sequence $G(n,3)$.

To do that, let $G(z) = \sum_{n \geq 3} G(n,3)z^n$. Then the previous equation leads to

$$G(z)(1-3z) = \frac{8z^3}{1-2z} + \frac{2z}{(1-z)^2} - 4z^2 - 2z - \frac{14z^3}{1-z},$$

$$G(z) = \frac{8z^3}{(1-2z)(1-3z)} + \frac{2z}{(1-z)^2(1-3z)} - \left(\frac{4z^2+2z}{1-3z} + \frac{14z^3}{(1-z)(1-3z)}\right).$$

Then we get $G(n,3)$ as the coefficient of z^n on the right-hand side, that is,

$$G(n,3) = 8(3^{n-2} - 2^{n-2}) + \frac{3^{n+1}-2n-3}{2} - 10 \cdot 3^{n-2} - 7(3^{n-2}-1)$$

$$= \frac{3^n + 11}{2} - n - 2^{n+1},$$

for all $n \geq 3$.

27. The identity states that the number of all surjections of domain $[n]$ that have an even-sized image and the number of surjections of domain $[n]$ that have an odd-sized image differ by $(-1)^n$.

In order to prove this, let $f : [n] \to [m]$ be a surjection. Let us first assume that it is not the case that $f^{-1}(m) = \{n\}$. Let $f(n) = i$. There are two cases.

(a) If n is the only element of $[n]$ that is mapped into i by f, then $i \neq m$, and we define $\overline{f}$ by setting $\overline{f}(x) = f(x) - 1$ if $f(x) > i$, and $\overline{f}(x) = f(x)$ otherwise. Note that $\overline{f}$ is a surjection from $[n]$ into $[m-1]$, and that n shares its image with at least one other element.

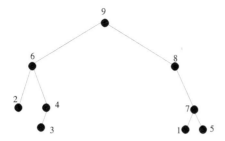

FIGURE 10.1
The decreasing binary tree of 263498175.

(b) If n is not the only element of $[n]$ that is mapped into i by f, then we define $\overline{f}$ by setting $\overline{f}(x) = f(x) + 1$ if $f(x) \geq i$ and $x \neq n$. Note that n is the unique element that is mapped to i by $\overline{f}$. Also note that $\overline{f}$ is a surjection from $[n]$ into $[m-1]$.

Note that the size of the images of f and $\overline{f}$ will always differ by one. If $f^{-1}(m) = \{n\}$, then we set $\overline{f}(n) = m$, and recursively proceed with this definition with $[n-1]$ and $[m-1]$ replacing $[n]$ and $[m]$. The only surjection that will not be matched to another one in this way is the identity map of $[n]$, proving our claim.

29. We prove the statement by induction on k, the initial case of $k = 0$ being trivially true. Now assume the statement is true for k, and prove it for $k+1$. We know from Theorem 1.7 that

$$A(n, k+1) - (k+1)A(n-1, k+1) = (n-k)A(n-1, k).$$

Here the right-hand side satisfies a polynomial recursion by the induction hypothesis, so the left-hand side will satisfy that same polynomial recursion. Rearranging that recurrence relation, we get a recurrence relation for $A(n, k)$.

31. For all $p \in S_n$, define the decreasing binary tree $T(p)$ as follows. The root of $T(p)$ is n, and the left (resp. right) child of n is the largest entry of p on the left (resp. right) of n. Then define the rest of the tree recursively. See Figure 10.1 for an example.

It is clear that T is a bijection from S_n to the set of all decreasing binary trees. Indeed the unique preimage of a decreasing binary tree can be read off the tree in order, that is, for each node, read the left subtree first, then the node itself, and then the right subtree.

33. If a permutation p has exactly k peaks then there are two possibilites.

(a) When $p_1 < p_2$. In this case, if $p_{n-1} > p_n$, then p has $2k$ alternating runs, and if $p_{n-1} < p_n$, then p has $2k + 1$ alternating runs.

(b) When $p_1 > p_2$. In this case, if $p_{n-1} > p_n$, then p has $2k + 1$ alternating runs, and if $p_{n-1} < p_n$, then p has $2k + 2$ alternating runs.

In other words, the set of n-permutations with k peaks consists of the entire set of n-permutations with $2k + 1$ alternating runs, half of the n-permutations with $2k$ alternating runs, and half of the n-permutations with $2k + 2$ alternating runs. This proves

$$Peak(n, k) = G(n, 2k + 1) + \frac{G(n, 2k) + G(n, 2k + 2)}{2}.$$

35. It is easy to prove, by induction, or otherwise, that the number of descents of p is equal to the number of *right edges* of $T(p)$, while the number of ascents of p is equal to the number of *left edges* of $T(p)$. Now it is clear that the symmetry of the sequence $A(n, k)_k$ can be proved by the simple bijection that reflects $T(p)$ through a vertical axis.

Proving unimodality is a more interesting task. Let T be a decreasing binary tree on n nodes. We define a total order of the nodes of T as follows. A node v of T is on *level* j of T if the distance of v from the root of T is j. Then our total order consists of listing the nodes on the highest level of T going from left to right, then the nodes on the second highest level left to right, and so on, ending with the root of T.

Let T_i be the subgraph of T induced by the first i vertices in this total order. Then T_i is either a tree or a forest with at least two components. In the first case, let $g(T_i)$ be the reflected image of T_i through a vertical axis. In the second case, if the components of T_i are $C_1, C_2, \cdots, C_t$, then let $g(T_i) = (g(C_1), g(C_2), \cdots g(C_2))$. Now prove that if $k \leq \lfloor (n-3)/2 \rfloor$, then i can always be chosen so that $f(T_i)$ has exactly one more right edge than T_i. Then an injection from the set of decreasing binary trees on n vertices with k right edges into the set of decreasing binary trees on n vertices with $k + 1$ right edges can be defined by finding the *smallest* such i, then replacing T_i by $g(T_i)$ in T.

37. (a) The first proof of this fact is due to Gábor Hetyei and Ethan Reiner [215], who used exponential generating functions and partial differential equations to get this result. The proof we present is combinatorial.

For any $i \leq n - 1$, either p_i is an ancestor of p_{i+1}, or p_{i+1} is an ancestor of p_i, or else p_i and p_{i+1} would have a common ancestor, which would put an index *between* i and $i + 1$. In particular, T_p^m cannot have both p_i and p_{i+1} as leaves. This makes the following definition meaningful. Let $i \in [n-2]$. Then the *ith local extremum* of a permutation p is the entry that is closest to the root of the minmax tree of p among p_i, p_{i+1}, and p_{i+2}. Denote this entry by e_i. Note that e_i always exists.

Let us assume without loss of generality that the entry 1 of p precedes the entry n of p. We can do that as the minmax tree of p^c is isomorphic to T_p^m.

First we are going to prove that p_1 is a leaf in $n!/3$ minmax trees. The simplest scenario is when the entry 1 is among the leftmost three entries, so in particular, $e_1 = 1$. This gives rise to three subcases:

- If $p_1 = 1$, then p_1 is the root of the minmax tree.
- If $p_2 = 1$, then p_2 is the root, and p_1 is a leaf.
- If $p_3 = 1$, then p_3 is the root, its left subtree has p_1 and p_2 as nodes, and among these, by definition, p_2 is a leaf, and p_1 is not.

Clearly, these cases are equally likely to occur, so p_1 will be a leaf with probability $1/3$ in this case.

Now let us assume that the entry 1 of the permutation is not among the first three elements. This entry is the root of T_p^m, and its left subtree has at least three nodes. Let these nodes be $b_1 < b_2 < \cdots < b_k$. Repeat the previous argument for this subtree, with b_1 playing the role of 1; now if b_1 is among the leftmost three elements of the permutation, p_1 is a leaf with probability $1/3$. Iterate this algorithm. It will eventually stop because we either get a left subtree of size three, or a subtree whose minimal entry is among the first three ones. This proves that there are $n!/3$ minmax trees on $[n]$ in which p_1 is a leaf.

The proof for general p_i is similar. Assume again that 1 is on the left of n in p, and let $i \in [n-2]$. If $1 \in \{p_i, p_{i+1}, p_{i+2}\}$, then there are three possibilities.

- If $p_i = 1$, then p_i is the root of the minmax tree.
- If $p_{i+1} = 1$, then p_{i+1} is the root, so p_i is the rightmost element of its left subtree, and as such, it is necessarily a leaf.
- If $p_{i+2} = 1$, then p_{i+2} is the root, p_i is the next-to-last element of the root's left subtree, and as such, it is always an internal node (having the leaf p_{i+1} for its only child).

Again each of these subcases occurs with probability $1/3$.

If $1 \notin \{p_i, p_{i+1}, p_{i+2}\}$, then we can proceed as above. That is, look for the entry 1 of p, then only consider the subtree that contains the positions i, $i+1$, and $i+2$. If 1 is not in any of these positions, then all three of them are in the same subtree. Iterating this algorithm, we eventually reach a subtree where we can apply the above method. The structures of the other subtrees do not influence whether p_i is a leaf or not, so p_i is a leaf with probability $1/3$. This completes the proof.

(b) Clearly, p_n is always a leaf because it cannot be the leftmost in any comparison, thus it cannot have descendants. Similarly, p_n is always the child of p_{n-1}, thus p_{n-1} is never a leaf.

39. Any lattice path included in our sum must end either in a horizontal step having label k or a vertical step having label $n - k + 1$. Then use induction and Theorem 1.7.

41. Similar to the solution of Exercise 39, just this time we have to use the recurrence proved in Exercise 17.

43. Note that when we insert two ns into an element of Q_{n-1}, we must insert them into consecutive positions. All the claims now follow by analyzing how many times this insertion increases the value of the statistic at hand.

45. Reversing the order of the last two entries of a permutation p changes as(p) by plus or minus one, and so it changes its parity.

10.2   Solutions for Chapter 2

1. There are two known solutions of these results. One [321] is by a combinatorial involution, and the other [142] is by a generating function argument. These solutions generalize in different directions.

3. As we discussed, $b(n, 3)$ is equal to the coefficient of x^3 in the polynomial $I_n(z) = (1+z)(1+z+z^2)\cdots(1+z+\cdots+z^{n-1})$; in other words, $b(n, 3)$ is the number of weak compositions of 3 into $n - 1$ parts, so that the first part is at most 1, and the second part is at most 2.

   There are $\binom{n+1}{3}$ weak compositions of 3 into $n-1$ parts. One has second part 3, one has first part 3, and $n-2$ of them have first part 2. Therefore, $b(n, 3) = \binom{n+1}{3} - n$.

5. The number of pairs of entries (x, y) so that $x > y+1$ is $\binom{n}{2} - (n - 1) = \binom{n-1}{2}$. Each such pair appears in reverse order in half of all permutations, so the answer is $n!\binom{n-1}{2}/2$.

7. The number of such trees with zero inversions is $n!$. Figure 10.2 shows the six trees with root 0, non-root vertex set [3], and no inversions. To prove this statement bijectively, let p be an n-permutation, and define $H(p)$ to be the tree on vertex set $[n] \cup 0$ defined as follows. If i is an entry of p, then the unique parent of i in $H(p)$ is the vertex j, where j is the closest entry in p that is on the left of i, and that is smaller than

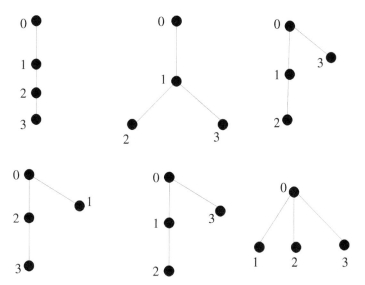

FIGURE 10.2

The six trees with no inversions.

i. If there is no such entry j, then the parent of i is 0. The reader is invited to verify that H is indeed a bijection.

9. The following proof was given by David Bressoud and Doron Zeilberger [99]. Let $\lambda = (\lambda_1, \lambda_2, \cdots, \lambda_t)$ be a partition of $n - a(j)$, for some even j. Define $\phi(\lambda)$ by

$$\phi(\lambda) = \begin{cases} (t + 3j - 1, \lambda_1 - 1, \lambda_2 - 1, \cdots \lambda_t - 1), & \text{if } t + 3j \geq \lambda_1, \\ \\ (\lambda_2 + 1, \cdots \lambda_t + 1, 1, 1, \cdots, 1) & \text{if } t + 3j < \lambda_1. \end{cases}$$

If the second rule is used, then there are $\lambda_1 - 3j - t - 1$ parts equal to 1 added at the end.

Note that the first rule sends λ into a partition of $n - a(j) + 3j = n - a(j-1)$, and the second rule sends λ into a partition of $n - a(j) - 3j - 2 = n - a(j+1)$. So in both cases, ϕ maps from $\cup_{j \text{ even}} Par(n - a(j))$ into $\cup_{j \text{ odd}} Par(n - a(j))$. Finally, note that applying ϕ twice is the identity, therefore ϕ has an inverse, and must be a bijection.

Another solution was given by Adriano Garsia and Steve Milne [188].

11. Yes. We prove the statement by induction on k, the initial case of $k = 0$ being trivial. Suppose we know that the statement is true for $k - 1$. Lemma 2.5 tells us that $b(n + 1, k) - b(n, k) = b(n + 1, k - 1)$ if $k \leq n$. From this, it is easy to see that $b(n, k)$ is also p-recursive. Finally, we mention that the constraint $k \leq n$ is not a reason for concern. Indeed,

it is obvious that if f and g are polynomially recursive functions, then so is $f + g$. On the other hand, for any fixed k, the difference of the functions $b(n + 1, k - 1)$ and $b(n + 1, k) - b(n, k)$ is nonzero for a finite number of values of n only, and is therefore certainly p-recursive.

13. There are $b(n, k)$ such n-tuples. Indeed, let $p = p_1 p_2 \cdots p_n$ be a permutation, and define $b(p)_i$ to be the number of indices $j < i$ so that $p_i < p_j$. Then clearly $0 \le b(p)_i \le i - 1$, and the map $b : S_n \to B_n$ defined by $b(p_1 p_2 \cdots p_n) = (b(p)_1, b(p)_2, \cdots, b(p)_n)$ is a bijection because the only preimage of $(b_1, b_2, \cdots, b_n)$ can be built up from left to right. The value of b_i reveals the relative size of p_i among the first i entries of p. Finally, note that $\sum_{i=1}^{n} b(p)_i = i(p)$ as $b(p)_i$ is the number of inversions whose first element is p_i. Therefore, the set of permutations with k inversions is mapped to the subset of B_n in which $\sum_{i=1}^{n} b_i = k$.

We note that $(b(p)_1, b(p)_2, \cdots, b(p)_n)$, or a trivial transformation thereof, is often called the *inversion table of* p.

15. In an n-permutation with k inversions, the entry n can be part of i inversions, with $0 \le i \le k$, so

$$b(n, k) = \sum_{i=0}^{k} b(n - 1, k - i).$$

17. Such a permutation is either alternating, or reverse alternating, but in both cases, it has k descents and k ascents. Therefore, it has to have at least k inversions and at least k noninversions. These are both attainable, with the permutations $13254 \cdots (2k+1)(2k)$ and its reverse.

19. The main idea of this solution can be found in [174]. Just as in the solution of Exercise 13, we are bijectively encoding a permutation by an n-tuple $(d_1, d_2, \cdots, d_n)$ of nonnegative integers satisfying $d_k \le k - 1$. This will prove that den is Mahonian. We define the d_k as follows. Let $p = p_1 p_2 \cdots p_n$ be an n-permutation. For fixed k, let

$$d_k = d_k(p) = \begin{cases} |\{l < k \text{ so that } p_k < p_l \le k\}| \text{ if } p_k \le k, \\[2mm] |\{l < k \text{ so that } p_l \le k\}| + |\{l < k \text{ so that } p_k < p_l\}| \\ \text{if } p_k > k. \end{cases}$$

It is easy to check that $\sum_{i=1}^{n} d_i(p) = den(p)$. All we have to show is that p can be recovered from its *Denert-table*, that is, from the n-tuple $(d_1(p), d_2(p), \cdots, d_n(p))$. Indeed, $p_n = n - d_n$. Now assume that the last $n - k$ elements of p have already been recovered. Then recover p_k as follows. Look at the list $k, k - 1, \cdots, 1, n, n - 1, \cdots, k + 1$, and delete all the entries that have already been assigned to a position in p. Then p_k has to be chosen so that there are exactly $d_k(p)$ entries on its left.

21. Let us multiply both sides of the identity by $[(n-k)]![k]!$ to get

$$[n]! = [n-1]![k]q^{n-k} + [n-1]![n-k].$$

Now let us divide both sides by $[n-1]!$ to obtain

$$[n] = [k]q^{n-k} + [n-k],$$

which follows directly from the definition of $[m]$.

23. This can be proved by repeated applications of the result of Exercise 21, but we prefer a combinatorial argument. The left-hand side provides the generating function for all k-subsets of $[m]$ according to their subset sums.

A typical term of the right-hand side is of the form $q^{m-k-j+1} \begin{bmatrix} m-j \\ k-1 \end{bmatrix}$, where $j \in [m-k]$. We claim that this term will provide the generating function for those k-subsets of $[m]$ whose largest element is equal to $m-j+1$. Indeed, the rest of such a subset is a $(k-1)$-subset of the set $[m-j]$. The term $q^{m-j-k+1}$ corrects the shift in the definitions of a_i and c_i, as seen in the proof of Theorem 2.27.

25. We claim that there is a bijection between the Ferrers shapes consisting of n boxes that fit within an $i \times k$ rectangle and k-subsets of $[i+k]$ whose sum of elements is $n + \binom{k+1}{2}$. Indeed, let F be the Ferrers shape of a partition of n into at most i parts of size at most k. Let the row lengths of F be $(f_1, f_2, \cdots, f_k)$ in nonincreasing order, where some f_i may be equal to 0. Then $\{f_1 + k, f_2 + k - 1, \cdots, f_k + 1\}$ is a k-element subset of $[i+k]$, and the sum of its elements is $n + \binom{k+1}{2}$. As this map is obviously a bijection, our statement is proved by Theorem 2.27.

27. See [333] for a bijective proof of this fact.

29. If n is odd, then we recall that $\begin{bmatrix} n \\ k \end{bmatrix}$ is the number of k-dimensional subspaces of an n-dimensional vector space over a q-element field. Matching each such subspace with its orthogonal complement, we get that $\sum_{k=0}^{n} (-1)^k \begin{bmatrix} n \\ k \end{bmatrix} = 0$.

If $n = 2m$ is even, then we claim that

$$\sum_{k=0}^{n} (-1)^k \begin{bmatrix} n \\ k \end{bmatrix} = (1-q)(1-q^3) \cdots (1 - q^{2m-1}).$$

A computational proof of this fact can be found in [13]. Recently, a more combinatorial proof was given in [139].

31. Consider the set of insertions that increase the number of descents of p and the insertions that increase the number of ascents of p separately. Show that both sets of insertions create a set of permutations with distint major indices, and that both sets of these major indices will form an interval.

33. Let G be a graph on five vertices, so that vertices A, B, C, and D form a complete graph except that BC is not an edge, and let BE be the last edge of G.

35. Let G be the inversion graph of the n-permutation p. Let P_p be a poset on vertex set $[n]$ so that $i <_P j$ if $i < j$ as integers and i precedes j in p. Then the comparability graph of P_p is precisely G. So all inversion graphs are comparability graphs.

For a comparability graph that is not an inversion graph, consider the poset Q with elements $\{A_1, \cdots, A_4, B_1, \cdots, B_4\}$, in which $A_i < B_j$ if $i \neq j$, and there are no other comparable pairs. Let G be the comparability graph of Q. Then G is a connected graph in which each vertex has degree three.

When we try to construct a permutation $p = p_1 \cdots p_8$ whose inversion graph is G, we quickly notice that the only way in which p_1 and p_8 can each be of three inversions is by $p_1 = 4$ and $p_8 = 5$. Similarly, $p_4 = 1$ and $p_5 = 8$ must hold. After this, it is straightforward to verify that the only candidate for p that satisfies the requirement that each p_i is part of three inversions is $p = 43218765$. However, the inversion graph of that permutation is the disjoint union of two complete graphs on four vertices, hence it is not isomorphic to G.

10.3   Solutions for Chapter 3

1. Interchanging the first two elements of a permutation p either increases or decreases the number of inversions of p. In either case, it changes the parity of p. So exactly half of all permutations are odd, and half are even.

3. Let (a, b) be an inversion of p. That means that looking at p as a function from $[n]$ to $[n]$, we have $p(i) = a$ and $p(j) = b$, with $a > b$ and $i < j$. Then, by the definition of the inverse, we also have $p^{-1}(a) = i$ and $p^{-1}(b) = j$, proving that (j, i) is an inversion of p^{-1}. This sets up a bijection between the inversions of p and p^{-1}.

5. Let $f(p) = 1$ if p is even, and let $f(p) = -1$ if p is odd. In other words, $f(p) = \det A_p$. Note that this map f is often called *sign*.

7. We claim that f must be the identity permutation. Let us assume that this is not the case, then f contains a k-cycle $(f_1 f_2 \cdots f_k)$, with $k > 1$. Let us first assume that $k > 2$. Let $g = (f_1 f_2)$. Then $f \cdot g \neq g \cdot f$. Indeed, $g(f(f_1)) = g(f_2) = f_1$, while $fg(f_1) = f(f_2) = f_3$. If, on the other hand,

f has no cycles longer than two (that is, $k = 2$), then let $g = (f_2 f_3)$, where f_3 is any element outside the cycle $(f_1 f_2)$. Then again $f \cdot g \neq g \cdot f$. Indeed, $g(f(f_1)) = g(f_2) = f_3$, whereas $f(g(f_1)) = f(f_1) = f_2$.

9. Such a permutation has a unique longest cycle C, of length k. We have $\binom{n}{k}(k-1)!$ choices for this cycle, then $(n-k)!$ choices for the rest of the permutation. Therefore, our total number of choices is

$$\binom{n}{k}(k-1)!(n-k)! = \frac{n!}{k}.$$

11. As each permutation is a product of its cycles, it suffices to prove our statement for permutations that consist of one cycle only. This is not difficult to do by induction as

$$(a_1 a_2 \cdots a_k) = (a_1 \cdots a_{k-1})(a_{k-1} a_k).$$

13. We claim that $n = 3$ is the only such value. Indeed, S_2 has two conjugacy classes of size 1. If $n \geq 5$, then S_n has two conjugacy classes consisting of $\frac{n!}{4(n-4)}$ elements. These are those that belong to types $(2, 1, 0, 0, \cdots, 0, 1, 0, 0, 0)$ and $(0, 0, 0, 1, \cdots, 0, 1, 0, 0, 0)$. For $n = 4$, these classes both have six elements.

We point out that much more is known. F. Markel [251] conjectured in 1973 that S_3 was the only finite *solvable* group that had no conjugacy classes of the same size. This conjecture stayed open for twenty years, and was then proved independently in [228] and [340]. However, it is not known at this time whether the condition that the group be solvable can be dropped.

15. Multiplying both sides by f_T from the right, we get that our formula is equivalent to

$$f_T C_T = (123 \cdots n) f_T.$$

Compute where each sides takes the node k. For the left-hand side, we get

$$f_T C_T(k) = C_T(f_T(k)) = C_T(C_T^{k-1}(1)) = C_T^k(1) = f_T(k+1).$$

For the right-hand side, we get

$$(123 \cdots n) f_T(k) = f_T((123 \cdots n)(k)) = f_T(k+1).$$

17. Clearly, $c(n, 1) = (n-1)!$ as such permutations have type $(0, 0, \cdots, 0, 1)$. We have $c(n, n-1) = \binom{n}{2}$ as all such permutations must have type $(n-2, 1, 0, \cdots, 0)$, and we have $\binom{n}{2}$ possibilities for the single 2-cycle.

19. We claim that

$$S(n, k) = \frac{n!}{k!} \sum_{r_1 + r_2 + \cdots + r_k = n} \frac{1}{r_1! r_2! \cdots r_k!},$$

where the r_i are positive integers. Indeed, order the elements of $[n]$ in one of $n!$ ways, then insert a bar after the first r_1 elements, then the next r_2 elements, and so on. This provides a partition counted by $S(n, k)$. However, each such partition will be obtained in $k! \cdot r_1! r_2! \cdots r_k!$ ways, as the order of the elements within each block does not matter, and the order of the blocks does not matter.

21. This approach to Stirling numbers was studied in [222]. Both polynomials have degree $2k$ and leading coefficient $\frac{(2k-1)!!}{(2k)!}$. This is straightforward to prove by induction, using the recurrence relations given in Exercise 8 of Chapter 1 and Lemma 3.19.

23. We will show that the right-hand side also counts all permutations of length $n + 1$ with $k + 1$ cycles. Let the entry $n + 1$ be part of an $(n - m + 1)$-cycle C. Then we have $\binom{n}{n-m}$ ways to choose the $n - m$ remaining elements of C, and we have $(n - m)!$ ways to choose C on these elements. Then, we have $c(n, m)$ ways to choose the rest of the permutation. Summing over m, we get the identity to be proved.

25. We claim that

$$c_3(n, k) = (n - 1)c_3(n - 1, k) + (n - 1)(n - 2)c_3(n - 3, k - 1),$$

with $c_3(0, 0) = 1$. Indeed, in a permutation enumerated by $c_3(n, k)$, the entry n is either in a 3-cycle, or in a larger cycle. There are $n - 1$ permitted ways to insert n into a gap position of a permutation counted by $c_3(n - 1, k)$ as the last gap position is forbidden (it would create a 1-cycle). The permutations obtained this way will contain n in a cycle longer than three.

Otherwise, there are $\binom{n-1}{2}$ ways to choose two entries that can share a 3-cycle with n, then there are two possible 3-cycles involving n and the two chosen elements.

27. Let $p(k, n)$ be the probability that we draw k white balls in n trials. That event can occur in two different ways, either we get the $k - 1$ white balls during the first $n - 1$ trials, or we get only k white balls during the first $n - 1$ trials, and the last one during the last trial. This leads to the recurrence relation

$$p(k, n) = \frac{k}{m} p(k, n - 1) + \frac{m - k + 1}{m} p(k - 1, n - 1).$$

Indeed, if the first a trials resulted in drawing b white balls, then there are $m - b$ white balls, and b black balls in the box.

This triangular recurrence is somewhat similar to that of the Stirling numbers of the second kind. To grasp this connection better, set

$$p(k, n) = d(n, k)\frac{(m)_k}{m^n}.$$

Then the numbers $d(n, k)$ satisfy the same recurrence as the numbers $S(n, k)$, and fulfill the same initial conditions. Therefore, $d(n, k) = S(n, k)$, and thus $p(k, n) = S(n, k)\frac{(m)_k}{m^n}$.

29. This is a simple application of Corollary 3.48. We take a partition of $[n]$ into k parts, but we do not put any structure on any of the blocks. The only requirement is that the blocks are not empty. Therefore, $F_i(u) = e^u - 1$ for all $i \in [k]$. Finally, unlike in Corollary 3.48, our partitions are *unordered*, so their number is only $1/k!$ times what it would be if they were ordered. This yields

$$g_k(u) = \sum_{n=k}^{\infty} S(n, k)\frac{u^n}{n!} = \frac{1}{k!}(\exp u - 1)^k.$$

31. By repeated applications of Proposition 3.12, we have

$$c(p, k) = \sum_{a_1+2a_2+\cdots+pa_p=p} \frac{p!}{a_1!a_2!\cdots a_p!1^{a_1}2^{a_2}\cdots p^{a_p}}.$$

As p is a prime, no positive integer smaller than p is divisible by p. Therefore, the denominator of no summand on the right-hand side is divisible by p. Indeed, we must always have $a_p = 0$, otherwise we would have a permutation with one cycle only, and that is not allowed. As $p!$ is divisible by p, the right-hand side is the sum of several integers, each of which is divisible by p.

33. Let $p > n > 1$, and let the matrices s and S be defined as in Theorem 3.30. As $s \cdot S = I$, the nondiagonal entries of this matrix are zero, so for $p \neq k$, we have

$$0 = \sum_{i=1}^{p} s(p, i)S(i, k) = \sum_{i=k}^{p} s(p, i)S(i, k).$$

We know from Exercises 31 and 32 that $s(p, i)$ is divisible by p unless $i = 1$ or $i = p$. Therefore, the only summand of the far-right-hand side in which the $s(p, i)$ term is not divisible by p is $s(p, p)S(p, k) = S(p, k)$. As the right-hand side is divisible by p, so too must be $S(p, k)$.

35. As $n \geq 1$, we can divide (3.3) by z to get

$$(z+1)\cdots(z+n-1) = \sum_{k=1}^{n} c(n,k)z^{k-1}.$$

Now equate the coefficients of z^{k-1} on the two sides.

37. Let A_i be the set of n-permutations in which i is a fixed point. Then $|A_{i_1} \cap \cdots \cap A_{i_k}| = (n-k)!$, and the result follows by the Principle of Inclusion–Exclusion.

39. One of these sets is always one larger than the other. For odd n, the set of n-permutations with exactly one fixed point is larger. For even n, the set of derangements of length n is larger.

To see this, let $G(n)$ be the number of n-permutations with exactly one fixed point, and note that $G(n) = nD_{n-1}$ holds for all $n \geq 3$. Indeed, to find a permutation counted by $G(n)$, first choose its only fixed point, then choose a derangement on the remaining $n-1$ entries. On the other hand, it follows from Corollary 3.57 that $D_n = nD_{n-1} + (-1)^n$. This proves our claim.

41. The notion of desarrangements, and the proofs below, are due to Jacques Désarmenien [141].

(a) Let p be a desarrangement of length n. Delete its last entry, and relabel the remaining entries accordingly. The obtained permutation p' is always a desarrangement of length $n - 1$, *except* when $p = n(n-1)\cdots 21$ and n is even. Conversely, each p' is obtained from n different n-desarrangements p this way, except for $p' = (n-1)(n-2)\cdots 21$ when n is odd. The latter is only obtained from $n-1$ desarrangements of length n as 1 could not be the last entry.

(b) A routine computation shows that the numbers D_n as given by Corollary 3.57 satisfy the same recurrence relation as the numbers $J(n)$ have been shown to satisfy in part [(a)] of this exercise. As $J(1) = D_1 = 0$, the proof follows.

43. The entry 1 of a derangement p of length n can be part of a 2-cycle or a larger cycle. There are $(n-1)$ other elements it can form a 2-cycle with, and then the remaining $n-2$ elements can form a derangement in D_{n-2} ways. On the other hand, if 1 is not to be in a 2-cycle, then we can just insert it in any gap position of any $(n-1)$-derangement (taken on the set $\{2, 3, 4, \cdots, n\}$) except the one that would put this entry into its own 1-cycle. This provides $(n-1)D_{n-1}$ additional derangements.

45. We will repeatedly use the triangular recurrence relation $c(m, k) = c(m-1, k-1)+(m-1)c(m-1, k)$. Subtract $n+k-1$ times the next-to-last row from the last row. Then the ith element of the last row becomes $c(n+k, i) - (n+k-1)c(n+k-1, i) = c(n+k-1, i-1)$. Now subtract $n+k-2$ times row $(k-2)$ from row $k-1$. This results in a row $k-1$ whose ith element is $c(n+k-1, i) - (n+k-2)c(n+k-2, i) = c(n+k-2, i-1)$. Continue in this way for all rows. We get the matrix

$$B = \begin{pmatrix} c(n+1, 1) & c(n+1, 2) & \cdots & c(n+1, k) \\ c(n+1, 0) & c(n+1, 1) & \cdots & c(n+1, k-1) \\ \cdots & \cdots & \cdots & \cdots \\ c(n+k-1, 0) & c(n+k-1, 1) & \cdots & c(n+k-1, k-1) \end{pmatrix},$$

that is, a matrix whose first row is identical to that of C_n, but in which the jth element of row i is $c(n+i-1, j-1)$, for $i > 1$. Expanding this matrix with respect to the first column and using induction (the $(n-1) \times (n-1)$ minor in the lower right corner is C_{k-1}), we get our claim.

47. List the elements of $[2n]$ in any of $(2n)!$ ways, then insert bars after every two elements. This will result in a fixed point-free involution. On the other hand, we obtain each such involution $n! \cdot 2^n$ times as we can change the order of the elements within each pair, and we can change the order of the pairs without changing the involution itself.

49. We use the Compositional formula, with $g(m) = m!$ and $f(k) = (k-1)!$ for $k \geq 1$. This yields $F(z) = -\ln(1-z)$, $G(z) = 1/(1-z)$, and so

$$H(z) = G(F(z)) = \frac{1}{1+\ln(1-z)}.$$

51. In such a permutation, each cycle length must be either one or three. Therefore, by Theorem 3.53, we get

$$G_n(z) = \exp\left(z + \frac{z^3}{3}\right) = \sum_{i \geq 0} \frac{z^i}{i!} \sum_{j \geq 0} \frac{z^{3j}}{j! 3^j}$$

$$= \sum_{n \geq 0} \frac{z^n}{n!} \sum_{j=1}^{[n/3]} \frac{n!}{3^j j!(n-3j)!},$$

showing that $g_n(C) = \sum_{j=1}^{[n/3]} \frac{n!}{3^j j!(n-3j)!}$. The reader is urged to find a combinatorial explanation for this formula.

53. By repeated applications of the method seen in Example 3.64, we get

$$G(z) = \cosh(z) \exp\left(\frac{z^2}{2}\right) \cosh\left(\frac{z^3}{3}\right) \exp\left(\frac{z^4}{4}\right) \cdots$$

$$= \exp\left(\sum_{n=1}^{\infty} \frac{z^{2n}}{2n}\right) \prod_{i\geq 1} \cosh\left(\frac{z^{2i-1}}{2i-1}\right)$$

$$= \sqrt{\frac{1}{1-z^2}} \prod_{i\geq 1} \cosh\left(\frac{z^{2i-1}}{2i-1}\right).$$

55. (This solution is due to Dennis White [53].) Let $p \in ODD(2m)$. Denote by $C_1, C_2, \cdots, C_{2k}$ the cycles of p in canonical order. We construct a bijection $\Phi : ODD(2m) \to EVEN(2m)$ as follows. For all i, $1 \leq i \leq k$, take the last element of C_{2i-1}, and put it at the end of C_{2i} to get $\Phi(p)$. For example, if $p = (4)(513)(726)(8)$, then $\Phi(p) = (5134)(72)(86)$. Note that if C_{2i-1} is a 1-cycle, it disappears, and that the canonical cycle structure of p is maintained.

To see that Φ is a bijection, it suffices to show that for all $\sigma \in S_n$, we can recover the only permutation $p \in ODD(2m)$ for which $\Phi(p) = \sigma$. While recovering p, we must keep in mind that it might have more than h cycles, because some of its 1-cycles might have been absorbed by the cycles immediately after them. If the last value in c_h is larger than the first value in c_{h-1}, then create a 1-cycle with this value, placing it in front of c_h and repeat the whole procedure using c_{h-2} and c_{h-1}. Otherwise, move this value from c_h to the end of c_{h-1} and repeat the whole procedure using c_{h-3} and c_{h-2}. If at any point only one cycle remains, create a 1-cycle with the last value in that cycle. It is then straightforward to check that the permutation p obtained this way fulfills $\Phi(p) = \sigma$. It also follows from the simple structure of Φ that at no point of the recovering procedure could we have done anything else.

57. By Theorem 3.53, the exponential generating function of these permutations is given by

$$G_C(x) = \exp\left(\sum_{n\neq kr} \frac{z^n}{n}\right).$$

The argument of exp on the right-hand side has to be computed a little bit differently from the special case of $k = 2$, which was covered in Corollary 3.59.

We have

$$\sum_{n\neq kr} \frac{z^n}{n} = \sum_{n\geq 1} \frac{z^n}{n} - \sum_{n} \frac{z^{kn}}{kn}$$

Therefore,

$$G_C(z) = \exp\left(\ln(1-z)^{-1} - \frac{1}{k}\ln(1-z^k)^{-1}\right) = \frac{(1-z^k)^{1/k}}{1-z}.$$

Now note that $1 - z^k = (1-z)(1 + z + z^2 + \cdots + z^{k-1})$, from which we get that

$$G_C(z) = (1 + z + z^2 + \cdots + z^{k-1})(1-z^k)^{(1-k)/k}. \qquad (10.3)$$

Let us assume for brevity that $n = mk$. Applying the binomial theorem, we get

$$(1-z^k)^{(1-k)/k} = \sum_{m\geq 0} (-1)^n z^{km} \binom{\frac{1-k}{k}}{m}$$

$$= \sum_{m\geq 0} (-1)^m z^{km} \frac{1}{m!} \cdot \frac{1-k}{k} \cdot \frac{1-2k}{k} \cdots \frac{1-(m-1)k}{k}$$

$$= \sum_{m\geq 0} z^{km} \frac{1}{m!} \cdot \frac{k-1}{k} \cdot \frac{2k-1}{k} \cdots \cdots \frac{(m-1)k-1}{k},$$

where $n = mk$.

So the coefficient of $\frac{z^n}{n!} = \frac{z^{km}}{(km)!}$ in $(1-z^k)^{(1-k)/k}$ is

$$a_n = \frac{(kn)!(k-1)(2k-1)\cdots((m-1)k-1)}{m!k^m}$$
$$= 1 \cdots (k-2)(k-1)^2(k+1)\cdots(2k-1)^2(2k+1)\cdots(n-1).$$

It follows immediately from (10.3) that a_n is also the coefficient of $a_n/n!$ in $G_C(x)$, in other words, $a_n = g_C(n)$. Finally, if n is not divisible by k, then the only difference is that the long product in our last formula has last term n, not $n-1$.

59. Take a pair $(\pi, k) \in ODD(2m+1) \times [2m+1]$, and insert $2m+2$ into the kth gap position. Note that this implies $2m+2$ cannot create a singleton cycle as it cannot go into the last gap position. Take away the cycle C containing $2m+2$, and run Φ of the solution of Exercise 55 through the remaining cycles. Then, together with C, we have a permutation in $EVEN(2m)$. Run it through Φ^{-1} to get $\tau(\pi, k) \in ODD(2m+2)$.

We claim that τ is indeed a bijection. To get the unique preimage of $\pi' \in ODD(2m+2)$ under τ, run π' first through Φ. This way $2m+2$ gets into the first position of an even cycle, and therefore it indicates a gap position, which is not the last one, and thus we recover k. Remove $2m+2$, leave its cycle intact, and run the remaining even cycles through Φ^{-1} to get $\tau^{-1}(\pi')$.

61. This argument is due to Dennis White [53]. In this solution, we are using the bijection $\Psi : ODD(2n) \times [2n+1] \to ODD(2n+1)$ of the solution of Exercise 58. If the reader has not solved that exercise yet, he is urged to do so now. As a hint, he may use the solution of Exercise 59. The maps Φ used in the solution of Exercise 58 and 59 are very similar.

Let $\mathbf{SQ(m)}$ denote the set of all m-permutations with a square root. We are going to construct a bijection κ from $\mathbf{SQ(2n)} \times [2n+1]$ onto $\mathbf{SQ(2n+1)}$. As the growth of $|\mathbf{SQ(n)}|$ is equal to that of $|ODD(n)|$ when passing from $2n$ to $2n+1$, we try to integrate the bijection $\Psi : ODD(2n) \times [2n+1] \to ODD(2n+1)$ into κ, by "stretching" the odd cycles part of our permutations. We proceed as follows.

Let $(\pi, k) \in \mathbf{SQ(2n)} \times [2n+1]$. Take π, and break it into even cycles part and odd cycles part, or, for brevity, odd part and even part. Let k mark a gap position in π. If this gap position is in the odd part, or at the end of π, then interpret the gap position as a gap position for the odd part only, and simply run the odd part and this gap position through Ψ to get $\kappa(\pi)$, together with the unchanged even parts. Note that $2n+1$ will appear in an odd cycle when we are done.

If the gap position marked by k is in one of the even cycles, say c, we can think of it as marking the member of c immediately following it, say x. Replace x by $2n+1$ in c. To keep the information encoded by x, we interpret x as a gap position in the odd part of π. Indeed, if x is larger than exactly $i-1$ entries in the odd part, then let us mark the ith gap position in the odd cycles part. So now we are in a situation like in the previous case, that is, the gap position is in the odd part.

Run the odd part and this gap position through Ψ. Instead of inserting $2n+1$ to the marked position, however, insert temporarily a symbol B, to denote a number larger than all entries in the odd part. Then decrease all entries in the odd part that are larger than x (including B) by one notch. The obtained odd cycles and the unchanged even cycles (except for the mentioned change in c) give us $\kappa(\pi)$. Note that $2n+1$ will be in an even cycle when we are done.

Now we show that κ is indeed a bijection. First, it is clear that κ maps into $\mathbf{SQ(2n+1)}$. Indeed, (π) and $\kappa(\pi)$ have the same number of cycles of each even length, so $\pi \in \mathbf{SQ(2n)}$ implies $\kappa(\pi) \in \mathbf{SQ(2n+1)}$.

For example, let $\pi = (31)(65)(742)(8)$ and let $k = 3$. Then k marks the entry 6. So we replace 6 by 9, get the new even part $(31)(95)$, and turn to the odd part, $(742)(8)$. In it, the entry 6 marks the third gap position as it is larger than two entries, 2 and 4. So we have to apply Ψ to $((742)(8), 3)$. When we do that, first we get $\Phi((742)(8)) = (74)(82)$, then we insert B into the third gap position to get $(74)(B82)$. Now we decrease the entries larger than 6 by one notch: B to 8, 8 to 7, 7 to 6, to get $(64)(872)$. Finally, we apply Φ^{-1} to (64) and complete the odd part of $\kappa(\pi)$, that is $(4)(6)(872)$. With the previously obtained even part, this yields that $\kappa(\pi) = (31)(4)(6)(872)(95)$.

To get the reverse of κ, take a permutation $\pi' \in \mathbf{SQ(2n+1)}$, and locate $2n+1$. If it is in an odd cycle, then run the odd cycles through Ψ^{-1}. This will yield an odd part one shorter, and an element of $[2n+1]$. Putting this together with the unchanged even part, we get $\kappa^{-1}(\pi')$.

If $2n+1$ is in an even cycle, then run the odd part through Ψ^{-1}. This will specify a gap position in the odd part, and so we recover the entry x. Increase entries larger than x by one notch in the odd part. To get the even part, put x back to the place of $2n+1$. The gap position immediately preceding $2n+1$ is our k in $\kappa^{-1}(\pi')$.

63. The trials of Exercise 6 of Chapter 1 are not independent in the sense that Lévy's theorem requires them to be. That is, for Lévy's theorem to be applicable, we have to define what a *success* is in these trials. Clearly, a success has to be defined as the event that the ball currently placed goes into a box that was previously empty (this is how the numbers of empty boxes will equal the Eulerian numbers). However, with that definition, the trials are not independent as the probability of success on trial i does depend on the number of successes on the previous trials if $i \geq 3$. Therefore, Lévy's theorem does not apply, except when $n \leq 2$.

65. Take the derivative of both sides, then set $z = 1$. We get the identity

$$\sum_{k=1}^{n} kc(n,k) = \sum_{i=0}^{n-1} \frac{n!}{1+i}.$$

Now divide both sides by $n!$ and notice that we obtain the equality that was to be proved.

67. Use the Exponential formula with $f(1) = u$ and $f(n) = (n-1)!$ for $n > 1$. Then

$$F(z,u) = uz + \sum_{n \geq 2} \frac{z^n}{n} = (u-1)z + \ln\left(\frac{1}{1-z}\right).$$

Therefore, by the Exponential formula, we have

$$H(z,u) = \exp F(z,u) = \frac{\exp(u-1)z}{1-z}.$$

69. Let q and S be as described in the exercise. Let M be the largest element of the set F consisting of all fixed points of q and all elements of $[n]$ that are not in S.

Now let $f(q,S)$ be obtained from (q,S) by simply switching the status of M. That is, if M is a fixed point of q, then move M outside S. If M

is outside S, move it to S and let M be a fixed point of the involution taken on S.

This sets up an involution on the set of all allowed pairs (q, S) that changes the parity of the length i of q. This involution is not defined if and only if M does not exist because the set F is empty. That happens when $i = n$ and q has no fixed points. This completes the proof.

71. It suffices to show that the map has an inverse. Let T be a non-plane tree on vertex set $\{0, 1, \cdots, n\}$ that satisfies the conditions on the labels. Then 0 must be the label of the root of T. We can reconstruct the unique inverse image of T as follows. The children $v_1 > v_2 > \cdots > v_k$ of 0 are the left-to-right minima of this inverse image p. As the left-to-right minima of any permutation are in decreasing order, the order of these left-to-right minima is given.

   We then recover the segment of p that starts at v_i and ends right before v_{i+1} by inductively applying this procedure to the subtree of T that is rooted at v_i.

73. Let us count all such trees with a neighbor of a leaf marked. Let $H(z)$ be the exponential generating function of such structures. Removing the root of such a tree, we get a collection of trees, one of which has a neighbor of a leaf marked, while other trees are simply increasing trees, and their set is in bijection with permutations. The only time when this does not happen is when the marked vertex was the root itself, and it has been the neighbor of a leaf. So in this case, after removing the (marked) root, at least one of the branches must have size one. This leads to the differential equation

$$H'(z) = H(z) \cdot \frac{1}{1-z} + \frac{1}{1-z} - \exp\left(\ln \frac{1}{1-z} - z\right),$$

   with initial condition $H(0) = 0$. Note that we obtained the number of trees that have a branch of size one by subtracting from the number of all trees the number of trees that do not have such a branch.

   Solving this differential equation, we get that $H(z) = -1 + \frac{\exp(-z)}{1-z}$. That shows that if $n > 0$, the number of such vertices is equal to the number of *derangments* of size n, which is an interesting fact.

75. Let us use the exponential formula with inside function $f(n) = (n-1)!$ for $n > 2$ and $f(2) = u$. Then the inside generating function is

$$F(z, u) = \sum_{n \geq 2} f(n) \frac{z^n}{n!} = \frac{uz^2}{2} - \frac{z^2}{2} + \sum_{n \geq 3} \frac{z^n}{n} = -z + \frac{z^2(u-1)}{2} + \ln\left(\frac{1}{1-z}\right).$$

   So the combined generating function is

$$H(z, u) = \exp(F(z, u)) = \frac{\exp\left(-z \frac{z^2(u-1)}{2}\right)}{1-z}.$$

Computing $dH/du(z, u)$, then setting $u = 1$, we get the generating function $h(z) = \frac{1}{2}\frac{z^2 e^{-z}}{1-z}$. From this, it follows that $h_n = \frac{n!}{2}\sum_{i=0}^{n-2}\frac{(-1)^i}{i!}$.

77. Let $i < j$. Write the entries of p around a circle so that p_i immediately follows i when moving clockwise. Let p' be the permutation obtained from p by interchanging i and j around the circle. If i and j are not in consecutive positions around the circle, then j precedes i in exactly one of p and p' when written in the one-line notation. Otherwise, if a, i, and j are in consecutive positions around the circle, examination of all cases shows that the only case when such a one-to-one match is not possible is when $i < a < j$. That happens for $j - i - 1$ values of a, and in those cases, j will precede i in both p and p'. This leads to the explicit formula

$$f(n) = n! \cdot \frac{3n - 1}{12},$$

for $n \geq 3$. See entry A227404 and its references in [286] for more information about this sequence.

10.4 Solutions for Chapter 4

1. Recall the notion of rank from Theorem 4.21. Generalize the Simion–Schmidt bijection of Lemma 4.4 as follows. Instead of fixing just the left-to-right minima, fix all entries that are of rank at most $k - 2$. Then proceed like in the Simion–Schmidt bijection.

3. Let p be a permutation enumerated by $\mathrm{Av}_n(132, 312) = a_n$. Let us say that n is in position i. Then all entries preceding n are larger than all entries that n precedes. Moreover, entries that n precedes are in decreasing order. This proves the recursive formula $a_n = \sum_{i=1}^{n} a_{i-1}$ for $n \geq 2$, with $a_1 = 1$. Solving this recursion, we get $a_n = 2^{n-1}$.

5. If p avoids 132, then all entries on the left of n are larger than all entries on the right of n. Furthermore, the subword on the left of n has to avoid 123 as well, while the subword on the right of n has to avoid 132 and 1234. These conditions together are sufficient.

The reader is invited to prove that $\mathrm{Av}_n(123, 132) = 2^{n-1}$. Now let $b_n = \mathrm{Av}_n(132, 1234)$. Let n be in position i of our permutation. Then it follows from the above that there are $2^{i-2}b_{n-i}$ ways for p to be $(132, 1234)$-avoiding if $i > 1$, and b_{n-1} ways if $i = 1$. This implies

$$b_n = b_{n-1} + \sum_{i=2}^{n} 2^{i-2}b_{n-i}.$$

Solving this recurrence relation, we get that $b_n = F_{2n}$, where F_i is the ith Fibonacci number, that is, $F_1 = 0$, $F_1 = 1$, and then $F_{i+1} = F_i + F_{i-1}$. This result was proved in [327] by a different argument. We point out that [327] has a catalog of results for sequences $\mathrm{Av}_n(p, q)$, where p is of length three, and q is of length four, as well as a general proof technique to obtain those results.

7. We show that for $n \geq 3$, we have $g(n) - g(n - 1) = n \cdot 2^{n-3}$. Indeed, there are $g(n - 1)$ permutations that avoid both 132 and 4231 in which n is in the last position. If n is not in the last position, then each entry of L is larger than each entry of R. Moreover, L is a permutation that avoids 132 and 312, and R is a permutation that avoids 132 and 231. These conditions are sufficient. It then follows from Exercise 3 that the number of such permutations for each position of n is equal to 2^{n-2} if n is not in the first position, and to 2^{n-1} if n is in the first position, proving our recursive formula. The statement of the exercise is then proved by induction.

9. (a) The numbers $r_{n-1} = \mathrm{Av}_n(3142, 2413)$ are the famous *large Schröder numbers*. See [326] for a proof of the recursive formula

$$r_n = \sum_{i=0}^{n} \binom{2n - i}{i} C_{n-i}.$$

It is easy to see from the above formula that the number r_n also counts subdiagonal lattice paths from $(0,0)$ to (n, n) that use steps $(0,1)$, $(1,0)$, and $(1,1)$. This proves the recurrence relation

$$r_n = r_{n-1} + \sum_{i=1}^{n} r_{j-1} r_{n-j}. \tag{10.4}$$

From (10.4) we see that

$$\sum_{n \geq 0} r_n x^n = \frac{1 - x - \sqrt{1 - 6x + x^2}}{2x}.$$

(b) There are nine other pairs (p, q) that are not trivially equivalent to $(3142, 2413)$, but are still enumerated by the same number of n-permutations. One of them is $(1324, 1423)$ as was proved in [326]. Another pair was found by Sophie Gire [196]. Finally, it was Darla Kremer [234] who gave an exhaustive list of all ten pairs. The remaining eight pairs are $(1234, 2134)$, $(1342, 2341)$, $(3124, 3214)$, $S(3142, 3241)$, $(3412, 3421)$, $S(2134, 1324)$, $(3124, 2314)$, and finally, $(2134, 3124)$. We point out that the Schröder numbers occur in many other contexts, and the interested reader should consult Chapter 6 of [297] for details.

11. (a) This result was proved in [49]. Denote by $h(n) = Av_n(1324, 2413)$ the number of the permutations to enumerate, for brevity. It is obvious that if n is the leftmost entry, then the number of such permutations is $h(n-1)$. Now let p be a (1324,2413)-avoiding n-permutation; suppose n is not the leftmost entry of p and let a be the smallest entry of p that precedes n. Then n precedes the entries $1, 2, \cdots, a - 1$. Furthermore, these $a - 1$ entries must occupy the last $a - 1$ positions (why?).

So the last $a - 1$ entries of p are the smallest ones, and so we can have $h(a - 1)$ different strings on them. Let $t(n - a + 1)$ be the number of possible substrings on the first $n - a + 1$ entries, in other words, $t(i)$ is the number of (1324,2413)-avoiding n-permutations in which the entry 1 precedes the entry n. In what follows, we are going to use this second interpretation of $t(n)$ so as to alleviate notation. Set $t(0) = 0$. Let $T(z) = \sum_{i \geq 1} t(n) z^n$. It follows from the above that permutations counted by $t(n)$ are precisely the indecomposable (1324,2413)-avoiding n-permutations. It is then clear that $H(z) = 1/(1 - T(z))$, and this includes even the case when n is the leftmost entry. Now we analyze the structure of permutations enumerated by the $t(i)$ in order to determine $T(z)$.

Call entries before the entry 1 *front* entries, entries after the entry n *back* entries, and entries between 1 and n *middle* entries. Say that an entry x *separates* two entries y and z written in increasing order if $y < x < z$.

The front entries must form a 132-avoiding permutation, the middle entries must form an increasing subsequence, and the back entries must form a 213-avoiding permutation. Similary, no front entry can separate two middle entries, or two back entries in increasing order; no middle entry can separate two front entries in increasing order or two back entries in increasing order; and no back entry can separate two middle entries or two front entries in increasing order.

Therefore, the only way for two entries of the same category to be in increasing order is when they relate to any entries of the other two categories in the same way. Such entries are said to form a *strong block*. The strong block subdivision of a permutation counted by $T(z)$ is shown in Figure 10.3.

As we said, each strong block between 1 and n consists of an increasing subsequence, while strong blocks in the front are 132-avoiding permutations, and strong blocks in the back are 231-avoiding permutations. Permutations satisfying all these conditions do avoid both 1324 and 2413.

Now for $i \geq 2$, let v_{i-1} be the number of those permutations counted by $t(i)$ containing no middle strong blocks, except for 1

n

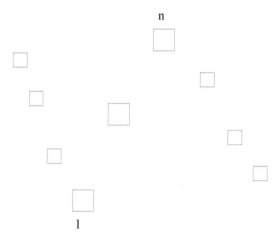

1

FIGURE 10.3

The strong block subdivision of p.

and i. So $v_1 = 1$, $v_2 = 2$, $v_3 = 6, \cdots$. Let $R(x) = \sum_{i\geq 1} v_i z^i$ be the generating function for the v_i. Then clearly $T(z) = \frac{z}{1-R(z)}$.

Note that v_{i-1} is just the number of ways to partition the interval $\{2, 3, \cdots, i-1\}$ into disjoint intervals, and then taking a 213-avoiding or a 132-avoiding permutation on each of them alternatingly. It is not hard to see by a lattice path argument that this means that $v_i = \binom{2i-2}{i-1}$, so $R(z) = \frac{z}{\sqrt{1-4z}}$. Therefore,

$$T(z) = \frac{z}{1 - R(z)} = \frac{z\sqrt{1 - 4z}}{\sqrt{1 - 4z} - z},$$

which implies

$$H(z) = \frac{1}{1 - T(z)} = \frac{(\sqrt{1 - 4z} - z)(\sqrt{1 - 4z}(1 - z) + z)}{(1 - 4z)(1 - z)^2 - z^2}$$

$$= \frac{1 - 5z + 3z^2 + z^2\sqrt{1 - 4z}}{1 - 6z + 8z^2 - 4z^3}.$$

(b) There are five nontrivially equivalent pairs (p, q) so that $Av_n(p, q) = h(n)$ for all n. Of the remaining four, the pair $(1324, 2143)$ was found in [209], though a more accessible reference is [89]. (We note that permutations avoiding this pair are called *smooth* permutations. The pair $(3214, 4123)$ was found in [288], and the pairs $(1342, 2314)$ and $(1342, 3241)$ were found in [49].

13. This problem has been solved in [207], where the authors showed that

$$\sum_{n\geq 0} I_n(2143)z^n = \frac{1 - z - \sqrt{1 - 2z - 3z^2}}{2z}. \tag{10.5}$$

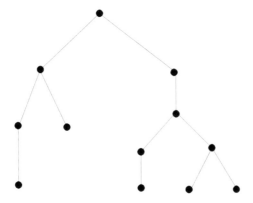

FIGURE 10.4
A 1-2 tree.

The numbers $I_n(2143)$ are called the *Motzkin* numbers, and occur in numerous combinatorial problems. See [297] for an extensive list of these problems.

The authors of [207] showed the above formula by finding a bijection between these involutions, and 1-2 trees. A 1-2 tree is a plane tree in which each vertex has 0, 1, or 2 children. However, in contrast to decreasing binary trees, if a vertex has a single child, that child is directly below its parent, not on its left or right. See Figure 10.4 for an example.

It is not difficult to see that the numbers of these trees satisfy the recurrence relation $M_n = M_{n-1} + \sum_{k=0}^{n-2} M_k M_{n-2-k}$, from which (10.5) follows for $I_n(2143) = M_n$.

15. As a 132-avoiding permutation is skew indecomposable if and only if it ends with its largest entry, the number of such k-permutations is C_{k-1}. Let our three blocks end in positions i, j, and n. With these restrictions, we clearly have $C_{i-1}C_{j-i-1}C_{n-j-1}$ permutations with the desired property. Now we have to sum this expression over all possible i and j to get the total number of 132-avoiding n-permutations having exactly three skew blocks. We do this in two steps. First, fix i, and compute the sum

$$C_{i-1} \sum_{j=i+1}^{n-1} C_{j-i-1}C_{n-j-1} = C_{i-1}C_{n-i-1}.$$

Second, we sum over all possible i to get that there exist

$$\sum_{i=1}^{n-2} C_{i-1}C_{n-i-1} = C_{n-1} - C_{n-2}$$

permutations with the desired property, as long as $n \geq 3$.

17. Let p be a 231-avoiding n-permutation. Let us call p *decomposable* if it can be cut into two parts so that each entry before the cut is smaller than each entry after the cut. For example, 21543 is decomposable. If p is decomposable, then say that its first cut is after i entries ($1 \leq i \leq n-1$), and that $p = LR$, where L is the substring of the first i entries. Then we define the northeastern lattice path $f(p)$ recursively, by taking the path $f(L)$ from $(0,0)$ to (i,i), and continuing it with a translated copy of the path $f(R)$ from (i,i) to (n,n).

We still have to define f for 231-avoiding permutations that are not dually decomposable. These permutations start in their entry n (otherwise they would have a cut immediately before the entry n). Let p be such a permutation, and let p' be the $(n-1)$-permutation obtained from p by omitting the entry n. Then define $f(p)$ as the concatenation of the step $(0,0)$ to $(1,0)$, a translated copy of $f(p')$ from $(1,0)$ to $(n, n-1)$, and the step $(n, n-1)$ from (n,n).

It is straightforward to prove again by induction that this recursively defined map f is indeed a bijection. By induction again, the number of ascents of p is equal to the number of north-to-east turns of $f(p)$.

19. Yes. If p is increasing and q is decreasing, then all permutations of length at least $|p| \cdot |q| + 1$ must contain at least one of p and q. This is a famous result of Erdős and Szekeres, and we proved it in Proposition 6.52.

21. As 231 and 312 are inverses of each other, and the inverse of an even permutation is even, the first equality is straightforward.

A permutation is 231-avoiding if and only if its reverse is 132-avoiding. On the other hand, reversing a permutation is the same as multiplying it with the transpositions $(1\,n), (2\,n-1), \cdots$. The number of these transpositions is $\lfloor n/2 \rfloor$, and our proof follows from the result of the previous exercise. This result first appeared in [284], in a slightly different form.

23. We claim that for $n \geq 1$, $\mathrm{Av}_n(123, 132, 213) = F_{n+1}$, where the F_n are the well-known Fibonacci numbers, starting with $F_0 = 0$, and $F_1 = 1$, and then given by $F_{n+1} = F_n + F_{n-1}$. This result was first mentioned in [284]. Indeed, in a permutation p enumerated by $\mathrm{Av}_n(123, 132, 213)$, the entry n must be in either the first or second position; otherwise, p could not avoid both of 123 and 213. If n is in the first place, then, by induction, we have F_n possibilities for the rest of p. If n is in the second place, then $n-1$ must be in the first place; otherwise, a 231-pattern is formed. Then, by induction again, we have F_{n-1} possibilities for the rest of p. This shows that $\mathrm{Av}_n(123, 132, 213) = F_{n+1}$. It is well-known,

and can be proved by routine generating function techniques, that

$$F_n = \frac{1}{\sqrt{5}} \cdot \left(\frac{1 + \sqrt{5}}{2} \right)^{n+1} - \frac{1}{\sqrt{5}} \cdot \left(\frac{1 - \sqrt{5}}{2} \right)^{n+1}.$$

As $\left(\frac{1-\sqrt{5}}{2} \right) < 1$, the first term is dominant, showing that we have $\sqrt[n]{\mathrm{Av}_n(123, 213, 132)} = \left(\frac{1+\sqrt{5}}{2} \right)$.

25. The pattern $132456 \cdots k$ can play the role of q. The proof is similar to that of Theorem 4.29. However, instead of simply defining a strong class by its left-to-right minima and right-to-left maxima, we also have to fix the value of the position of entries that become right-left maxima if the original right-to-left maxima are removed. We then have to iterate this procedure $n - 4$ more times.

27. (a) Let p be a 132-avoiding n-permutation in which the entry n is in position i. Then the binary plane tree $T(p)$ will have a left subtree of $i - 1$ nodes and a right subtree of $n - i$ nodes. The rest of the tree is constructed recursively by the same rule. The map T is a bijection as the position of n in p can be read off $T(p)$ as the size of the left subtree of $T(p)$ plus one. The position of the other vertices can then be found recursively, noting that if n is in position i, then the set of entries of the left of n must be $\{i + 1, i + 2, \cdots, n\}$, and the set of entries on the right of n must be $[i - 1]$. It is here that we use the fact that p is 132-avoiding.

Note that while $T(p)$ is an unlabeled tree, each node of $T(p)$ is naturally associated to an entry of p. Nevertheless, $T(p)$ is unlabeled as writing these entries to the nodes would not carry any additional information.

(b) It is straightforward to prove by induction that $d(p)$ is the number of right edges (that is, edges that go down and right, or, if you like, from northwest to southeast).

(c) We claim that $p_i > p_{i+1}$ if and only if the vertex corresponding to p_i appears on a higher level in $T(p)$ than the vertex corresponding to p_{i+1}. This is obvious if $n = 2$. Now assume our claim is true for all integers less than n. Then if p_i and p_{i+1} are on the same side of the entry n, then the corresponding vertices are in the same (left or right) subtree of $T(p)$, and our claim follows by induction. Consecutive entries cannot be on two different sides of n, so the only remaining case is when one of p_i and p_{i+1} is equal to n. That case is trivial, however, as that vertex will correspond to the root itself.

29. The number of 1234-avoiding n-permutations is equal to the number of strong classes of n-permutations. On the other hand, the number of

1324-avoiding n-permutations is asymptotically more than that. Indeed, for $n > 7$, it is very easy to construct strong classes that end in the class $3 * 1 * 7 * 5$ by concatenation. These classes contain at least two 1324-avoiding permutations. On the other hand, they constitute at least a constant factor of all strong classes, implying our claim.

31. Let p be a pattern of length k starting with 1 and ending with k so that $\text{Av}_n(p) < C^n$ holds for all n, for some constant C. Then we claim that $\text{Av}_n(q') < (4cC)^n$, thus we can set $K = 4cC$. Take an n-permutation π which avoids q'. Suppose it contains q (this will only exclude c^n permutations). Then consider all copies of q in our permutation and consider their entries x. Color these entries red. Clearly, the red entries must form a permutation which does not contain p. Suppose they do, and denote x_1 and x_k the first and last elements of that purported copy of p. Then the initial segment of the copy of q which contains x_1 and the ending segment of the copy of q which contains x_k would form a copy of q'.

Now remove all the red entries from π, to get π'. Then π' must be q-avoiding as all copies of q in π lost their entries playing the role of x. Therefore, there are at most c^n possibilities for the permutation of the non-red entries. There are at most 2^{n-1} choices for the positions of the red entries, at most 2^{n-1} choices for the values of the red entries, and at most C^{n-1} choices for the permutation of the red entries. This shows that less than $(4C)^{n-1} \cdot c^n + c^n < (4Cc)^n$ permutations of length n can avoid q'.

33. It suffices to prove that every q-avoiding n-permutation p can be extended into a q-avoiding $(n + 1)$-permutation by prepending it with a new first entry, in $k - 1$ different ways. Let i be the first entry of q. Then we can prepend p with any one of the entries $1, 2, \cdots, i - 1$, as well as any one of the entries $n, n - 1, \cdots, n - k + i + 1$ without creating a copy of q. Indeed, the first entry of the obtained $(n + 1)$-permutation would be either too small or too large to play the role of i in any copy of q, and, being the leftmost entry, it certainly cannot play the role of any other entry of q.

35. These will be the trees in which going from the leaves up, each label is as large as it can possibly be, that is, the sum of its children plus 1. Indeed, in 123-avoiding permutations, the entries that are not left-to-right minima form a decreasing sequence, meaning that each vertex will contribute to its own label.

Note that this observation provides a bijection from the set of indecomposable 123-avoiding n-permutations to that of *unlabeled* plane trees on n vertices.

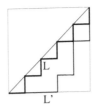

FIGURE 10.5
The area between L and $L' = f(53124)$ is equal to $i(53124) = 6$.

37. The number of inversions is translated to an *area*. More precisely, let L be $f(123\cdots n)$, the staircase lattice path. Let $f(p) = L'$ for some 231-avoiding n-permutation p. Then $i(p)$ is equal to the area between L and L'. See Figure 10.5 for an example. This can be proved by induction, using the block decomposition of p explained in the solution of Exercise 17.

We mention that the area statistic of these permutations leads to an interesting open problem. Let n be fixed, and let a_k be the number of 231-avoiding n-permutations p so that the area below $f(p)$ is equal to k. Then it is conjectured in [304] that the sequence $\{a_k\}$ is unimodal. It is also conjecture in [304] that unimodality remains true if the staircase Ferrers shape that $f(p)$ is not allowed to enter is replaced by another self-conjugate Ferrers shape.

39. Yes. We claim that for $k \geq 2$, we have $\mathrm{Av}_n(123, (k-1)k\cdots 21) \leq n^{2(k-2)}$. We prove this claim by induction on k. For $k = 2$, the statement is trivial, and for $k = 3$, the statement is true as $\mathrm{Av}_n(123, 231) = \binom{n}{2} + 1 \leq n^2$.

Now let us assume that the statement is true for k, and prove it for $k+1$. It is clear that if a permutation p avoids both 123 and $(k-1)k\cdots 21$, then the substring p' obtained from p by omitting its right-to-left minima must avoid both 123 and $(k-2)(k-1)\cdots 21$. Note that p can have at most two right-to-left minima as p is the union of two decreasing sequences. One of these two right-to-left minima must be the entry 1, and the other is the rightmost entry. We have at most n choices for the position of n, at most n choices for the rightmost entry, and, by induction, at most $n^{2(k-3)}$ choices for p'. Therefore, we have at most $n \cdot n \cdot n^{2(k-3)} = n^{2(k-2)}$ choices for p as claimed.

41. It is proved in [3], along with many similar results on multiset permutations avoiding patterns of length three, that

$$B_1(a_1, a_2, a_3) = \binom{a_1 + a_2 + a_3}{a_3} - \binom{a_2 + a_3}{a_3} + \binom{a_1 + a_2 + a_3}{a_2}.$$

43. Let a_n be the number of 1s contained in A_n. Then $a_0 = 1$, and $a_{n+1} = 2a_n + 2^n$ for $n \geq 0$. So, if $A(z) = \sum_{n \geq 0} a_n z^n$, then we have

$$A(z) = \frac{1}{1 - 2z} + \frac{z}{(1 - 2z)^2} = 1 + \sum_{n \geq 1} (n + 2)2^{n-1} z^n.$$

So $a_n = (n + 2)2^{n-1}$ if $n \geq 1$ (and the rule works even for $n = 0$). Note that A_n does not contain B. If $m = 2^n$, then this means that there exists a matrix of size $m \times m$ that contains more than $(m/2) \cdot \log_2 m$ entries equal to 1 that avoids B.

45. This result is due to Sergi Elizalde and Marc Noy [158].

47. If $n = 2m + 1$, then this number is the Catalan number C_m, and if $n = 2k$, then this number is the Catalan number C_k. These facts can be proved by considering the position of the entry 1 of our permutations, and then recognizing that the numbers of our permutations satisfy the same recurrence as the Catalan numbers.

49. If p avoids q, then $p^{-1} \ominus p$ is an involution that avoids q.

51. Let p avoid H_k, and let p_i be the leftmost entry of p that is of rank $k-1$. Then the string $p_1 p_2 \cdots p_{i-1}$ is a $12 \cdots (k-1)$-avoiding permutation, while the string $p_i \cdots p_n$ is an increasing sequence. This yields the chain of inequalities

$$\mathrm{Av}_n(H_k) \leq \sum_{i=1}^{n} \binom{n}{n-i-1} ((k-2)^2)^{i-1}$$

$$\leq \sum_{i=1}^{n} \binom{n}{n-i-1} ((k-2)^2)^{i+1}$$

$$\leq ((k-1)^2 + 1)^n.$$

We used the binomial theorem in the last step.

53. (a) Let skew indecomposable 132-avoiding permutations correspond to Dyck paths that never touch the horizontal axis, except in their starting and ending point. Then define f recursively, noting that skew indecomposable 132-avoiders must end in their largest entry, while their images have the property that removing the original $(1,1)$-step and the ending $(1,-1)$-step, we get a Dyck path of semilength $n - 1$.

(b) It is easy to prove that p has a ULIS if and only if $f(p)$ has a unique peak of maximal length. Furthermore, p is an involution if and only if $f(p)$ is symmetric to the vertical line $x = n$. So if $p \in UI_n(132)$, then $f(p)$ is symmetric to the line $x = n$ and has a unique peak of maximal length (which is necessarily on that line).

55. Let p be a permutation of length $n \geq (k-1)^3 + 1$ that avoids the pattern $12 \cdots k$. Then p is the union of $k-1$ decreasing subsequences. Color these subsequences with colors $1, 2, \ldots, k-1$ so that a maximum-length subsequence gets the color 1. Then there are at least $(k-1)^2 + 1$ entries of color 1, say in positions $i_1 < i_2 < \cdots < i_m$, and they are $p(i_1) > p(i_2) > \cdots > p(i_m)$. Let $P_j = p(i_j)$.

Consider the $m \geq (k-1)^2 + 1$ positions $P_m < P_{m-1} < \cdots < P_1$ in p. There will be a set S of at least k positions among them so that entries in these positions are of the same color, that is, that form a decreasing subsequence S. Those same k entries form an increasing sequence in p^2. Indeed, let positions P_a and P_b contain two entries of S, with $a < b$, so $P_a > P_b$. That means that position $p(i_a)$ contains a smaller entry than position $p(i_b)$, that is, $p(p(i_a)) < p(p(i_b))$. Repeating this argument for every pair of entries in positions that belong to S proves that in p^2, the k entries that are in positions in S form an increasing subsequence. See [79] for more on this problem.

57. It is proved in [80] that $\mathrm{Cyc}_n(123, 132) = 2^{\lfloor (n-1)/2 \rfloor}$.

59. Let $p = p_1 p_2 \cdots p_n$ be any cyclic permutation of length n that avoids q. Insert the entry $n+1$ to the next-to-last position of p. Then p is still q-avoiding, since $n+1$ is too far back in p to be part of any copies of q. Furthermore, the obtained permutation p' is still cyclic, since $p_i = p'_i$ for all $i \leq n-1$, and p maps n to x, while p' maps n to $n+1$, and then $n+1$ to x. So, we get the cyclic diagram of p' by simply inserting the entry $n+1$ between n and x in the cyclic diagram of p.

Doing this for all $\mathrm{Cyc}_n(q)$ cyclic, q-avoiding permutations of length n yields a set S of $\mathrm{Cyc}_n(q)$ cyclic q-avoiding permutations of length $n+1$, each of which contains the entry $n+1$ in the nth position. As q is an involution, the inverse r^{-1} of any q-avoiding permutation r is also q-avoiding. So taking the inverse of each permutation in S yields a set T of $\mathrm{Cyc}_n(q)$ cyclic q-avoiding permutations of length $n+1$, each of which contains the entry n in the $n+1$st position. Finally, S and T are disjoint sets, since a cyclic permutation that is longer than 2 cannot contain the 2-cycle $(n\ n+1)$. This result was first proved in [80].

61. Let $\mathcal{C} = Av(12 \cdots (k+1), 132, 213)$. If $p \in \mathcal{C}$, Then the entry 1 can be only in one of the last k positions. Furthermore, all entries on the right of 1 must be smaller than all entries on the left of 1, and the entries on the right of 1 must form an increasing sequence. It is now easy to set up a bijection from $\mathcal{C}_n$ to the set of all compositions of n into parts at most k. The length of that last increasing subsequence starting at 1 will correspond to the last part of a composition, then the other parts are recursively defined by this same process. (Alternatively, the numbers of

both kinds of objects satisfy the recurrence relation $f(0) = 0$, $f(i) = 2^i$ for $i \in [k]$, and $f(n) = \sum_{i=1}^{k} f(n-i)$ if $n \geq k$.)

10.5   Solutions for Chapter 5

1. We claim that $S_{132,1}(n) = \binom{2n-3}{n-3}$. See [47] for a proof. The main idea is the following. In a permutation enumerated by $S_{132,1}(n)$, there is either one or no front entry that is smaller than a back entry. If there is one such front entry, then its position and size is very restricted. If there is no such front entry, then the single 132-pattern of the permutation is formed either by front entries only, or by back entries only. This leads to a recurrence relation involving the numbers $S_{132,1}(n)$ and the Catalan numbers, and solving that recurrence relation, we obtain the above explicit formula.

3. This classic problem was first solved by Cayley [111], who proved that

$$f(n,d) = \frac{1}{n+d+2}\binom{n+d+2}{d+1}\binom{n-1}{d}.$$

Recently, Richard Stanley [296] gave a proof based on a bijection between these polygon dissections and Standard Young Tableaux. We will revisit that proof in Exercise 22 of Chapter 7.

5. We prove the statement by induction on k. For $k = 1$, we have $S(n,1) = 1$, which is obviously P-recursive. Now let us assume the statement is true for $k-1$, and prove it for k. We have seen in Exercise 8 of Chapter 1 that

$$S(n,k) - kS(n-1,k) = S(n-1,k-1).$$

Let $S_k(z) = \sum_{n \geq k} S(n,k)z^n$. Multiplying both sides of the above equation by z^n, and summing over all $n \geq k$, we get

$$S_k(z)(1 - kz) = zS_{k-1})(z)$$

$$S_k(z) = \frac{zS_{k-1}(z)}{1-kz},$$

so $S_k(z)$ is the product of the d-finite generating functions $zS_{k-1}(z)$ and $\frac{1}{1-kz}$, and is therefore d-finite.

Note that it follows from the above that

$$S_k(z) = \frac{z^k}{(1-z)(1-2z)\cdots(1-kz)}.$$

7. Let $A(z)$ be algebraic of degree a, and let $B(z)$ be algebraic of degree b. Then $A(z)^a$ can be obtained as a linear combination of the power series $1, A(z), A(z)^2, \cdots, A(z)^{a-1}$ in which the coefficients are polynomials. The analogous statement holds for $B(z)^b$. That means that any expression involving $A(z)$ and $B(z)$ and algebraic operations on them can be obtained using linear combinations of the $A(z)^i$ and $B(z)^j$, with $i \leq a-1$ and $j \leq b-1$, and the same algebraic operations that are used in the desired expression.

9. On one hand, $f''(z) = -f(z)$, so the dimension of the vector space spanned by the derivatives of f is at most of dimension two; therefore, f is d-finite. On the other hand, assume that f is algebraic of degree d, that is,

$$P_0(z) + P_1(z) \sin(z) + \cdots + P_d(z) \sin^d(z) = 0,$$

with d being minimal. As $P_1(z) \sin(z) + \cdots + P_d(z) \sin^d(z) = 0$ for infinitely many values of z, it follows that $P_0(z)$ must be the zero polynomial. So the above equation reduces to

$$\sin(z)(P_1(z) + \cdots + P_d \sin^{d-1}(z)) = 0.$$

As $\sin(z)$ is not identically zero, it follows that $P_1(z) + \cdots + P_d \sin^{d-1}(z) = 0$ as a function, contradicting the assumption that the degree of $\sin(z)$ was d.

Note that this proof depended on the fact that $\sin(z)$ has infinitely many roots. See [297], Chapter 6, for two proofs of the fact that e^z is not algebraic. On the whole, however, it is often difficult to prove that a series is not algebraic.

11. It is proved in Exercise 11 of Chapter 4 that

$$H(z) = \sum_{n \geq 0} \mathrm{Av}_n(1342, 2431) z^n = \frac{1 - 5z + 3z^2 + z^2\sqrt{1 - 4z}}{1 - 6z + 8z^2 - 4z^3}.$$

In other words, the generating function $H(z)$ of our sequence is algebraic. Indeed, if $a(z) = 1 - 5z + 3z^2$ and $b(z) = 1 - 6z + 8z^2 - 4z^3$, then $(b(z)H(z) - a(z))^2 = z^4(1 - 4z)$, and the algebraic property of $H(z)$ follows. Therefore, $H(z)$ is d-finite, proving that our sequence is P-recursive.

13. A permutation is layered if and only if it avoids both 312 and 231. On one hand, the condition is necessary, since in a layered permutation, any purported copy of any of those two patterns would have to contain the entries playing the roles of 3 and 1 within the same descending run, and that would make it impossible to find an entry to play the role of 2.

On the other hand, in permutations avoiding 312, the entries after the maximal entry have to be in decreasing order, and in permutations

$k^2{-}k{+}1$	$k^2{-}k{+}2$		k^2
$k{+}1$	$k{+}2$		$2k$
1	2		k

FIGURE 10.6
A grid for a k-superpattern.

avoiding 231, the entries after the maximal entry have to be larger than entries before the maximal entry. So, in permutations avoiding both 312 and 231, entries weakly after the maximal entry must form the subsequence $n(n-1)\cdots(n-k)$ for some k. Iterating this argument completes the proof.

15. Induction on n. The claim is clearly true if $n \leq 2$. Assuming it is true for $n-1$, first consider $c \leq \binom{n-1}{2}$ and let $p \in S_{n-1}$ satisfy the lemma. Then $pn \in S_n$ works for such c. On the other hand, if $\binom{n-1}{2} < c \leq \binom{n}{2}$ then consider $c' = c - (n-1) \leq \binom{n-1}{2}$. Pick $p \in S_{n-1}$ with c' copies of 21 and none of 132. Then $np \in S_n$ is the desired permutation.

17. A 321-avoiding n-permutation consists of two increasing subsequences, of sizes a and $n-a$. Therefore, the number of inversions in p is at most $a(n-a) \leq \frac{n^2}{4}$. This is indeed attainable, for permutations like 456123.

19. (a) We claim that $sp(3) = 5$. On one hand, 41352 is a 3-superpattern. On the other hand, there is no 3-superpattern of length four as a pattern of length four can contain at most four different patterns of length three.

   (b) The permutation 1 3 6 10 2 5 9 4 8 7 is a 4-superpattern.

   (c) Consider the grid shown in Figure 10.6.

   Then read the columns one after another starting with the leftmost one, going from the bottom up in each column. The obtained permutation is clearly a k-superpattern. Indeed, if we want to find a copy of $q = q_1 \cdots q_k$, all we need to do is to choose the q_ith entry of the ith column for each i.

21. (a) A permutation of length six has $\binom{6}{4} = 15$ subwords of length four, so it cannot contain all 24 different patterns of length four.

   (b) A 4-superpattern has to contain both the increasing and the decreasing pattern of length four. A permutation p of length seven can only do that if it is the union of a 1234-pattern and a 4321-pattern, which intersect in exactly one entry. If p has this decomposition property, then all of its subwords can be decomposed into an increasing and a decreasing sequence. Therefore, 3412 cannot be a subword of p.

23. For any k, the number of different k-element patterns contained in p is at most $\min\left(\binom{8}{k}, k!\right)$, as there are $\binom{8}{k}$ subwords of p that have k elements, and there are $k!$ different possibilities for the pattern of these subwords. One checks that $\binom{8}{k} \leq k!$ if and only if $k \leq 5$, so we get that the number of different patterns in p is at most $1+1+2+6+24+56+28+8+1 = 127$.

25. No, this is not always true. As is shown in [268], the case of $k = 6$ provides the smallest counterexample. In that case, the layered permutation having seven layers of length $n/7$ each will have more copies of q than the permutation having six layers of length $n/6$ each.

27. This result and its proof can be found in [4].

29. Set $M_0 = 1$, and $M(z) = \sum_{n \geq 0} M_n z^n$. The number of such paths that first return to the line $y = 0$ at the point $(k, 0)$ is clearly $M_{k-2}M_{n-k}$ if $2 \leq k \leq n$, and M_{n-1} if $k = 1$. This leads to the functional equation

$$M(z) = 1 + M(z)z + M^2(z)z^2,$$

hence

$$M(z) = \frac{1 - z - \sqrt{1 - 2z - 3z^2}}{2z}.$$

In other words, $M(z)$ is algebraic, and therefore, d-finite.

31. On one hand $S = 1232123$ contains all six permutations of length three, hence, allperm(3) ≤ 7. On the other hand, let us assume that there exists a sequence Z of length six of that property. If there is an entry, say 1, that occurs in Z only once, then there would have to be at least three entries before that entry 1, and at least three entries after that entry 1 in order for Z to contain all four permutations starting or ending in 1. That is impossible, since Z has only six entries. So Z must contain two copies of each of 1, 2, and 3. Let us assume without loss of generality that Z starts with 1. Then the other occurrence of Z cannot be in the second, third, fourth, or sixth positions, for obvious reasons. So the second occurence of 1 is in the fifth position. Then Z cannot contain both 312 and 213, since the sixth entry of Z cannot be both 2 and 3. This proves that allperm(3) = 7.

33. Let Z be a sequence of length allperm(n) that contains all $n!$ permutations of length n as a subword. Let i be the element of $[n]$ that shows up *last* in Z. That means that the leftmost copy of i in Z is in position

n or later. Note that Z contains all n-permutations, so in particular, Z contains all n-permutations starting with i. So if we remove all elements of Z that either precede the first occurrence of i in Z, or are equal to i, we get a shorter sequence Z' whose elements come from an $(n-1)$-element set and that contains all $(n-1)!$ permutations of that set as a subword. As Z' is at least n letters shorter than Z, this implies that allperm$(n) \geq$ allperm$(n-1)+n$, after which our claim is routine to prove by induction.

35. In fact, the generating function of the sequence is not only d-finite, but also *rational*. This is easy to prove by induction on k.

37. This result has been first proved by Daniel Ashlock and Jenett Tillotson [18]. The proof is by induction.

Take an n-superpermutation p. Start moving left to right. When you find the first factor q that is equal to a permutation of length n, replace that copy of q by the string $q(n+1)q$. This adds $n+1$ new letters to p. Continue reading, from the first digit of the *second* copy of q in that newly added string $q(n+1)q$. When you find an n-permutation that you did not encounter before, stop, and repeat the above procedure.

As p contained all $n!$ permutations of length n, we will carry out the above procedure $n!$ times, adding a total of $(n+1)n! = (n+1)!$ new letters.

The obtained long string is an $(n+1)$-superpermutation because it contains the permutation $L(n+1)R$ as a factor. Indeed, at some point during the above procedure, the permutation RL will be replaced by the string $RL(n+1)LR$.

For instance, for $n = 2$, we can start with the 2-superpermutation 121, and get the 3-superpermutation 123121321.

Note that this argument proves that if spp$(n) \leq m$, then spp$(n+1) \leq m+(n+1)!$

10.6   Solutions for Chapter 6

1. (a) The choice of i, j, k, and ℓ is clearly insignificant. Applying the Transition Lemma, we see that the four entries $n-3$, $n-2$, $n-1$, and n are in the same cycle of p if and only if n is the leftmost of the four of them in $f(p)$. This obviously happens in $1/4$ of all n-permutations, so the probability in question is $1/4$.

   (b) The entries $n-3$, $n-2$, $n-1$, and n belong to different cycles of p if and only if they are in increasing order in $f(p)$. The latter happens in $1/24$ of all n-permutations, so the probability we are looking for is $1/24$.

3. We can again assume that the four entries in question are $n - 3$, $n - 2$, $n - 1$, and n.

   (a) This will happen if and only if the pattern formed by our four entries has two left-to-right maxima. There are $c(4, 2) = 11$ such permutations, so the probability we are looking for is $11/24$.

   (b) Similarly, this will happen if the pattern of our four entries has three left-to-right maxima. The number of such permutations is $c(4, 3) = 6$, so the probability in question is $\frac{6}{24} = \frac{1}{4}$.

5. This exercise is very similar to Example 6.27. Let S be a k-element subset of $[n]$, and let $X(p)$ be the number of k-cycles of p. Then the probability that the entries belonging to S form a k-cycle is $(k - 1)! \cdot \frac{(n-k)!}{n!}$; therefore, this is the expectation of the corresponding indicator variable X_S. As there are $\binom{n}{k}$ such indicator variables, we get that

$$E(X) = \sum_S E(X_S) = \binom{n}{k} \cdot (k - 1)! \cdot \frac{(n - k)!}{n!} = \frac{1}{k}.$$

7. For obvious symmetry reasons, we have $E(Z) = \binom{n}{2}/2$. To compute $E(Z^2)$, introduce the indicator variables $Z_{i,j}$ defined for $p = p_1 p_2 \cdots p_n$ and for all pairs of elements $i < j$ by

$$Z_{i,j}(p) = \begin{cases} 1 \text{ if } p_i > p_j, \\ 0 \text{ if not.} \end{cases}$$

It is then clear that $Z = \sum_{i<j} Z_{i,j}$. Furthermore, $E(Z_{i,j}) = E(Z_{i,j}^2) = 1/2$. There are several cases to consider when computing the expectations $E(Z_{i,j} Z_{k,\ell})$. The simplest is the case of the $\binom{n}{2}\binom{n-2}{2}/2$ pairs when $\{i, j\}$ and $\{k, \ell\}$ are disjoint. In that case, clearly $E(Z_{i,j} Z_{k,\ell}) = 1/4$. If $i = k$, but $j \neq \ell$, then $E(Z_{i,j} Z_{k,\ell}) = 1/3$ as two of the six possible patterns for the triple $p_i p_j p_\ell$ are favorable (the ones in which p_i is the largest of the three entries). Similarly, we have $E(Z_{i,j} Z_{k,\ell}) = 1/3$ if $j = \ell$ but $i \neq k$. Each of these possibilities occurs in $2\binom{n}{3}$ cases. Finally, it can also happen that $j = k$, in which case $E(Z_{i,j} Z_{k,\ell}) = 1/6$ (the pattern $p_i p_j p\_ell$ has to be decreasing) or that $i = \ell$, in which case again, $E(Z_{i,j} Z_{k,\ell}) = 1/6$ (the pattern $p_k p_i p_j$ has to be decreasing). Each of these scenarios occurs $\binom{n}{3}$ times. Therefore, we have

$$E(Z^2) = \sum_{i<j} E(Z_{i,j}^2) + \sum_{(i,j) \neq (k,\ell)} E(Z_{i,j} Z_{k,\ell})$$

$$= \frac{1}{2}\binom{n}{2} + \frac{1}{4}\binom{n}{2}\binom{n-2}{2} + \frac{5}{3}\binom{n}{3},$$

and consequently,

$$\mathrm{Var}(Z) = E(Z^2) - E(Z)^2$$

$$= \frac{1}{2}\binom{n}{2} + \frac{1}{4}\binom{n}{2}\binom{n-2}{2} + \frac{5}{3}\binom{n}{3} - \frac{n^2(n-1)^2}{16}$$

$$= \binom{n}{2}\frac{2n+5}{36}.$$

9. No, there is not. If there were, applying the hooklength formula for such a Ferrers shape, the numerator, being 20!, would be divisible only by 7^2, while the denominator would contain three factors equal to 7.

11. We prove that we can even create a tree that consists of a single path that satisfies at least one-third of the constraints.

These trees correspond to permutations of $[n]$. The generic constraint $\{(a, b), (c, d)\}$ will be satisfied by eight of the possible 24 relative orders of these four elements as we can swap entries within the pairs, or we can swap the pairs. Let p be a random n-permutation, and let $X(p)$ be the indicator variable of the event that the tree defined by p satisfies the constraint X. Then $E(X) = 1/3$, where the expectation is taken over all n-permutations p. Using the linearity of expectation for the indicator variables of all constraints, we obtain our claim.

13. Let R be the range of X. By the definition of expectation, we have

$$\mu = E(X) = \sum_{i \in R} iP[X = i] = \sum_{i > \alpha\mu} iP[X = i] + \sum_{i \le \alpha\mu} iP[X = i]$$

$$\ge \sum_{i > \alpha\mu} iP[X = i] > \alpha\mu \cdot P[X > \alpha\mu].$$

Dividing by $\alpha\mu$, we get Markov's inequality.

15. As p is an involution, its cycles are all of length one or length two. Let I_n denote the number of involutions of length n. Then the number of involutions of length n in which the entry 1 is fixed is I_{n-1}, whereas the number of involutions of length n in which the entry 1 is part of a 2-cycle is $(n-1)I_{n-2}$. This shows that

$$E(Y) = \frac{I_{n-1} + 2(n-1)I_{n-2}}{I_n}.$$

Finally, as it is easy to compute by the Exponential formula, or directly, $I_n = \sum_{i=0}^{\lfloor n/2 \rfloor} \binom{n}{2i}(2i-1)!!$, where we set $(-1)!! = 1$.

17. By symmetry, we have

$$P\left[i(p) - \frac{1}{2}\binom{n}{2} > a\binom{n}{2}\right] = \frac{1}{2}P\left[\left|i(p) - \frac{1}{2}\binom{n}{2}\right| > a\binom{n}{2}\right].$$

We have seen in the solution of Exercise 7 that if $Z(p) = i(p)$, then $\text{Var}(Z) = \binom{n}{2}\frac{2n+5}{36}$. Therefore, $\sigma = \sqrt{\text{Var}(Z)} = O(n^{3/2})$, and our statement follows from Chebyshev's inequality by setting $\lambda = ca\sqrt{n}$ for an appropriate constant c.

19. If n is in a cycle of length more than 2, then deleting n, we get a derangement of length $n - 1$. If n is in a 2-cycle, then deleting its cycle, we get a derangement of length $n - 2$. This leads to the formula

$$E(Y_n) = \frac{(n-1)D(n-1)}{D(n)} \cdot E(Y_{n-1} + 1) + \frac{(n-1)D(n-2)}{D(n)}E(Y_{n-2}).$$

21. This fact was published in [125] without proof. Let r be an n-permutation that consists of one cycle, and let Z_r be the indicator variable of the event that the vertices of $G_{p,q}$ form a Hamiltonian cycle in the order given by r. Then

$$E(Z_r) = \sum_{k=0}^{n} \binom{n}{k} \frac{(n-k)!\,k!}{n!} \frac{k!}{n!} = \sum_{k=0}^{n} \frac{1}{n!}$$
$$= \frac{n+1}{n!}.$$

Indeed, first choose the k edges that will come from the permutation p, which specifies the values of $p(i)$ for k distinct values of i. Then choose the remaining $n - k$ values of p in $(n - k)!$ ways. Similarly, the $n - k$ edges of the Hamiltonian cycle r that come from q specify the values of $q(i)$ for k distinct values of i, then choose the remaining values in $k!$ distinct ways. Using linearity of expectation, we get that

$$E(Z) = (n-1)!E(Z_r) = \frac{n+1}{n}.$$

23. The number of total cycles of p is equal to the number of its 1-cycles, plus the number of its 2-cycles, and so on. If $X_k(p)$ denotes the number of k-cycles of $p \in S_n$ and $X(p)$ denotes the numbers of all cycles of p, then it follows from the result of Example 6.3 that

$$E(X) = \sum_{k=1}^{n} E(X_k) = \sum_{k=1}^{n} \frac{1}{n}.$$

25. Our proof will be by induction on n. For $n = 1$ and $n = 2$, the statement is vacuous, and it is straightforward to check that the statement is true for $n = 3$. Indeed, in the only nontrivial case, we get $A_{1,\lambda} = A_{2,\lambda} = 1/2$. Now assume that the statement is true for $n - 1$, and prove it for n. Then in any SYT on λ, the entry n is in one of the inner corners of λ. The presence of λ does not have any influence on the occurrence of

$A_{i,\lambda}$ if $i < n-1$, so by the induction hypothesis (after removing the box containing n), we get that $A_{1,\lambda} = A_{2,\lambda} = \cdots = A_{n-2,\lambda}$. So our claim will be proved if we show that $A_{n-2,\lambda} = A_{n-1,\lambda}$.

Let X be the set of SYT on λ in which $n-2$ is a descent but $n-1$ is not, and let Y be the set of SYT on λ in which $n-1$ is a descent but $n-2$ is not. It clearly suffices to show that X and Y are equinumerous; we do so by presenting a bijection $g : X \to Y$.

Let $T \in X$. We consider three separate cases.

(a) When $n-1$ and n are in the same row, then we define $g(T)$ by swapping $n-1$ and $n-2$ in T. Note that $n-2$ and n will be in the same row of $g(T)$.

(b) When $n-1$ is in a row that is strictly below the row containing n, and $n-2$ is in the same row as n, then define $g(T)$ by swapping $n-1$ and n in T. Note that $n-2$ and $n-1$ will be in the same row of $g(T)$.

(c) In all other cases, $n-2$, $n-1$, and n are in all different rows.

    (c1) First, consider the case when $n-2$ is the north neighbor of $n-1$. In this case, swap $n-1$ and n to get $g(T)$. So in $g(T)$, the entry $n-2$ will be the north neighbor of n.

    (c2) If $n-2$ is the north neighbor of n, then swap $n-1$ and n again to get $g(T)$. So $n-2$ is the north neighbor of n in $g(T)$.

    (c3) If $n-2$ is an inner corner, then from top to bottom, the order of our three maximal entries is either n, $n-2$, $n-1$, or $n-2$, n, $n-1$. We then obtain $g(T)$ by ordering them, respectively, $n-1$, $n-2$, n, and $n-1$, n, $n-2$. Note that all three entries are in inner corners in $g(T)$.

Note that in all three subcases, the three maximal entries are in different rows of $g(T)$.

This completely defines the map $g : X \to Y$. To see that g is a bijection, we show that it has an inverse. Let $U \in Y$. By our remarks at the end of each case, we can establish (from the positions of the three maximal entries in U) which rule was used to create U, and our statement is proved.

We mention that [297] contains a non-induction proof of the result we have just proved.

27. Note that p has ri rising sequences if and only if p^{-1} has $ri-1$ descents, or ri ascending runs. So replacing $ri(p)$ by $d(p) + 1$ in the result of the previous exercise, we get the probability that p^{-1} is obtained by our shuffle. If we sum that equation over all $p \in S_n$, then we get 1 on the left-hand side. Multiplying both sides by a^n, we get the identity of Theorem 1.8.

29. As for any $X \subseteq [n]$, all entries of X must have the same chance to be the index of the minimum element of $f(X)$, the size of F must be divisible by the size of X, and this has to hold for all possible X. Therefore, $|F|$ has to be divisible by the least common multiple of the numbers $1, 2, 3, \cdots, n$. This argument is from [100]. The authors then mention the well-known number theoretical fact [14] that this least common divisor is of size $e^{n-o(n)}$.

31. Let X_i be the indicator random variable of the event that $p_i + 1 = p_{i+1}$. Then $E(X_i) = \frac{n-1}{n(n-1)} = \frac{1}{n}$. Therefore, $E(X) = \sum_{i=1}^{n-1} E(X_i) = (n - 1)/n$.

33. This follows from the fact, proved in Exercise 43 of Chapter 1, that ascents, descents, and plateaux satisfy the same recurrence relation, and therefore are equidistributed.

35. Let $f(z) = \sum_{k=1}^{n} c(n, k) z^k$, and compute $f''(1) = \sum_{k=1}^{n} k(k-1)c(n, k)$. (Note that $f'(1)$ was computed in Exercise 65, when we computed $E(X) = \sum_{k=1}^{n} \frac{1}{k}$.) Using the product rule of derivatives, we get that

$$f''(1) = \sum_{i \neq j} \frac{n!}{i \cdot j},$$

where $1 \leq i, j \leq n$. This implies that

$$E(X(X-1)) = \frac{f''(1)}{n!}$$

$$= \left(\sum_{k=1}^{n} \frac{1}{k} \right)^2 - \sum_{k=1}^{n} \frac{1}{k^2}.$$

So

$$\text{Var}(X) = E(X^2) - E(X)^2$$
$$= E(X(X-1)) + E(X) - E(X)^2$$
$$= \sum_{k=1}^{n} \frac{1}{k} - \sum_{k=1}^{n} \frac{1}{k^2}.$$

37. (a) Let $T(z)$ denote the ordinary generating function for the total number of leaves of all such trees, so $T(z) = z + z^2 + 3z^3 + \cdots$. In other words, $T(z)$ counts all such trees with a leaf marked. Removing the root of such a structure, we get a sequence of rooted plane trees, one of which has a marked leaf. (With the exception of the case when the original tree had just one vertex, that vertex was both the root and the marked leaf.) The obtained trees without a marked vertex are rooted plane unlabeled trees, and such trees are counted

by the generating function $C_1(z) = zC(z) = (1 - \sqrt{1 - 4z})/2$. This leads to the functional equation

$$T(z) = z + T(z) \sum_{k \geq 0} (k+1)(C_1(z))^k = z + \frac{T(z)}{(1 - C_1(z))^2}$$

$$= z + \frac{4zT(z)}{(1 + \sqrt{1 - 4z})^2}.$$

Solving this equation for $T(z)$, we get

$$T(z) = \frac{z}{2} + \frac{1}{2} \frac{z}{\sqrt{1 - 4z}}.$$

This shows that for $n \geq 2$, the total number of leaves is $\frac{1}{2}\binom{2n-2}{n-1}$. The total number of vertices in all such trees of size n is of course $nc_{n-1} = \binom{2n-2}{n-1}$. So $a_n = 1/2$, meaning that half of all vertices are leaves.

(b) Such trees on n vertices are in natural bijection with 132-avoiding permutations of length $n - 1$. Leaves correspond to left-to-right minima. So for $n > 1$, half of all entries are left-to-right minima in such permutations.

For more information about this topic, the reader may consult the paper [126] of Keith Copenhaver.

39. Let $T(z)$ be the exponential generating function of the numbers T_n, with $T(0) = 0$. Removing the root of such a tree, we get a sequence of such trees. This leads to the differential equation $T'(z) = 1/(1 - T(z))$. This yields $T(z) = 1 - \sqrt{1 - 2z}$, and that leads to the formula $T_n = (2n - 3)!!$ for $n > 1$, and $T_1 = 1$.

41. Let A_n denote the total number of *leaves* in all non-plane 1-2 trees on vertex set $[n]$. Let $A(z) = \sum_{n \geq 0} A_n \frac{z^n}{n!}$. Then we claim that

$$A(z) = \frac{z - 1 + \cos z}{1 - \sin z}.$$

Indeed, let (v, T) be an ordered pair in which T is a non-plane 1-2 tree on vertex set $[n]$ and v is a leaf of T. Then $A(z)$ is the exponential generating function counting such pairs. Let us first assume that $n > 1$, and let us remove the root of T. On the one hand, this leaves a structure that is counted by $A'(z)$. On the other hand, this leaves an ordered pair consisting of a non-plane 1-2 tree with a leaf marked, and a non-plane 1-2 tree. By the Product formula of exponential generating functions, such ordered pairs are counted by the generating function $A(z)E(z)$, where $E(z) = \tan z + \sec z$ is the exponential generating function of the

Euler numbers, equivalently, of decreasing nonplane 1-2 trees. Finally, if $n = 1$, then no such ordered pair is formed, while $A'(z)$ has constant term 1. This leads to the linear differential equation

$$A(z) = E(z)A(z) + 1,$$

with the initial condition $A_0(0) = 0$. Solving this equation we get the formula we claimed.

Now we need the following lemma from Complex Analysis.

LEMMA 10.1

Let $H(z) = \frac{f(z)}{g(z)}$ be a function so that f and g are analytic functions at z_0, and $f(z_0) \neq 0$, while $g(z_0) = g'(z_0) = 0$, and $g''(z) \neq 0$. Then

$$H(z) = \frac{2f(z_0)}{g''(z_0)} \cdot \frac{1}{(z - z_0)^2} + \frac{h_{-1}}{z - z_0} + h_0 + \cdots .$$

PROOF   The conditions directly imply that g has a double root, and hence H has a pole of order two, at z_0. In order to find the coefficient that belongs to that pole, let $g(z) = q(z)(z - z_0)^2$. Now differentiate both sides twice with respect to z, to get

$$g''(z) = q''(z)(z - z_0)^2 + 4q'(z)(z - z_0) + 2q(z).$$

Setting $z = z_0$, we get

$$g''(z_0) = 2q(z_0). \tag{10.6}$$

By our definitions, in a neighborhood of z_0, the function $H(z)$ behaves like

$$\frac{f(z)}{q(z)(z - z_0)^2},$$

and our claim follows by (10.6).   ∎

Applying this to the dominant term of $A(z)$ with $D = \pi - 2$ and $a = \pi/2$, we get that

$$\frac{A_n}{n!} \sim (n + 1)(\pi - 2) \cdot \left(\frac{2}{\pi}\right)^{n+2}. \tag{10.7}$$

So now we have the number of all leaves. On the other hand, the total number of all vertices is nE_n, where E_n, the Euler number. We know that $E(z) = \tan z + \sec z$ has a simple pole at $z_0 = \pi/2$. Using that observation, it is straightforward to compute that

$$\frac{E_n}{n!} \sim \frac{4}{\pi} \cdot \left(\frac{2}{\pi}\right)^n. \tag{10.8}$$

Comparing (10.7) and (10.8), we get that

$$\lim_{n \to \infty} a_n = 1 - \frac{2}{\pi} \approx 0.3633802278.$$

43. Let $A(z)$ be the ordinary generating function for the number a_n of all inversions in all such permutations, and let $C(z)$ be the generating function of Catalan numbers. Then, it follows from the structure of 312-avoiding permutations that the differential equation

$$A(z) = 2zA(z)C(z) + z^2C(z)C'(z)$$

holds, with initial condition $A(0) = 0$. Solving this equation, and then computing the coefficients of the obtained power series $A(z)$, we get that

$$a_n = 2^{2n-1} - \binom{2n+1}{n} + \binom{2n-1}{n-1}.$$

Therefore,

$$E(X) = \frac{a_n}{C_n} \sim \frac{\sqrt{\pi}}{2} \cdot n^{3/2}.$$

10.7   Solutions for Chapter 7

1. We know that *permutations* of length n that avoid $12 \cdots k$ correspond to *pairs* of SYT of the same shape that have at most $k - 1$ columns. Because of Theorem 7.11, *involutions* of length n correspond to such pairs in which the two elements are identical. In other words,

$$I_n(123 \cdots k) = \sum_F f_F,$$

where F runs through SYT on n boxes having at most $k - 1$ columns. The proof is then immediate.

3. These SYT bijectively correspond to northeastern lattice paths from $(0,0)$ to (n,n) that never go above the main diagonal. The bijection is given by reading the entries in an SYT in increasing order, and if entry i is in row 1, then taking step i of the corresponding path to the east, otherwise taking it to the north.

5. We claim that $P(\pi)$ must be of rectangular shape (10×8). Indeed, $P(\pi)$ has ten rows, and its rows are of length eight or less; otherwise, the first row would be of length nine, implying that π contains an increasing subsequence of length nine. Therefore, the fifth row of $P(\pi)$ is also of length eight.

7. By the argument seen in the solution of Exercise 1, it suffices to find a formula for the number of all SYT on n boxes that have at most two columns. In such SYT, the entry n must be at the end of either column. If n is even, and our SYT is rectangular, then n must be at the end of the second column. This implies the recursive formulae

$$
I_n(123) = \begin{cases} 2I_{n-1}(123) \text{ if } n \text{ is even,} \\[2mm] 2I_{n-1}(123) - C_{(n-1)/2} \text{ if } n \text{ is odd,} \end{cases}
$$

where in the last step we use the result of Exercise 3. The result is now straightforward by induction as for $n = 2m + 2$, we get $2 \cdot \binom{2m+1}{m} = \binom{2m+2}{m+1}$ and for $n = 2m + 1$, we get $2\binom{2m}{m} - (\binom{2m}{m} - \binom{2m}{m-1}) = \binom{2m+1}{m}$.

9. Theorem 7.5, with its notation, shows that necessarily $a_1 = k$ and $a_1 + a_2 + \cdots + a_r = r \cdot k$. Thus necessarily $a_1 = a_2 = \cdots = a_r = k$ and $a_{r+1} < k$, otherwise there would be $r + 1$ increasing subsequences of length k that are disjoint. This means that the size of the last column is $m_k = r$. Applying (7.1) with k variables instead of $k - 1$ and fixing $m_k = r$ we get the proof exactly as we got the proof of Theorem 7.4.

11. In any SYT of shape F, the entry n has to be in one of the inner corners. So removing n from any such tableaux, we get an SYT of shape F' for some F' that is part of the summation.

13. Note that with any reductions, the value $q(a, b)$ never increases. So no series of reductions can turn q into p.

15. Yes, there is. Let $a_1 = 13\ 12\ 10\ 14\ 8\ 11\ 6\ 9\ 4\ 7\ 3\ 2\ 1\ 5$. Then let a_{i+1} be obtained from i by simply inserting two consecutive elements right after the maximum element m of a_i, and giving them the values $(m - 4)$ and $(m - 1)$, and of course, relabeling the other elements naturally. It is not difficult to see that the a_i consist of two decreasing subsequences, and that they form an antichain. This construction is due to Miklós Bóna and Daniel Spielman [52].

17. An isomorphism between the two posets can be constructed using the idea given in the proof of Exercise 3 of Chapter 3.

19. The statement can be proved by induction on the size of P, the initial case being obviously true. See [279], pages 95–97 for the details of the inductive step.

21. This result is based on an observation from [145]. It is clear that the right-hand side is the number of 321-avoiding n-permutations. We claim that the left-hand side is the same, counted by the length k of the longest increasing subsequence. Indeed, if π is such a permutation, then $risk(\pi)$ has at most two rows, and exactly k columns. There is exactly

one Ferrers shape F satisfying these criteria, and then the hooklength formula shows that

$$f^F = \frac{2k - n + 1}{n + 1}\binom{n + 1}{n - k},$$

and the proof follows.

23. Consider the fraction on the right-hand side of (7.1). Its numerator is clearly $n!$, as it should be. In the denominator, the term $(m_{k-1})!$ is the product of the hooklengths of all boxes in the last row of F. How about the term $(m_{k-2} + 1)!$ in the denominator? It is *almost* the product of the hooklengths of all boxes in the next-to-last row of F. We must say "almost" because if $m_{k-1} < m_{k-2}$, then there will be one term missing from the $m_{k-2} + 1$ terms (hooklengths) whose product is $(m_{k-2} + 1)!$. This is because the rightmost box of row $k - 2$ that has a southern neighbor will have a hooklength that is larger by 2 than the hooklength of its right neighbor. Therefore, the missing hooklength will be $m_{k-2} - m_{k-1} + 1$. However, this is corrected by the appropriate term of the product in the brackets, that is, the term given by $i = k - 2$ and $j = k - 1$.

Continuing this way, we see that the product of the hooklengths in row i is

$$\left[\prod_{i \leq j \leq k-1} (m_i - m_j + j - i)\right] \cdot (m_i + k - i - 1)!,$$

and the proof follows by taking products for $i \in [k - 1]$.

25. Yes, P'_n is a lattice. We prove this by induction on n, the initial case of $n = 1$ being trivial. Let us assume that the statement is true for $n - 1$, and prove it for n. Let $x = x_1 x_2 \cdots x_n$ and $y = y_1 y_2 \cdots y_n$ be two elements of P'_n. Let x' (resp. y') be the $(n - 1)$-permutation obtained from x (resp. y) by removing the maximal entry n. Let $v = x' \wedge y'$, and let $z = x' \vee y'$. Let $x_i = n$, and $y_j = n$, and assume without loss of generality that $i < j$. It is then easy to verify that inserting n into the jth position of v results in $x \wedge y$, and inserting n into the ith position of z results in $x \vee y$.

27. No, if $n > 2$, then P'_n is not complemented. For instance, let $x = 2134 \cdots n$. Then $y = n(n - 1) \cdots 4132$ and $y' = n(n - 1) \cdots 4312$ both satisfy the requirements.

29. No, I_n is not self-dual in general. For instance, if $n = 4$, then there are three elements in I_n that cover the minimum element, namely 1243, 1324, and 2134. At the same time, there are only two elements in I_n that are covered by the maximum, namely 4231 and 3412.

31. It is proved in [332] that the number of these posets is
$$\frac{(1 + o(1))n!^2}{2\sqrt{e}}.$$

33. We have seen that i is a descent in $Q(p)$ if and only if i is a descent of p. However, now p is an involution, so $P(p) = Q(p)$. Therefore, the question is reduced to asking what the probability is that i is a descent of a randomly selected SYT on n boxes. As we have seen in Exercise 24 of Chapter 6, this probability is $1/2$.

35. Let f_F be the number of Standard Young Tableaux of shape F. Then by the Robinson–Schensted correspondence, and Theorem 7.11, we have

$$n! = \sum_{|F|=n} f_F^2 \leq \left(\sum_{|F|=n} f_F \right)^2 = a_n^2.$$

On the other hand, the Cauchy–Schwarz inequality implies that for any positive real numbers $x_1, x_2, \cdots, x_n$, we have

$$\frac{1}{n} \left(\sum_{i=1}^{n} x_i \right)^2 \leq \sum_{i=1}^{n} x_i^2.$$

Applying this inequality for the $p(n)$ positive real numbers f_F where $|F| = n$, we get the second part of the claim.

37. Consider the elements of S as strings of zeros and ones, given by the parity of each entry. Then the elements of S are 0-1 strings of length n that contain exactly $\lfloor n/2 \rfloor$ ones. Therefore, if S has more than $\binom{n}{\lfloor n/2 \rfloor}$ elements, then S has two elements p and q whose associated 0-1 strings are identical. Then p and q cannot be colliding.

This elegant argument is due to János Körner and Claudia Malvenuto [231]. In that paper, the authors mention that for $n \geq 7$, a set of $\binom{n}{\lfloor n/2 \rfloor}$ pairwise colliding n-permutations actually exist. It is not known if that remains true if $n > 7$.

39. Both sides count layered permutations. Indeed, the left-hand side counts layered permutations, and a layered permutation is automatically an involution.

10.8 Solutions for Chapter 8

1. This is a classic algorithm for generating permutations, which was found independently by Johnson [221] and Trotter [312]. The proof is not

difficult by induction on n, the initial case of $n = 2$ being obvious. Now let us assume that the statement is true for $n - 1$. Then one proves from the definition that the algorithm will list the $n!$ permutations of length n so that the first n, the next n, the following n, and so on, will only differ in the position of n, while the subsequence of the entries from $[n-1]$ will be unchanged within each of these n-tuples of permutations. Among the n-tuples, these subwords will be changed according to the list of $(n-1)!$ permutations of length $n - 1$, generated by this same algorithm.

3. Such an algorithm can be found in [154].

5. We have seen that for any n-permutation p, the image $s(p)$ ends in the entry n. Iterating this argument (to the shorter string preceding n in $s(p)$), the image $s(s(p))$ ends in the string $(n - 1)n$, the image $s^3(p)$ ends in $(n - 2)(n - 1)n$, and so on, the image $s^{n-1}(p)$ ends in $23 \cdots n$, so must be the identity permutation.

7. We claim that for $n \geq 3$, these are the $(n-2)!$ permutations of the form $Sn1$, where S is any permutation of the set $\{2, 3, \cdots, n-1\}$. We prove this statement by induction on n. The initial case of $n = 3$ is obvious. Let us assume that we know the statement for n, and let p be an $(n+1)$-permutation that is not $(n-1)$-stack-sortable. Let $p = L(n+1)R$. Then $s(p) = s(L)s(R)(n + 1)$ is not $(n - 2)$-stack-sortable. By our induction hypothesis, this means that $s(L)s(R) = 23 \cdots n1$. As $s(R)$ has to end in its largest entry, we must have $s(R) = R = 1$, showing that p is indeed of the form $L(n + 1)1$ as claimed.

9. We have seen in Proposition 8.17 that a t-stack-sortable permutation must always avoid the pattern $23 \cdots (t+2)1$. On the other hand, observe that Theorem 4.23 implies that

$$\mathrm{Av}_n(234 \cdots (t + 2)1) = \mathrm{Av}_n(123 \cdots (t + 2)) < (t + 1)^{2n},$$

where the last inequality was proved in Theorem 4.21.

11. (a) The image of 231 is 123 as shown in Figure 10.7.

   (b) These are precisely the t-stack-sortable permutations. Even more strongly, $s_t(p) = s^t(p)$ for all p. Indeed, it is easy to see by induction on i that the entries of p will leave stack i in the order identical to $s^i(p)$.

13. An n-permutation is $(n - 2)$-stack-sortable except when it ends in the string $n1$. So $W_{n-2}(n, k)$ is just the number of $(n - 2)$-permutations with $k - 1$ descents, that is, $W_{n-2}(n, k) = A(n - 2, k)$. As Eulerian polynomials are log-concave, our statement is proved.

Output	Stack 2	Stack 1	Input
			231
		2	31
	2		31
	2	1 3	
	1 2	3	
1	2	3	
12	3		
123			

FIGURE 10.7
Passing 231 through two stacks.

15. Let $p = LnR$, when L and R are allowed to be empty. If neither L nor R are empty, then let $f(p) = f(L)nf(R)$. If L is empty, that is, $p = nR$, then let $f(p) = f(R)n$. Finally, if R is empty, that is, $p = Ln$, then let $f(p) = nf(L)$. Use this same rule recursively to compute $f(L)$ and $f(R)$.

17. Yes. Take the antichain A from the solution of Exercise 15 of Chapter 7, then take the reverse of all the permutations in it, to get the antichain A'. Then A' consists of 321-avoiding permutations, that is, permutations containing an increasing subsequence of length at least $n/2$. Now affix the entry 1 to the end of each entry of A' to get the new antichain A''.

19. The condition that $pr = qr$ means that going through the stack has the same effect on p and on q. In other words, the *movement sequences* associated to the two permutations are the same. By this we mean the following. To each permutation of length n, we can associate a *sequence of parentheses* of length $2n$, consisting of n copies of "(" (left parenthesis) and n copies of ")", (right parenthesis) describing how the permutation passes through the stack. Each time an entry goes DOWN in the stack, we write a left parenthesis, and each time an entry comes UP, we write a right parenthesis.

For example, sequence of parentheses of $x = 123$ is ()()(), the sequence of parentheses of $y = 321$ is ((())), while the sequence of parentheses of both $p = 132$ and $q = 231$ is ()(()).

So p and q satisfy the conditions of this exercise if and only if they have the same sequence of parentheses. We claim that this, in turn, is equivalent to the condition that $T(p)$ and $T(q)$ are the same as unlabeled trees. This is straightforward by induction on n, if we note that the size of the right subtree of $T(p)$ and $T(q)$ is k, where there are $2k$ parentheses (k left, k right) *inside* the pair of parentheses that ends last. Indeed, this is just the number of entries that were on the right of n; therefore, they entered the stack after n but exited the stack before n.

21. We have seen in Exercise 19 that b_p just describes what effect the stack has on p. We have also seen that this effect only depends on the sequence of parentheses associated to p, or, in other words, the unlabeled tree obtained from $T(p)$ by omitting its labels. As there are C_n such trees, we get that B has $C_n = \binom{2n}{n}/(n+1)$ elements.

23. We have seen in Exercise 21 that going through a stack can effect an n-permutation in at most C_n ways, so an n-permutation can have at most C_n preimages. On the other hand, the identity permutation does have C_n preimages, namely all the 231-avoiding permutations. We claim this is the only permutation with that property.

To see that no other n-permutation can have C_n preimages, we apply induction on n. For $n \leq 3$, the statement is clearly true. Now let us assume the statement is true for all positive integers less than n. If $s(q) = p$, and $q = LnR$, then we have $p = s(L)s(R)n$. Now keep n fixed in q, in position k, and change L and R so that $s(q)$ does not change. Note that this means that the *set* of entries in L and the set of entries of R cannot change, as otherwise the set of the first $k - 1$ entries in $s(q)$ would change. Similarly, $s(L)$, and $s(R)$ have to remain unchanged.

By our induction hypothesis, there are at most C_{k-1} ways we can permute the entries of L without changing $s(L)$, and there are at most C_{n-k} ways we can do this with R, with equality holding only if both $s(L)$ and $s(R)$ are monotone increasing, that is, only if p is increasing. As k can range from 1 to n, this means that the number of preimages of p is at most

$$\sum_{k=1}^{n} C_{k-1}C_{n-k} = C_n,$$

with equality holding only if $p = 123\cdots n$.

25. (a) These are the permutations that avoid both 2431 and 4231.

(b) It follows from part (a) and the characterization of ir-sortable permutations that p is ir-sortable if and only if $((p^r)^{-1})^r$ is or-sortable, proving our claim.

27. An n-permutation p is separable if and only if it avoids both 2413 and 3142. First we show that this is necessary. Indeed, the statement is obvious for $n \leq 4$, and follows by induction on n for larger n.

Now let p avoid both 2413 and 3142. Then clearly, the reverse of p also avoids these patterns, so we can assume without loss of generality that 1 precedes n in p. Then all entries on the right of n have to be larger than all entries on the left of 1, or a 3142 pattern would be formed. If there is no entry on the left of n that is larger than an entry on the right of n, then p is separable, and we are done. Otherwise, let d be the leftmost entry on the left of n that is larger than the smallest entry c on the right of n. Then dnc is a 231-pattern, meaning that there cannot be any entry $b < c$ located between d and n, for $dbnc$ would be a 3142 pattern. So all entries located between d and n are larger than c, whereas all entries on the left of d are smaller than c, proving that p is separable. Indeed, $p = LR$, where the split occurs immediately before d. Then both L and R avoid 2413 and 3142, and therefore, are separable.

The concept of separable permutations was introduced in [85].

29. These are the permutations avoiding the pattern $q_t = (t + 1)t \cdots 1$. It is clear by the Pigeonhole Principle that these permutations cannot be t-queue-sortable. On the other hand, if p avoids q_t, then let us define the co-rank of an entry to be the length of the longest decreasing subsequence ending at that entry. Then p is t-queue-sortable by sending all entries of co-rank i to queue i.

31. We claim that this number is

$$P(n, k) = 2^{n-1-2k} \binom{n-1}{2k} \frac{\binom{2k}{k}}{k+1}.$$

The result was first announced in this form by Petter Brändén [94], with a computational proof. A combinatorial proof can be given as follows. Let us identify stack-sortable permutations with their decreasing binary tree which, in this case, can be viewed as an unlabeled plane tree. Let us take an unlabeled binary plane T tree on k vertices. If a vertex v of T has less than two children, add new children as needed until all the vertices of T have two children. Let T' be the tree obtained in this way. Then T' has k peaks and $k+1$ leaves; hence T' has $2k$ edges. Now subdivide some of the edges of T' by new (blue) vertices until the total number of edges of the obtained tree T'' is $n - 1$. (The new vertices can also be added on the left of the leftmost vertex of T'' and on the right of the rightmost vertex of T''.) Now T'' has $n - 1 - 2k$ vertices that have only one child; there are 2^{n-1-2k} ways to rearrange the left-right orientation of these edges without changing the structure of the non-blue vertices. As there are $\binom{2k}{k}/(k+1)$ choices for T, there are $\binom{n-1}{2k}$ ways to turn T' into T'', and the statement is proved.

33. Setting $a_0 = 0$, we have $A(z) = \sum_{n \geq 0} a_n z^n = z/(1 - 2z - z^2)$. The key observation is that in permutations counted by a_n, the entry n has to be in the first, second, or last position. For details, and many similar results, see [156].

35. Let γ have fertility A, and let τ have fertility B. Then we claim that the permutation

$$\pi = (\gamma \ominus \tau) \oplus 1$$

has fertility AB.

For instance, 213 has fertility one, while 123 has fertility five, so our claim in this case is that 5461237 has fertility five.

It is clear that π has fertility at least AB, since if $s(\rho) = \gamma$ and $s(\zeta) = \tau$, and $\phi = (\rho \oplus 1) \ominus \zeta$, then $s(\phi) = \pi$. In the above example, $\pi = 5467123$.

Now we show that no other permutations are preimages of π. Note that in π, all entries coming from γ precede all entries coming from τ. As the former are larger, it follows that in any preimage of π, there must be an entry between the leftmost of the entries that belong to γ and the rightmost of the entries that belong to τ. By exclusion, that entry must be the maximum entry, and our claim is proved.

10.9 Solutions for Chapter 9

1. First, let n be odd. Taking an alternating path that starts with the gray edge from 0 to 1, then continues with the black edge from 1 to 2, and so on, we reach the vertex n with a gray edge, after which we must go to 0 using a black edge, completing an alternating cycle. The remaining edges form the other alternating cycle.

   Now let n be even. Take the alternating path that starts as the one in the previous paragraph. This path will reach the vertex n with a black edge, after which we must go to the vertex $n + 1$ using a gray edge, then to 1 using a black edge. After this, we walk through the vertices $1, 2, \cdots, n$ again, using the edges that have not been used yet, finishing with the black edge from 1 to 0. This shows that all of $G(p)$ is one alternating cycle.

3. We claim that the canonical block interchange defined in the proof of Theorem 9.9 has this property, with $x = p_i$. Recall that x is defined as the smallest entry of p for which there exists an entry y in p that precedes x and is larger than x.

Indeed, if p does not start with the entry 1, then $x = 1$, and x has the property required in this exercise. If $p = p_1 p_2 \cdots p_n$, and $p_j = j$ for $j < k$, but $p_k \neq k$, then the smallest entry p_i for which $p_i \neq i$ is k. This entry k is not in the kth position, so the entry in the kth position is larger than k. So k is eligible for the role of x, and no entry less than k is eligible for the role of x, since the entries that are smaller than k are in increasing order at the beginning of p.

5. A block transposition removes three black edges in $G(p)$, and replaces them by three new black edges. The old black edges could be part of one, two, or three cycles, of which none, one, two, or three could have been odd, and the new black edges are also part of one, two, or three cycles, of which none, one, two, or three could be odd. The only way in which this would result in $c_{odd}(G(p))$ growing by more than two would be if the three new black edges would be in three separate odd cycles, while the three old black edges would have been in three separate even cycles. However, that is impossible for parity considerations.

7. Let us assume without loss of generality that the alternating path P from u to v starts with the gray edge from u to $u + 1$ and ends in the gray edge from $v - 1$ to v. We show that P can be completed to an alternating cycle. At each vertex v_i starting at $v_1 = v$, use the edge leaving v_i whose color is different from the edge that we used to arrive at v_i. This creates an alternating path that eventually gets back at u with a black edge since the graph is finite and no vertex is the endpoint of more than one edge of each color.

9. There is a natural bijection between the cycles of the two graphs. If C is a cycle of $B(p)$, consisting of the vertices

$$L_{a_1} R_{a_1-1} L_{a_2} R_{a_2-1} \cdots L_{a_k} R_{a_k-1},$$

then $R_{a_i-1} L_{a_{i+1}}$ is a black edge in $B(p)$, implying that $a_i - 1$ is the entry immediately preceding a_{i+1} in p. However, that means that there is a black edge from a_{i+1} to $a_i - 1$ in $G(p)$, and so

$$a_k \, (a_{k-1} - 1) \, a_{k-1} \, (a_{k-2} - 1) \, a_{k-2} \cdots a_1 \, (a_k - 1)$$

is a cycle of $G(p)$.

11. Create the permutation $p' = 897564231$ with the first move. Then note that due to three entries followed by entries one larger, sorting p' is equivalent to sorting the decreasing permutation of length six, and we have seen in the text how to achieve that with four block transpositions.

13. Create $p' = 963785241$, then $p'' = 967852341$ with the first two moves. Then proceed as in the solution of Exercise 11.

15. If p is not the increasing permutation, then the value of is(p) can be increased by one by a block transposition. Indeed, find the longest increasing subsequence s of p, find an entry $x \notin s$, and insert x into the right position of s. An example for permutations for which this bond is tight is

$$p = 2\ 4\ 6 \cdots 2n\ 1\ 3\ 5 \cdots (2n-1).$$

References

[1] W. Aitken, Total relative displacement of permutations. *J. Combin. Theory A*, **87** (1989), no. 1, 1–21.

[2] S. B. Akers, D. Harel, B. Krishnamurthy, The star graph: An attractive alternative to the n-cube. Proceedings of the Fourth International Conference on Parallel Processing (ICPP), 393–400, Pennsylvania State University Press, 1987.

[3] M. H. Albert, R. Aldred, M. D. Atkinson, C. C. Handley, D. A. Holton, Permutations of a multiset avoiding permutations of length 3. *European J. Combin.*, **22** (2001), no. 8, 1021–1031.

[4] M. H. Albert, M. D. Atkinson, C. C. Handley, D. A. Holton, W. Stromquist, On packing densities of permutations. *Electron. J. Combin.*, **9** (2002), no. 1.

[5] M. H. Albert, M. Elder, A. Rechnitzer, P. Westcott, M. Zabrocki, On the Stanley-Wilf limit of 4231-avoiding permutations and a conjecture of Arratia. *Adv. in Appl. Math.* **36** (2006), no. 2, 96–105.

[6] M. H. Albert, M. D. Atkinson, V. Vatter, Counting 1324, 4231-avoiding permutations. *Electronic J. Combin* **16** (2009) # R136.

[7] M. H. Albert, M. Coleman, R. Flynn, I. Leader, Permutations containing many patterns. *Annals of Combinatorics* **11** (2007), 265–270.

[8] M. H. Albert, M. Engen, J. Pantone, V. Vatter, Universal layered permutations Electronic J. Combin **25** (2018), no. 3, P3.23.

[9] M. H. Albert, S. Linton, Growing at a Perfect Speed. *Combinatorics, Probability, and Computing* **18** (2009), 301–308.

[10] M. H. Albert, V. Vatter, How many pop-stacks does it take to sort a permutation? *Preprint*, available at https://arxiv.org/abs/2012.05275.

[11] D. Aldous, P. Diaconis, Longest increasing subsequences: From patience sorting to the Baik-Deift-Johansson theorem. *Bull. Amer. Math. Soc. (N.S.)*, **36** (1999), no. 4, 413–432.

[12] N. Alon, E. Friedgut, On the number of permutations avoiding a given pattern. *J. Combin. Theory Ser. A*, **89** (2000), no. 1, 133–140.

[13] G. Andrews, *The Theory of Partitions*. Addison-Wesley, Reading MA, 1976.

[14] T. Apostol, *Introduction to Analytical Number Theory*. Springer Verlang, NY, 1976.

[15] K. Archer, S. Elizalde, Cyclic permutations realized by signed shifts. *J. Comb.* **5** (2014), no. 1, 1–30.

[16] K. Archer, A. Gregory, B. Pennington, S. Slayden, Pattern restricted quasi-Stirling permutations. *Australas. J. Combin.* **74** (2019), 389–407.

[17] R. Arratia, On the Stanley-Wilf conjecture for the number of permutations avoiding a given pattern. *Electronic J. Combin.*, **6** (1999), no. 1, N1.

[18] D. Ashlock, J. Tillotson, Construction of small superpermutations and minimal injective superstrings. *Congr. Numer.* **93** (1993) 91–98.

[19] M. D. Atkinson, Permutations which are the union of an increasing and a decreasing subsequence. *Electronic J. Combin.*, **5** (1998), no. 1, R6.

[20] M. D. Atkinson, Restricted permutations. *Discrete Math.*, **195** (1999), no. 1-3, 27–38.

[21] M. D. Atkinson, Pop-stacks in parallel. *Inform. Process. Lett.*, **70** (1999) 63–67.

[22] M. D. Atkinson, M. M. Murphy, N. Ruskuc, Sorting with two ordered stacks in series. *Theoret. Comput. Sci.*, **289** (2002), no 1., 205–223.

[23] D. Avis, M. Newborn, On pop-stacks in series. *Utilitas Math.*, **19** (1981), 129–140.

[24] E. Babson, J. West, The permutations $123p_4 \cdots p_m$ and $321p_4 \cdots p_m$ are Wilf-equivalent. *Graphs Combin.*, **16** (2000), no. 4, 373–380.

[25] E. Babson, E. Steingrímsson, Generalized permutation patterns and a classification of the Mahonian statistics. *Séminaire Lotharingien de Combinatoire.*, **44** (2000).

[26] J. Backelin, J. West, G. Xin, Wilf equivalence for singleton classes. *Adv. in Appl. Math.* **38** (2007), no. 2, 133--148.

[27] V. Bafna, P. Pevzner, Sorting by transpositions. *SIAM J. Discrete Math.* **11** (1998), no. 2, 224–240 (electronic).

[28] J. Baik, P. Deift, K. Johansson, On the distribution of the length of the longest increasing subsequence of random permutations. *J. Amer. Math. Soc.*, **12** (1999), no. 4, 1119–1178.

[29] H. Barcelo, S. Sundaram, On some submodules of the action of the symmetric group on the free Lie algebra. *J. Algebra,* **154** (1993), no. 1, 12–26.

[30] H. Barcelo, R. Maule, S. Sundaram, On counting permutations by pairs of congruence classes of major index. *Electron. J. Combin.*, **9** (2002), no. 1.

[31] M. Bayer, P. Diaconis, Trailing the dovetail shuffle to its lair. *Ann. Appl. Prob.*, **2** (1992), no. 2, 294–313.

[32] J. S. Beissinger, Similar constructions for Young tableaux and involutions, and their application to shiftable tableaux. *Discrete Math.*, **67** (1987), no. 2, 149–163.

[33] E. Bender, R. Canfield, Log-concavity and related properties of the cycle index polynomials. *J. Combin. Theory Ser. A*, **74** (1996), no. 1, 57–70.

[34] E. A. Bertram, B. Gordon, Counting special permutations. *European J. Combin.*, **10** (1989), no. 3, 221–226.

[35] O. Bernardi, Bijective counting of Kreweras walks and loopless triangulations. *J. Combin. Theory Ser. A*, **114** (2007), no. 5, 931–956.

[36] O. Bernardi, A. Morales, R. Stanley, R. X. Du, Separation probabilities for products of permutations. *Combin. Probab. Comput.* **23** (2014), no. 2, 201–222.

[37] D. Bevan, Intervals of permutation class growth rates. *Combinatorica* **38** (2018), no. 2, 279–303.

[38] D. Bevan, Permutations avoiding 1324 and patterns in Łukasiewicz paths. *J. Lond. Math. Soc. (2)* **92** (2015), no. 1, 105–122.

[39] D. Bevan, R. Brignall, A. E. Price, J. Pantone, A structural characterisation of Av(1324) and new bounds on its growth rate. *European J. Combin.* **88** (2020), 103115, 29 pp.

[40] A. Björner, F. Brenti, *Combinatorics of Coxeter Groups*. Springer Verlag, New York, 2004.

[41] J. Bloom, S. Elizalde, Pattern avoidance in matchings and partitions. *Electron. J. Combin.* **20** (2013), no. 2, Paper 5, 38 pp.

[42] J. Blum, Enumeration of the square permutations in S_n. *J. Combin. Theory A*, **17** (1974), 156–161.

[43] E. D. Bolker, A. M. Gleason, Counting permutations. *J. Combin. Theory A*, **29** (1980), no. 2, 236–242.

[44] M. Bóna, Permutations avoiding certain patterns; The case of length 4 and generalizations. *Discrete Math.*, **175** (1997), no. 1–3, 55–67.

[45] M. Bóna, A combinatorial proof of a result of Hetyei and Reiner on Foata-Strehl-type permutation trees. *Ann. Comb.*, **1** (1997), no. 2, 119–122.

[46] M. Bóna, Exact enumeration of 1342-avoiding permutations; A close link with labeled trees and planar maps. *J. Combin. Theory A*, **80** (1997), 257–272.

[47] M. Bóna, Permutations with one or two 132-subsequences. *Discrete Mathematics*, **175** (1997), 55–67.

[48] M. Bóna, The number of permutations with exactly r 132-subsequences is P-recursive in the size! *Adv. Appl. Math*, **18** (1997), 510–522.

[49] M. Bóna, The permutation classes equinumerous to the Smooth class. *Electronic J. Combin.*, **5** (1998).

[50] M. Bóna, Partitions with k crossings. *The Ramanujan Journal*, **3** (1999), no. 2, 215–220.

[51] M. Bóna, R. Ehrenborg, A combinatorial proof of the log-concavity of the numbers of permutations with k runs. *J. Combin. Theory A*, **90** (2000), no. 2, 293–303.

[52] M. Bóna, D. A. Spielman, An infinite antichain of permutations. *Electronic J. of Combin.*, **7** (2000).

[53] M. Bóna, A. MacLennan, D. White, Permutations with roots. *Random Structures and Algorithms*, **17** (2000), no. 2, 157–167.

[54] M. Bóna, R. Simion, A self-dual poset on objects counted by the Catalan numbers and a type-B analogue. *Discrete Mathematics* **220** (2000), no. 1-3, 35–49.

[55] M. Bóna, B. E. Sagan, V. Vatter, Frequency sequences with no internal zeros. *Adv. Appl. Math*, **28** (2002), 395–420.

[56] M. Bóna, A simplicial complex of 2-stack sortable permutations. *Adv. Appl. Math*, **29** (2002), 499–508.

[57] M. Bóna, The limit of a Stanley-Wilf sequence is not always rational, and layered patterns beat monotone patterns. *J. Combin. Theory Ser. A*, **110** (2005), no. 2, 223–235.

[58] M. Bóna, A combinatorial proof of the log-concavity of a famous sequence counting permutations. *Electronic J. Combin.* **11** (2) (2005), Note 2.

[59] M. Bóna, On a balanced property of derangements. *Electron. J. Combin.* **13** (2006), no. 1, Research Paper 102, 12 pp. (electronic).

[60] M. Bóna, New records in Stanley-Wilf limits. *European J. Combin.* **28** (2007), no. 1, 75–85.

[61] M. Bóna, Where the monotone pattern (mostly) rules. *Discrete Mathematics*, **308** (2008), no. 23, 5782–5788.

[62] M. Bóna, Generalized descents and normality. *Electron. J. Combin.* **15** (2008), no. 1, Note 21, 8 pp.

[63] M. Bóna, Real zeros and normal distribution for statistics on Stirling permutations defined by Gessel and Stanley. *SIAM J. Discrete Math.* **23** (2008/09), no. 1, 401–406.

[64] M. Bóna, On two related questions of Wilf concerning standard Young tableaux. *European J. Combin.* **30** (2009), no. 5, 1318–1322.

[65] M. Bóna, R. Flynn, The average number of block interchanges needed to sort a permutation and a recent result of Stanley. *Inform. Process. Lett.* **109** (2009), no. 16, 927–931.

[66] M. Bóna, The absence of a pattern and the occurrences of another. *Discrete Math. Theor. Comput. Sci.* **12** (2010), no. 2, 89–102.

[67] M. Bóna, On three different notions of monotone subsequences. *Permutation patterns*, 89–114, London Math. Soc. Lecture Note Ser., **376**, Cambridge University Press, Cambridge, UK, 2010.

[68] M. Bóna, Non-overlapping permutation patterns. *Pure Mathematics and Its Applications*, **22** (2011), no. 2.

[69] M. Bóna, Surprising symmetries in objects counted by Catalan numbers. *Electronic Journal of Combinatorics*, **19** (2012), P62.

[70] M. Bóna, On a Family of Conjectures of Joel Lewis on Alternating Permutations. *Graphs Combin.* 30 (2014), no. 3, 521–526.

[71] M. Bóna, k-protected vertices in binary search trees. *Adv. in Appl. Math.* **53** (2014), 1–11.

[72] M. Bóna, A new record for 1324-avoiding permutations. *Eur. J. Math.* **1** (2015), no. 1, 198–206.

[73] M. Bóna, editor, *Handbook of Enumerative Combinatorics.* CRC Press – Chapman & Hall, Boca Raton, FL, 2015.

[74] M. Bóna, *A Walk Through Combinatorics, 4th Edition.* World Scientific, River Edge, NJ, 2016.

[75] M. Bóna, *Introduction to Enumerative and Analytic Combinatorics, 2nd Edition.* CRC Press – Chapman & Hall, Boca Raton, FL, 2016.

[76] M. Bóna, B. Pittel, On a random search tree: asymptotic enumeration of vertices by distance from leaves. *Adv. in Appl. Probab.* **49** (2017), no. 3, 850–876.

[77] M. Bóna, M.-L Lackner, B. Sagan, Longest Increasing Subsequences and Log Concavity. *Annals of Combinatorics* **21** (2017) , 535–549.

[78] M. Bóna, B. Pittel, On the cycle structure of the product of random maximal cycles. *Sém. Lothar. Combin.* **80** (2019), 1–37.

[79] M. Bóna, R. Smith, Pattern avoidance in permutations and their squares. *Discrete Math.* **342** (2019), no. 11, 3194–3200.

[80] M. Bóna, M. Cory, Cyclic permutations avoiding pairs of patterns of length three. *Discrete Math. Theor. Comput. Sci.* **21** (2019), no. 2, Paper No. 8, 15 pp.

[81] M. Bóna, E. DeJonge, Pattern avoiding permutations with a longest increasing subsequence. *Electronic J. Combin.* **27** (4), (2020), P.4.44.

[82] M. Bóna, Generating Functions of Permutations with Respect to Their Alternating Runs. Séminaire Lotharingien de Combinatoire, **B85b** (2021), 5 pp.

[83] M. Bóna, A. Burstein, Permutations with exactly one copy of a decreasing pattern of length k. Preprint, available at `https://arxiv.org/abs/2101.00332`.

[84] M. Bóna, J. Pantone, Permutations avoiding sets of patterns with long monotone subsequences. Preprint, available at `https://arxiv.org/abs/2103.06918`.

[85] P. Bose, J. F. Buss, A. Lubiw, Pattern matching for permutations. *Inform. Process. Lett.,* **65** (1998), 277–283.

[86] M. Bousquet-Mélou, C. Chauve, G. Labelle, P. Leroux, A bijective proof for the arborescent form of the multivariable Lagrange inversion formula. *Mathematics and Computer Science, (Versailles, 2000), Trends Math.,* Birkhäuser, Basel, 2000, 89–100.

[87] M. Bousquet-Mélou, Sorted and/or sortable permutations. *Formal Power Series and Algebraic Combinatorics, (Toronto, ON, 1998). Discrete Math.,* **225** (2000), no. 1-3, 25–50.

[88] M. Bousquet-Mélou, Counting walks in the quarter plane. *Mathematics and Computer Science: Algorithms, Trees, Combinatorics and Probabilities, Trends in Mathematics,* Birkhäuser, 2002, 49–67.

[89] M. Bousquet-Mélou, S. Butler, Forest-like permutations. *Annals of Combinatorics* **11** (2007), 335–354.

[90] M. Bousquet-Mélou, Ponty, Y, Culminating paths. *Discrete Math. Theor. Comput. Sci.* **281** (2008), no. 1-2, 125–152.

[91] P. Brändén, q-Narayana numbers and the flag h-vector of $J(\mathbf{2} \times \mathbf{n})$. *Discrete Math.* **281** (2004), no. 1–3, 67–81.

[92] P. Brändén, Counterexamples to the Neggers-Stanley conjecture. *Electron. Res. Announc. Amer. Math. Soc.* **10** (2004), 155–158.

[93] P. Brändén, On linear transformations preserving the Pólya frequency property. *Trans. Amer. Math. Soc.* **358** (2006), no. 8, 3697–3716.

[94] P. Brändén, Actions on permutations and unimodality of descent polynomials. *European J. Combin.* **29** (2008), no. 2, 514–531.

[95] F. Brenti, Hilbert polynomials in combinatorics. *J. Algebraic Combin.*, **7** (1998), no. 2, 127–156.

[96] F. Brenti, Combinatorics and total positivity. *J. Combin. Theory A*, **71** (1995), no. 2, 175–218.

[97] F. Brenti, Permutation enumeration, symmetric functions, and unimodality. *Pacific J. Mathematics*, **157** (1993), no. 1, 1–28.

[98] F. Brenti, Log-concave and unimodal sequences in algebra, combinatorics, and geometry: an update. *Jerusalem Combinatorics '93, Contemp. Math.* **178** (1994), *Amer. Math. Soc.*, Providence, RI, 71–89.

[99] D. Bressoud, D. Zeilberger, Bijecting Euler's partitions-recurrence. *American Math. Monthly*, **92** (1985), no. 1, 54–55.

[100] A. Broder, M. Charikar, A. Frieze, M. Mitzenmacher, Min-wise independent permutations. *30th Annual ACM Symposium on Theory of Computing (Dallas, TX, 1998). J. Comput. System Sci.*, **60** (2000), no. 3, 630–659.

[101] W. G. Brown, J. W. Tutte, On the enumeration of rooted non-separable planar maps. *Canad. J. Math.*, **16** (1964), 572–577.

[102] S. Brunetti, A. DelLungo, F. DelRistoro, A cycle lemma for permutation inversions. *Discrete Math.*, **257** (2002), no. 1, 1–13.

[103] A. Burstein, A short proof for the number of permutations containing the pattern 321 exactly once. *Elec. J. Combin.* **18** (2), 2001, P21.

[104] A. Burstein, J. Pantone, Two examples of unbalanced Wilf-equivalence. *J. Comb.* **6** (2015), no. 1-2, 55–67.

[105] L. Butler, The q-log-concavity of q-binomial coefficients. *J. Combin. Theory A*, **54** (1990), no. 1, 54–63.

[106] L. Butler, A unimodality result in the enumeration of subgroups of a finite Abelian group. *Proc. Amer. Math. Soc.*, **101** (1987), vol. 4, 771–775.

[107] T. W. Cai, Average length of the longest k-alternating subsequence. *J. Combinatorial Theory A,* **134** (2015), 51–57.

[108] L. Cangelmi, Factorizations of an n-cycle into two n-cycles. *Eur. J. Combin.* **24** (2003), 849–853.

[109] C. Charalambides, *Enumerative Combinatorics.* Chapman & Hall / CRC, Boca Raton, FL, 2002.

[110] L. Carlitz, Eulerian numbers and polynomials. *Math. Mag.*, **32** (1958-1959), 247–260.

[111] A. Cayley, On the partitions of a polygon. *Proc. London Math. Soc.*, **22** (1890–1891), no. 1, 237–262.

[112] R. Chapman, An involution on derangements. 17th British Combinatorial Conference (Canterbury, 1999). *Discrete Math.* **231** (2001), no. 1–3, 121–122.

[113] R. Chapman, L. K. Williams, A conjecture of Stanley on alternating permutations. *Electron. J. Combin.* **14** (2007), no. 1, Note 16, 7 pp. (electronic).

[114] L. Chua, K. R. Sankar, Equipopularity Classes of 132-Avoiding Permutations. *Electronic J. Combin.* **21** no. 1, (2014), P1.59.

[115] J. Y. Choi, J. D. H. Smith, On the unimodality and combinatorics of Bessel numbers. *Discrete Math.* **264** (2003), no. 1–3, 45–53.

[116] D. A. Christie, Sorting Permutations By Block Interchanges, *Inform. Process. Lett.* **60** (1996), 165–169.

[117] Z. Chroman, M. Kwan, M. Singhal, Lower bounds for superpatterns and universal sequences. *J. Combin. Theory Ser. A* **182** (2021), 105467.

[118] J. Cibulka, On constants in the Füredi-Hajnal and the Stanley-Wilf conjecture. *Journal of Combinatorial Theory, Series A* **116** (2009) 209–302.

[119] A. Claesson, V. Jelinek, E. Steingrímsson, Upper bounds for the Stanley-Wilf limit of 1324 and other layered patterns. *J. Combin. Theory Ser. A* **119** (2012), no. 8, 1680–1691.

[120] A. Claesson, Generalized pattern avoidance. *Europ. J. Combinatorics*, **22** (2001), no. 7, 961–971.

[121] A. Claesson, S. Kitaev, Classification of bijections between 321- and 132-avoiding permutations. *Sém. Lothar. Combin.* **60** (2008/09), Art. B60d, 30 pp.

[122] M. Coleman, An answer to a question by Wilf on packing distinct patterns in a permutation. *Electronic J. Combin.* **11** (2004), N#8.

[123] L. Comtet, *Advanced Combinatorics. The Art of Finite and Infinite Expansions.* D. Reidel Publishing Co., Dordrecht, The Netherlands, Revised and Enlarged Edition, 1974.

[124] M. Conger, A refinement of the Eulerian numbers, and the joint distribution of $\pi(1)$ and $\mathrm{Des}(\pi)$ in S_n. *Ars Combin.* **95** (2010), 445–472.

[125] C. Cooper, The union of two random permutations does not have a directed Hamilton cycle. *Random Structures Algorithms*, **18** (2001), no. 1, 95–98.

[126] K. Copenhaver, k-protected vertices in unlabeled rooted plane trees *Graphs Combin.* **33** (2017), no. 2, 347–355.

[127] R. Cori, B. Jacquard, G. Schaeffer, Description trees for some families of planar maps. *Proceedings of the 9th Conference on Formal Power Series and Algebraic Combinatorics*, Vienna, Austria, 1997, 196–208.

[128] R. Cori, G. Schaeffer, Description trees and Tutte formulas. *Theoret. Comput. Sci*, **292** (2003), no. 1, 165–183.

[129] D. W. Cranston, I. H. Sudborough, D. B. West, Short Proofs for Cut-and-Paste Sorting of Permutations. *Discrete Math* **307** (2007), no. 22, 2866–2870.

[130] J. N. Darroch, On the distribution number of successes in independent trials. *Ann. Math. Stat.*, **35** (1964), 1317–1321.

[131] F. N. David, D. E. Barton, *Combinatorial Chance*. Charles Griffin and Company, Ltd. London, UK, 1962.

[132] C. Defant, *Preimages under the stack-sorting algorithm. Graphs Combin.* **33** (2017), no. 1, 103–122.

[133] C. Defant, *Counting 3-stack-sortable permutations. J. Combin. Theory Ser. A* **172** (2020), 105209, 26 pp.

[134] C. Defant, Catalan intervals and uniquely sorted permutations. *J. Combin. Theory Ser. A* **174** (2020), 105250, 43 pp.

[135] C. Defant, Fertility numbers. *J. Comb.* **11** (2020), no. 3, 527–548.

[136] C. Defant, Descents in t-sorted permutations. *J. Comb.* **11** (2020) no. 3, 511–526.

[137] C. Defant, Fertility monotonicity and average complexity of the stack-sorting map. *Eur. J. Combin.* **93** (2021) 103276.

[138] C. Defant, A. E. Price, A. J. Guttmann, Asymptotics of 3-stack-sortable permutations. *Electronic J. Combin.* **28** (2021), no. 2, Paper No. 2.49, 29 pp.

[139] R. S. Deodhar, M. K. Srinivasan, A statistic on involutions. *J. Algebraic Combin.*, **13** (2001), 187–198.

[140] J. Dénes, The representation of a permutation as the product of a minimal number of transpositions, and its connection with the theory of graphs. *A Magyar Tudományos Akadémia Matematikai Intézetének közleményei* [Communications of the Mathematical Institute of the Hungarian Academy of Sciences], **4** (1959), 63–70.

[141] J. Désarmenien, Une autre interprétation du nombre de dérangements. *Sém. Lotharing. Combin.*, **19** (1984), 11–16.

[142] J. Désarmenien, D. Foata, The signed Eulerian numbers. *Discrete Math.*, **99** (1992) no. 1–3, 49–58.

[143] J. Désarmenien, M. Wachs, Descentes des dérangements et mots circulaires. *Sém. Lotharing. Combin.*, **19** (1984), 13–21.

[144] J. Désarmenien, M. Wachs, Descent classes on permutations. *J. Combin. Theory A*, **64** (1993), no. 2, 311–328.

[145] E. Deutsch, A. J. Hildebrand, H. S. Wilf, Longest increasing subsequences in pattern-restricted permutations. *Electronic J. Combin.*, **9** (2003), R12.

[146] E. Deutsch, A. Robertson, D. Saracino, Refined restricted involutions. *European J. Combinatorics*, **28** (2007), no. 1, 481–498.

[147] P. Diaconis, M. MacGrath, J. Pitman, Riffle shuffles, cycles, and descents. *Combinatorica*, **15** (1995), no. 1, 11–29.

[148] J-P. Doignon, A. Labarre, On Hultman numbers. *Journal of Integer Sequences*, **10** (2007), Article 07.6.2.

[149] S. Dulucq, S. Gire, J. West, Permutations with forbidden subsequences and nonseparable planar maps. *Discrete Math.*, **153** (1996), no. 1–3, 85–103.

[150] P. H. Edelman, Inversions and cycles in permutations. *Europ. J. Combinatorics*, **8** (1987), 269–279.

[151] P. H. Edelman, C. Greene, Combinatorial correspondences for Young tableaux, balanced tableaux, and maximal chains in the weak Bruhat order of S_n. *Combinatorics and Algebra (Boulder, CO, 1983), Contemp. Math.*, **34** (1984), *Amer. Math. Soc.*, Providence, RI, 155–162.

[152] P. H. Edelman, C. Greene, Balanced tableaux. *Adv. in Math.*, **63** (1987), no. 1, 42–99.

[153] P. H. Edelman, V. Reiner, h-shellings and h-complexes. *Adv. Math.*, **106** (1994), no. 1, 36–64.

[154] S. Effler, F. Ruskey, A CAT algorithm for generating permutations with a fixed number of inversions. *Inform. Process. Lett.*, **86** (2003), no. 2, 107–112.

[155] E. Egge, Restricted permutations related to Fibonacci numbers and k-generalized Fibonacci numbers. *Integers* **5** (2005), no. 1, A1, 12 pp.

[156] E. Egge, T. Mansour, 132-avoiding two-stack sortable permutations, Fibonacci numbers and Pell numbers. *Disc. Appl. Math* **143** (2004) 72–83.

[157] I. Elias, T. Hartman, A 1.375-approximation algorithm for sorting by transpositions. *IEEE-ACM Transactions on Computational Biology and Bioinformatics*, **3** (2006) no. 4, 369–379.

[158] S. Elizalde, M. Noy, Consecutive patterns in permutations. Formal power series and algebraic combinatorics (Scottsdale, AZ, 2001), *Adv. in Appl. Math.* **30** (2003), no. 1–2, 110–125.

[159] S. Elizalde, Asymptotic enumeration of permutations avoiding generalized patterns. *Adv. Appl. Math.* **36** (2006), 138–155.

[160] S. Elizalde, The most and the least avoided consecutive patterns. *Proc. Lond. Math. Soc. (3)* **106** (2013), no. 5, 957–979.

[161] S. Elizalde, Descents on quasi-Stirling permutations. *J. Combin. Theory Ser. A*, **180** (2021), 105429, 35 pp.

[162] S. Elizalde, J. Troyka, Exact and asymptotic enumeration of cyclic permutations according to descent set. *J. Combin. Theory Ser. A*, **165** (2019), 360–391.

[163] M. El-Zahar, N.W. Sauer, Asymptotic enumeration of two-dimensional posets. *Order* **5** (1988), no. 3, 239–244.

[164] M. Engen, V. Vatter, Containing all permutations. *The American Mathematical Monthly*, **128** (2021), no. 1, 4–24.

[165] P. Erdős, On a conjecture of Hammersley. *J. London Math. Soc.*, **28** (1953), 232–236.

[166] H. Eriksson, K. Eriksson, S. Linusson, J. Wästlund, Dense packing of patterns in a permutation. *Ann. Comb.* **11** (2007), no. 3–4, 459–470.

[167] H. Eriksson, K. Eriksson, J. Karlander, L. Svensson, J. Wästlund, Sorting a bridge hand. *Disc. Math.* **241** (2001), 289–300.

[168] P. Flajolet, Analytic models and ambiguity of context-free languages. *Theoret. Comput. Sci.*, **49** (1987), 282–309.

[169] P. Flajolet, R. Sedgewick, *Analytic Combinatorics*. Cambridge University Press, Cambridge, UK, 2009.

[170] D. Foata, On the Netto inversion number of a sequence. *Proc. Amer. Math. Soc.*, **19** (1968), no. 1, 236–240.

[171] D. Foata, Distributions Eulériennes et Mahoniennes sur le groupe des permutations. In *Higher Combinatorics*, Martin Aigner, editor, D. Reidel, Dordrecht, The Netherlands, 1977.

[172] D. Foata, M. P. Schützenberger, *Théorie Geometrique des Polynômes Eulériens*. In *Lecture Notes in Mathematics*, **138** (1970), Springer-Verlag, Berlin, Germany.

[173] D. Foata, M. P. Schützenberger, Major index and inversion number of permutations. *Math. Nachr.*, **83** (1978), 143–159.

[174] D. Foata, D. Zeilberger, Denert's permutation statistic is indeed Euler-Mahonian. *Stud. Appl. Math.*, **83** (1990), 31–59.

[175] D. Foata, G-N. Han, Fix-Mahonian calculus, I: Two transformations. *European J. Combin.* **29** (2008), 1721–1732.

[176] S. Fomin, Finite, partially ordered sets and Young diagrams. (Russian) *Dokl. Akad. Nauk SSSR,* **243** (1978), no. 5, 1144–1147.

[177] H. O. Foulkes, A nonrecursive rule for Eulerian numbers. *J. Combin. Theory Ser. A,* **22** (1977), no. 2, 246–248.

[178] J. Fox, Stanley-Wilf limits are typically exponential. Preprint, arXiv:1310.8378v1.

[179] A. Frank, On chain and antichain families of a partially ordered set. *J. Combin. Theory Ser. B,* **29** (1980), no. 2, 176–184.

[180] A. Frieze, Hamilton cycles in the union of random permutations. *Random Structures Algorithms*, **18** (2001), no. 1, 83–94.

[181] G. Frobenius, Über die Charaktere der symmetrischen Gruppe. *Preuss. Akad. Wiss. Sitz.*, 1900, 516–534.

[182] G. Frobenius, Über die charakteristischen Einheiten der symmetrischen Gruppe. *Preuss. Akad. Wiss. Sitz.*, 1903, 328–358.

[183] G. Frobenius, Über die Bernoullischen Zahlen und die Eulerischen Polynome. *Sits. Ber. Preuss. Akad. Wiss.*, 1910, 808–847.

[184] M. Fulmek, Enumeration of permutations containing a prescribed number of occurrences of a pattern of length three. *Adv. in Appl. Math.*, **30** (2003), no. 4, 607–632.

[185] W. Fulton, Young tableaux. With applications to representation theory and geometry. *London Mathematical Society Student Texts,* **35** (1997), Cambridge University Press, Cambridge, 1997.

[186] Z. Füredi, P. Hajnal, Davenport-Schinzel theory of matrices. *Discrete Math.*, **103** (1992), 233–251.

[187] S. Garrabrant, I. Pak, Pattern avoidance is not *P*-recursive. Preprint, arXiv:1505.06508

[188] A. Garsia, S. Milne, A Rogers-Ramanujan bijection. *J. Combin. Theory Ser. A,* **31** (1981), 289–339.

[189] V. Gasharov, On the Neggers-Stanley conjecture and the Eulerian polynomials, *J. Combin. Theory Ser. A,* **82** (1998), no. 2, 134–146.

[190] I. Gessel, Personal communication, 1997.

[191] I. Gessel, Symmetric functions and P-recursiveness. *J. Combin. Theory Ser. A*, **53** (1990), no. 2, 257–285.

[192] I. Gessel, G. Labelle, Lagrange inversion for species. *J. Combin. Theory Ser. A*, **72** (1995), no. 1, 95–117.

[193] I. Gessel, C. Reutenauer, Counting permutations with given cycle structure and descent set. *J. Combin. Theory Ser. A*, **64** (1993), 189–215.

[194] I. Gessel, R. Stanley, Stirling polynomials. *J. Combin. Theory Ser. A*, **24** (1978), 24–33.

[195] I. Gessel, Counting forests by descents and leaves. *Electronic J. Combin.*, **3** (1996), no. 2, R8.

[196] S. Gire, Arbres, permutations à motifs exclus et cartes planaires: quelque problèmes algorithmiques et combinatoires. PhD thesis, University Bordeaux I, 1993.

[197] D. Gouyou-Beauchamps, Standard Young tableaux of height 4 and 5. *European J. Combin.*, **10** (1989), no. 1, 69–82.

[198] H. Gordon, Discrete probability. *Undergraduate Texts in Mathematics*, Springer-Verlag, New York, 1997.

[199] C. Greene, An extension of Schensted's theorem. *Advances in Math.*, **14** (1974), 254–265.

[200] C. Greene, D. Kleitman, The structure of Sperner k-families. *J. Combinatorial Theory Ser. A*, **20** (1976), 41–68.

[201] C. Greene, A. Nijenhuis, H. S. Wilf, A probabilistic proof of a formula for the number of Young tableaux of a given shape. *Adv. in Math.*, **31** (1979), no. 1, 104–109.

[202] C. Greene, A. Nijenhuis, H. S. Wilf, Another probabilistic method in the theory of Young tableaux. *J. Combin. Theory Ser. A*, **37** (1984), no. 2, 127–135.

[203] H. W. Gould, Evaluation of sums of convolved powers using Stirling and Eulerian numbers. *Fibonacci Quart.*, **16** (1978), no. 6, 488–497.

[204] I. P. Goulden, J. West, Raney paths and a combinatorial relationship between rooted nonseparable planar maps and two-stack-sortable permutations. *J. Combin. Theory Ser. A*, **75** (1996), no. 2, 220–242.

[205] O. Guibert, T. Mansour, Restricted 132-involutions and Chebysev polynomials. *Ann. Comb.* **6** (2002), no. 3-4, 349–374.

[206] O. Guibert, Combinatoire des permutations à motifs exclus en liason avec des mots, cartes planaires, et Tableaux de Young. PhD thesis, University Bordeaux I, 1995.

[207] O. Guibert, R. Pinzani, E. Pergola, Vexillary involutions are enumerated by Motzkin numbers. *Ann. Comb.*, **5** (2001), no. 2, 153–174.

[208] B. Hackl, C. Heuberger, H. Prodinger, S. Wagner, Analysis of bidirectional ballot sequences and random walks ending in their maximum. *Ann. Comb.* **20** (2016), no. 4, 775–797.

[209] M. Haiman, *Smooth Schubert varieties.* Unpublished manuscript, 1992.

[210] J. Hammersley, A few seedlings of research. *Proc. Sixth Berkeley Symposium Math. Statist. and Probability*, Volume **1**, 345–394, University of California Press, Berkeley, CA, 1972.

[211] S. Hannenhalli, P. Pevzner, Transforming cabbage into turnip: Polynomial algorithm for sorting signed permutations by reversals. *J. ACM* **46** (1999), no. 1, 1–27.

[212] G-H. Nan, G. Xin, Permutations with extremal number of fixed points. *J. Combin. Theory Series A* **116** (2009) 449–459.

[213] T. E. Harris, *A Theory of Branching Processes.* Prentice Hall, Englewood Cliffs, NJ, 1963.

[214] P. Hästö, On descents in Standard Young Tableaux. *Electronic J. Combin*, **7** (2000), R59.

[215] G. Hetyei, E. Reiner, Permutation trees and variation statistics. *European J. Combin.*, **19** (1998), no. 7, 847–866.

[216] C. Homberger, Expected Patterns in Permutation Classes. Electronic J. Combin. **19** no 3. (2012), P43.

[217] R. Houston, Tackling the Minimal Superpermutation Problem. Preprint, available at `arxiv.org/pdf/1408.5108.pdf`

[218] B. Huang, An upper bound on the number of (132,213)-avoiding cyclic permutations. *Discrete Math.* **342** (2019), no. 6, 1762–1771.

[219] D. M. Jackson, R.C. Read, A note on permutations without runs of given length. *Aequationes Mathematicae*, **17** (1978), 336–343.

[220] A. D. Jaggard, Prefix exchanging and pattern avoidance by involutions. *Elec. J. Combin.*, **9** (2003). no. 2, R16, 2003.

[221] S. M. Johnson, Generation of permutations by adjacent transpositions. *Mathematics of Computation*, **17** (2003), 282–285.

[222] C. Jordan, *Calculus of Finite Differences.* Chelsea, New York, NY, 2nd edition, 1960.

[223] T. Kaiser, M. Klazar, On growth rates of closed permutation classes, *Electr. J. Combinatorics* **9** (2003), vol. 2, R10.

[224] M. Klazar, A general upper bound in extremal theory of sequences. *Comment. Math. Univ. Carolin.*, **33** (1992), 737–746.

[225] M. Klazar, P. Valtr, Generalized Davenport-Schinzel sequences. *Combinatorica* **14** (1994), 463–476.

[226] M. Klazar, The Füredi-Hajnal conjecture implies the Stanley-Wilf conjecture. *Formal Power Series and Algebraic Combinatorics*, Springer Verlag, Berlin, Germany, 2000, 250–255.

[227] D. J. Kleitman, D. J. Kwiatkowski, A lower bound on the length of the sequence containing all permutations as subsequences. *J. Combin. Theory (A)*, **21** (1976), 129–136.

[228] R. Knörr, W. Lempken, B. Thielcke, The S_3-conjecture for solvable groups. *Israel J. Math.*, **91** (1995), no. 1-3, 61–76.

[229] D. E. Knuth, Permutations, matrices, and generalized Young tableaux. *Pacific J. Math.*, **34** (1970), 709–727.

[230] D. E. Knuth, *The Art of Computer Programming, Volume 3,* Addison-Wesley Publishing Co., Reading MA, 1973.

[231] J. Körner, C. Malvenuto, Pairwise colliding permutations and the capacity of infinite graphs. *SIAM J. Discrete Mathematics*, **20** (2007), 203–212.

[232] J. Körner, C. Malvenuto, G. Simonyi, Graph-different permutations. *SIAM J. Discrete Mathematics*, **22** (2008), 489–499.

[233] C. Krattenthaler, The enumeration of lattice paths with respect to their number of turns. *Advances in Combinatorial Methods and Applications to Probability and Statistics*, N. Balakrishnan, ed., Birkhäuser, Boston, MA 1997, 29–58.

[234] D. Kremer, Permutations with forbidden subsequences and a generalized Schröder number. *Discrete Mathematics*, **218** (2000), no. 1-3, 121–130.

[235] D. Kremer, W. C. Shiu, Finite transition matrices for permutations avoiding pairs of length four patterns. *Discrete Mathematics*, **268** (2003), no. 1-3, 171–183.

[236] G. Kreweras, Une famille de polynômes ayant plusieurs propriétés énumeratives. *Period. Math. Hungar.*, **11** (1980), no. 4, 309–320.

[237] G. Kreweras, Sur une classe des problèmes liés au treillis des partitions d'entiers. *Cahiers du B.U.R.O*, **6** (1965), 5–105.

[238] A. Labarre, S. Grusea, The distribution of cycles in breakpoint graphs of signed permutations. *Discrete Appl. Math.* **161** (2013), no. 10-11, 1448–1466.

[239] A. Labarre, A new tight upper bound on the transposition distance. *Algorithms in bioinformatics*, 216–227, *Lecture Notes in Comput. Sci.*, **3692**, Springer, Berlin, 2005.

[240] A. Labarre, Edit distances and factorisations of even permutations. *Proceedings of the Sixteenth Annual European Symposium on Algorithms*, 635–646. *Lecture Notes in Comput. Sci.*, **5193**, Springer, Berlin, 2008.

[241] G. Labelle, Une nouvelle démonstration combinatoire des formules d'inversion de Lagrange. *Adv. in Math.*, **42** (1981), no. 1, 217–247.

[242] R. Laver, Well-quasi orderings and sets of finite sequences. *Math. Proc. of the Cambr. Philos. Soc.* **79** (1976), 1–10.

[243] J. B. Lewis, Generating trees and pattern avoidance in alternating permutations. *Elec. J. Combin.* **19** (2012) #P21.

[244] N. Linial, Graph coloring and monotone functions on posets. *Discrete Math.* **58** (1986), 97–98.

[245] B. F. Logan, L. A. Schepp, A variational problem for random Young tableaux. *Advances in Math.* **26** (1977), no. 2, 206–222.

[246] L. Lovász, *Combinatorial Problems and Exercises.* Second edition, North Holland, Amsterdam, The Netherlands, 1994.

[247] L. Lu, Y. Yang, A lower bound on the transposition diameter. *SIAM J. Discrete Math.* **24** (2010), no. 4, 1242–1249.

[248] P. MacMahon, Two applications of general theorems in combinatory analysis. *Proc. London Math. Soc.* **15** (1916), 314–321.

[249] T. Mansour, A. Vainshtein, Counting occurrences of 132 in a permutation. *Adv. Appl. Math.*, **28** (2002), no. 2, 185–195.

[250] A. Marcus, G. Tardos, Excluded permutation matrices and the Stanley-Wilf conjecture. *J. Combin. Theory Ser. A*, **107** (2004), no. 1, 153–160.

[251] F. Markel, Groups with many conjugate elements. *J. Algebra* **26** (1973) 69–74.

[252] A. Miller, Asymptotic bounds for permutations containing many different patterns. *J. Combin. Theory Ser. A* **116** (2009) 92–108.

[253] P. Moszkowski, A solution to a problem of Dénes: A bijection between trees and factorizations of cyclic permutations. *Europ. J. Combinatorics*, **10** (1989), no. 1, 13–16.

[254] E. Netto, *Lehrbuch der Combinatorik.* Chelsea, New York, NY, 1901.

[255] J. Noonan, The number of permutations containing exactly one increasing subsequence of length three. *Discrete Math.*, **152** (1996), no. 1-3, 307–313.

[256] J. Noonan, D. Zeilberger, The enumeration of permutations with a prescribed number of forbidden patterns. *Adv. in Appl. Math.*, **17** (1996), no. 4, 381–407.

[257] J. C. Novelli, I. Pak, A. Stoyanovskii, A direct bijective proof of the hook-length formula. *Discrete Math. Theor. Comput. Sci.*, **1** (1997), no. 1, 53–67.

[258] K. O'Hara, Unimodality of Gaussian coefficients: A constructive proof. *J. Combin. Theory A*, **53** (1990), no. 1, 29–52.

[259] J. Pantone, The enumeration of permutations avoiding 3124 and 4312. *Ann. Comb.* **21** (2017), no. 2, 293–315.

[260] J. Pantone, V. Vatter, Growth rates of permutation classes: categorization up to the uncountability threshold. *Israel J. Math.* **236** (2020), no. 1, 1–43.

[261] T. K. Petersen, Eulerian Numbers. Birkhäuser, Basel, Switzerland, 2015.

[262] J. Pitman, Probabilistic bounds on the coefficients of polynomials with only real zeros. *J. Combin. Theory A*, **77** (1997), no. 2, 279–303.

[263] H. Kestel, B. Pittel, A Local Limit Theorem for the Number of Nodes, the Height, and the Number of Final Leaves in a Critical Branching Process Tree. *Random Structures Algorithms* **8** (1996), no. 4, 243–299.

[264] V. R. Pratt, Computing permutations with double-ended queues, parallel stacks and parallel queues. *Proc. ACM Symp. Theory of Computing* **5** (1973), 268–277.

[265] C. B. Presutti, Determining lower bounds for packing densities of non-layered patterns using weighted templates. *Electron. J. Combin.* **15** (1) Research paper 50, 10, 2008.

[266] C. B. Presutti, W. Stromquist, Packing rates for measures and a conjecture for the packing density of 2413. *Permutation Patterns, London Mathematical Society Lecture Note Series,* **376** (2010) 287–316.

[267] V. Pratt, Computing permutations with double-ended queues. Parallel stacks and parallel queues. Fifth Annual ACM Symposium on Theory of Computing (Austin, Tex., 1973), pp. 268–277. Assoc. Comput. Mach., New York, 1973.

[268] A. Price, Packing densities of layered patterns. PhD Thesis, University of Pennsylvania, 1997.

[269] R. Proctor, Solution of two difficult combinatorial problems with linear algebra. *Amer. Math. Monthly*, **89** (1982), no. 10, 721–734.

[270] S. Radomirović, A construction of short sequences containing all permutations of a set as subsequences. *Electron. J. Combin.* **19** (2012), no. 4, Paper 31, 11 pp.

[271] D. Rawlings, The r-Major index. *J. Combin. Theory A*, **31** (1981), no. 2, 175–183.

[272] A. Regev, Asymptotic values for degrees associated with strips of Young diagrams. *Advances in Mathematics*, **41** (1981), 115–136.

[273] A. Reifegerste, On the diagram of Schröder permutations. *Electronic J. Combin*, **9** (2003), no. 2, R8.

[274] A. Reifegerste, On the diagram of 132-avoiding permutations. *European J. Combin*, **24** (2003), no. 6, 759–776.

[275] J. Remmel, A note on a recursion for the number of derangements. *European J. Combin.*, **4** (1984), no. 4, 371–374.

[276] A. Robertson, D. Saracino, D. Zeilberger, Refined restricted permutations. *Annals of Combinatorics*, **6** (2003), 427–444.

[277] K. Rudolph, Pattern Popularity in 132-Avoiding Permutations. *Electonic J. Combin.* **20** (2013), no. 1, P8.

[278] B. E. Sagan, Inductive and injective proofs of log concavity results. *Discrete Math.*, **68** (1998), no. 2–3, 281–292.

[279] B. E. Sagan, *The Symmetric Group*, Second Edition. Springer Verlag, New York, NY, 2001.

[280] C. Schensted, Longest increasing and decreasing subsequences. *Canad. J. Math.*, **13** (1961), 179–191.

[281] O. Schlömilch, Recherches sur les coefficients des facultés analytiques. *Crelle* **44** (1852), 344–355.

[282] M. P. Schützenberger, Quelques remarques sur une construction de Schensted. *Math. Scand.*, **12** (1963), 117–128.

[283] J. Shareshian, M. Wachs, q-Eulerian polynomials: Excedance and major index. *Electron. Res. Announc. Amer. Math. Soc.* **13** (2007), 33–45 (electronic).

[284] R. Simion, F. W. Schmidt, Restricted permutations. *European Journal of Combinatorics*, **6** (1985), 383–406.

[285] M. Skandera, An Eulerian partner for inversions. *Sém. Lothar. Combin.*, **46** 2001, B46d.

[286] N. P. Sloane, The Online Encyclopedia of Integer Sequences, Internet database, http://oeis.org/.

[287] R. Smith, Comparing algorithms for sorting with t stacks in series. *Ann. Comb.*, **8** (2004), no. 1, 113–121.

[288] Z. Stankova, Forbidden subsequences. *Discrete Math.*, **132** (1994), no. 1-3, 291–316.

[289] Z. Stankova, J. West, A new class of Wilf-equivalent permutations. *J. Algebraic Combin.*, **15** (2002), no. 3, 271–290.

[290] R. Stanley, Binomial posets, Möbius inversion, and permutation enumeration. *J. Combin. Theory Ser. A* **20** (1976), 336–356.

[291] R. Stanley, Eulerian partitions of a unit hypercube. In *Higher Combinatorics*, Martin Aigner, editor, D. Reidel, Dordrecht, The Netherlands, 1977.

[292] R. Stanley, Differentiably finite power series. *European J. Combin.*, **1** (1980), 175–188.

[293] R. Stanley, Factorization of permutations into n-cycles. *Discrete Math.* **37** (1981), 255–262.

[294] R. Stanley, On the number of reduced decompositions of elements of Coxeter groups. *European J. Combin.*, **5** (1984), no. 4, 359–372.

[295] R. Stanley, Log-concave and unimodal sequences in algebra, combinatorics, and geometry. *Graph Theory and Its Applications: East and West. Ann. NY Acad. Sci.* , **576** (1989), 500–535.

[296] R. Stanley, Polygon dissections and standard Young tableaux. *J. Combin. Theory Ser. A*, **76** (1996), no. 2, 175–177.

[297] R. Stanley, *Enumerative Combinatorics, Volume 2*. Cambridge University Press, Cambridge UK, 1999.

[298] R. Stanley, Alternating permutations and symmetric functions. *J. Combin. Theory Ser. A*, **114** (2007) 436–460.

[299] R. Stanley, Longest alternating subsequences of permutations. *Michigan Math. J.*, **57** (2008), 675–687.

[300] R. Stanley, A survey of alternating permutations. *Contemporary Mathematics*, **531** (2010), 165–196.

[301] R. Stanley, Personal communication, August 2010.

[302] R. Stanley, *Enumerative Combinatorics, Volume 1*, Third Edition. Cambridge University Press, Cambridge UK, 2011.

[303] R. Stanley, Two enumerative results on cycles of permutations. *European J. Combin.*, **32** (2011), no. 6, 937–943.

[304] D. Stanton, Unimodality and Young's lattice. *J. Combin. Theory Ser. A*, **54** (1990), no. 1, 41–53.

[305] V. Strehl, Symmetric Eulerian distributions for involutions. *Actes 2e Séminaire Lotharingien de Combinatoire, Premiere Session, Publ.* 140/S-02, IRMA, Strasbourg, 1981, 12.

[306] V. Strehl, Inversions in 2-ordered permutations—a bijective counting. *Bayreuth. Math. Schr.,* **28** (1989), 127–138.

[307] S. Sundaram, The homology of partitions with an even number of blocks. *J. Algebraic Combin.,* **4** (1995), 69–92.

[308] L. Takács, A generalization of the Eulerian numbers. *Publ. Math. Debrecen,* **26** (1979), no. 3-4, 173–181.

[309] S. Tanimoto, An operator on permutations and its application to Eulerian numbers. *Europ. J. Combinatorics,* **22** (2001), no. 4, 569–576.

[310] R. E. Tarjan, Sorting using networks of queues and stacks. *J. Assoc. Comput. Mach,* **19** (1972), 341–346.

[311] I. Tomescu, Graphical Eulerian numbers and chromatic generating functions. *Discrete Math.,* **66** (1987), 315–318.

[312] H. F. Trotter, Algorithm 115. *Comm. Assoc. Comput. Mach* **5** (1962), 434–435.

[313] J. W. Tutte, A census of planar maps. *Canadian Journal of Mathematics* **33**, 249–271, 1963.

[314] P. Ungar, 2N Noncollinear Points Determine at Least 2N Directions *J. Combinatorial Theory A,* **33** (1982), 343–347.

[315] V. R. Vatter, Permutations avoiding two patterns of length three. *Electronic J. Combin.,* **9** (2003), no. 2, R6.

[316] V. Vatter, Small permutation classes. *Proc. London Math. Soc.,* (3) **103** (2011), 879–921.

[317] V. Vatter, Permutation classes of every growth rate above 2.48188. *Mathematika,* **56** (2010) 182–192.

[318] V. Vatter, Growth rates of permutations classes: from countable to uncountable. *Proceedings of the London Mathematical Society,* **103** (2011), 879–921.

[319] A. M. Vershik, S. V. Kerov, Asymptotic behavior of the Plancherel measure of the symmetric group and the limit form of Young tableaux. *Dokl. Akad. Nauk SSSR* **233** (1977), no. 6, 1024–1027, 1977. (Russian, English translation: *Soviet Math. Dokl.* **233** (1977), no. 1–6, 527–531.)

[320] X. G. Viennot, Une forme géométrique de la correspondance de Robinson-Schensted. *Combinatoire et représentation du groupe symétrique (Actes Table Ronde CNRS,* Univ. Louis-Pasteur Strasbourg,

Strasbourg, 1976), 29–58. *Lecture Notes in Math.*, Vol. **579**, Springer, Berlin, 1977.

[321] M. Wachs, An involution for signed Eulerian numbers. *Discrete Math.*, **99** (1992), no. 1-3, 59–62.

[322] D. Wagner, Enumeration of functions from posets to chains. *European J. Combin.* **13** (1992), no. 4, 313–324.

[323] R. Warlimont, Permutations avoiding consecutive patterns. *Ann. Univ. Sci. Budapest. Sect. Comput.* **22** (2003), 373–393.

[324] R. Warlimont, Permutations avoiding consecutive patterns. II. *Arch. Math.* (Basel) **84** (2005), no. 6, 496–502.

[325] J. West, Permutations with forbidden subsequences; and, stack sortable permutations. PhD thesis, Massachusetts Institute of Technology, 1990.

[326] J. West, Generating trees and the Catalan and Schröder numbers. *Discrete Math.*, **146** (1995), no. 1-3, 247–262.

[327] J. West, Generating trees and forbidden subsequences. *Discrete Math.*, **157** (1996), no. 1-3, 363–374.

[328] H. S. Wilf, A bijection in the theory of derangements. *Mathematics Magazine*, **57** (1984), no. 1, 37–40.

[329] H. S. Wilf, Ascending subsequences of permutations and the shapes of tableaux. *J. Combin. Theory Ser. A*, **60** (1992), no. 1, 155–157.

[330] H. S. Wilf, Real zeroes of polynomials that count runs and descending runs. *Unpublished manuscript*, 1998.

[331] H. S. Wilf, The variance of the Stirling cycle numbers. *Preprint*, available at `arXiv:math/0511428v2[math.CO]`.

[332] P. Winkler, Random orders of dimension 2. *Order* **7** (1991), 329–339.

[333] L. Yen, A note on multiset permutations. *SIAM J. Discrete Math.* **7** (1994), no. 1, 152–155.

[334] A. Young, On quantitative substitutional analysis II. *Proc. London. Math. Soc.* **34** (1902), no. 1, 361–397.

[335] E. Zălinescu, Shorter strings containing all k-element permutations. *Inform. Process. Lett.* **111** (2011), no. 12, 605–608.

[336] D. Zeilberger, Kathy O'Hara's constructive proof of the unimodality of the Gaussian polynomials. *Amer. Math. Monthly*, **96** (1990), no. 7, 590–602.

[337] D. Zeilberger, A holonomic systems approach to special functions identities. *J. Computational and Applied Mathematics*, **32** (1990), no. 3, 321–368.

[338] D. Zeilberger, A proof of Julian West's conjecture that the number of two-stack-sortable permutations of length n is $2(3n)!/((n+1)!(2n+1)!)$. *Discrete Math.*, **102** (1992), no. 1, 85–93.

[339] D. Zeilberger, Alexander Burstein's lovely combinatorial proof of John Noonan's beautiful theorem that the number of n-permutations that contain the pattern 321 exactly once equals $(3/n)(2n)!/((n-3)!(n+3)!)$. *Pure Math. Appl.* (PU.M.A.), **22** (2011), no. 2, 297–298.

[340] J. P. Zhang, Finite groups with many conjugate elements. *J. Algebra*, **170** (1994), no. 2, 608–624.

List of Frequently Used Notation

- $A(n,k)$ — number of n-permutations with $k-1$ descents

- $c(n,k)$ — number of n-permutations with k cycles

- $d(p)$ — number of descents of the permutation p

- $i(p)$ — number of inversions of the permutation p

- $I_n(q)$ — number of involutions of length n avoiding the pattern q

- $n!$ — $n(n-1)\cdots 1$

- $\binom{n}{k}$ — $\frac{n!}{k!(n-k)!}$

- $[n]$ — $\{1,2,\cdots,n\}$

- $[\mathbf{n}]$ — $q^{n-1} + q^{n-2} + \cdots + 1$

- $[\mathbf{n}]!$ — $[\mathbf{n}][\mathbf{n-1}]\cdots[\mathbf{1}]$

- $\begin{bmatrix} \mathbf{n} \\ \mathbf{k} \end{bmatrix}$ — $\frac{[\mathbf{n}]!}{[\mathbf{k}]![\mathbf{n-k}]!}$

- $(n)_m$ — $n(n-1)\cdots(n-m+1)$

- $p(n)$ — number of partitions of the integer n

- S_n — set of all n-permutations

- $S_n(q)$ — number of n-permutations avoiding the pattern q

- $S_{n,r}(q)$ — number of n-permutations containing exactly r copies of q

- $S(n,k)$ — number of partitions of the set $[n]$ into k blocks

- $s(n,k)$ — $(-1)^{n-k}c(n,k)$

- $s(p)$ — image of the permutation p under the stack sorting operation

- $c(G(p))$ — number of alternating cycles in the alternating cycle decomposition of the cycle graph $G(p)$ of the permutation p

- $c(\Gamma(p))$ — number of cycles of the permutation p

- $btd(p)$ — number of block transpositions needed to turn the permutation p into the identity permutation

Index

Printed in the USA
CPSIA information can be obtained
at www.ICGtesting.com
LVHW011825041124
795688LV00003B/363

9 780367 222581